Perspectives in Biophysical Plant Ecophysiology

Park S. Nobel (1998) at Agave Hill,
Deep Canyon Reserve, in South Eastern California.
(Photograph was kindly provided by Catherine Goodman)

Perspectives in Biophysical Plant Ecophysiology

A Tribute to Park S. Nobel

Edited by

Erick De la Barrera
Centro de Investigaciones en Ecosistemas
Universidad Nacional Autónoma de México
Campus Morelia

William K. Smith
Department of Biology
Wake Forest University
Winston-Salem, North Carolina

UNIVERSIDAD NACIONAL AUTÓNOMA DE MÉXICO

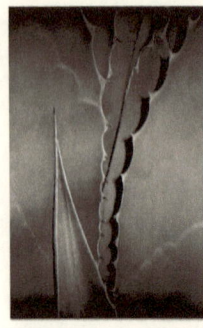

Cover:
Detail of *Agave* from the gardens of Park S. Nobel.
Photograph and design by Eric A. Graham.

Universidad Nacional Autónoma de México
Ciudad Universitaria, Delegación Coyoacán
México, D.F. C. P 04510, México
www.unam.mx

Centro de Enseñanza para Extranjeros,
Escuela Permanente de Extensión en San Antonio
600 Hemisfair Park
San Antonio, TX 78283-0426, U.S.A.
www.usa.unam.edu

Centro de Investigaciones en Ecosistemas, UNAM Campus Morelia
Antigua Carretera a Pátzcuaro 8701
Col. Ex-Hacienda de San José de la Huerta
Morelia, Michoacán 58190, México
www.oikos.unam.mx

ISBN: 978-0-578-00421-1 (Hardcover)
        978-0-578-00676-5 (Paperback)

CONTENTS

I. SUCCULENT PLANTS AND CRASSULACEAN ACID METABOLISM (CAM) PHOTOSYNTHESIS

II. PLANT ECOPHYSIOLOGY FROM GRASSLANDS TO ALPINE ENVIRONMENTS

III. ECOSYSTEM PROCESSES AND CLIMATE CHANGE

IV. PERSPECTIVES

PREFACE

Park S. Nobel pioneered the coupling of cellular physical chemistry with plant physiology, providing a physicochemical interpretation of the laws of diffusion to a rapidly expanding field of plant ecophysiology. For the first time, the commonly applied Fick's and Ohm's Laws for diffusion were explained on a thermodynamic basis using a sound analytical approach with exceptional clarity. His classical textbook, *Physicochemical and Environmental Plant Physiology* (Academic Press/Elsevier) is still unique today in providing a comprehensive array of quantitative problems and solutions from the molecular to the ecological level. Accompanying Park's production of such a prestigious and widely-read textbook are an extraordinary number of refereed publications and four other books on the physiological ecology of desert plants.

Back in 1973, when one of the editors here was the first ecological type to join the Nobel laboratory and the other one was just born, Park was awarded a sabbatical leave to the Australian National University (Guggenheim Fellowship). Despite careful planning, his project involving stomatal guard cells was recognized, after only the first few weeks in Canberra, as impossible to complete in the allotted time period. With nearly the entire sabbatical remaining, Park drew from his engineering physics background and, using an available low-speed wind tunnel, developed boundary layer equations for bluff objects which are still used today. This was his inaugural voyage into the field of environmental biophysics. After returning to UCLA, he made a radical change in his research program and began a remarkable journey of research in the field of environmental biophysics and ecophysiology of desert plants. In particular, he began to unravel the biophysics of plant form and structure that could accomplish the same adaptations as found in biochemistry.

Coming with only laboratory research experience, Park immersed himself in field work, recognizing his need to understand the environment from a first-hand perspective. At the infamous Agave Hill near Palm Desert, California, he would lay out his sleeping bag within arms distance of his spine-laden research plants and measure photosynthetic gas exchange every few minutes throughout the entire night. He would not hesitate to stay up consecutive nights and then drive 3-hrs back to his office to analyze data. We surmised that he must have started writing manuscripts just as quickly. The oldest editor and his wife will always remember an initial visit to Park's home and his detailed description of the mechanics of his solar heating system to our children— they were 2 and 3 yrs-old at the time. This kind of excitement, enthusiasm, and dedication to research and teaching inspired us to produce this volume. Yet, at the same time, we will always remember Park's humor during particularly slow walks to lunch on warm summer days in California, as well as his overall penchant for always "marching to his own drummer". A marvelous overview of Park as a laboratory biophysicist turned field researcher is found in Betsy Gladfelter's *Agassiz's Legacy* (Oxford University Press, 2002).

In honor of Park's 70th birthday, former graduate students, postdocs, and colleagues have contributed a series of papers that, while covering a broad spectrum of modern ecophysiology, illustrate some of the broad influence of Park's prolific career.

Chapters 1 through 6 focus on the ecophysiology of succulent CAM plants, a favorite topic of Park's. In Chapter 1, Ed Bobich and Gretchen North present the biomechanical implications of the anatomy and architecture of succulent plants. Many of such plants can be found in deserts and Paul Schulte explains in Chapter 2 the processes for water transport for these water limited plants. In addition to water limitations such desert plants have to deal with extremely high temperature. The strategies that CAM succulents from arid and semi-arid environments utilize to cope with high temperatures are discussed by Pippa Drennan in Chapter 3. In addition to being prevalent in arid lands, CAM can be found ubiquitously among epiphytic plants. The microenvironments in which these plants can be found are discussed by José Luis Andrade, Carlos Cervera, and Eric Graham in Chapter 4. In Chapter 5, Casandra Reyes and Howard Griffiths discuss with further detail the strategies for the bromeliads from a tropical dry forest. One of Park's research foci has been the prickly pear cactus, *Opuntia ficus-indica*, a CAM crop whose cultivation spans more than one million hectares distributed in forty countries. In Chapter 6, Paolo Inglese, Giuseppe Barbera, Giovanni Gugliuzza, and Giorgia Liguori, explain the ecophysiology of this CAM crop, focusing on economic history and agricultural aspects of fruit production.

Chapters 7 through 10 deal with the ecophysiology of plants from various ecosystems and of different life forms. First, Michelle DaCosta and Bingru Huang present in Chapter 7 the adaptations that perennial grasses have developed to deal with drought, focusing on the production of turf grass. In Chapter 8, Augusto Franco discusses the various functional strategies that trees from the *cerrado* employ in that semi-arid Brazilian ecosystem. Also distributed in semi-arid environments are species of trees of the genus *Prosopis*. Peter Felker summarizes his more than two decades of work with this genus around the world in Chapter 9, discussing physiological adaptations as well as the economic botany for this multi-use tree. A longstanding question in plant ecophysiology— gas exchange by plants from high altitudes and under unique biophysical challenges are discussed in Chapter 10 by Bill Smith and Dan Johnson.

The quantitative approach to the study of plant ecophysiology that Park developed throughout his career represents a conceptual bridge between classical ecophysiology and modern ecosystem physiology, which is the focus of Chapters 11 through 14. In Chapter 11, Eric Graham presents a critical view of how digital photography can help in answering emerging questions in plant science. Beyond the organismic level, at which interactions with the physical environment are crucial, during their reproductive cycle, plants interact with each other and with other organisms. In Chapter 12, Erick De la Barrera, Eulogio Pimienta, and Jorge Schondube, highlight some aspects of our current understanding of plant reproductive ecophysiology, with special attention to the origin and evolution of nectar, as a driver of plant-animal interactions. In Chapter 13, Enrico Yepez and David Williams, discuss gas exchange at the ecosystem level, with particular emphasis of the way semi-arid environments manage precipitation pulses. Finally, after presenting the various physiological responses of plants to elevated CO_2, Stan Smith, David Tissue, Travis Huxman, and Michael Loik, discuss in Chapter 14 the specific response characteristics of plant functional groups found in the desert.

We sincerely thank all of the authors and referees who participated in the production of this volume, and our attempt to maintain the high quality that we grew accustomed to in Park's lab. The support of Drs. José Antonio Vela, José Luis Palacio, and Ken Oyama, directors of UNAM's Escuela Permanente de Extensión en San Antonio, Centro de Enseñanza para Extranjeros, and Centro de Investigaciones en Ecosistemas, as well as Dirección General de Asuntos del Personal Académico (PAPIIT IN221407), were instrumental for the publication of this festschrift. During the initial stages of this project Dr. Chuck Crumly, a former editor for various versions of Park's textbook at Academic Press (currently at UC Press), provided valuable advice and encouragement. Following in Park's tradition of involving students in the editorial process, we would like to thank Alejandra González, Whaleeha Gudiño, Roberto Sáyago, Cynthia Armendáriz, Ana Moreno, Isadora Torres, and, especially, Iván Camargo and Fernando Pineda, who participated in a graduate seminar course about plant ecophysiology and thoroughly revised each one of the chapters.

Finally, we want to express emphatically our respect, admiration, and, most of all, our ongoing appreciation for our mentor, Distinguished Professor, Park S. Nobel.

Erick De la Barrera

William K. Smith

November 4, 2008

A SALUTE TO PARK S. NOBEL

Arthur C. Gibson

Park S. Nobel fashioned a career at UCLA that few in the sciences could ever match. Surely there must be parallels within and outside biology, but it was our good fortune to have witnessed this one up close. Park set a pace of productivity, coupled with the highest quality product, which impressed even our most distinguished, senior faculty members, and wowed faculty promotion committees so as to place his applications on a plane separate from standard submissions. Here was a role model that was impossible to mimic, because he crammed more into a long workday than anyone else—correspondence, researching and preparing research papers, submitting reports and grants, reviewing, editing, teaching preparation and lecturing, committee service, and always reserving time for regularly scheduled meetings with his students and postdocs, especially to keep focus on the task and scientific data flowing through the pipeline. The S in his name must stand for self-disciplined.

Clearly his first book, *Plant Cell Physiology* (Nobel 1970), written and published while still an assistant professor, defined the approaches that Park would use throughout his career to treat and explain biological processes. I still recall from 1973 in a hallway at the University of Arizona when a plant physiologist flashed a copy of that book and heaped praise on the author. Using the techniques of a physicist, Park had chosen the equations that best explain how the physical world interacts with a cell or with tissues, what became known as the physicochemical approach, which bridged the gulf between physical chemistry, cell and organelle physiology, and biochemistry. This was at a time when many in plant biology were rushing to interpret new observations on ultrastructure of plant cells using transmission electron microscopy and while many in the physiologic and biochemical communities were focused on metabolic pathways, before the meteoric rise of molecular plant biology. Although the book jacket advertised this book as an introduction for advanced college students, perhaps more importantly at that time, it actually became a textbook for the professors. Why? Park Nobel was able to explain, using equations and understandable terms, what factors or parameters must be measured to understand structural design and fluxes of liquids and gases, among other things.

The expanded version of that textbook, *Introduction to Biophysical Plant Physiology* (Nobel 1974), was exactly what was needed as the era of gas exchange studies was preparing to explode. Portable instruments to measure photosynthesis and transpiration in the field were becoming available. Park added two chapters. One addressed the biophysical factors affecting leaf physiology; the other was concerned with higher level issues of the plant canopy interacting with the environment, delivery of water to leaves from soil via roots and xylem, and translocation of sugars from leaves to other parts of the plant. Collectively, on these topics Park would spend the bulk of his experimental effort until retirement, leaving behind the spinach and peas of the early years when he investigated diffusion properties of chloroplasts. Park has told us that his interests shifted dramatically to environmental physiology when, by a quirk of fate in 1973 while a Guggenheim Fellow in Canberra, Australia, he began studies on bluff bodies in a wind tunnel.

To most of us who followed his contributions closely, Park Nobel has been a person with an extremely broad range of interests, covering processes from cells to whole plants, from surface to interior, from freezing and low temperatures to very high temperatures, from wet times to the driest of times, from full sun to darkness, from soil and roots to plant and atmosphere, from vegetative to reproductive. To outsiders, Park may have given the impression that he was narrow in scope, because most often he used succulent, CAM species of agaves and cacti (pronounced by him kak'-tee) for addressing these topics. To his credit, Park single-handedly created new model systems in biology, with *Opuntia ficus-indica* and *Agave*, taking the topic of Crassulacean acid metabolism, one of only three known photosynthetic carbon assimilation cycles (C_3, C_4, and CAM), in its infancy to one where one has to think long and hard to find a topic that has not already been addressed by Park with his students. This will be evident from other chapters in this festschrift. Park was way ahead of the curve on the study of energy budgets of plants in natural and agricultural situations, effects on plant productivity and distribution under global warming, reactions of roots to dry and drying soils, and water relations between vegetative and reproductive structures, to name a few.

Having a sharp focus on his favorite experimental species made it difficult to divert Park's attention to other creatures. I tried a few times. Not so distant, Park agreed to coauthor *The Cactus Primer* (Gibson and Nobel 1986), in which we were able to review the known biology, designs, ecology, and evolution of that interesting family of dicotyledons, and from that also resulted a 26-minute film entitled *Adaptations of the American Cacti* (1987). I would like to think that writing the popular book aroused his enthusiasm to author his own books on agaves and cacti (Nobel 1988, 1994). Also during the 1980s, while Park was nurturing his interest in hydraulic properties of plants, we combined with students and postdocs to investigate how water flows through plants, beginning with ferns and progressing to seed plants with woody stems and to plants with vessels rather than tracheids to determine how well water flow in xylem can be predicted by Hagen-Poiseuille relationships (*e.g.*, Gibson, Calkin, and Nobel 1984, Calkin, Gibson, and Nobel 1985, 1986; Schulte, Gibson, and Nobel 1987, 1989). This was during a period when plant-water relations began to take on new importance in ecology and evolution of structure.

There are some contributions by Park that may go unnoticed if not stated here categorically, and hopefully not incorrectly. Through his textbook, his research papers, his co-editing of the *Encyclopedia of Plant Physiology* (Lange *et al.* 1981), and his reviewing and editing of countless journal manuscripts, Park did as much as anyone to standardize metric units for expressing fluxes, and much of the credit for the popularization of Fick's laws of diffusion can be traced to Park. Indeed, it still amazes me how beautifully simple these processes are when explained in this manner, but I remain frustrated that general biology textbooks still tend to avoid the subject for freshman biology. Park found a simple way to introduce Fick's laws when he taught freshman.

Teaching was something that Park assumed very seriously, because he wanted his presentations to be carefully structured, as a physicist would expect. Once I taught general biology with Park, and he assigned me all of the animal stuff (except the physiology, including Fick's law applications). Inside his own sphere of expertise, in physicochemical plant physiology and environmental physiology, he taught his book, making it likely the most intellectually challenging course in the life sciences at UCLA because of the computational component. The course was not only a powerful experience for undergraduates but a great testing and proving ground for the parade of teaching assistants, his graduate students plus five other exceptional ones. What has to be said is that so many around the world have learned the subject also from his textbooks, but without having to suffer the low scores on his exams. So it is that the versions of his textbooks are dog-eared on our shelves and with notes in the margins, while many of the others still crack when we open them.

Literature cited

Calkin, H.W., A.C. Gibson, and P.S. Nobel. 1985. Xylem water potentials and hydraulic conductances in eight species of ferns. *Canadian Journal of Botany* 63:631–637.

Calkin, H.W., A.C. Gibson, and P.S. Nobel. 1986. Biophysical model of xylem conductance in tracheids of *Pteris vittata*. *Journal of Experimental Botany* 37:1054–1064.

Gibson, A.C., H.W. Calkin, and P.S. Nobel. 1984. Xylary anatomy, water flow, and hydraulic conductance in the fern *Cyrtomium falcatum*. *American Journal of Botany* 71:564–574.

Gibson, A.C. and P.S. Nobel. 1986. *The Cactus Primer*. Harvard University Press, Cambridge, MA.

Lange, O.L., P.S. Nobel, C.B. Osmond, and H. Ziegler (eds.). 1981. *Physiological Plant Ecology, Encyclopedia of Plant Physiology, New Series*, vol. 12A–D. Springer-Verlag, Berlin.

Nobel, P.S. 1970. *Plant Cell Physiology*. W.H. Freeman and Company, San Francisco.

Nobel, P.S. 1974. *Introduction to Biophysical Plant Physiology.* W.H. Freeman and Company, San Francisco.
Nobel, P.S. 1988. *Environomental Biology of Agaves and Cacti.* Cambridge University Press, New York.
Nobel, P.S. 1994. *Remarkable Agaves and Cacti.* Oxford University Press, New York.
Schulte, P.S., A.C. Gibson, and P.S. Nobel. 1987. Xylem anatomy and hydraulic conductance of *Psilotum nudum. American Journal of Botany* 74:1438–1445.
Schulte, P.S., A.C. Gibson, and P.S. Nobel. 1989. Water flow in vessels with simple or compound perforation plates. *Annals of Botany* 64:171–178.

PUBLICATIONS BY PARK S. NOBEL

Books

P.S. Nobel. 1970. *Plant Cell Physiology: A Physicochemical Approach.* W.H. Freeman, San Francisco. 267 pp.

P.S. Nobel. 1974. *Introduction to Biophysical Plant Physiology.* W.H. Freeman, San Francisco. 488 pp.

P.S. Nobel. 1983. *Biophysical Plant Physiology and Ecology.* W.H. Freeman, San Francisco/New York. 608 pp.

A.C. Gibson and P.S. Nobel. 1986. *The Cactus Primer.* Harvard University Press, Harvard, Cambridge. 286 pp.

P.S. Nobel. 1988. *Environmental Biology of Agaves and Cacti.* Cambridge University Press, New York. 270 pp.

P.S. Nobel. 1991. *Physicochemical and Environmental Plant Physiology.* Academic Press, San Diego. 635 pp.

P.S. Nobel. 1994. *Remarkable Agaves and Cacti.* Oxford University Press, New York. 166 pp. Spanish Translation by E. García Moya. 1998. *Los Incomparables Agaves y Cactos.* Editorial Trillas, Mexico City. 211 pp.

P.S. Nobel. 1999. *Physicochemical and Environmental Plant Physiology, 2nd Ed.* Academic Press, San Diego. 474 pp.

P.S. Nobel. 2005. *Physicochemical and Environmental Plant Physiology, 3rd Ed.* Elsevier/Academic Press, San Diego. 567 pp.

P.S. Nobel. 2009. *Physicochemical and Environmental Plant Physiology, 4th Ed.* Elsevier/Academic Press, San Diego.

Edited books

O.L. Lange, P.S. Nobel, C.B. Osmond, and H. Ziegler, eds. 1981. *Physiological Plant Ecology I: Responses to the Physical Environment. Encyclopedia of Plant Physiology, New Series, Vol. 12A,* Springer-Verlag, Berlin, 625 pp.

O.L. Lange, P.S. Nobel, C.B. Osmond, and H. Ziegler, eds. 1982. *Physiological Plant Ecology II: Water Relations and Carbon Assimilation. Encyclopedia of Plant Physiology, New Series, Vol. 12B,* Springer-Verlag, Berlin, 747 pp.

O.L. Lange, P.S. Nobel, C.B. Osmond, and H. Ziegler, eds. 1983. *Physiological Plant Ecology III: Responses to the Chemical and Biological Environment. Encyclopedia of Plant Physiology, New Series, Vol. 12C,* Springer-Verlag, Berlin, 799 pp.

O.L. Lange, P.S. Nobel, C.B. Osmond, and H. Ziegler, eds. 1983. *Physiological Plant Ecology IV: Ecosystem Processes: Mineral Cycling, Productivity, and Man's Influence. Encyclopedia of Plant Physiology, New Series, Vol. 12D,* Springer-Verlag, Berlin, 644 pp.

P.S. Nobel, ed. 2002. *Cacti: Biology and Uses.* University of California Press, Berkeley, California. 280 pp.

Research Articles and Reviews

1. Nobel, P.S., and Packer, L. 1964. Energy-dependent ion uptake in spinach chloroplasts. *Biochim. Biophys. Acta* **88**:453-455.
2. Bearden, A.J., Mattern, P.L., and Nobel, P.S. 1964. Mössbauer-effect apparatus for an advanced teaching laboratory. *Amer. J. Physics* **32**:109-119.
3. Beeler, G.W., Fender, D.H., Nobel, P.S., and Evans, C.R. 1964. Perception of pattern and colour in the stabilized retinal image. *Nature* **203**:1200.
4. Packer, L., Siegenthaler, P-A., and Nobel, P.S. 1965. Light-induced high-amplitude swelling of spinach chloroplasts. *Biochem. Biophys. Res. Comm.* **18**:474-477.
5. Nobel, P.S., and Packer, L. 1965. Light-dependent ion translocation in spinach-chloroplasts. *Plant Physiol.* **40**:633-640.
6. Packer, L., Siegenthaler, P-A., and Nobel, P.S. 1965. Light-induced volume changes in spinach chloroplasts. *J. Cell Biol.* **26**:593-599.
7. Nobel, P.S., and Mel, H.C. 1966. Electrophoretic studies of light-induced charge in spinach chloroplasts. *Arch. Biochem. Biophys.* **113**:695-702.
8. Packer, L., Nobel, P.S., Gross, E.L., and Mel, H.C. 1966. Fractionation of spinach chloroplasts by flow sedimentation-electrophoresis. *J. Cell Biol.* **28**:443-448.
9. Nobel, P.S., Murakami, S., and Takamiya, A. 1966. Localization of light-induced strontium accumulation in spinach chloroplasts. *Plant Cell Physiol.* **7**:263-275.
10. Nobel, P.S., Murakami, S., and Takamiya, A. 1966. Localization of light-induced barium accumulation in spinach chloroplasts. *Proc. 6th Intl. Congr. Electron Microscopy*, Kyoto, Japan. Pp. 373-374.
11. Nobel, P.S., and Murakami, S. 1967. Electron microscopic evidence for the location and amount of ion accumulation by spinach chloroplasts. *J. Cell Biol.* **32**:209-211.
12. Nobel, P.S. 1967. Relation of swelling and photophosphorylation to light-induced ion uptake by chloroplasts *in vitro*. *Biochim. Biophys. Acta* **131**:127-140.
13. Nobel, P.S. 1967. Calcium uptake, ATPase and photophosphorylation by chloroplasts *in vitro*. *Nature* **214**:875-877.
14. Nobel, P.S. 1967. A rapid technique for isolating chloroplasts with high rates of endogenous photophosphorylation. *Plant Physiol.* **42**:1389-1394.
15. Murakami, S., and Nobel, P.S. 1967. Lipids and light-dependent swelling of isolated spinach chloroplasts. *Plant Cell Physiol.* **8**:657-671.
16. Nobel, P.S. 1968. Chloroplast shrinkage and increased photophosphorylation *in vitro* upon illuminating intact plants of *Pisum sativum*. *Biochim. Biophys. Acta* **153**:170-182.
17. Nobel, P.S. 1968. Light-induced chloroplast shrinkage *in vivo* detectable after rapid isolation of chloroplasts from *Pisum sativum*. *Plant Physiol.* **43**:781-87.
18. Nobel, P.S. 1968. Energetic basis of the light-induced chloroplast shrinkage *in vivo*. *Plant Cell Physiol.* **9**:499-509.
19. Nobel, P.S. 1969. Light-induced changes in the ionic content of chloroplasts in *Pisum sativum*. *Biochim. Biophys. Acta* **172**:134-143.
20. Nobel, P.S., Chang, D.T., Wang, C-t., Smith, S.S., and Barcus, D.E. 1969. Initial ATP formation, NADP reduction, CO_2 fixation, and chloroplast flattening upon illuminating pea leaves. *Plant Physiol.* **44**:655-661.
21. Nobel, P.S. 1969. The Boyle-Van't Hoff relation. *J. Theor. Biol.* **23**:375-379.
22. Nobel, P.S. 1969. Light-dependent potassium uptake by *Pisum sativum* leaf fragments. *Plant Cell Physiol.* **10**:597-605.
23. Nobel, P.S. 1969. Density of pea chloroplasts determined by four different methods. *Biochim. Biophys. Acta* **189**:452-454.
24. Nobel, P.S. 1970. Increased CO_2 fixation by *Pisum sativum* chloroplasts *in vitro* reflecting a change in coupling caused by illuminating the plants. *Plant Cell Physiol.* **11**:380-388.

25. Nobel, P.S., and Wang, C-t. 1970. Amino acid permeability of pea chloroplasts as measured by osmotically determined reflection coefficients. *Biochim. Biophys. Acta* **211**:79-87.

26. Nobel, P.S. 1970. Relation of light-dependent potassium uptake by pea leaf fragments to the Pk of the accompanying organic acid. *Plant Physiol.* **46**: 491-493.

27. Wang, C-t., and Nobel, P.S. 1971. Permeability of pea chloroplasts to alcohols and aldoses as measured by reflection coefficients. *Biochim. Biophys. Acta* **241**:200-212.

28. Lin, D.C., and Nobel, P.S. 1971. Control of photosynthesis by Mg^{2+}. *Arch. Biochem. Biophys.* **145**:622-632.

29. Nobel, P.S., and Craig, R.L. 1971. Relative anion permeabilities and concentrations in leaf cells of *Pisum sativum* determined using electrical measurements and the Goldman equation. *Plant Cell Physiol.* **12**:653-656.

30. Miller, M.M., and Nobel, P.S. 1972. Light-induced changes in the ultrastructure of pea chloroplasts *in vivo*. *Plant Physiol.* **49**:535-541.

31. Nobel, P.S., and Cheung, Y.-N.S. 1972. Two amino-acid carriers in pea chloroplasts. *Nature, New Biology* **237**:207-208.

32. Nobel, P.S. 1973. Mitochondrial permeability for alcohols, aldoses, and amino acids. *J. Memb. Biol.* **12**:287-299.

33. Nobel, P.S., and Wang, C-t. 1973. Ozone increases the permeability of isolated pea chloroplasts. *Arch. Biochem. Biophys.* **157**:388-394.

34. Cheung, Y.-N.S., and Nobel, P.S. 1973. Amino acid uptake by pea leaf fragments. *Plant Physiol.* **52**:633-637.

35. Nobel, P.S. 1973. Review: *The Quantitative Analysis of Plant Growth* by C.C. Evans, Univ. of Calif. Press, Berkeley, 1972. *Madroño* **22**:215-216.

36. Nobel, P.S. 1974. Temperature dependence of the permeability of chloroplasts from chilling-sensitive and chilling-resistant plants. *Planta* **115**:369-382.

37. Nobel, P.S. 1974. Ozone effects on chlorophylls *a* and *b*. *Die Naturwissenshaften* **61**:80-81.

38. Nobel, P.S. 1974. Boundary layers of air adjacent to cylinders. Estimation of effective thickness and measurements on plant material. *Plant Physiol.* **54**:177-181.

39. Nobel, P.S. 1974. Free energy in biology. *In*: N. Calder, ed., *Nature in the Round: A Guide to Environmental Science*, Weidenfeld and Nicolson, London. Pp. 157-167.

40. Nobel, P.S. 1974. Rapid isolation techniques for chloroplasts. *In*: S. Fleischer and L. Packer, eds., *Methods in Enzymology*, Vol XXXI, Biomembranes, Academic Press, New York. Pp. 600-606.

41. Nobel, P.S. 1975. Chloroplast reflection coefficients: Influence of partition coefficients, carriers, and membrane phase transitions. *In*: U. Zimmermann and J. Dainty, eds., *Membrane Transport in Plants*, Springer-Verlag, Berlin. Pp. 289-295.

42. Nobel, P.S. 1975. Effective thickness and resistance of the air boundary layer adjacent to spherical plant parts. *J. Exp. Bot.* **26**:120-130.

43. Nobel, P.S., Zaragoza, L.J., and Smith, W.K. 1975. Relation between mesophyll surface area, photosynthetic rate, and illumination level during development for leaves of *Plectranthus parviflorus* Hanckel. *Plant Physiol.* **55**:1067-1070.

44. Nobel, P.S. 1975. Chloroplasts. *In*: D.A. Baker and J.L. Hall, eds., *Ion Transport in Plant Cells and Tissues*, Elsevier, Amsterdam. Pp. 101-124.

45. Nobel, P.S. 1976. Photosynthetic rates of sun versus shade leaves of *Hyptis emoryi* Torr. *Plant Physiol.* **58**:218-223.

46. Hartsock, T.L., and Nobel, P.S. 1976. Watering converts a CAM plant to daytime CO_2 uptake. *Nature* **262**:574-576.

47. Nobel, P.S. 1976. Water relations and photosynthesis of a desert CAM plant, *Agave deserti*. *Plant Physiol.* **58**:576-582.

48. Nobel, P.S. 1976. Review: *Mathematical Models in Plant Physiology* by J.H.M. Thornley, Academic Press, London. *Plant Science Bulletin* **22**:45-46.

49. Nobel, P.S. 1977. Water relations and photosynthesis of a barrel cactus, *Ferocactus acanthodes*, in the Colorado Desert. *Oecologia* **27**:117-133.
50. Nobel, P.S. 1977. Water relations of flowering of Agave deserti. *Bot. Gaz.* **138**:1-6.
51. Smith, W.K., and Nobel, P.S. 1977. Temperature and water relations for sun and shade leaves of a desert broadleaf, *Hyptis emoryi*. *J. Exp. Bot.* **28**:169-183.
52. Nobel, P.S. 1977. Internal leaf area and cellular CO_2 resistance: Photosynthetic implications of variations with growth conditions and plant species. *Physiol. Plant.* **40**:137-144.
53. Lewis, D.A., and Nobel, P.S. 1977. Thermal energy exchange model and water loss of a barrel cactus, *Ferocactus acanthodes*. *Plant Physiol.* **60**:609-616.
54. Smith, W.K., and Nobel, P.S. 1977. Influences of seasonal changes in leaf morphology on water-use efficiency for three desert broadleaf shrubs. *Ecology* **58**:1033-1043.
55. Nobel, P.S. 1978. Microhabitat, water relations, and photosynthesis of a desert fern, *Notholaena parryi*. *Oecologia* **31**:293-309.
56. Nobel, P.S., and Hartsock, T.L. 1978. Resistance analysis of nocturnal carbon dioxide uptake by a Crassulacean acid metabolism succulent, *Agave deserti*. *Plant Physiol.* **61**:510-514.
57. Smith, W.K., and Nobel, P.S. 1978. Influence of irradiation, soil water potential, and leaf temperature on leaf morphology of a desert broadleaf, *Encelia farinosa* Gray (Compositae). *Amer. J. Bot.* **65**:429-432.
58. Nobel, P.S., Longstreth, D.J., and Hartsock, T.L. 1978. Effect of water stress on the temperature optima of net CO_2 exchange for two desert species. *Physiol. Plant.* **44**:97-101.
59. Nobel, P.S. 1978. Surface temperatures of cacti — Influences of environmental and morphological factors. *Ecology* **59**:986-996.
60. Nobel, P.S., and Hartsock, T.L. 1979. Environmental influences on open stomates of a Crassulacean acid metabolism plant, *Agave deserti*. *Plant Physiol.* **63**:63-66.
61. Longstreth, D.J., and Nobel, P.S. 1979. Salinity effects on leaf anatomy. Consequences for photosynthesis. *Plant Physiol.* **63**:700-703.
62. Jordan, P.W., and Nobel, P.S. 1979. Infrequent establishment of seedlings of *Agave deserti* (Agavaceae) in the northwestern Sonoran Desert. *Amer. J. Bot.* **66**:1079-1084.
63. Longstreth, D.J., and Nobel, P.S. 1980. Nutrient influences on leaf photosynthesis. *Plant Physiol.* **65**:541-543.
64. Longstreth, D.J., Hartsock, T.L., and Nobel, P.S. 1980. Mesophyll cell properties for some C_3 and C_4 species with high photosynthetic rates. *Physiol. Plant.* **48**:494-498.
65. Nobel, P.S. 1980. Leaf anatomy and water-use efficiency. *In*: N.C. Turner and P.H. Kramer, eds., *Adaptations of Plants to Water and High Temperature Stress*, Wiley, New York. Pp. 43-55.
66. Nobel, P.S. 1980. Interception of photosynthetically active radiation by cacti of different morphology. *Oecologia* **45**:160-166.
67. Nobel, P.S. 1980. Morphology, surface temperatures, and northern limits of columnar cacti in the Sonoran Desert. *Ecology* **61**:1-7.
68. Nobel, P.S. 1980. Water vapor conductance and CO_2 uptake for leaves of a C_4 desert grass, *Hilaria rigida*. *Ecology* **61**:252-258.
69. Nobel, P.S. 1980. Morphology, nurse plants, and minimum apical temperatures for young *Carnegiea gigantea*. *Bot. Gaz.* **141**:188-191.
70. Woodhouse, R.M., Williams, J.G., and Nobel, P.S. 1980. Leaf orientation, radiation interception, and nocturnal acidity increases by the CAM plant *Agave deserti* (Agavaceae). *Amer. J. Bot.* **67**:1179-1185.
71. Nobel, P.S. 1980. Productivity of selected plant species adapted to arid regions. *Proc. IV Inter. Symp. Alcohols Fuels Technology*, Vol. 1. Guaruja, São Paulo, Brazil. Oct. 5-8, 1980. Pp. 131-138.
72. Nobel, P.S. 1980. Influences of minimum stem temperatures on ranges of cacti in southwestern United States and central Chile. *Oecologia* **47**:10-15.

73. Nobel, P.S., and Hartsock, T.L. 1981. Development of leaf thickness for *Plectranthus parviflorus* — Influence of photosynthetically active radiation. *Physiol. Plant.* **51**:163-166.
74. Nobel, P.S. 1981. Influence of freezing temperatures on a cactus, *Coryphantha vivipara. Oecologia* **48**:194-198.
75. Longstreth, D.J., Hartsock, T.L., and Nobel, P.S. 1981. Light effects on leaf development and photosynthetic capacity of *Hydrocotyle bonariensis* Lam. *Photosynthesis Res.* **2**:95-104.
76. Jordan, P.W., and Nobel, P.S. 1981. Seedling establishment of *Ferocactus acanthodes* in relation to drought. *Ecology* **62**:901-906.
77. Nobel, P.S. 1981. Influences of photosynthetically active radiation on cladode orientation, stem tilting, and height of cacti. *Ecology* **62**:982-990.
78. Lange, O.L., Nobel, P.S., Osmond, C.B., and Ziegler, H. 1981. Introduction: Perspectives in ecological plant physiology. *In*: O.L. Lange, P.S. Nobel, C.B. Osmond, and H. Ziegler, eds., *Physiological Plant Ecology, Encyclopedia of Plant Physiology, New Series*, Vol. 12A, Springer-Verlag, Berlin. Pp. 1-9.
79. Nobel, P.S. 1981. Wind as an ecological factor. *In*: O.L. Lange, P.S. Nobel, C.B. Osmond, and H. Ziegler, eds., *Physiological Plant Ecology, Encyclopedia of Plant Physiology, New Series*, Vol. 12A, Springer-Verlag, Berlin. Pp. 475-500.
80. Nobel, P.S. 1981. Spacing and transpiration of various sized clumps of a desert grass, *Hilaria rigida. J. Ecol.* **69**:735-742.
81. Nobel, P.S., and Longstreth, D.J. 1981. Effects of environmental factors on leaf anatomy, mesophyll cell conductance, and photosynthesis. *In*: G. Akoyunoglou, ed., *Photosynthesis VI. Photosynthesis and Productivity, Photosynthesis and Environment*, Balban International Science Services, Philadelphia. Pp. 245-254.
82. Nobel, P.S., and Hartsock, T.L. 1981. Shifts in the optimal temperature for nocturnal CO_2 uptake caused by changes in growth temperatures for cacti and agaves. *Physiol. Plant.* **51**:163-166.
83. Nobel, P.S. 1981. Review: *Biophysical Ecology* by D.M. Gates. Springer-Verlag, Berlin. *Amer. Sci.* **69**:459.
84. Woodhouse, R.M., and Nobel, P.S. 1982. Stipe anatomy, water potentials, and xylem conductances in seven species of ferns (Filicopsida). *Amer. J. Bot.* **69**:135-140.
85. Didden-Zopfy, B., and Nobel, P.S. 1982. High temperature tolerance and heat acclimation of *Opuntia bigelovii. Oecologia* **52**:176-180.
86. Nobel, P.S. 1982. Interaction between morphology, PAR interception, and nocturnal acid accumulation in cacti. *In*: I.P. Ting and M. Gibbs, eds., *Crassulacean Acid Metabolism*, American Society of Plant Physiologists, Rockville, Maryland. Pp. 260-277.
87. Nobel, P.S. 1982. Orientations of terminal cladodes of platyopuntias. *Bot. Gaz.* **143**:219-224.
88. Nobel, P.S. 1982. Orientation, PAR interception, and nocturnal acidity increases for terminal cladodes of a widely cultivated cactus, *Opuntia ficus-indica. Amer. J. Bot.* **69**:1462-1469.
89. Nobel, P.S. 1982. Low-temperature tolerance and cold hardening of cacti. *Ecology* **63**:1650-1656.
90. Jordan, P.W., and Nobel, P.S. 1982. Height distributions of two species of cacti in relation to rainfall, seedling establishment, and growth. *Bot. Gaz.* **143**:511-517.
91. Nobel, P.S., and Hartsock, T.L. 1983. Relationships between photosynthetically active radiation, nocturnal acid accumulation, and CO_2 uptake for a Crassulacean acid metabolism plant, *Opuntia ficus-indica. Plant Physiol.* **71**:71-75.
92. Nobel, P.S. 1983. Low and high temperature influences on cacti. *In*: R. Marcelle, H. Clijsters, and M. van Pouke, eds., *Effects of Stress on Photosynthesis*, Proc. Inter. Conf., Limburgs Universitair Centrum, Diepenbek, Belgium, August 22-27, 1982. Pp. 165-174.
93. Nobel, P.S. 1983. Spine influences on PAR interception, stem temperature, and nocturnal acid accumulation by cacti. *Plant Cell Environ.* **6**:153-159.

94. Robberecht, R., and Nobel, P.S. 1983. A fibonacci sequence in rib number for a barrel cactus. *Ann. Bot.* **51**:153-155.

95. Woodhouse, R.M., Williams, J.G., and Nobel, P.S. 1983. Simulation of plant temperature and water loss by the desert succulent, *Agave deserti*. *Oecologia* **57**:291-297.

96. Acevedo, E., Badilla, I., and Nobel, P.S. 1983. Water relations, diurnal acidity changes, and productivity of a cultivated cactus, *Opuntia ficus-indica*. *Plant Physiol.* **72**:775-780.

97. Nobel, P.S. 1983. Nutrient levels in cacti — Relation to nocturnal acid accumulation and growth. *Amer. J. Bot.* **70**:1244-1253.

98. Nobel, P.S., and Jordan, P.W. 1983. Transpiration stream of desert species: Resistances and capacitances for a C_3, a C_4, and a CAM plant. *J. Exp. Bot.* **34**:1379-1391.

99. Smith, S.D., Hartsock, T.L., and Nobel, P.S. 1983. Ecophysiology of *Yucca brevifolia*, an aborescent monocot of the Mojave Desert. *Oecologia* **60**:10-17.

100. Robberecht, R., Mahall, B.E., and Nobel, P.S. 1983. Experimental removal of intraspecific competitors — Effects on water relations and productivity of a desert bunchgrass, *Hilaria rigida*. *Oecologia* **60**:21-24.

101. Nobel, P.S., and Smith, S.D. 1983. High and low temperature tolerances and their relationships to distribution of agaves. *Plant Cell Environ.* **6**:711-719.

102. Nobel, P.S., and Hartsock, T.L. 1984. Physiological responses of *Opuntia ficus-indica* to growth temperature. *Physiol. Plant.* **60**:98-105.

103. Nobel, P.S. 1984. PAR and temperature influences on CO_2 uptake by desert CAM plants. Proceedings, IV International Congress on Photosynthesis. *Adv. Photosynthesis Res. IV.* **3**:193-200.

104. Barcikowski, W., and Nobel, P.S. 1984. Water relations of cacti during desiccation: Distribution of water in tissues. *Bot. Gaz.* **145**:110-115.

105. Smith, S.D., Didden-Zopfy, B., and Nobel, P.S. 1984. High temperature responses of North American cacti. *Ecology* **65**:643-651.

106. Nobel, P.S., and Sanderson, J. 1984. Rectifier-like activities of roots of two desert succulents. *J. Exp. Bot.* **35**:727-737.

107. Gibson, A.C., Calkin, H.W., and Nobel, P.S. 1984. Xylary anatomy, water flow, and hydraulic conductance in the fern *Cyrtomium falcatum*. *Amer. J. Bot.* **71**:564-574.

108. Nobel, P.S., Lüttge, U., Heuer, S., and Ball, E. 1984. Influence of applied NaCl on Crassulacean acid metabolism and ionic levels in a cactus, *Cereus validus*. *Plant Physiol.* **75**:799-803.

109. Lüttge, U., and Nobel, P.S. 1984. Day-night variations in malate concentration, osmotic pressure, and hydrostatic pressure in *Cereus validus*. *Plant Physiol.* **75**:804-807.

110. Nobel, P.S. 1984. Extreme temperatures and thermal tolerances for seedlings of desert succulents. *Oecologia* **62**:310-317.

111. Nobel, P.S. 1984. Productivity of *Agave deserti*: Measurements by dry weight and monthly prediction using physiological responses to environmental parameters. *Oecologia* **64**:1-7.

112. Jordan, P.W., and Nobel, P.S. 1984. Thermal and water relations of roots of desert succulents. *Ann. Bot.* **54**:705-717.

113. Nobel, P.S., Calkin, H.W., and Gibson, A.C. 1984. Influences of PAR, temperature, and water vapor concentration on gas exchange by ferns. *Physiol. Plant.* **62**:527-534.

114. Geller, G.N., and Nobel, P.S. 1984. Cactus ribs: Influence on PAR interception and CO_2 uptake. *Photosynthetica* **18**:482-494.

115. Nobel, P.S. 1984. Review: *CO_2 and Plants* by E.R. Lemon, ed. AAAS, Westview Press, Boulder, CO. *Quart. Rev. Biol.* **59**:328-329.

116. Gibson, A.C., Calkin, H.W., Raphael, D.O., and Nobel, P.S. 1985. Water relations and xylem anatomy of ferns. *Proc. Roy. Soc. Edinburgh* **86B**:81-92.

117. Calkin, H.W., Gibson, A.C., and Nobel, P.S. 1985. Xylem water potentials and hydraulic conductances in eight species of ferns. *Can. J. Bot.* **63**:631-637.

118. Nobel, P.S. 1985. PAR, water, and temperature limitations on the productivity of cultivated *Agave fourcroydes* (henequen). *J. Appl. Ecol.* **22**:157-173.

119. Garcia de Cortázar, V., Acevedo, E., and Nobel, P.S. 1985. Modeling of PAR interception and productivity by *Opuntia ficus-indica*. *Agric. Forest Meteor.* **34**:145-162.

120. Nobel, P.S., and Berry, W.L. 1985. Element responses of agaves. *Amer. J. Bot.* **72**:686-694.

121. Berry, W.L., and Nobel, P.S. 1985. Influence of soil and mineral stresses on cacti. *J. Plant Nutrition* **8**:679-696.

122. Nobel, P.S. 1985. Desert succulents. *In*: B.F. Chabot and H.A. Mooney, eds., *Physiological Ecology of North American Plant Communities*, Chapman & Hall, London. Pp. 181-197.

123. Nobel, P.S., and Walker, D.B. 1985. Structure of leaf photosynthetic tissue. *In*: J. Barber, and N.R. Baker, eds., *Topics in Photosynthesis, Vol. 6, Photosynthetic Mechanisms and the Environment*. Elsevier, Amsterdam. Pp. 501-536.

124. Nobel, P.S. 1985. Environmental responses of agaves — A case study with *Agave deserti*. *In*: C. Cruz, L. del Castillo, M. Robert, and R.N. Ondarza, eds., *Biologia y Aprovechamiento Integral del Henequen y otros Agaves*. Centro de Investigacion Cientifica de Yucatan, A.C., Merida, Mexico. Pp. 55-66.

125. Nobel, P.S. 1985. Water relations and carbon dioxide uptake of *Agave deserti* — Special adaptations to desert climates. *Desert Plants* **7**:51-56.

126. Gibson, A.C., Calkin, H.W., and Nobel, P.S. 1985. Hydraulic conductance and xylem structure in tracheid-bearing plants. *IAWA Bull.* **6**:293-302.

127. Chuan Kee, S., and Nobel, P.S. 1985. Fatty acid composition of chlorenchyma membrane fractions from three desert succulents grown at moderate and high temperatures. *Biochim. Biophys. Acta* **820**:100-106.

128. Nobel, P.S., and Long, S.P. 1985. Canopy structure and light interception. *In*: J. Coombs, D.O. Hall, S.P. Long, and J. Scurlock, eds., *Techniques in Bioproductivity and Photosynthesis*, 2nd ed., Pergamon, London. Pp. 41-49.

129. Nobel, P.S., and Meyer, S.E. 1985. Field productivity of a CAM plant, *Agave salmiana*, estimated using daily acidity changes under various environmental conditions. *Physiol. Plant.* **65**:397-404.

130. Nobel, P.S., and Quero, E. 1986. Environmental productivity indices for a Chihuahuan Desert CAM plant, *Agave lechuguilla*. *Ecology* **67**:1-11.

131. Garcia de Cortázar, V., and Nobel, P.S. 1986. Modeling of PAR interception and productivity of a prickly pear cactus, *Opuntia ficus-indica* L., at various spacings. *Agron. J.* **78**:80-85.

132. Nobel, P.S., and Hartsock, T.L. 1986. Temperature, water, and PAR influences on predicted and measured productivity of *Agave deserti* at various elevations. *Oecologia* **68**:181-185.

133. Kee, S.C., and Nobel, P.S. 1986. Concomitant changes in high temperature tolerance and heat-shock proteins in desert succulents. *Plant Physiol.* **80**:596-598.

134. Nobel, P.S. 1986. Relation between monthly growth of *Ferocactus acanthodes* and an environmental productivity index. *Amer. J. Bot.* **73**:541-547.

135. Nobel, P.S., and Hartsock, T.L. 1986. Leaf and stem CO_2 uptake in the three subfamilies of the Cactaceae. *Plant Physiol.* **80**:913-917.

136. Young, D.R., and Nobel, P.S. 1986. Predictions of soil-water potentials in the north-western Sonoran Desert. *J. Ecol.* **74**:143-154.

137. Nobel, P.S., Geller, G.N., Kee, S.C., and Zimmerman, A.D. 1986. Temperatures and thermal tolerances for cacti exposed to high temperatures near the soil surface. *Plant Cell Environ.* **9**:279-287.

138. Raphael, D.O., and Nobel, P.S. 1986. Growth and survivorship of ramets and seedlings of *Agave deserti*: influences of parent-ramet connections. *Bot. Gaz.* **147**:78-83.

139. Nobel, P.S. 1986. Form and orientation in relation to PAR interception by cacti and agaves. *In*: T.J. Givnish, ed., *On the Economy of Plant Form and Function*. Cambridge University Press, Cambridge. Pp. 83-103.

140. Nobel, P.S., and Hartsock, T.L. 1986. Influence of nitrogen and other nutrients on the growth of *Agave deserti. J. Plant Nutrition* 9:1273-1288.
141. Geller, G.N., and Nobel, P.S. 1986. Branching patterns of columnar cacti: Influences on PAR interception and CO_2 uptake. *Amer. J. Bot.* 73:1193-1200.
142. Smith, S.D., and Nobel, P.S. 1986. Deserts. *In*: N.R. Baker and S.P. Long, eds., *Photosynthesis in Contrasting Environments*. Elsevier Science Publishers B.V., Amsterdam. Pp. 13-62.
143. Smith, J.A.C., and Nobel, P.S. 1986. Water movement and storage in a desert succulent: Anatomy and rehydration kinetics for leaves of *Agave deserti. J. Exp. Bot.* 37:1044-1053.
144. Calkin, H.W., Gibson, A.C., and Nobel, P.S. 1986. Biophysical model of xylem conductance in tracheids of the fern *Pteris vittata. J. Exp. Bot.* 37:1054-1064.
145. Nobel, P.S., and Hartsock, T.L. 1986. Short-term and long-term responses of Crassulacean acid metabolism plants to elevated CO_2. *Plant Physiol.* 82:604-606.
146. Calkin, H.W., and Nobel, P.S. 1986. Nonsteady-state analysis of water flow and capacitance for *Agave deserti. Can. J. Bot.* 64:2556-2560.
147. Nobel, P.S., and Hartsock, T.L. 1986. Environmental influences on the productivity of three desert succulents in the south-western United States. *Plant Cell Environ.* 9:741-749.
148. Nobel, P.S., and Franco, A.C. 1986. Annual root growth and intraspecific competition for a desert bunchgrass. *J. Ecol.* 74:1119-1126.
149. Osmond, C.B., Austin, M.P., Berry, J.A., Billings, W.D., Boyer, J.S., Dacey, J.W.H., Nobel, P.S., Smith, S.D., and Winner, W.E. 1987. Stress physiology and the distribution of plants. *BioScience* 37:38-48.
150. Nobel, P.S. 1987. Water relations and plant size aspects of flowering for *Agave deserti. Bot. Gaz.* 148:79-84.
151. Nobel, P.S., and Geller, G.N. 1987. Temperature modelling of wet and dry desert soils. *J. Ecol.* 75:247-258.
152. Nobel, P.S., and Valenzuela, A.G. 1987. Environmental responses and productivity of the CAM plant, *Agave tequilana. Agric. Forest Meterol.* 39:319-334.
153. Hunt, E.R., Jr., and Nobel, P.S. 1987. A two-dimensional model for water uptake by desert succulents: Implications of root distribution. *Ann. Bot.* 59:559-569.
154. Hunt, E.R., Jr., and Nobel, P.S. 1987. Allometric root/shoot relationships and predicted water uptake for desert succulents. *Ann. Bot.* 59:571-577.
155. Nobel, P.S., Russell, C.E., Felker, P., Galo, J.M., and Acuña, E. 1987. Nutrient relations and productivity of prickly pear cacti. *Agron. J.* 79:550-555.
156. Geller, G.N., and Nobel, P.S. 1987. Comparative cactus architecture and PAR interception. *Amer. J. Bot.* 74:998-1007.
157. Nobel, P.S., and Hartsock, T.L. 1987. Drought-induced shifts in daily CO_2 uptake patterns for leafy cacti. *Physiol. Plant.* 70:114-118.
158. Hunt, E.R., Jr., Rock, B.N., and Nobel, P.S. 1987. Measurement of leaf relative water content by infrared reflectance. *Remote Sensing Environ.* 22:429-435.
159. Chetti, M.B., and Nobel, P.S. 1987. High-temperature sensitivity and its acclimation for photosynthetic electron transport reactions of desert succulents. *Plant Physiol.* 84:1063-1067.
160. Hunt, E.R., Jr., Zakir, N.J.D., and Nobel, P.S. 1987. Water costs and water revenues for established and rain-induced roots of *Agave deserti. Funct. Ecol.* 1:125-129.
161. Nobel, P.S. 1987. Photosynthesis and productivity of desert plants. *In*: L. Berkofsky and M.G. Wurtele, eds., *Progress in Desert Research*, Rowman & Littlefield, Totowa, New Jersey. Pp. 41-66.
162. Schulte, P.J., Gibson, A.C., and Nobel, P.S. 1987. Xylem anatomy and hydraulic conductance of *Psilotum nudum. Amer. J. Bot.* 74:1438-1445.
163. Hunt, E.R., Jr., and Nobel, P.S. 1987. Non-steady-state water flow for three desert perennials with different capacitances. *Aust. J. Plant Physiol.* 14:363-375.

164. Nobel, P.S. 1987. Transpiration analysis using resistances and capacitances. *In*: D.W. Newman and K.G. Wilson, eds., *Models in Plant Physiology and Biochemistry*, Volume III. CRC Press, Boca Raton, Florida. Pp. 37-39.

165. Smith, J.A.C., Schulte, P.J., and Nobel, P.S. 1987. Water flow and water storage in *Agave deserti*: Osmotic implications of Crassulacean acid metabolism. *Plant Cell Environ.* **10**:639-648.

166. Quero, E., and Nobel, P.S. 1987. Predictions of field productivity for *Agave lechuguilla*. *J. Appl. Ecol.* **24**:1053-1062.

167. Nobel, P.S., and Garcia de Cortázar, V. 1987. Interception of photosynthetically active radiation and predicted productivity for *Agave* rosettes. *Photosynthetica* **21**:261-272.

168. Nobel, P.S. 1988. Productivity of desert succulents. *In*: E.E. Whitehead, C.F. Hutchinson, B.N. Timmermann, and R.G. Varady, eds., *Arid Lands: Today and Tomorrow*. Westview Press, Boulder, Colorado. Pp. 137-147.

169. Silverman, E.P., Young, D.R., and Nobel, P.S. 1988. Effects of applied NaCl on *Opuntia humifusa*. *Physiol. Plant.* **72**:343-348.

170. Tissue, D.T., and Nobel, P.S. 1988. Parent-ramet connections in *Agave deserti*: Influences of carbohydrates on growth. *Oecologia* **75**:266-271.

171. Denison, R.F., and Nobel, P.S. 1988. Growth of *Agave deserti* without current photosynthesis. *Photosynthetica* **22**:51-57.

172. Nobel, P.S., and McDaniel, R.G. 1988. Low temperature tolerances, nocturnal acid accumulation, and biomass increases for seven species of agave. *J. Arid Environ.* **15**:147-155.

173. Geller, G.N., and Nobel, P.S. 1988. Cactus morphology: Effect on the interception of photosynthetically active radiation and CO_2 uptake. *In*: J.E. Keeley and G. Sibley, eds., *Desert Ecology 1986: A Research Symposium*. Southern California Academy of Sciences and the Southern California Desert Studies Consortium, Los Angeles. Pp. 149-161.

174. Nobel, P.S., Quero, E., and Linares, H. 1988. Differential growth responses of agaves to nitrogen, phosphorus, potassium, and boron applications. *J. Plant Nutrition* **11**:1683-1700.

175. Chetti, M.B., and Nobel, P.S. 1988. Recovery of photosynthetic reactions after high-temperature treatments of a heat-tolerant cactus. *Photosynthesis Res.* **18**:277-286.

176. Franco, A.C., and Nobel, P.S. 1988. Interactions between seedlings of *Agave deserti* and the nurse plant *Hilaria rigida*. *Ecology* **69**:1731-1740.

177. Nobel, P.S. 1988. Principles underlying the prediction of temperature in plants, with special reference to desert succulents. *In*: S.P. Long and F.I. Woodward, eds., *Plants and Temperature*. Society for Experimental Biology, Company of Biologists, Cambridge, U.K. Pp. 1-23.

178. Schulte, P.J., and Nobel, P.S. 1989. Responses of a CAM plant to drought and rainfall: Capacitance and osmotic pressure influences on water movement. *J. Exp. Bot.* **40**:61-70.

179. Nobel, P.S. 1989. Influence of photoperiod on growth for three desert CAM species. *Bot. Gaz.* **150**:9-14.

180. Palta, J.A., and Nobel, P.S. 1989. Root respiration for *Agave deserti*: Influence of temperature, water status, and root age on daily patterns. *J. Exp. Bot.* **40**:181-186.

181. Nobel, P.S. 1989. Productivity and water-use efficiency of prickly pear cacti. *In*: C.W. Hanselka and J.C. Paschal, eds., *Developing Prickly Pear as Forage, Fruit and Vegetable Resource*. Texas A & I University, Kingsville, Texas. Pp. 1-5.

182. Palta, J.A., and Nobel, P.S. 1989. Influences of water status, temperature, and root age on daily patterns of root respiration for two cactus species. *Ann. Bot.* **63**:651-662.

183. Palta, J.A., and Nobel, P.S. 1989. Influence of soil O_2 and CO_2 on root respiration for *Agave deserti*. *Physiol. Plant.* **76**:187-192.

184. Schulte, P.J., Gibson, A.C., and Nobel, P.S. 1989. Water flow in vessels with simple or compound perforation plates. *Ann. Bot.* **64**:171-178.

185. Nobel, P.S. 1989. Shoot temperatures and thermal tolerances for succulent species of *Haworthia* and *Lithops*. *Plant Cell Environ.* **12**:643-651.
186. Nobel, P.S. 1989. A nutrient index quantifying productivity of agaves and cacti. *J. Appl. Ecol.* **26**:635-645.
187. Nobel, P.S. 1989. Temperature, water availability, and nutrient levels at various soil depths — Consequences for shallow-rooted desert succulents, including nurse plant effects. *Amer. J. Bot.* **76**:1486-1492.
188. Nobel, P.S., and Palta, J.A. 1989. Soil O_2 and CO_2 effects on root respiration of cacti. *Plant Soil* **120**:263-271.
189. Franco, A.C., and Nobel, P.S. 1989. Effect of nurse plants on the microhabitat and growth of cacti. *J. Ecol.* **77**:870-886.
190. Schulte, P.J., Smith, J.A.C., and Nobel, P.S. 1989. Water storage and osmotic pressure influences on the water relations of a dicotyledonous desert succulent. *Plant Cell Environ.* **12**:831-842.
191. Nobel, P.S., Quero, E., and Linares, H. 1989. Root versus shoot biomass: Responses to water, nitrogen, and phosphorus applications for *Agave lechuguilla*. *Botan. Gaz.* **150**:411-416.
192. Nobel, P.S. 1989. Productivity of desert succulents. *Excelsa* **14**:21-28.
193. Tissue, D.T., and Nobel, P.S. 1990. Carbon relations of flowering in a semelparous clonal desert perennial. *Ecology* **71**:273-281.
194. Franco, A.C., and Nobel, P.S. 1990. Influences of root distribution and growth on predicted water uptake and interspecific competition. *Oecologia* **82**:151-157.
195. Garcia de Cortázar, V., and Nobel, P.S. 1990. Worldwide environmental productivity indices and yield predictions for a CAM plant, *Opuntia ficus-indica*, included effects of doubled CO_2 levels. *Agric. Forest Meteorol.* **49**:261-279.
196. Nobel, P.S., Schulte, P.J., and North, G.B. 1990. Water influx characteristics and hydraulic conductivity for roots of *Agave deserti* Engelm. *J. Exp. Bot.* **41**:409-415.
197. Nobel, P.S., and Hartsock, T.L. 1990. Diel patterns of CO_2 exchange for epiphytic cacti differing in succulence. *Physiol. Plant.* **78**:628-634.
198. Nobel, P.S. 1990. Low-temperature tolerance and CO_2 uptake for platyopuntias — A laboratory assessment. *J. Arid. Environ.* **18**:313-324.
199. Garcia-Moya, E., and Nobel, P.S. 1990. Leaf unfolding rates and responses to cuticle damaging for pulque agaves in Mexico. *Desert Plants* **10**:55-57.
200. Nobel, P.S., and Loik, M.E. 1990. Thermal analysis, cell viability, and CO_2 uptake of a widely distributed North American cactus, *Opuntia humifusa*, at subzero temperatures. *Plant Physiol. Biochem.* **28**:429-436.
201. Nobel, P.S. 1990. Soil O_2 and CO_2 effects on apparent cell viability for roots of desert succulents. *J. Exp. Bot.* **41**:1031-1038.
202. Nobel, P.S. 1990. Photosynthesis and field environmental productivity indices. *In*: M. Baltscheffsky, ed., *Current Research in Photosynthesis, Vol. IV*. (*Proc. VIIIth International Congr. Photosynthesis*, Stockholm, Sweden). Kluwer Academic Publishers, Dordrecht, The Netherlands. Pp. 821-825.
203. Nobel, P.S. 1990. Low temperature responses of cacti: A review. *In*: P. Felker, ed., *Proceedings, First Annual Texas Prickly Pear Council*. Caesar Kleberg Wildlife Research Institute, Kingsville, Texas. Pp. 38-48.
204. Nobel, P.S. 1990. Productivity of agaves and cacti. *In*: H. Elattir, ed., *Proc. Intern. Symp. on Drought Tolerant Species for Pre-Saharan Areas*, Association Iligh pour le Développment et la Coopération, Agadir, Morocco. Pp. 1-4.
205. Tissue, D.T., and Nobel, P.S. 1990. Carbon translocation between parents and ramets of a desert perennial. *Ann. Bot.* **66**:551-557.
206. Nobel, P.S. 1990. Environmental influences on CO_2 uptake by agaves, CAM plants with high productivities. *Econ. Bot.* **44**:488-502.
207. Nobel, P.S. 1990. Orientation, EPI, and productivity for *Opuntia ficus-indica*. *In*: J.J. Lopez Gonzalez and M.J. Ayala Ortega, eds., *El Nopal: Su Conocimiento y Aprovechamiento. III Reunion Nacional y I Internacional*. Universidad Autonoma Agraria "Antonio Narro," Saltillo, Coahuila, Mexico. Pp. 4-9.

208. Goldstein, G., Ortega, J.K.E., Nerd, A., and Nobel, P.S. 1991. Patterns of water potential components for the Crassulacean acid metabolism plant *Opuntia ficus-indica* when well-watered or droughted. *Plant Physiol.* **95**:274-280.

209. Nobel, P.S., and Garcia de Cortázar, V. 1991. Growth and predicted productivity of *Opuntia ficus-indica* for current and elevated carbon dioxide. *Agron. J.* **83**:224-230.

210. Alm, D.M., and Nobel, P.S. 1991. Root system water uptake and respiration for *Agave deserti*: Observations and predictions using a model based on individual roots. *Ann. Bot.* **67**:59-65.

211. Lopez, F.B., and Nobel, P.S. 1991. Root hydraulic conductivity of two cactus species in relation to root age, temperature, and soil water status. *J. Exp. Bot.* **235**:143-149.

212. Goldstein, G., Andrade, J.L., and Nobel, P.S. 1991. Differences in water relations parameters for the chlorenchyma and parenchyma of *Opuntia ficus-indica* under wet *versus* dry conditions. *Aust. J. Plant Physiol.* **18**:95-107.

213. Nobel, P.S. 1991. Ecophysiology of roots of desert plants, with special emphasis on agaves and cacti. *In*: Y. Waisel, A. Eshel, and U. Kafkafi, eds., *Plant Roots: The Hidden Half*. Marcel Dekker, New York. Pp. 839-866.

214. Nerd, A., and Nobel, P.S. 1991. Effects of drought on water relations and nonstructural carbohydrates in cladodes of *Opuntia ficus-indica*. *Physiol. Plant.* **81**:495-500.

215. Nobel, P.S., and Lee, C.H. 1991. Variations in root water potentials: Influence of environmental factors for two succulent species. *Ann. Bot.* **67**:549-554.

216. North, G.B., and Nobel, P.S. 1991. Changes in hydraulic conductivity and anatomy caused by drying and rewetting roots of *Agave deserti* (Agavaceae). *Amer. J. Bot.* **78**:906-915.

217. Tissue, D.T., Yakir, D., and Nobel, P.S. 1991. Diel water movement between parenchyma and chlorenchyma of two desert CAM plants under dry and wet conditions. *Plant Cell Environ.* **14**:407-413.

218. Rundel, P.W., and Nobel, P.S. 1991. Structure and function in desert root systems. *In*: D. Atkinson, ed., *Plant Root Growth — An Ecological Perspective*. Blackwell Scientific, Oxford. Pp. 349-378.

219. Nobel, P.S., and Meyer, R.W. 1991. Biomechanics of cladodes and cladode-cladode junctions for *Opuntia ficus-indica* (Cactaceae). *Amer. J. Bot.* **78**:1252-1259.

220. Nobel, P.S. 1991. Environmental productivity indices and productivity for *Opuntia ficus-indica* under current and elevated atmospheric CO_2 levels. *Plant Cell Environ.* **14**:637-646.

221. Garcia de Cortázar, V., and Nobel, P.S. 1991. Prediction and measurement of high annual productivity for *Opuntia ficus-indica*. *Agric. Forest Meteorol.* **56**:261-272.

222. Nobel, P.S., Lopez, F.B., and Alm, D.M. 1991. Water uptake and respiration for root systems of two cacti: Observations and predictions based on individual roots. *J. Exp. Bot.* **42**:1215-1223.

223. Nobel, P.S. 1991. Tansley Review 32. Achievable productivities of certain CAM plants: Basis for high values compared with C_3 and C_4 plants. *New Phytol.* **119**:183-205.

224. Nobel, P.S., Loik, M.E., and Meyer, R.W. 1991. Microhabitat and diel tissue acidity changes for two sympatric cactus species differing in growth habit. *J. Ecol.* **79**:167-182.

225. Goldstein, G., and Nobel, P.S. 1991. Changes in osmotic pressure and mucilage during low-temperature acclimation of *Opuntia ficus-indica*. *Plant Physiol.* **97**:954-961.

226. Loik, M.E., and Nobel, P.S. 1991. Water relations and mucopolysaccharide increases for a winter hardy cactus during acclimation to subzero temperatures. *Oecologia* **86**:340-346.

227. Nobel, P.S., and Goldstein, G. 1992. Desiccation and freezing phenomena for plants with large water capacitance — Cacti and espeletias. *In*: G.N. Somero, C.B. Osmond, and C.L. Bolis, eds., *Water and Life: Comparative Analysis of*

Water Relationships at the Organismic, Cellular, and Molecular Levels. Springer-Verlag, Berlin. Pp. 240-257.

228. Nobel, P.S., Alm, D.M., and Cavelier, J. 1992. Growth respiration, maintenance respiration, and structural-carbon costs for roots of three desert succulents. *Funct. Ecol.* **6**:79-85.

229. Nobel, P.S. 1992. Annual variations in flowering percentage, seedling establishment, and ramet production for a desert perennial. *Int. J. Plant Sci.* **153**:102-107.

230. North, G.B., and Nobel, P.S. 1992. Drought-induced changes in hydraulic conductivity and structure in roots of *Ferocactus acanthodes* and *Opuntia ficus-indica. New Phytol.* **120**:9-19.

231. Alm, D.M., Cavelier, J., and Nobel, P.S. 1992. A finite-element model of radial and axial conductivities for individual roots: Development and validation for two desert succulents. *Ann. Bot.* **69**:87-92.

232. Nobel, P.S., and Cui, M. 1992. Hydraulic conductances of the soil, the root-soil air gap, and the root: Changes for desert succulents in drying soil. *J. Exp. Bot.* **43**:319-326.

233. Nobel, P.S., Garcia-Moya, E., and Quero, E. 1992. High annual productivity of certain agaves and cacti under cultivation. *Plant Cell Environ.* **15**:329-335.

234. Nobel, P.S., Cavelier, J., and Andrade, J.L. 1992. Mucilage in cacti: Its apoplastic capacitance, associated solutes, and influence on tissue water relations. *J. Exp. Bot.* **43**:641-648.

235. Garcia de Cortázar, V., and Nobel, P.S. 1992. Biomass and fruit production for the prickly pear cactus, *Opuntia ficus-indica. J. Amer. Soc. Hort. Sci.* **117**:558-562.

236. Ewers, F.W., North, G.B., and Nobel, P.S. 1992. Root-stem junctions of a desert monocotyledon and a dicotyledon: Hydraulic consequences under wet conditions and during drought. *New Phytol.* **121**:377-385.

237. North, G.B., Ewers, F.W., and Nobel, P.S. 1992. Main root-lateral root junctions of two desert succulents: Changes in axial and radial components of hydraulic conductivity during drought. *Amer. J. Bot.* **79**:1039-1050.

238. Nobel, P.S., and Huang, B. 1992. Hydraulic and structural changes for lateral roots of desert succulents in response to soil drying and rewetting. *Int. J. Plant Sci.* **153**:S163-S170.

239. Nobel, P.S., and Cui, M. 1992. Prediction and measurement of gap water vapor conductance for roots located concentrically and eccentrically in air gaps. *Plant Soil* **145**:157-166.

240. Nobel, P.S., Miller, P.M., and Graham, E. 1992. Influence of rocks on soil temperature, soil water potential, and rooting patterns for desert succulents. *Oecologia* **92**:90-96.

241. Huang, B., and Nobel, P.S. 1992. Hydraulic conductivity and anatomy for lateral roots of *Agave deserti* during root growth and drought-induced abscission. *J. Exp. Bot.* **43**:1441-1449.

242. Nobel, P.S., and Cui, M. 1992. Shrinkage of attached roots of *Opuntia ficus-indica* in response to lowered water potentials — Predicted consequences for water uptake or loss to soil. *Ann. Bot.* **70**:485-491.

243. Luo, Y., and Nobel, P.S. 1992. Carbohydrate partitioning and compartmental analysis for a highly productive CAM plant, *Opuntia ficus-indica. Ann. Bot.* **70**:551-559.

244. Cui, M., and Nobel, P.S. 1992. Nutrient status, water uptake and gas exchange for three desert succulents infected with mycorrhizal fungi. *New Phytol.* **122**:643-649.

245. Nobel, P.S., Forseth, I.W., and Long, S.P. 1993. Canopy structure and light interception. *In*: D.O. Hall, J.M.O. Scurlock, H. Bolhàr-Nordenkampf, R.C. Leegood, and S.P. Long, eds., *Photosynthesis and Production in a Changing Environment: A Field and Laboratory Manual.* Chapman & Hall, London. Pp. 79-90.

246. Nobel, P.S., Huang, B., and García-Moya, E. 1993. Root distribution, growth, respiration, and hydraulic conductivity for two highly productive agaves. *J. Exp. Bot.* **44**:747-754.

247. Luo, Y., and Nobel, P.S. 1993. Growth characteristics of newly initiated cladodes of *Opuntia ficus-indica* as affected by shading, drought and elevated CO_2. *Physiol. Plant.* **87**:467-474.

248. Huang, B., and Nobel, P.S. 1993. Hydraulic conductivity and anatomy along lateral roots of cacti: Changes with soil water status. *New Phytol.* **123**:499-507.

249. North, G.B., Huang, B., and Nobel, P.S. 1993. Changes in structure and hydraulic conductivity for root junctions of desert succulents as soil water status varies. *Bot. Acta* **106**:126-135.

250. Loik, M.E., and Nobel, P.S. 1993. Freezing tolerance and water relations of *Opuntia fragilis* from Canada and the United States. *Ecology* **74**:1722-1732.

251. Nobel, P.S., and North G.B. 1993. Rectifier-like behaviour of root-soil systems: New insights from desert succulents. *In*: J.A.C. Smith and H. Griffiths, eds. *Water Deficits: Plant Responses From Cell to Community*. BIOS Scientific, Oxford. Pp. 163-176.

252. Gersani, M., Graham, E.A., and Nobel, P.S. 1993. Growth responses of individual roots of *Opuntia ficus-indica* to salinity. *Plant Cell Environ.* **16**:827-834.

253. Cui, M., Miller, P.M., and Nobel, P.S. 1993. CO_2 exchange and growth of the Crassulacean acid metabolism plant *Opuntia ficus-indica* under elevated CO_2 in open-top chambers. *Plant Physiol.* **103**:519-524.

254. Huang, B., North, G.B., and Nobel, P.S. 1993. Soil sheaths, photosynthate distribution to roots, and rhizosphere water relations for *Opuntia ficus-indica*. *Int. J. Plant Sci.* **154**:425-431.

255. Nobel, P.S., and Alm, D.M. 1993. Root orientation versus water uptake simulated for monocotyledonous and dicotyledonous succulents by a root-segment model. *Funct. Ecol.* **7**:600-609.

256. Loik, M.E., and Nobel, P.S. 1993. Exogenous abscisic acid mimics cold acclimation for cacti differing in freezing tolerance. *Plant Physiol.* **103**:871-876.

257. Nobel, P.S. 1993. Water conservation and productivity of certain CAM plants *In*: T.J. Mabry, H.T. Nguyen, R.A. Dixon, and M.S. Bonness, eds., *Biotechnology for Aridland Plants*. IC² Institute, The University of Texas at Austin, Austin, Texas. Pp. 59-71.

258. Nobel, P.S., Cui, M., Miller, P.M., and Luo, Y. 1994. Influences of soil volume and an elevated CO_2 level on growth and CO_2 exchange for the Crassulacean acid metabolism plant *Opuntia ficus-indica*. *Physiol. Plant.* **90**:173-180.

259. North, G.B., and Nobel, P.S. 1994. Changes in root hydraulic conductivity for two tropical epiphytic cacti as soil moisture varies. *Amer. J. Bot.* **81**:46-53.

260. Nobel, P.S. 1994. Root-soil responses to water pulses in dry environments. *In*: M.M. Caldwell and R.W. Pearcy, eds., *Exploitation of Environmental Heterogeneity by Plants: Ecophysiological Processes Above- and Belowground*. Academic Press, San Diego, California. Pp. 285-304.

261. Goldstein, G., and Nobel, P.S. 1994. Water relations and low-temperature acclimation for cactus species varying in freezing tolerance. *Plant Plysiol.* **104**:675-681.

262. Nobel, P.S., and Israel, A.A. 1994. Cladode development, environmental responses of CO_2 uptake, and productivity for *Opuntia ficus-indica* under elevated CO_2. *J. Exp. Bot.* **45**:295-303.

263. Pimienta-Barrios, E., and Nobel, P.S. 1994. Pitaya (*Stenocereus* spp., Cactaceae): An ancient and modern fruit crop of Mexico. *Econ. Bot.* **48**:76-83.

264. Nobel, P.S. 1994. Physiology and productivity of CAM plants such as *Opuntia ficus-indica*. *In*: Actas del II Congreso Internacional de la Tuna y Cochinilla. Universidad de Chile, Santiago, Chile. Pp. 105-112.

265. Cui, M., and Nobel, P.S. 1994. Water budgets and root hydraulic conductivity of opuntias shifted to low temperatures. *Int. J. Plant Sci.* **155**:167-172.

266. Inglese, P., Israel, A.A., and Nobel, P.S. 1994. Growth and CO_2 uptake for cladodes and fruits of the Crassulacean acid metabolism species *Opuntia ficus-indica* during fruit development. *Physiol. Plant.* **91**:708-714.

267. Israel, A.A., and Nobel, P.S. 1994. Activities of carboxylating enzymes in the CAM species *Opuntia ficus-indica* grown under current and elevated CO_2 concentrations. *Photosyn. Res.* **40**:223-229.

268. Nobel, P.S., and Loik, M.E. 1994. Low-temperature tolerance of prickly-pear cacti. *In*: P. Felker and J.R. Moss, eds. *Proceedings, Fourth Annual Texas Prickly Pear Council*. Caesar Kleberg Wildlife Research Institute, Kingsville, Texas. Pp. 1-9.

269. Cui, M., and Nobel, P.S. 1994. Gas exchange and growth responses to elevated CO_2 and light levels for the CAM species *Opuntia ficus-indica*. *Plant Cell Environ.* **17**:935-944.

270. Nobel, P.S., Cui, M., and Israel, A.A. 1994. Light, chlorophyll, carboxylase activity and CO_2 fixation at various depths in the chlorenchyma of *Opuntia ficus-indica* (L.) Miller under current and elevated CO_2. *New Phytol.* **128**:315-322.

271. Huang, B., and Nobel, P.S. 1994. Root hydraulic conductivity and its components, with emphasis on desert succulents. *Agron. J.* **86**:767-774.

272. Nobel, P.S., Andrade, J.L., Wang, N., and G.B. North. 1994. Water potentials for developing cladodes and fruits of a succulent plant, including xylem-versus-phloem implications for water movement. *J. Exp. Bot.* **45**:1801-1807.

273. North, G.B., Moore, T.L., and Nobel, P.S. 1995. Cladode development for *Opuntia ficus-indica* (Cactaceae) under current and doubled CO_2 concentrations. *Amer. J. Bot.* **82**:159-166.

274. Raveh, E., Gersani, M., and Nobel, P.S. 1995. CO_2 uptake and fluorescence responses for a shade-tolerant cactus *Hylocereus undatus* under current and doubled CO_2 concentrations. *Physiol. Plant.* **93**:505-511.

275. Nobel, P.S., and Pimienta-Barrios, E. 1995. Monthly stem elongation for *Stenocereus queretaroensis*: Relationships to environmental conditions, net CO_2 uptake, and seasonal variations in sugar content. *Environ. Exp. Bot.* **35**:17-24.

276. Israel, A.A., and Nobel, P.S. 1995. Growth temperature versus CO_2 uptake, Rubisco and PEPCase activities, and enzyme high-temperature sensitivities for a CAM plant. *Plant Physiol. Biochem.* **33**:345-351.

277. Wang, N., and Nobel, P.S. 1995. Phloem exudate collected via scale insect stylets for the CAM species *Opuntia ficus-indica* under current and doubled CO_2 concentrations. *Ann. Bot.* **75**:525-532.

278. North, G.B., and Nobel, P.S. 1995. Hydraulic conductivity of concentric root tissues of *Agave deserti* Engelm. under wet and drying conditions. *New Phytol.* **130**:47-57.

279. Nobel, P.S., Wang, N., Balsamo, R.A., Loik, M.E., and Hawke, M.A. 1995. Low-temperature tolerance and acclimation of Opuntia spp. after injecting glucose or methylglucose. *Int. J. Plant Sci.* **156**:496-504.

280. Zhang, H., Sharifi, M.R., and Nobel, P.S. 1995. Photosynthetic characteristics of sun versus shade plants of *Encelia farinosa* as affected by photosynthetic photon flux, intercellular CO_2 concentration, leaf water potential, and leaf temperature. *Aust. J. Plant Physiol.* **22**:833-841.

281. Nobel, P.S. 1995. Avances recientes en la ecofisiologia de *Opuntia ficus-indica* y otras cactaceas. *In*: E. Pimienta-Barrios, C. Neri Luna, A. Munoz Urias, and F.M. Huerta Martinez, eds. *Conocimiento y Aprovechamiento del Nopal. 6to. Congresso Nacional y 4to. Congreso Internacional*. Universidad de Guadalajara, Guadalajara, Jalisco, Mexico. Pp. 77-83.

282. Pimienta-Barrios, E., and Nobel, P.S. 1995. Reproductive characteristics of pitayo (*Stenocereus queretaroensis*) and their relationships with soluble sugars and irrigation. *J. Amer. Soc. Hort. Sci.* **120**:1082-1086.

283. Nerd, A., and Nobel, P.S. 1995. Accumulation, partitioning, and assimilation of nitrogen in *Opuntia ficus-indica*. *J. Plant Nutrition* **18**:2533-2549.

284. Nobel, P.S. 1995. Environmental biology. *In*: G. Barbera, P. Inglese, and E. Pimienta-Barrios, eds. *Agroecology, Cultivation and Uses of Cactus Pear*. Food and Agriculture Organization of the United Nations, Rome, Italy. Pp. 36-48.

285. Nobel, P.S. 1996. High productivity of certain agronomic CAM species. *In*: K. Winter and J.A.C. Smith, eds. *Crassulacean Acid Metabolism: Biochemistry, Ecophysiology and Evolution*. Springer, Berlin. Pp. 255-265.

286. Nobel, P.S., and North, G.B. 1996. Features of roots of CAM plants. *In*: K. Winter and J.A.C. Smith, eds. *Crassulacean Acid Metabolism: Biochemistry, Ecophysiology and Evolution.* Springer, Berlin. Pp. 266-280.
287. Nobel, P.S. 1996. Ecophysiology of roots of desert plants, with special emphasis on agaves and cacti. *In*: Y. Waisel, A. Eshel, and U. Kafkafi, eds. *Plant Roots: The Hidden Half,* 2nd Ed. Marcel Dekker, New York. Pp. 823-844.
288. Graham, E.A., and Nobel, P.S. 1996. Long-term effects of a doubled atmospheric CO_2 concentration on the CAM species *Agave deserti. J. Exp. Bot.* **47**:61-69.
289. Wang, N., and Nobel, P.S. 1996. Doubling the CO_2 concentration enhanced the activity of carbohydrate-metabolism enzymes, source carbohydrate production, photoassimilate transport, and sink strength for *Opuntia ficus-indica. Plant Physiol.* **110**:893-902.
290. North, G.B., and Nobel, P.S. 1996. Radial hydraulic conductivity of individual root tissues of *Opuntia ficus-indica* (L.) Miller as soil moisture varies. *Ann. Bot.* **77**:133-142.
291. Drennan, P.M., and Nobel, P.S. 1996. Temperature influences on root growth for *Encelia farinosa* (Asteraceae), *Pleuraphis rigida* (Poaceae), and *Agave deserti* (Agavaceae) under current and doubled CO_2 concentrations. *Amer. J. Bot.* **83**:133-139.
292. Andrade, J.L., and Nobel, P.S. 1996. Habitat, CO_2 uptake, and growth for the CAM epiphytic cactus *Epiphyllum phyllanthus* in a Panamanian tropical forest. *J. Tropical Ecol.* **12**:291-306.
293. Zhang, H., and Nobel, P.S. 1996. Photosynthesis and carbohydrate partitioning for the C_3 desert shrub *Encelia farinosa* under current and doubled CO_2 concentrations. *Plant Physiol.* **110**:1361-1366.
294. Nobel, P.S., Israel, A.A., and Wang, N. 1996. Growth, CO_2 uptake, and responses of the carboxylating enzymes to inorganic carbon for two highly productive CAM species at current and doubled CO_2 concentrations. *Plant Cell Environ.* **19**:585-592.
295. Nobel, P.S. 1996. Recent ecophysiological advances for *Opuntia ficus-indica* and other cacti. *In*: P. Felker and J. Moss, eds. Proceedings, First Annual Conference, Professional Association for Cactus Development, Dallas, Texas. Pp. 1-11.
296. Zhang, H., and Nobel, P.S. 1996. Dependency of c_i/c_a and leaf transpiration efficiency on the vapor pressure deficit. *Aust. J. Plant Physiol.* **23**:561-568.
297. Nobel, P.S. 1996. Responses of some North American CAM plants to freezing temperatures and doubled CO_2 concentrations: Implications of global climate change for extending cultivation. *J. Arid Environ.* **34**:187-196.
298. Nobel, P.S. 1996. Shading, osmoticum, and hormone effects on organ development for detached cladodes of *Opuntia ficus-indica. Int. J. Plant Sci.* **157**:722-728.
299. Mizrahi, Y., Nerd, A., and Nobel, P.S. 1997. Cacti as crops. *Hort. Rev.* **18**:291-319.
300. North, G.B., and Nobel, P.S. 1997. Root-soil contact for the desert succulent *Agave deserti* Engelm. in wet and drying soil. *New Phytol.* **135**:21-29.
301. Wang, N., Zhang, H., and Nobel, P.S. 1997. Phloem-xylem water flow in developing cladodes of *Opuntia ficus-indica* during sink-to-source transition. *J. Exp. Bot.* **48**:675-682.
302. Nobel, P.S. 1997. Root distribution and seasonal production in the northwestern Sonoran Desert for a C_3 subshrub, a C_4 bunchgrass, and a CAM leaf succulent. *Amer. J. Bot.* **84**:949-955.
303. Drennan, P.M., and Nobel, P.S. 1997. Frequencies of major C_3, C_4 and CAM perennials on different slopes in the northwestern Sonoran Desert. *Flora* **192**:297-304.
304. Nobel, P.S. 1997. Recientes descubrimentos ecosfisiológicos en *Opuntia ficus-indica.* Memorias de VII Congreso Nacional y V Internacional Sobre el Conocimiento y Aprovechamiento del Nopal, Monterrey, Nuevo Leon, México. Pp. 11-20.

305. Andrade, J.L., and Nobel, P.S. 1997. Microhabitats and water relations of epiphytic cacti and ferns in a lowland neotropical forest. *Biotropica* **29**:261-270.
306. North, G.B., and Nobel, P.S. 1997. Drought-induced changes in soil contact and hydraulic conductivity for roots of *Opuntia ficus-indica* with and without rhizosheaths. *Plant and Soil* **191**:249-258.
307. Nobel, P.S., and Zhang, H. 1997. Photosynthetic responses of three codominant species from the north-western Sonoran Desert — a C_3 deciduous subshrub, a C_4 deciduous bunchgrass, and a CAM evergreen leaf succulent. *Aust. J. Plant Physiol.* **24**:787-796.
308. Nobel, P.S., and Linton, M.J. 1997. Frequencies, microclimate, and root properties for three codominant perennials in the northwestern Sonoran Desert on north- versus south-facing slopes. *Ann. Bot.* **80**:731-739.
309. Nobel, P.S. 1997. Recent ecophysiological findings for *Opuntia ficus-indica*. *J. Professional Assoc. Cactus Development* **2**:89-96.
310. Pimienta, E., Hernandez, G., Domingues, A., and Nobel, P.S. 1998. Growth and development of the arborescent cactus *Stenocereus queretaroensis* in a subtropical semiarid environment, including the effects of gibberellic acid. *Tree Physiol.* **18**:59-64.
311. Nobel, P.S., and Castañeda, M. 1998. Seasonal, light, and temperature influences on organ initiation for unrooted cladodes of the prickly pear cactus *Opuntia ficus-indica*. *J. Amer. Soc. Hort. Sci.* **123**:47-51.
312. Wang, N., and Nobel, P.S. 1998. Phloem transport of fructans in the Crassulacean acid metabolism species *Agave deserti*. *Plant Physiol.* **116**:709-714.
313. Dubrovsky, J.G., North, G.B., and Nobel, P.S. 1998. Root growth, developmental changes in the apex, and hydraulic conductivity for *Opuntia ficus-indica* during drought. *New Phytol.* **138**:75-82.
314. Raveh, E., Wang, N., and Nobel, P.S. 1998. Gas exchange and metabolite fluctuations in green and yellow bands of variegated leaves of the monocotyledonous CAM species *Agave americana*. *Physiol. Plant.* **103**:99-106.
315. Nobel, P.S., Zhang, H., Sharifi, R., Castañeda, M., and Greenhouse, B. 1998. Leaf expansion, net CO_2 uptake, Rubisco activity, and efficiency of long-term biomass gain for the common desert subshrub *Encelia farinosa*. *Photosyn. Res.* **56**:67-73.
316. Nobel, P.S., Castañeda, M., North, G., Pimienta-Barrios, E., and Ruiz, A. 1998. Temperature influences on leaf CO_2 exchange, cell viability, and cultivation range for *Agave tequilana*. *J. Arid Environ.* **39**:1-9.
317. North, G.B., and Nobel, P.S. 1998. Water movement and structural plasticity along roots of a desert monocot during and after prolonged drought. *Plant Cell Environ.* **21**:705-713.
318. Wang, N., Zhang, H., and Nobel, P.S. 1998. Carbon flow and carbohydrate metabolism during sink-to-source transition for developing cladodes of *Opuntia ficus-indica*. *J. Exp. Bot.* **49**:1835-1843.
319. Drennan, P., and Nobel, P.S. 1998. Root growth dependence on soil temperature for *Opuntia ficus-indica*: Influences of air temperature and a doubled CO_2 concentration. *Funct. Ecol.* **12**:959-964.
320. Pimienta-Barrios, E., and Nobel, P.S. 1998. Vegetative, reproductive, and physiological adaptations to aridity of pitayo (*Stenocereus queretaroensis*, Cactaceae). *Econ. Bot.* **52**:401-411.
321. Raveh, E., and Nobel, P.S. 1999. CO_2 uptake and water loss accompanying vernalization for detached cladodes of *Opuntia ficus-indica*. *Int. J. Plant Sci.* **160**:92-97.
322. Nobel, P.S., and Loik, M.E. 1999. Form and function of cacti. *In*: R.H. Robichaux, ed. *Ecology of Sonoran Desert Plants and Plant Communities*. University of Arizona Press, Tucson. Pp. 143-163.
323. Graham, E.A., and Nobel, P.S. 1999. Root water uptake, leaf water storage and gas exchange of a desert succulent: Implications for root system redundancy. *Ann. Bot.* **84**:213-223.
324. Linton, M.J., and Nobel, P.S. 1999. Loss of water transport capacity due to xylem cavitation in roots of two CAM succulents. *Amer. J. Bot.* **86**:1538-1542.

325. Nobel, P.S. 1999. Photosynthetic characteristics of CAM succulents with high productivity. *In*: G. Garab, ed. Photosynthesis: Mechanisms and Effects, Vol. V. Kluver Academic Publishing, Dordrecht, The Netherlands. Pp. 3955-3960.

326. North, G.B., and Nobel, P.S. 2000. Heterogeneity in water availability alters cellular development and hydraulic conductivity along roots of a desert succulent. *Ann. Bot.* **85**: 247-255.

327. Nobel, P.S., and De la Barrera, E. 2000. Carbon and water balances for young fruits of platyopuntias. *Physiol. Plant.* **109**: 160-166.

328. Nobel. P.S. 2000. Crop ecosystem responses to climate change: Crassulacean acid metabolism crops. *In*: K.R. Reddy and H.F. Hodges, eds. *Climate Change and Global Crop Productivity.* CAB International, Oxford, UK. Pp. 315-331.

329. Pimienta-Barrios, Eu., Zañudo, J., Yepez, E., Pimienta-Barrios, En., and Nobel, P.S. 2000. Seasonal variations of net CO_2 uptake for cactus pear (*Opuntia ficus-indica*) and pitayo (*Stenocereus queretaroensis*) in a semi-arid environment. *J. Arid Environ.* **44**: 73-83.

330. Drennan, P.M., and Nobel, P.S. 2000. Responses of CAM species to increasing atmospheric CO_2 concentrations. *Plant Cell Environ.* **23**: 767-781.

331. Nerd, A., and Nobel, P.S. 2000. Water relations during ripening for fruit of well-watered versus water-stressed *Opuntia ficus-indica*. *J. Amer. Soc. Hort. Sci.* **125**: 653-657.

332. Bobich, E.G., and Nobel, P.S. 2001. Biomechanics and anatomy of cladode junctions for two *Opuntia* (Cactaceae) species and their hybrid. *Am. J. Bot.* **88**: 391-400.

333. Bobich, E.G., and Nobel, P.S. 2001. Vegetative reproduction as related to biomechanics, morphology and anatomy of four cholla cactus species in the Sonoran Desert. *Ann. Bot.* **87**: 485-493.

334. Martre, P., North, G.B., and Nobel, P.S. 2001. Hydraulic conductance and mercury-sensitive water transport for roots of *Opuntia acanthocarpa* in relation to soil drying and rewetting. *Plant Physiol.* **126**: 352-362.

335. Linton, M.J., and Nobel, P.S. 2001. Hydraulic conductance, xylem cavitation and water potential for succulent leaves of *Agave deserti* and *A. tequilana*. *Int. J. Plant Sci.* **162**: 747-754.

336. Pimienta-Barrios, E., Robles-Murguia, C., and Nobel, P.S. 2001. Net CO_2 uptake for *Agave tequilana* in a warm and a temperate environment. *Biotropica* **33**: 312-318.

337. Nobel, P.S. 2001. Ecophysiology of *Opuntia ficus-indica*. In: C. Mondragon-Jocobo and S. Perez-Gonzalez, eds, Cactus (Opuntia spp.) As Forage, FAO Plant Production and Protection Paper 169, FAO, Rome, Pp. 13-20.

338. Pimienta-Barrios, E., Gonzalez del Castillo-Aranda, M.E., and Nobel, P.S. 2002. Ecophysiology of a wild platyopuntia exposed to prolonged drought. *Environ. Exp. Bot.* **47**: 77-86.

339. Nobel, P.S., and Bobich, E.G. 2002. Plant frequency, stem and root characteristics, and CO_2 uptake for *Opuntia acanthocarpa*: Elevational correlates in the northwestern Sonoran Desert. *Oecologia* **130**: 165-172.

340. Nobel, P.S. 2002. Ecophysiology of roots of desert plants, with special reference to agaves and cacti. In: Y. Waisel, A. Eshel, and U. Kafkafi, eds. *Plant Roots: The Hidden Half,* 3rd ed. Marcel Dekker, New York. Pp. 961-973.

341. Nobel, P.S., Pimienta-Barrios, E., Zanudo Hernandez, J., Ramirez-Hernandez, B. 2002. Historical aspects and net CO_2 uptake for cultivated CAM plants in Mexico. *Ann. Appl. Biol.* **140**: 133-142.

342. Pimienta-Barrios, E., Pimienta-Barrios, En., Salas-Galvan, M.E., Zanudo-Hernandez, J., and Nobel, P.S. 2002. Growth and reproductive characteristics for the columnar cactus *Stenocereus queretaroensis* and their relationships with environmental factors and colonization by arbuscular mycorrhizae. *Tree Physiol.* **21**: 667-674.

343. Nobel, P.S., and Bobich, E.G. 2002. Environmental biology. *In*: P.S. Nobel, ed. *Cacti: Biology and Uses*, University of California Press, Berkeley, California. Pp. 57-74.

344. Nobel, P.S., and De la Barrera, E. 2002. High temperatures and net CO_2 uptake, growth, and stem damage for the hemiepiphytic cactus *Hylocereus undatus*. *Biotropica* **34**: 225-231.

345. Bobich, E.G., and Nobel, P.S. 2002. Cladode junction regions and their biomechanics for arborescent platyopuntias. *Int. J. Plant Sci.* **163**: 507-517.

346. Nobel, P.S. 2002. Physiological ecology of columnar cacti. In: T.H. Fleming and A. Valiente-Banuet, eds. *Evolution, Ecology and Conservation of Columnar Cacti and Their Mutualists*, University of Arizona Press, Tucson, Arizona. Pp. 189-204.

347. Nobel, P.S., and De la Barrera, E. 2002. Stem water relations and net CO_2 uptake for a hemiepiphytic cactus during short-term drought. *Environ. Exp. Bot.* **48**: 129-137.

348. Nobel, P.S., and Bobich, E.G. 2002. Initial net CO_2 uptake responses and root growth for a CAM community placed in a closed environment. *Ann. Bot.* **90**: 593-598.

349. Nobel, P.S. 2002. Cactus physiological ecology, emphasizing gas exchange of cactus fruits. *Acta Hort.* **581**: 143-150.

350. Martre, P., North, G.B., Bobich, E.G., and Nobel, P.S. 2002. Root deployment and shoot growth for two desert species in response to soil rockiness. *Am. J. Bot.* **89**: 1933-1939.

351. Nobel, P.S., and De la Barrera, E. 2002. Nitrogen relations for net CO_2 uptake by the cultivated hemiepiphytic cactus, *Hylocereus undatus*. *Scientia Hort.* **96**: 281-292.

352. Martre, P., Morillon, R., Barrieu, F., North, G.B., Nobel, P.S., and Chrispeels, M.J. 2002. Plasma membrane aquaporins play a significant role during recovery from water deficit. *Plant Physiol.* **130**: 2101-2110.

353. Nobel, P.S., De la Barrera, E., Beilman, D.W., Doherty, J.H., and Zutta, B.R. 2002. Temperature limitations for cultivation of edible cacti in California. *Madroño* **49**: 228-236.

354. De la Barrera, E., and Nobel, P.S. 2003. Physiological ecology of seed germination for the columnar cactus *Stenocereus queretaroensis*. *J. Arid Environ.* **53**: 297-306.

355. Nobel, P.S., and De la Barrera, E. 2003. Tolerances and acclimation to low and high temperatures for cladodes, fruits and roots of a widely cultivated cactus, *Opuntia ficus-indica*. *New Phytologist* **157**: 271-279.

356. Nobel, P.S. 2003. Water relations of plants/Basic water relations. In: D. Murphy and B. Murray, eds. *Encyclopedia of Applied Sciences*, Academic Press, London. Pp. 1435-1440.

357. Pimienta-Barrios, E., Gonzalez del Castillo-Aranda, M.E., Muñoz-Urias, A., and Nobel, P.S. 2003. Effects of benomyl and drought on the mycorrhizal development and daily net CO_2 uptake of a wild platyopuntia in a rocky semi-arid environment. *Ann. Bot.* **92**: 239-245.

358. Nobel, P.S. and De la Barrera, E. 2004. CO_2 uptake by the cultivated hemiepiphytic cactus, *Hylocereus undatus*. *Ann. Appl. Biol.* **144**: 1-8.

359. North, G.B., Martre, P., and Nobel, P.S. 2004. Aquaporins account for variations in hydraulic conductance for metabolically active root regions of *Agave deserti* in wet, dry, and rewetted soil. *Plant Cell Environ.* **27**: 219-228.

360. De la Barrera, E. and Nobel, P.S. 2004. Carbon and water relations for developing fruits of *Opuntia ficus-indica* (L.) Miller, including effects of drought and gibberellic acid. *J. Exp. Bot.* **55**: 719-729.

361. De la Barrera, E. and Nobel, P.S. 2004. Nectar: properties, floral aspects, and speculations on origin. *Trends Plant Sci.* **9**: 65-69.

362. Pimienta-Barrios, E., Pimienta-Barrios, En., and Nobel, P.S. 2004. Ecophysiology of the pitayo de Queretaro (*Stenocereus queretaroensis*). *J. Arid Environ.* **59**: 1-17.

363. Smith, W.K., Nobel, P.S., Reiners, W.A., Vogelmann, T.C., and Chritchley, C. 2004. Summary and future perspectives. In: W.K. Smith, T.C. Vogelmann, and C. Critchley, eds. *Photosynthetic Adaptation: Chloroplast to Landscape*. Springer, New York. Pp. 297-309.

364. Graham, E.A. and Nobel, P.S. 2005. Daily changes in stem thickness and related gas exchange patterns for the hemiepiphytic cactus *Hylocereus undatus*. *Int. J. Plant Sci.* **166:** 13-20.

365. Pimienta-Barrios, E., Zañudo-Hernandez, J., Rosas-Espinosa, V.C., Valenzuela-Tapia, A, and Nobel, P.S. 2005. Young daughter cladodes affect CO_2 uptake by mother cladodes of *Opuntia ficus-indica*. *Ann. Bot.* **95:** 363-369.

366. Nobel, P.S. and Zutta, B.R. 2005. Morphology, ecophysiology, and seedling establishment for *Fouquieria splendens* in the northwestern Sonoran Desert. *J. Arid Environ.* **62:** 251-265.

367. Pimienta-Barrios, E., Zañudo-Hernandez, J., and Nobel, P.S. 2005. Effects of young cladodes on the gas exchange of basal cladodes of *Opuntia ficus-indica* (Cactaceae) under wet and dry conditions. *Int. J. Plant Sci.* **166:** 961-968.

368. Nobel, P.S. 2006. Parenchyma–chlorenchyma water movement during drought for the hemiepiphytic cactus *Hylocereus undatus*. *Ann. Bot* **97:** 469-474.

369. Ben-Asher, J., Nobel, P.S., Yossov, E., Mizrahi, Y. 2006. Net CO_2 uptake rates for *Hylocereus undatus* and *Selenicereus megalanthus* under field conditions: Drought influence and a novel method for analyzing temperature dependence. *Photosynthetica* **44:** 181-186.

370. Nobel, P.S., Zutta, B.R. 2007. Rock associations, root depth, and temperature tolerances for the "rock live-forever" (*Dudleya saxosa*) at three elevations in the northwestern Sonoran Desert. *J. Arid Environ.* **69:** 15-28.

371. Nobel, P.S., Zutta, B.R. 2007. Carbon dioxide uptake, water relations and drought survival for *Dudleya saxosa*, the 'rock live-forever', growing in small soil volumes. *Funct. Ecol.* **21:** 698-704.

372. Pimienta-Barrios, E., Castillo-Cruz, I., Zañudo-Hernández, J., Méndez-Morán, L., Nobel, P.S. 2007. Effects of shade, drought and daughter cladodes on the CO_2 uptake by cladodes of *Opuntia ficus-indica*. *Ann. Appl. Biol.* **151:** 137-144.

373. Nobel, P.S., Zutta, B.R. 2008. Temperature tolerances for stems and roots of two cultivated cacti, *Nopalea cochenillifera* and *Opuntia robusta*: Acclimation, light, and drought. *J. Arid Environ.* **72:** 633-642.

CONTRIBUTORS

José Luis Andrade
Unidad de Recursos Naturales
Centro de Investigación Científica de
Yucatán, A.C.
Apartado Postal 87, Cordemex
Mérida, Yucatán 97310
MEXICO

Giuseppe Barbera
Dipartimento di Colture Arboree
Università degli Studi di Palermo
ITALY

Edward G. Bobich
Biological Sciences Department
California State Polytechnic Univer-
sity
Pomona, California 91768
U.S.A.

J. Carlos Cervera
Unidad de Recursos Naturales
Centro de Investigación Científica de
Yucatán, A.C.
Apartado Postal 87, Cordemex
Mérida, Yucatán 97310
MEXICO

Michelle DaCosta
Department of Plant, Soil, and Insect
Sciences
University of Massachusetts
Amherst, Massachusetts 01003
U.S.A.

Erick De la Barrera
Centro de Investigaciones
  en Ecosistemas
Universidad Nacional Autónoma
  de México
Antigua Carretera a Pátzcuaro 8701
Morelia, Mich. 58190
MEXICO

Philippa M. Drennan
Department of Biology
Loyola Marymount University
1 LMU Drive
Los Angeles, California 90045
U.S.A.

Peter Felker
D'Arrigo Bros.
P.O. Box 850
Salinas, California 939902
U.S.A.

Augusto C. Franco
Departamento de Botanica
Universidade de Brasilia
Caixa postal 04457
Brasilia, D.F. 70919-970
Brazil

Arthur C. Gibson
Department of Ecology and Evolu-
tionary Biology
University of California
Los Angeles, California 90095
U.S.A.

Eric A. Graham
Center for Embedded Networked
Sensing
University of California, Los Angeles
3563 Boelter Hall
Los Angeles, California 90095-1596
U.S.A.

Howard Griffiths
Physiological Ecology Group
Department of Plant Sciences
University of Cambridge
Cambridge CB2 3EA
U.K.

Giovanni Gugliuzza
Dipartimento di Colture Arboree
Università degli Studi di Palermo
ITALY

Bingru Huang
Department of Plant Biology and
Pathology
Rutgers University
New Brunswick, New Jersey 08901
U.S.A.

Travis E. Huxman
Department of Ecology and Evolu-
tionary Biology
University of Arizona
Tucson, Arizona
U.S.A.

Paolo Inglese
Dipartimento di Colture Arboree
Università degli Studi di Palermo
ITALY

Daniel M. Johnson
Department of Biology
Wake Forest University
Winston-Salem, NC 27109-7325,
U.S.A.

Michael E. Loik
Department of Environmental Stud-
ies
University of California
Santa Cruz, California
U.S.A.

Giorgia Liguori
Dipartimento di Colture Arboree
Università degli Studi di Palermo
ITALY

Gretchen B. North
Department of Biology
Occidental College
Los Angeles, California 91141
U.S.A.

Eulogio Pimienta-Barrios
Departamento de Ecología
Centro Universitario de Ciencias
Biológicas y Agropecuarias
Universidad de Guadalajara
Km 15.5 Carretera Guadalajara-
Nogales
Zapopan, Jalisco 45110
MEXICO

Casandra Reyes-García
Physiological Ecology Group
Department of Plant Sciences
University of Cambridge
Cambridge CB2 3EA
U.K.
Present address:
Unidad de Recursos Naturales,
Centro de Investigación Científica de
Yucatán, A.C.
Apartado Postal 87, Cordemex
Mérida, Yucatán 97310
MEXICO

Jorge E. Schondube
Centro de Investigaciones
  en Ecosistemas
Universidad Nacional Autónoma
  de México
Antigua Carretera a Pátzcuaro 8701
Morelia, Mich. 58190
MEXICO

Paul J. Schulte
Department of Biological Sciences
University of Nevada
Las Vegas, Nevada
U.S.A.

Stanley D. Smith
School of Life Sciences
University of Nevada
Las Vegas
U.S.A.

William K. Smith
Department of Biology
Wake Forest University
Winston-Salem, NC 27109-7325,
U.S.A.

David T. Tissue
Department of Biological Sciences
Texas Tech University
U.S.A.

Centre for Plant and Food Science
University of Western Sydney
Hawkesbury Campus
Bourke Street
Building S12, Room 1-005
Richmond NSW 2753
AUSTRALIA

David G. Williams
Departments of Renewable Re-
sources and Botany
University of Wyoming
Laramie, Wyoming 82071
U.S.A.

Enrico A. Yepez
Departamento de Agua y
  Medio Ambiente
Instituto Tecnológico de Sonora
5 de Febrero 818 Sur
Ciudad Obregón, Sonora 85000
MEXICO

REFEREES

F. J. Adamsen, U.S. Arid-Land Agricultural Research Center, U.S.A.

R. Bello-Bedoy, Instituto de Ecología, UNAM, Mexico

S. D. Davis, Pepperdine University, U.S.A.

J. Flores Rivas, Instituto Potosino de Investigación Científica y Tecnológica, Mexico

J. G. García-Franco, Instituto de Ecología, A.C., Mexico

F. J. González, Universidad Autónoma de San Luis Potosí, Mexico

Y. Jiang, Purdue University, U.S.A.

C. Kleier, Regis University, U.S.A.

Ch. Körner, Universität Basel, Switzerland

J. López-Portillo, Instituto de Ecología, Mexico

H. Maherali, University of Guelph, Canada

R. Méndez-Alonzo, Centro de Investigaciones en Ecosistemas, UNAM, Mexico

A. Muñoz Urias, Universidad de Guadalajara, Mexico

F. Pineda, Centro de Investigaciones en Ecosistemas, UNAM, Mexico

W. T. Pockman, University of New Mexico, U.S.A.

L. Santiago, University of California, Riverside, U.S.A.

G. Solís Garza, Universidad de Sonora, Mexico

S. Schwinning, Texas State University, U.S.A.

T. Terrazas Salgado, Colegio de Postgraduados, Mexico

F. Volaire, Institut National de la Recherche Agronomique, France

K. Winter, Smithsonian Tropical Research Institute, Panama

I. Succulent Plants and Crassulacean Acid Metabolism (CAM) Photosynthesis

Chapter 1

STRUCTURAL IMPLICATIONS OF SUCCULENCE: ARCHITECTURE, ANATOMY, AND MECHANICS OF PHOTOSYNTHETIC STEM SUCCULENTS, PACHYCAULS, AND LEAF SUCCULENTS

Edward G. Bobich and Gretchen B. North

Perspectives in Biophysical Plant Ecophysiology: A Tribute to Park S. Nobel, pp. 3-37
Edited by: E. De la Barrera and W.K. Smith
© 2009 by The Authors
Book Compilation © 2009 Universidad Nacional Autónoma de México

Introduction

For plants, succulence is defined as the presence of thickened tissues in plant organs for which the primary function is water-storage and, consequently, drought avoidance (Gibson 1982; Gibson and Nobel 1986; von Willert 1992). Because they store water, organs with succulent tissues tend to have large volume to surface area ratios compared with non-succulent organs on the same plant (Gibson and Nobel 1986; Rowley 1987). Such increases in volume:surface area result in organs with relatively large cross-sectional areas, which leads to greater resistance of bending stresses (Niklas 1992). However, succulent tissue is comprised primarily of thin-walled parenchyma with highly extensible cell walls that have a low modulus of elasticity compared with the cell walls of chlorenchyma and those of other tissues, such as sclerenchyma and secondary xylem (Goldstein *et al.* 1991; Niklas and Buchman 1994). The low modulus of elasticity of the cell walls of succulent tissue allows the cells to swell and contract with changes in water availability; as succulent plants lose water, the water in the succulent tissue is translocated to other tissues that are metabolically more active, such as chlorenchyma in stem and leaf succulents (Nobel 1988; Goldstein *et al.* 1991). The implications of such water movement have been discussed in terms of ecophysiology more than in terms of biomechanics.

In addition to being less stiff than other plant tissues, succulent tissues are also usually denser due to their relatively high water content (Gibson and Nobel 1986). The water storage organs of woody stem succulents and leaf succulents are often 90-95% water (Nobel 1988) compared with 40-70% for non-succulent wood, and 70-90% for herbaceous leaves (Rowley 1987). Succulents can have densities of 800-1,100 kg m^{-3} (Niklas *et al.* 1999), whereas woody stems typically have densities of 350-850 kg m^{-3} (Hacke *et al.* 2001). The combination of the lower elastic modulus of supporting tissues and the increased weight due to water storage places great limitations on the architecture of succulents and significant stresses on their tissues, and has probably led to the evolution of the following modifications in succulent plants: reduced branching, reduced stems or leaves, and roots with contractile function (Gibson and Nobel 1986; Rowley 1987; Bell 1991). Anatomical modifications associated with succulence are also numerous, including parenchymatization of wood, presence of a thick hypodermis, and delayed formation of periderm (Gibson 1982; Gibson and Nobel 1986), as well as root modifications described later in this chapter.

The morphological and anatomical adaptations of the succulent habit have been studied extensively over the last 150 years, but the mechanical influences behind the evolution of the structural modifications have been studied primarily during the last two decades. The overall goal of this chapter is to synthesize the literature on the morphology, anatomy, and biomechanics of the shoots and roots of photosynthetic stem succulents, pachycauls, and leaf succulents, with a primary focus on North American representatives, in order to explain how the evolution of succulent tissue led to further above- and belowground modifications.

Plant architecture and stem morphology

Photosynthetic stem succulents

Succulents with photosynthetic stems form more of a functional group than a structural group because, other than the fact that they have functioning stomates and chlorenchyma in their stems, they almost all perform Crassulacean acid metabolism (Gibson 1982). Photosynthetic stem succulents actually include a wide range of growth habits, from geophytes and caespitose (cushion) forms to frutescent (shrubby) and arborescent (tree-like) forms (Fig. 1.1; Gibson and Nobel 1986; Anderson 1998; Anderson 2001). Photosynthetic stem succulents are most notably represented by the Cactaceae of the Americas and stem succulent Euphorbiaceae of the tropics and subtropics in Africa, Madagascar, and the Americas, but are also represented by members of the Asclepiadaceae (Anderson 1998), and some salt tolerant members of the Chenopodiaceae, a family in which succulence is an adaptation to high soil salt concentrations.

The basic shoot construction of photosynthetic stem succulents varies from long-lived monopodial axes, like the main trunks and branches of globose, barrel, and columnar forms to forms with sympodial axes, like opuntioid cacti of the Cactaceae, for which each branch is made up of several stems. Both types of stem construction are present in low-growing, epiphytic, lianoid, shrubby, and arborescent forms (Fig. 1.1). Perhaps what is so interesting is that large sympodial forms, as for leptocaulous (thin-stemmed) plants with the same construction, have trunks similar to plants with monopodial axes, like *Ulmus rubra* (Fisher and Stevenson 1981). However, most plants with sympodial axes, especially opuntioids like prickly pears and chollas, retain the distinction between many of their younger stems. The mechanics of the junctions between these stems—cladodes (flattened in cross section) in prickly pears and joints in chollas (approximately circular in cross section)—influences the overall form of these plants (Evans *et al.* 2004a, b; Bobich and Nobel 2001a, b). The number of stems or branches of opuntioids is related both to the maximum bending stress that causes the failure of the tissue at the junction between terminal and subterminal stems and to the rooting of terminal stems after detachment (Evans *et al.* 2004a). Overall clone size can also be influenced by the mechanics of the junctions, with hybrid prickly pears in Southern California forming extensive clonal populations, in which one individual can cover tens of square meters. These large clones form either by dropping stems or, more commonly, by having the branches gradually bend down and come into contact with the soil as a result of the biomechanics and anatomy of their stem junctions (Benson and Walkington 1965; Bobich and Nobel 2001a). Chollas like *Cylindropuntia bigelovii* and *C. fulgida* also have biomechanically weak stem junctions (Bobich and Nobel 2001b; Kahn-Jetter *et al.* 2001; Bobich 2004), a trait that is related to their repro-

Figure 1.1. Growth forms of photosynthetic stem succulents: A) cushion (*Euphorbia tubiglans* at California State Polytechnic University, Pomona), B) branching globose (*Mammillaria dioica* in Palm Desert, California); C) low shrub (*Euphorbia horrida* at California State Polytechnic University, Pomona); D) expansive shrub (*Opuntia englemannii* near Tucson, Arizona); E) arborescent (*Carnegiea gigantea* near Tucson, Arizona).

Figure 1.2. Cereus repandus stems with prominent ribs at California State Polytechnic University, Pomona.

ductive strategy and form. Both species primarily reproduce by dropping their stems as they grow, which is why branches are restricted to the upper third of the plant.

Cacti are constructed of long photosynthetic shoots and inconspicuous short shoots. The long shoot composes the overall structure of the stem, and the short shoots are the areoles, which are specialized axillary buds that consist of meristematic tissue and the short shoots with leaf spines (Gibson and Nobel 1986). The spines on a cactus are analogous to those on other photosynthetic stem succulents, like the stipular spines of euphorbias. In addition to providing protection with leaf spines, long shoot-short shoot architecture in cacti also minimizes the carbon allocated to branches.

Although photosynthetic stem succulents include a diverse array of growth habits, they share many of the same morphological characteristics, one of which is a general reduction in the production of photosynthetic branches compared with other vascular plants (Gibson and Nobel 1986; Price and Enquist 2006). Other than caespitose species, such as cacti in *Maihuenia*, *Echinocereus*, and *Echinopsis* and small succulent Euphorbiaceae (Fig. 1.1A), stem succulents typically display an apical dominance that either completely eliminates branching or suppresses branching so that branches do not greatly shade apical meristems or other stems (Nobel 1988). The height where branching occurs can also be affected by the surrounding vegetation, with taller surrounding vegetation leading to greater branching heights (Racine and Downhower 1974; Cody 1984).

The frequency of branching and the orientation of stems, such as the flattened stems of prickly pears, are typically explained in terms of carbon uptake and partitioning (Geller and Nobel 1986; Price and Enquist 2006). However, branching in stem succulents is also probably influenced by their stem biomechanics. Succulent stems are relatively heavy due to their high density and relatively small wood cross-sectional area/stem cross-sectional area (Gibson and Nobel 1986). One way that columnar cacti and large Euphorbiaceae reduce the bending moments due to branching is by orienting their branches vertically. A bending moment is a force that causes any structure to bend and its magnitude is calculated as the force × the length of the moment arm. Thus for a stem, the bending moment due to gravity is related to the length or vector of the stem perpendicular to vertical times the force of the mass of the stem. Columnar cacti such as *Pachycereus pringlei* and *Carnegiea gigantea*, as well as *Euphorbia ingens* appear to decrease the bending moments due to branching by minimizing the horizontal length of their branches (Fig. 1.1E).

Most of the genera of the Cactoideae subfamily of the Cactaceae and the succulent Euphorbiaceae and Asclepiadaceae have ribs (Gibson and Nobel 1986; Anderson 1998), which allow plants to increase surface area for greater light absorption and greater CO_2 uptake (Geller and Nobel 1984). Ribs also allow the plants to swell and shrink like an accordion depending on the water content of the plant (Nobel 1988) and are an ancestral trait for the Cactoideae (Anderson 2001). In globose and caespitose Cactoideae, ribs tend be less pronounced or even flattened, but for large columnar species the ribs can be relatively large in cross-section (Fig. 1.2) and are important for support, especially in the younger portions of the plant near the apex (Gibson and Nobel 1986; Niklas *et al.* 1999). Near the apex of *Pachycereus pringlei* stems, the tissues of the ribs are just as stiff as the vascular tissue because the vascular tissues near the apex have a small cross-sectional area, especially compared with the ground tissues (Niklas *et al.* 1999). The ribs of columnar forms may also stiffen stems because of their triangular shape in cross-section, which results in a progressively increasing ratio of collenchyma to parenchyma per cross-sectional area towards the ridge (periphery).

For an individual photosynthetic stem succulent, the structural properties of the overall form can differ with plant age. *Carnegiea gigantea* and other columnar cacti appear to be overbuilt when young, meaning that their stem diameter is more than adequate to support the overall aboveground structure (Niklas and Buchman 1994; Cornejo and Simpson 1997). As the plants age their height becomes greater relative to their diameter, which is reflected in greater overall surface area to volume ratios (Cornejo and Simpson 1997). However, as is the case with saguaro, mature plants are typically well within the safety factors for their height (Niklas and Buchman 1994). In fact, cacti tend to have greater ratios of trunk diameter to stem height than do palms, cycads, and leptocaulous trees (Niklas *et al.* 2006).

Pachycauls

Succulent pachycauls, or sarcocaulescents, all possess one basic trait: their trunks store large amounts of water and are greatly enlarged compared to the rest of their body (Figs. 1.3 and 1.4). Unlike photosynthetic stem succulents, the primary photosynthetic organs of pachycauls are usually seasonal leaves (Rowley 1987). The pachycaul form has arisen numerous times in a variety of tropical and subtropical habitats, including deserts, thornscrub, and tropical deciduous forests, all ecosystems with seasonal precipitation (Rowley 1987; Turner *et al.* 1995). In addition, pachycaulesence has evolved numerous times on islands from a variety of leptocaulous forms (Carlquist 1980). Pachycauls also occur in an incredible number of dicotyledon and monocotyledon families, including Apocynaceae, Asteraceae, Campanulaceae, Dracaenaceae, Goodeniaceae, and Sterculiaceae (Carlquist 1960; Rowley 1987). In arid habitats of North America, pachycauls are primarily in the Agavaceae, Anacardiaceae, Burseraceae, Fouquieriaceae, and Nolinaceae (Asparagaceae, The Angiosperm Phylogeny Group II 2003; Rowley 1987).

The massive trunks of pachycauls are often upright, especially in the Apocynaceae, Didiereaceae, and Fouquieriaceae, which reduces the tensile stresses within the trunk (Fig. 1.3). The stems of pachycauls should be able to resist compressive forces better than tensile forces because of the large amounts of parenchyma in the tissues (see below; Niklas 1992). In addition, a pachycaul trunk has a relatively large second moment of area, which

Figure 1.3. Plants displaying the pachycaulous form: A) *Brighamia insignis* at California State Polytechnic University, Pomona, B) *Beaucarnea recurvata* at California State Polytechnic University, Pomona, C) *Fouquieria columnaris* at Rancho Santa Ana Botanic Garden, Claremont California; D) *Dracaena draco* at California State Polytechnic University, Pomona.

Figure 1.4. Close-up of the swollen trunks of *Beaucarnea recurvata* at California State Polytechnic University, Pomona.

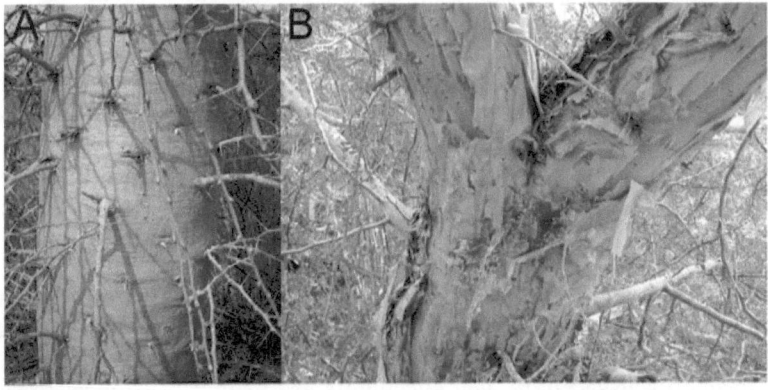

Figure 1.5. Bark of A) *Fouquieria columnaris* (boojum) at Rancho Santa Ana Botanical Garden and B) *Pachycormus discolor* at Biosphere 2 Center, Oracle, Arizona. Note the leptocaulis lateral branches of *F. columnaris* and the flaking phellem of *P. discolor*.

allows the plant to spread the stresses out over a larger area, reducing the stresses experienced in any one region (Niklas 1992).

Pachycauls also reduce stresses associated with bending due to gravity by having fewer branches than leptocaulous species (Rowley 1987). In addition, the branches of pachycauls usually have small diameters compared with their succulent trunks. For *Dendrosycios* on the Island of Socotra off the coast of Arabia, the branches are actually decumbent and relatively weak, possibly owing this trait to its proposed derivation from caudiciform vines (Olson 2003). The reduction in branching can once again be related to the massiveness of sarcocaulescent stems due to their water storage and wood anatomy (see below). When branching does occur, it often occurs after terminal inflorescences; thus, the axillary buds produce new stems (Mabberley 1974b). As the plant grows and produces more flowers, branches proliferate; however their branches, especially dendrosenecios of the Asteraceae and members of the Apocynaceae and Dracaenaceae (Fig. 1.3) have progressively smaller and smaller diameters, which may limit plant height (Mabberley 1974a, b).

Reduction in branch size is even more pronounced for the pachycauls of North America such as *Bursera hindsiana*, *B. microphylla*, *Fouquieria columnaris* (Fig. 1.3C), *Jatropha cuneata*, and *Pachycormus discolor*, all of which exhibit long shoot-short shoot architecture, which allows trees to reduce the amount of carbon allocated to stems versus leaves (Turner *et al.* 1995). Succulent Fouquieriaceae take reduction of stem size further by having leptocaulous lateral branches, which bear short shoots with leaves (Henrickson 1969a, b). These branches allow for increased photosynthetic area due to their architecture and minimize carbon investment into construction and resistance of bending moments.

In addition to their interesting architecture, the trunks of pachycauls of North America also have interesting bark. The bark is often flakey, exposing a green photosynthetic stem, which may function in the fixation of respired CO_2 within the plant, but does not function in exogenous CO_2 uptake (Fig. 1.5; Franco-Vizcaíno *et al.* 1990; Nilsen *et al.* 1990). Such a flaky bark covering a photosynthetic phelloderm and cortex with thin walls probably provides little in terms of support, but allows plants like *Fouquieria columnaris* and *Pachycormus discolor* to respond quickly to the infrequent rains of Baja California and Sonora, Mexico (Franco-Vizcaíno *et al.* 1990).

Leaf succulents

Leaf succulents include a wide variety of forms, such as geophytes, basal rosettes, arborescent rosettes, prostrate creeping perennials, drought deciduous perennials, and annuals (Fig. 1.6). Just as for the two stem succulent forms discussed here, leaf succulents are primarily tropical to subtropical and are well represented in a number of different families, including: Agavaceae, Aizoaceae, Crassulaceae, Portulacaceae, Ruscaceae (includes *Sansevieria*; the family is now included in the Asparagaceae; Angiosperm Phylogeny Group II 2003), and Asphodelaceae (inclues *Aloe*).

Figure 1.6. Growth forms of leaf succulents: A) shrubby (*Agave shawii* in San Diego County, California), B) arborescent (*Aloe arborescens* at California State Polytechnic University Pomona), C) small acaulescent rosette (*Dudleya pulverulenta* at Rancho Santa Ana Botanic Garden), and D) annual (*Mesembryanthemum crystallinum* in the University of California, San Diego Scripps Coastal Reserve).

In the arid regions of North America the most prominent leaf succulents are in the Agavaceae and, to a lesser extent, the Crassulaceae.

Much like the lower-growing photosynthetic stem succulents, geophytic and other leaf succulent forms that are short in stature are supported by their cell turgor pressure, especially those in the Aizoaceae and Crassulaceae (Fig. 1.6). Larger leaf succulents often have a rosette form. The rosette form is convergent for many different families growing in a variety of harsh ecosystems or microclimates and appears to serve a variety of functions, such as funneling rain water to roots (Gentry 1982) and freezing prevention (Beck 1994). As stated above, rosette leaf succulents can vary from acaulescent (basal) to caulescent forms, with some, such as arborescent aloes, having an almost pachycaulous stem (Anderson 1998). For acaulescent rosettes and other lower forms, the leaf is the main structural component; individual leaves can vary in length from a few millimeters to well over 2 m (Gentry 1982). In the larger rosettes, such as those in the Agavaceae and *Sansevieria*, the leaves are usually stiff due to the fact that they have a large number of phloem and cortical fibers (Smith and Nobel 1986; Koller and Rost 1988). Some leaves are so long, such as those of *Agave americana*, that the leaves actually bend down from vertical near the tip (Gentry 1982), indicating that the leaf decreases in stiffness acropetally.

Branching for leaf succulents is often absent, sparse, or by way of rhizomes. Perhaps one reason for the minimal branching among leaf succulents is the preponderance of the rosette form; branches would shade a considerable amount of the canopy because the leaves are tightly compacted near the apex of the stem. Species of agaves that do branch within their rosette, like *Agave shawii* (Fig. 1.6), tend to have elongated stems compared with other agaves (Gentry 1982; Tony Burgess pers. comm.). Another possible reason for the reduced branching in leaf succulents is that the stems of many of the plants of the Crassulaceae and *Aloe* have small diameters or are sarcocaulescent and not very stiff.

Anatomical and mechanical properties of shoots

Photosynthetic stem succulents

The gross anatomy of most stem succulents is similar (Mauseth 2004a). In general, stem succulents have a single-layer (uniseriate) epidermis (Gibson and Nobel 1986; Mauseth 2004a). The outer tangential and radial walls of epidermal cells can be cutinized but are not lignified or suberized, at least before periderm development, which is delayed due to the fact that the stem is the main photosynthetic organ (Gibson and Nobel 1986).

Most cacti and certain succulent Euphorbiaceae and Asclepiadaceae have a multilayered, collenchymatous hypodermis to the inside of the epidermis (Gibson 1982; Mauseth 2004a). Collenchyma is a very plastic supporting tissue that often provides support in growing tissues due to its ability to deform (Fahn 1990); along with the epidermis it serves an important structural role in the support of many succulents (Niklas *et al.* 1999, 2003). In fact, for the cacti *Stenocereus gummosus* and *S. eruca*, collenchyma, based on its exterior location, offers the greatest resistance to bending stresses of any other aboveground tissues (Niklas *et al.* 2003). Not only can the collenchyma and the epidermis resist bending stresses that occur due to applied forces and gravity, they can also resist the internal forces exerted by the turgid parenchyma of the plant body (Niklas 1992). The "hoop stress" or reinforcement applied by the epidermis and hypodermis actually pushes back on the parenchyma to the stem interior, allowing for more rigid, taller structures (Niklas 1992) and helps explain why some barrel cacti can reach heights of 2 m without producing a woody skeleton (Cornejo and Simpson 1997; Fig. 1.7A).

Either the cortex or the pith is typically the water storage region in photosynthetic stem succulents, with the cortex being the main storage region in the subfamily Cactoideae of the cactus family (Gibson and Nobel 1986); this subfamily includes globose, columnar (Fig. 1.1), and barrel (Fig. 1.7A), as well as vining and epiphytic forms (Fig. 1.7B). The cortex is also the main water-storing region for a few known succulent Euphorbiaceae and Asclepiadaceae (Mauseth 2004a). The pith is the main water-storing region for the cactus subfamily Opuntioideae and for the majority of succulent Euphorbiaceae and Asclepiadaceae (Gibson and Nobel 1986; Mauseth 2004a).

Figure 1.7. A) A 1.9 m tall barrel cactus, *Ferocactus wislizeni*, near Tucson, Arizona and B) a vining hemiepiphyte, *Hylocereus triangularis*, in Quintana Roo, Mexico.

The significance of the location of the water-storing region has to do with the distribution of stresses in structures. For caespitose and globose forms, the importance of the location of the water storage tissue is reduced because the plants and/or their branches are not long enough to experience a significant bending moment for most static and applied forces, like wind. However, for barrel and arborescent columnar forms, bending moments due to gravity and wind can be significant (see Niklas *et al.* 1997). For any object, the stress resulting from bending increases from the center, where the stresses are minimal, to the outside of the object (Niklas 1992). Having a thick cortex with large intercellular air spaces and easily deformed water-storage parenchyma can lead to stem deformation, especially if there is little vascular tissue.

For most perennial shrubs and trees wood is the primary support tissue (Fahn 1990). However, this may not be the case for many photosynthetic stem succulents. In general the smaller cactus growth forms have a reduction in the number of fibers in their wood, with fibers being totally replaced by parenchyma or wide-band (vascular) tracheids in many species (Gibson 1973, 1977a, 1978); thus, smaller forms are supported primarily by the hydrostatic pressure of their chlorenchyma and water-storage parenchyma (Gibson and Nobel 1986). For sprawling forms like *Stenocereus eruca* and certain morphs of *S. gummosus*, wood provides little support because it has a relatively small cross-sectional area and is centrally located in the stem, surrounded by a thick cortex (Niklas *et al.* 2003). In many smaller growth forms and in the junctions of stems for prickly pear cacti (platyopuntias), which have flattened stems (cladodes), the wood is polymorphic, meaning that wood production within a plant can developmentally change from fiberless woods often with wide-band (vascular) tracheids to wood with fibers (Gibson 1973, 1977b; Mauseth and Plemons 1995; Bobich and Nobel 2001a, 2002), or in the case of cactus epiphytes, change from producing fibrous to more succulent wood (Gibson 1973; Gibson and Nobel 1996; Mauseth 1993). Production of fibers is extremely important because they can resist tensile stresses much better than parenchyma (Kahn-Jetter *et al.* 2000, 2001; Evans *et al.* 2004). In fact, the length and diameter of individual fibers are positively correlated with plant height in both platyopuntias (Gibson 1978; Bobich and Nobel 2001a) and Cactoideae (Gibson 1973), indicating that larger fibers may provide more resistance to bending and gravitational forces in large growth forms.

Wide-band tracheids are particularly interesting cells because of their function and shape and because their presence may be ancestral for cacti (Mauseth 2004b). They are imperforate and have either annular (typically in Opuntioideae) or helical secondary wall thickenings (Gibson 1973, 1978; Mauseth and Plemons 1995; Bobich and Nobel 2001a, b). The function of wide-band tracheids may be to increase water-storage in cacti while preventing cell collapse when water is lost (Gibson 1973, 1977a, 1978; Mauseth 1995), but their structure may also be significant for another mechanical reason (Bobich and Nobel 2001). When a stem is subjected to bending forces, the middle lamellae may not be able to hold the cells together, resulting in shearing between cells. Because their secondary thickenings alternate with those of adjacent wide-band tracheids cells, the cells often "interlock", especially in Opuntioideae, thus reducing shear and increasing stem stiffness (Bobich and Nobel 2001a).   Because they have helically thickened secondary walls in Cactoideae (Gibson 1973; Gibson and Nobel 1986; Mauseth *et al.* 1995), wide-band tracheids should also provide for stiffer stems than would parenchyma alone.

In woody plants, branches that are not vertical or that experience uneven stresses due to gravity or applied forces may produce reaction wood (Fisher and Stevenson 1981; Fahn 1990).   In dicotyledons, reaction wood can have one to all of the following traits: increased wood deposition on upper side of the stem, which experiences tension, compared with the lower side, which experiences compression; gelatinous fibers, which have a cellulose-rich layer in their secondary cell walls and little lignin and may act in reorienting stems toward vertical by contracting; and a greater density than "normal wood" (Fahn 1990; Fisher and Stevenson, 1981; Fournier *et al.* 1994).

In cacti, reaction wood occurs within cladode junctions of *Opuntia* (Fig. 1.8; Bobich and Nobel 2001a, 2002) and the joint junctions of one species of *Cylindropuntia* (Bobich and Nobel 2001b).   Cladode junctions (the regions between two successive cladodes) can have less than 10% of the cross-sectional area of the widest regions of their subtending cladodes and are the most flexible regions of a branch (Nobel and Meyer 1991). Anatomically, cladode junctions are comprised of a ring of parenchyma surrounding a central core of phloem, wood, and pith (Fig. 1.8; Bobich and Nobel 2001a, 2002). The wood within relatively young cladode junctions of large platyopuntias is asymmetrical, with greater wood accumulation on the side of the cladode under compression (Fig. 1.8).   This asymmetry may be because the wood is comprised primarily of thin walled parenchyma,

Figure 1.8. A three-year-old cladode junction of *Opuntia ficus-indica* stained with phloroglucinol to show lignification in wood. Note the greater accumulation of wood on the lower side of the stem and the paucity of lignification in the early-formed wood. Stained areas primarily represent the lignified middle lamellae and primary walls of libriform fibers; most libriform fibers have a gelatinous layer in their secondary wall.

which is more effective in resisting compressive than tensile forces (Niklas 1992; Bobich and Nobel 2002). In general, fiber production does not begin until the cladode junction is supporting a total of three cladodes (the sub-tending upper cladode plus two more distal cladodes; Bobich and Nobel 2001a, 2002). The majority of the fibers are gelatinous, but in contrast to most dicotyledons, gelatinous fibers occupy the greatest area in the lateral regions of the junction, near the neutral plane of bending in terms of gravitational forces. It is believed that gelatinous fiber production is a response to bending forces caused by wind; thus the gelatinous fibers may prevent excessive swaying of the cladode at that stage of branch development (Bobich and Nobel 2002). Although reaction wood has not been described for other stem succulents, it may be present in the branches of large columnar forms, like *Carnegiea gigantea*, as well as in forms with stems that tilt towards the equator, like certain species of *Copiapoa* in South America and *Ferocactus* in North America (Ehleringer *et al.* 1980).

The anatomical modifications for stem succulence have a great effect on the elastic modulus of the stems (Table 1.1). The elastic modulus, which is most simply defined as the applied stress/the elastic deformation of any material regardless of its geometry, for cactus stems and probably for other stem succulents, can be one to three orders of magnitude less than that of other woody plants, which range from 5,000-15,000 MPa (Table 1.1; Niklas 1992). The lower elastic modulus values for cactus stems are because of the increased amount of chlorenchyma and water-storage parenchyma in their stems compared with non-succulent woody plants. In general, the elastic modulus of cactus stems also appears to be correlated with plant height, as are the anatomical features of the wood of Opuntioideae and Cactoideae (Gibson 1973, 1978; Table 1.1). Tall columnar forms (*Carnegiea gigantea* and *Pachycereus pringlei*) have the highest elastic moduli, followed in order by large arborescent and shrubby *Opuntia*, and sprawling (*Stenocereus gummosus*) and prostrate forms (*S. eruca*; Table 1.1).

Changes in elastic modulus occur along the stem axis for the innermost tissues of *Pachycereus pringlei* stems (Niklas *et al.* 1999). In general, the elastic modulus of the wood and pith in *P. pringlei* increases basipetally, as would be expected because of the development of secondary xylem and the shear mass of a succulent of that size. However, near the base of the stem the elastic modulus of the wood greatly decreases (Niklas *et al.* 1999). The geometry of the stem makes up for this decrease though, as the cross-sectional area is much greater near the base of the stem and more than compensates for the lower values for the elastic modulus, making this the stiffest region of the stem. Differences in elastic modulus not only exist within individual cacti, but also among different morpho- or ecotypes of certain species, especially *Stenocereus gummosus*, which ranges in form from sprawling to very large shrubs with stems over 4 m tall (Nobel 1988).

Table 1.1. Elastic moduli (E) for aboveground organs of cacti.

Species	E (MPa)	Tissue/location	Reference
Carnegiea gigantea	1,600	Stem	Niklas and Buchman (1994)
Opuntia cochenillifera	682	Cladode junction	Bobich and Nobel (2002)
O. ficus-indica	375	Cladode junction	Bobich and Nobel (2002)
O. robusta	179	Cladode junction	Bobich and Nobel (2002)
O. undulata	364	Cladode junction	Bobich and Nobel (2002)
Pachycereus pringlei	1,670[a]	Wood	Niklas *et al.* (2000)
P. pringlei	1,090[b]	Wood	Niklas *et al.* (2000)
Stenocereus eruca	4[c]	Stem	Niklas *et al.* (2003)
S. gummosus	7[c]	Stem	Niklas *et al.* (2003)

[a] Maximal radial value along stem axis.
[b] Maximal tangential value along stem axis.
[c] Measured in tension.

Pachycauls

Succulent pachycauls have a variety of traits that impact the mechanics of their tissues. In species with photosynthetic tissue in their trunks (Fig. 1.3), the outer stem tissues are interesting both structurally and physiologically. For the succulent Fouquieriaceae, specifically *F. fasciculata* and *F. purpusii*, the stems retain their epidermis throughout the life of the plant and increase in epidermal surface area by anticlinal divisions (Henrickson 1969c). The stem of *Fouquieria columnaris*, or boojum, is covered with a thin layer of suberized phellem, which surrounds the cortex (Henrickson 1969c).

The cortex of the succulent Fouquieriaceae is very large in cross-sectional area and performs different functions in relation to position within the stem axis. The outer cortex is composed of chlorenchyma and probably does not afford much support (Henrickson 1969c). To the inside of the chlorenchyma are large groups of thick-walled brachysclereids; the groups can be as large as 5 mm in diameter in *F. columnaris*. The nests are closely packed and essentially form a cylinder within the trunk of *F. columnaris*, *F. fasciculata*, and *F. purpusii*. Further inside the cortex are networks of water-storage parenchyma cells with thin cell walls, which, much like the photosynthetic cells, probably do not provide much support, and a network of starch-storing parenchyma (Henrickson 1969c). Because of their location and their assumed structural properties, the groups of brachysclereids stiffen the stems of pachycauls in the Fouquieriaceae and probably keep the stems from collapsing on a regular basis, which they can do during extended drought (Henrickson 1972). In *F. fasciculata* and *F. purpusii*, a continuous ring of primary phloem fibers occurs just to the inside of the cortex and may significantly aid in resisting bending moments even though the fibers are more centrally located within the stem than are the brachysclereids.

As stated previously, the trunks of *Pachycormus discolor* and *Bursera* both have a flaky periderm that peels off in large sheets (Fig. 1.5; Gibson 1981; Nilsen *et al.* 1990). The periderm of *P. discolor* has several layers of phellem, and copious amounts of photosynthetic phelloderm. The amount of parenchyma in the phelloderm, the large amount of secondary phloem, which contains numerous large parenchyma cells, and a thick cortex indicate that the bark of *P. discolor* is a major water storage organ and provides support via turgor (Gibson 1981). The outer tissues of *Dendrosycios socotrana* also have abundant parenchyma, specifically in the phelloderm and cortex (Olson 2003); the cortex in older regions of the trunk actually appears to be a result of secondary growth.

The anatomy of the wood of North American pachycauls is diverse in terms of water storage and mechanics. In mature trunks of *F. fasciculata* and *F. purpusii*, the wood contains a large amount of water storage parenchyma due to either cambial or intrusive growth and cell divisions by cells lower in the trunk (Henrickson 1969b; Carlquist 2001). The axial regions of the wood have concentric bands of primarily lignified tissue that alternate with concentric bands of unlignified parenchyma; the lignified bands appear to stiffen the parenchymatous bands (Henrickson 1969b). In addition, the rays become very wide with age, increasing the water-storage capacity of the plant substantially. The axial regions of the wood also have parenchyma within the lignified regions (Carlquist 2001).

Another feature of succulent *Fouquieria* wood is the presence of gelatinous fibers throughout much of the trunk; gelatinous fibers are conspicuously absent from the lateral branches of *F. columnaris* (Carlquist 2001). The presence of gelatinous fibers indicates that the wood in the trunks of all three succulent species and in some smaller branches of *F. fasciculata* and *F. purpusii* is reaction wood. In the case of the succulent Fouquieriaceae the fibers may be a response to gravitational stresses resulting from the high density of the tissue. However, the gelatinous fibers may function in stiffening the main trunks, which can reach 16 m in height, by contracting longitudinally and compressing the various tissues of the trunk.

One of the most extreme cases of wood succulence among pachycauls is that of *Pachycormus discolor*. The wood of *P. discolor* is diffuse porous with wide-diameter vessels and wide, short libriform fibers that lack lignin in their cell wall (Gibson 1981). Parenchyma is located in multiseriate rays, but is scanty and paratracheal in axial regions and also unlignified. In fact, only the vessels are lignified in the wood, meaning that the stem is supported almost exclusively by turgor pressure (Gibson 1981).

Many pachycauls develop trunks with large cross-sectional areas without developing typical secondary tissues. *Dendrosycios socotrana* has anomalous secondary growth, in which individual plates of xylem form from discontinuous cambia and are often associated with small regions of interxylary phloem (Olson 2003). Axially, the plates of xylem bifurcate and reconnect and are separated by extremely wide rays. Parenchyma also alternates with concentric bands of rays and xylem plates as conjunctive tissue, further increasing the water and starch storage of the trunk (Olson 2003). Furthermore, the water-storage parenchyma cells have undulating tangential walls, indicating that they may function in water-storage and shearing prevention in the same manner as the wide-band tracheids of cacti with annularly or helically thickened secondary walls (Mauseth 1995a; Bobich and Nobel 2001a).

Monocotyledonous pachycauls have an understandably different tissue organization from that of the dicotyledonous pachycauls. Like most monocots, the primary body of *Beaucarnea recurvata* (Figs. 1.3 and 1.4) consists of numerous vascular bundles in a matrix of ground tissue (Stevenson 1980). The primary body of *B. recurvata*, like that of several other arborescent monocots of the Asparagaceae and Xanthorrhoeaceae, is produced not only by the apical meristem, but also by a primary thickening meristem (Stevenson 1980); both of these meristems appear to be continuous in the primary body of this species. A secondary thickening meristem develops within the outer cortex of the stem and produces conjunctive tissue (parenchyma), which is eventually permeated by amphivasal vascular (specifically, desmogen) strands, centripetally (Fisher 1975; Stevenson 1980). Very early in development, phellogen develops to the outside of the secondary thickening meristem and produces phellem. The conjunctive tissue in *B. recurvata* and other similar forms, like *Yucca elephantipes*, is generally unlignified except when adjacent to the secondary vascular bundles (Fisher 1975). Thus, much of the strength of the stems of monocotyledonous pachycauls is probably due to hydrostatic pressure, although the phellem and vascular bundles probably do significantly stiffen the conjunctive and primary ground tissues.

An interesting aspect of the secondary tissues of monocotyledonous pachycauls is that they display eccentric growth with respect to gravity (Fisher 1975). Both *B. recurvata* and *Y. elephantipes* accumulate more secondary tissue on the lower side of horizontal stems; the accumulation on the lower side can be over 12 times greater than that on the upper side (Fisher 1975). The parenchyma cells and the vascular bundles are also larger in the lower than in the upper sides of stems. The greater growth on the lower side of the stem appears to be similar to the responses of prickly pear junctions that have greater wood accumulation on the lower sides of the stems and the majority of the cells are parenchyma, which are best at resisting compressive forces.

Leaf succulents

The leaves of many leaf succulents are essentially large hydrostats because they lack fibers (Niklas 1992; Gibson 1996). The epidermis and cuticle, like the epidermis and hypodermis of cactus stems, provide a hoop stress for a succulent leaf (Niklas 1992) and allow the leaf to have a considerable mass. The most common plants native to arid lands in North America that have leaves that act as hydrostats are in the Crassulaceae. In general, leaves within this family are relatively flexible (Gibson 1996). The basic structure of a Crassulaceaen leaf is very simple, with a uniseriate, less commonly two-layered, thin-walled epidermis (Fig. 1.9A). Except for vascular bundles, which are located near the middle of the leaf and have few lignified cells, the majority of the leaf cross-section is composed of large, thin-walled storage parenchyma cells that are usually circular in cross-section (Fig. 1.9A). The location of the vascular bundles in the center of the leaves, surrounded by relatively soft tissue, leads to leaves that are "soft."

Hydrostatic leaves of other leaf succulents have more reinforcement. Unlike typical Crassulaceae, the outer and sometimes inner periclinal walls of the epidermis of African *Aloe* leaves are thickened (Fig. 1.9B; Gibson 1996). Also, *Aloe* leaves have a definitive palisade mesophyll with long slender cells and most of the cells appear to be smaller in cross-sectional area than those of Crassulaceae; smaller cells tend to provide more support because they have more cell wall material per area than do larger cells (Niklas 1988).

Aizoaceae of Africa also stiffen their epidermis, with most xeriphytic species having thick outer periclinal walls and a thick cuticle (Gibson 1996). One subfamily of the Aizoaceae, the Rushioideae, also has wideband tracheids in the xylem of secondary and tertiary leaf veins; the leaf midrib lacks these cells (Landrum 2001). The presence of wide-band tracheids, which are the majority of the cells in secondary and tertiary veins, allows the leaf to maintain its structure during drought, which is important because the leaves of this subfamily are usually the main aboveground structures (Landrum 2001).

Other leaf succulents have leaves that are not hydrostats; the most recognized plants with these leaves in North America are the agaves (Fig. 1.6; Gibson 1996). Agave leaves tend to have very thick cuticles (usually over 20 µm thick) and epidermal cells with a great deal of secondary thickening in their outer periclinal cells walls and, less often, on their anticlinal and inner periclinal walls (Fig. 1.9C; Gibson 1996). Although they have palisade parenchyma, which may provide more stiffness than isodiametric parenchyma, it is the vascular bundles that provide much of the support. For *Agave deserti* the number of vascular bundles increases linearly with leaf surface area (Smith and Nobel 1986). The vascular bundles do not appear to branch within the leaf and are relatively evenly distributed throughout the storage parenchyma of a leaf; there are never more than 8 storage parenchyma cells separating the bundles in *A. deserti* (Smith and Nobel 1986). The palisade mesophyll completely lacks vascular bundles. The bundles themselves are relatively simple, with a few to several tracheids in the xylem and numerous phloem fibers capping the bundle. It is the fibers that are commercially important and their presence allows agave leaves to become several times thicker than those of other leaf succulents.

Another group of plants that have reinforced succulent leaves are those in the genus *Sansevieria*, which are native to tropical and subtropical Africa and Madagascar. *Sansevieria* leaves are similar to those of *Agave* in that they have a uniseriate epidermis and contain numerous vascular bundles throughout their water storage parenchyma (Koller and Rost 1988). The vascular bundles can be quite large, especially in the more xeromorphic leaves, and typically have lignified phloem cap fibers (Koller and Rost 1988). In addition to the vascular bundles, support is also provided throughout the entire leaf by fiber bundles (Koller and Rost 1988). The location of the fiber bundles appears to greatly stiffen the outer tissues of the leaf and minimizes bending. Resistance to collapse appears to be further provided by the presence of bands of thickened cell wall on specialized dead water storage cells to the inside of the chlorenchyma in many *Sansevieria* species (Koller and Rost 1988).

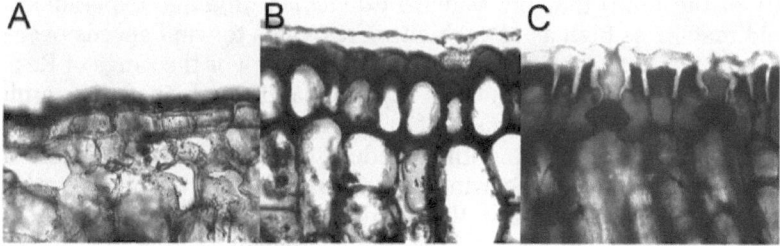

Figure 1.9. Cross sections of leaves of A) *Dudleya virens*, B) *Aloe arborescens*, and C) *Agave shawii*, stained with toluidine blue O. Note the thickening of the outer periclinal walls of the epidermis and the thickened cuticles for both *Aloe arborescens* and *Agave shawii*.

Wind

Wind is known to have a great effect on plant morphology, resulting in shrubby to prostrate growth habits of plants that are often trees and flag trees, which orient their stems' growth downwind to minimize the amount of force experienced. Because of their mass, relatively low stem elastic moduli, shallow rooting depth, and the fact that they act as bluff bodies because they can change characteristics of airflow (Nobel 2005), succulents, especially those with taller growth habits, should be greatly affected by wind in terms of morphology. On the island of Piedra in the Laguna Ojo de Liebre (Eye of the Jackrabbit Lagoon) of the Vizcaíno region of Baja California Sur, wind, which averages 1.8 m s^{-1}, has apparently resulted in all of the *Ferocactus fordii* on the island to be tilted or prostrate, with their apices oriented to the southeast; the prevailing winds come from the northwest (Ortega-Rubio *et al.* 1995). Perhaps the most striking effect of wind on the morphology of a succulent is that on the pachycaul *Pachycormus discolor* (Turner *et al.* 1995; Peinado *et al.* 2005; Arizona-Sonora Desert Museum web site, http://www.desertmuseum.org/programs/succulents_Pacdis.html). On more protected sites inland, *P. discolor* can reach heights of 8 m, but on Punta Eugenia on the Vizcaíno Peninsula, *P. discolor* is completely prostrate and is shrubby in other locations in the region, where the wind is constant throughout the year.

As discussed previously, ribs may act in reinforcing large photosynthetic stem succulents in addition to increasing the photosynthetic area of the plant. There is also evidence that ribs, along with spines, decrease wind effects experienced by cacti (Talley *et al.* 2001; Talley and Mungal 2002). The Reynolds number (Re; defined as the magnitude of the fluid velocity times the characteristic dimension of an object divided by the kinematic velocity of the fluid) that the saguaro *Carnegiea gigantea* experiences in the field may be as high as 10^6, which corresponds to wind speeds over 30 m s^{-1} (Talley *et al.* 2001; Talley and Mungal 2002). For the range of Re a *C. gigantea* may experience, cylinders with ribs and rough surfaces simulating spines have a lower coefficient of drag and experience lower fluctuating side forces compared to smooth cylinders, which means that ribs and spines actually decrease the wind forces experienced by photosynthetic stem succulents. Furthermore, the hemispherical apex of a succulent like *C. gigantea* also appears to decrease the coefficient of drag (Talley *et al.* 2001; Talley and Mungal 2002). Arborescent stem succulents, like *C. gigantea*, can be uprooted during high winds because of their shallow root systems and their behavior as bluff bodies (Pierson and Turner 1998); however, it appears that arborescent stem succulents would be far more susceptible to uprooting if they lacked ribs.

Roots of desert succulents

Because of the importance of the hydrostatic skeleton in supporting the shoots of desert succulents, the primary biomechanical contribution of their roots may be to supply shoots with water nearly as soon as it becomes available. This ability has been documented extensively for agaves and cacti (Kausch 1965; Nobel, 1988; Dubrovsky *et al.* 1998) and is associated with architectural, anatomical, and physiological features that characterize the root systems of many desert succulents. In warm deserts, the combination of infrequent and limited rainfall, high temperatures, and poorly developed rocky soils often underlain by a layer of hardpan has selected for plants with comparatively shallow root systems (with the exception of phreato-phytes; Canadell *et al.* 1996; Schenk and Jackson 2002). Succulents in the Agavaceae and the Cactaceae, two families widely represented in North American deserts, tend to have roots restricted to the top 0.5 m of soil, with most roots growing within 0.2 m of the soil surface (Nobel 2002). Such shallow roots may be well positioned to take advantage of limited rainfall, but they would seem less well suited to provide anchorage, particularly for massive succulent shoots. Constrained by their shallow distribution, the roots of desert succulents have evolved a number of features that can improve plant establishment, stability, and endurance in arid soils.

Root system architecture

One apparent adaptation exhibited by a number of arborescent and co-lumnar cacti, such as saguaro (*Carnegiea gigantea*; Table 1.2) is a dimor-phic root system, consisting primarily of shallow, far-reaching lateral roots branching from a much deeper taproot. A plant anchored by a taproot sys-tem is less likely to become uprooted than one with a system of lateral roots only (or a "plate-type system;" Stokes 2002). Thus, it is not surprising that the tallest cacti (Table 1.2) have relatively deep taproots. The root system of *Pachycereus pringlei* has been described as having a combination plate, tap, and fibrous root system; yet even with a taproot, older plants of this species are more susceptible than younger ones to uprooting due to a dis-proportionate increase in stem height with age relative to depth of the tap-root ("bayonet root;" Niklas *et al.* 2002). At an unspecified site in the Sono-ran Desert, one observer noted that half of the saguaros appeared to have been blown over in one direction (Benson 1982). Although not nearly as tall as saguaro, two heavy-stemmed barrel cacti, *Ferocactus cylindraceus* and *F. wislizenii*, are perhaps even more prone to uprooting by wind or other forces due to their lack of deep taproots (Cannon 1911; Barbour 1973). Medium-height, branching cylindropuntias, such as *Cylindropuntia arbuscula*, do tend to have taproots (Table 1.2), as does the unrelated oco-tillo (*Fouquieria splendens*), but the relatively small diameters and shallow depths of their taproots correspond to the modest succulence of their shoots.

Table 1.2. Average shoot height and maximum root depth and root length recorded in the field.

Family Species	Shoot height (m)	Root depth (max; m)	Root length (max; m)	References
Agavaceae				
Agave deserti				Rundel and
	0.3-0.5[a]	0.3	-	Nobel, 1991
Dasylirion				Cannon, 1911
texanum	1.0	0.4	2.2	
Hesperoyucca				Hellmers *et al.*,
whipplei	0.6	0.6	4.0	1955
Yucca elata	0.6	0.7	0.7	Sisson, 1983
Y. schidigera	1-5	0.5	-	Cody, 1986
Cactaceae				
Carnegiea				Cannon, 1911;
gigantea				Turner *et al.*,
	12	0.8	9.7	1995
Cereus				Lüttge *et al.*,
horrispinus	6.5	0.5	-	1989
Cylindropuntia				Cannon, 1911
arbuscula	0.6-1.2	0.1	3.0	
C. fulgida				Cannon, 1911;
	1.0-3.5	0.4	3.0	Preston, 1900
C. imbricata	3.5	0.7	4.6	Dittmer, 1959
C. leptocaulis	0.5-0.7	0.2	2.8	Cannon, 1911
C. versicolor	3.5	0.7	4.6	Dittmer, 1959
Ferocactus				Rundel and
cylindraceus	1-3	0.3	-	Nobel, 1991
F. wislizeni	0.4	0.3	2.4	Cannon, 1911
Pachycereus				Turner *et al.*,
pringlei				1995; Niklas *et*
	15-20	1.2	3.9-5.2	*al.*, 2002
Fouquieriaceae				
Fouquieria				Cannon, 1911
splendens	2-10	0.4	>2.0	

[a] All heights given as ranges are from Hickman (1993). Otherwise, individual heights are from references in table.

An architectural feature common to nearly all desert succulents is the wide-ranging lateral extent of their roots. Specifically, in a global survey of root system characteristics, succulents had the largest root lateral spread: root depth ratio of all plant functional groups (Schenk and Jackson 2002). Although not strictly succulent nor a true desert species, chaparral yucca (*Hesperoyucca whipplei*) had the second largest root spread:shoot spread ratio (4.4) of 18 chaparral species examined (Hellmers *et al.* 1955). A widely spreading root system has a double payoff for arid-land plants: it maximizes the radius and thereby the mass and stability of the root-soil "plate," and it maximizes root surface area in the soil horizon that is most likely to be wetted by rainfall. In addition, the warmer temperatures near the surface of the soil are conducive to root growth for several desert succulents, at least in late winter and early spring when warm (but not excessively high) temperatures and soil moisture availability are likely to coincide (Cannon 1916; Drennan and Nobel 1996). The prevalence of rocks on and near the surface of desert soils not only helps prevent local soil temperatures from becoming too high for root growth but also prolongs the availability of soil water (Nobel *et al.* 1992). In addition, roots that grow through rock crevices and under and between rocks are likely to be better anchored than roots in rock-free soil.

Several cacti exhibit determinate root growth, a developmental feature with architectural consequences that can improve both water uptake and plant anchorage in desert soils. The primary roots of several cacti in the genera *Ferocactus*, *Pachycereus*, and *Stenocereus* cease to elongate due to exhaustion of the root apical meristem within a few days of root initiation (Dubrovsky and North 2002), resulting in a more rapid proliferation of lateral roots than occurs for roots that continue to elongate. Accelerated initiation of lateral roots occurs in root systems of epiphytic cacti (North and Nobel 1994; Dubrovsky and North 2002) and in roots of *Ferocactus cylindraceus* and *Opuntia ficus-indica* in response to rewetting after drought (North *et al.* 1993; Dubrovsky *et al.* 1998). In addition to increasing root surface area for water and nutrient absorption, the rapid production of lateral roots can help provide anchorage, particularly for epiphytic cacti such as *Epiphyllum phyllanthus* and *Rhipsalis baccifera* that become established on bare tree branches (Andrade and Nobel 1997).

Root anatomy

At first glance, roots of desert succulents appear to lack the distinctive anatomical specializations seen in shoots. In an otherwise painstaking investigation of the anatomy of ocotillo, Flora Murray Scott dismissed the roots in a few sentences, stating, "The structure of the root is in no way unusual" (Scott 1932). The roots of ocotillo and other succulent species, however, have a number of anatomical features related to the biomechanical demands of anchoring succulent shoots. The anatomical properties involved in resisting uprooting are most often associated with secondary xylem, or wood, in eudicots such as ocotillo and cacti. Though lacking wood, the roots of monocots such as agaves and yuccas also show anatomical modifications that help maintain or improve plant position in the soil.

One of the most basic root modifications to improve shoot anchorage is an increase in girth, because roots with larger diameters are less likely to deform or break (the stiffness of a beam with a circular outline, such as a taproot, is related to the radius to the fourth power; Stokes 2002). Moreover, the large volume of soil pressing on the surface of a large taproot helps prevent it from pivoting or becoming dislodged (Niklas 1992). In a survey of enlarged roots of cacti (Stone-Palmquist and Mauseth 2002) and an investigation of the biomechanics of the branching columnar cactus *Pachycereus pringlei* (Niklas *et al.* 2002), the increased girth of the taproot is shown to be due to an increased volume fraction of wood, and not, as for shoots, to increases in water-storing parenchyma in ground tissues or a primary cortex. Thus, most cactus taproots cannot properly be called succulent, although water-storing roots of smaller globe-or disc-shaped cacti can perform the traditional role of succulence in supplying water to maintain shoot function. The wood of cactus roots, like most wood, is composed of axial tissues, which are longitudinally oriented vessels, tracheids and fibers, and ray tissues, which are radially oriented files of cells that are usually parenchymatous and associated with storage (Fig. 1.10A). For taproots of *Pachycereus pringlei*, resistance to breakage increases with the increasing volume fraction of axial tissue, and a corresponding reduction in ray tissue or secondary cortex (Niklas *et al.* 2002). Roots highly modified for storage of water and starch would thus not be well suited as taproots for columnar or arborescent cacti, which require strong anchoring roots.

As stated earlier, the succulent stems of Fouquieriaceae have gelatinous fibers throughout their wood (Carlquist 2001). The root wood also has gelatinous fibers (Fig. 1.10B), with some species having only gelatinous fibers in their taproots (Carlquist 2001). The presence of gelatinous fibers in aerial roots of leptocaulous trees (*Ficus benjamina*; Zimmermann *et al.* 1968) suggests that fibers may help the far-reaching lateral roots function as guy ropes to keep the short taproot and root crown of succulent Fouquieriaceae from being pulled out of the soil by the movement of their large stems.

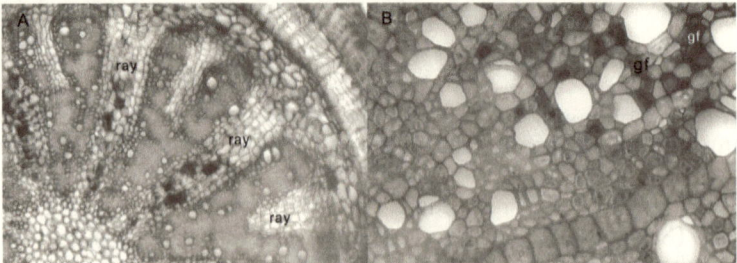

Figure 1.10. A) Cross-section of a root of *Opuntia basilaris*, stained with toluidine blue O, with parenchymatous rays indicated, and B) cross-section of root of *Fouquieria splendens*, stained with toluidine blue O, with gelatinous fibers (gf) indicated.

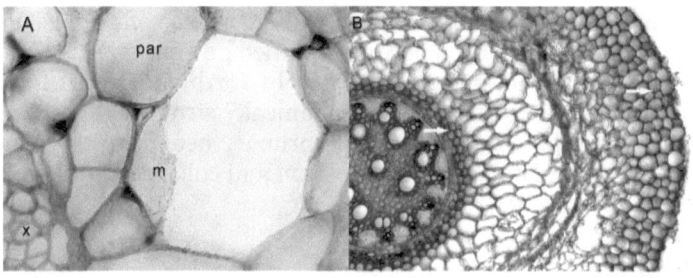

Figure 1.11. A) Cross-section of a root of *Cylindropuntia acanthocarpa*, stained with toluidine blue O, with mucilage cell (m), xylem (x), and parenchyma (par) indicated, and B) cross-section of root of *Agave deserti*, stained with phloroglucinol-HCl, with arrows indicating lignified cells outside endodermis and inside exodermis.

Although lignin does help to stabilize the water content of cell walls (Niklas 1992), an important consideration for plants in arid soils, cactus roots may tolerate unlignified cells in their wood due to two other anatomical features. One is the early and extensive development of a suberized, lignified periderm that can help retard the loss of water from roots to a drier soil (Dubrovsky and North 2002). The second is mucilage-containing cells or ducts, which are produced in the secondary tissues of the roots of many cacti, including *Opuntia ficus-indica*, *Cylindropuntia acanthocarpa* (Fig. 1.11A), the tropical epiphytes *Epiphyllum phyllanthus* and *Rhipsalis baccifera* (North and Nobel 1992, 1994), and the geophytic *Ariocarpus fissuratus* (Anderson 2001). In roots as in shoots, mucilage can act as a capacitor and help moderate changes in the water potential of surrounding tissues (Nobel *et al.* 1992).

As monocots, leaf succulents in the Agavaceae generally lack secondary tissues in their roots, although several species produce rhizomes and stems with some secondary thickening. No agaves and few yuccas attain shoot heights comparable to those of saguaro and other columnar cacti, with the exception of Joshua tree (*Yucca brevifolia*) and a few closely related species. Reliable information on the rooting depth of Joshua tree is not available, despite some references to its relatively deep roots or taproot (Cody 1986). While not nearly as massive as the bayonet root of *Pachycereus pringlei*, the proximal region of main roots of three to four-year old plants of *Y. baccata* and *Y. schidigera* can exceed 5 cm in diameter. Such roots have been described as succulent, as have the roots of many cactus species (Stone-Palmquist and Mauseth 2002). However, in only one species of agave and yucca examined, *Y. rostrata*, does water content approach that of succulent shoots (88%, as compared to 90-95% for shoots; Nobel 1988). A substantial proportion of the mass and volume of mature roots of agaves and yuccas is represented by sclerenchyma. As roots of these species age, cells in the mid-cortex die and collapse while cortical cells adjacent to the exodermis and the endodermis become lignified and suberized (Fig. 1.11B), processes that are accelerated in drying soil (North and Nobel 1991). The resulting core-and-rind structure is mechanically stronger than roots lacking sclerenchyma (Niklas 1992), but the primary benefit to the plant may be to increase root resistance to compression and collapse and thereby help maintain root-soil contact.

Contractile roots

Rimbach (1922) described the roots of *Agave americana* as being capable of contraction (longitudinal shortening), and most agaves and yuccas examined to date show varying degrees of contractile function (G. North, unpublished observations). Morphological evidence of contraction is transverse wrinkling of the outermost tissues (Fig. 1.12A); anatomical evidence is the compression of vascular tissues and the radial rather than axial elongation of inner cortical cells, which leads to longitudinal shortening (Fig. 1.12B; Pütz 2002). For most of these species, the contractile region is the proximal 10-15 cm of adventitious roots, the oldest root region that originates from the stem just below the root-shoot junction. The roots of Joshua tree are exceptional in that they are contractile for most of their length, at least in the case of young plants. Contractile roots have been said to occur in cacti, and longitudinal sections through wrinkled taproots of the small geophytic cacti *Ariocarpus fissuratus* and *Leuchtenbergia principis* show compression of the vascular tissues (G. North, unpublished observations). When the roots contract, shoots are pulled further down in the soil because the distal root regions are firmly anchored. For example, seedlings of *Agave mckelveyana* moved 4-5 cm down further into the soil over a three-month period (G. North, unpublished observations). Contractile roots have been studied most intensively for bulbous plants and geophytes such as gladiolus and hyacinth (Cyr *et al.* 1988), in which root contraction pulls the shoot apical meristem well below the soil surface where it is protected from unfavorable temperatures, dry conditions, and perhaps herbivory.

Figure 1.12. A) Basal region of contracted roots of *Yucca schidigera*, showing transverse wrinkling, and B) cross-section of root of *Agave deserti*, stained with toluidine blue O, showing radially elongated inner cortical cells (c) in the basal contractile region.

For agaves and yuccas, contractile roots could confer several benefits. From a biomechanical perspective, root contraction would help provide anchorage, particularly for seedlings in sandy soils subject to disturbance by wind and flash floods. Contractile roots help restore the position of partially dislodged or excavated shoots of *A. deserti* (G. North, unpublished observations), a service that could be particularly useful for plants such as *A. deserti*, *Hesperoyucca whipplei*, and *Nolina parryi* that grow on rocky slopes. For Joshua tree, contractile roots pull seedlings well below the soil surface; for older plants with tall stems, contraction along nearly the entire length of lateral roots could improve their function as guy ropes. A final biomechanical consequence is that the proximal root region becomes radially enlarged before it contracts, which may help establish or re-establish root-soil contact. Other developmental and anatomical features associated with root contraction, such as delayed senescence of cortical cells, continued aquaporin activity in the proximal root region, and reduced suberization of the endodermis, are associated with unusually high rates of hydraulic conductance for proximal roots of *A. deserti* (North *et al.* 2004) and a few species of yucca. Thus, contractile roots provide not only additional anchorage for desert succulents but also play an important role in the rapid uptake of ephemeral soil water.

Conclusions and future research

The evolution of succulence has led to numerous morphological and anatomical adaptations, many of which have been outlined here, including enlarged organs, reduced branching, long-shoot short-shoot architecture, parenchymatization of primary and secondary tissues, reaction wood, epidermal modifications, shallow root systems, and contractile roots. Needless to say, succulents have received a great deal of attention in terms of their structure. However, other than a few studies (*e.g.*, Fisher 1975), experiments testing hypotheses concerned with the mechanics associated with the anatomical and structural adaptations of succulents, except for cacti, are lacking. Future studies of interest could examine the function of the secondary thickenings in the walls of wide-band tracheids and the thickened walls of water-storage parenchyma of the various succulents covered here. Questions pertaining to these cells are: Do wide-band tracheids and reinforced water-storage parenchyma really function in providing stiffness to dehydrating tissues? And, do wide-band tracheids and reinforced water-storage parenchyma have similar mechanical properties? Also, the reasons for the existence of reaction wood in succulent dicots and asymmetrical growth in pachycaulis monocots need to be better explained. Finally, research into whether and how contractile regions are maintained throughout the life of roots of large pachycauls and leaf succulents, let alone small species of cacti and mesembryanthemums, is certainly necessary. Clearly, the possibilities for further research in the structure and mechanics of succulent plants are numerous and the increased availability of plants through commercial means should aid in furthering the research in this field in the future.

Acknowledgements

The authors thank Park Nobel for, among other things, calling their attention to biomechanics and supporting their research in functional anatomy and morphology. Financial support from the California State Polytechnic University, Pomona Research Scholarship and Creative Activities program to EGB and the National Science Foundation grant # 0517740 to GBN is also appreciated.

Literature cited

Anderson EF. 2001. *The Cactus Family*. Timber Press, Portland, Oregon.

Anderson M. 1998. *The World Encyclopedia of Cacti & Succulents*. Hermes House, New York.

Andrade JL and Nobel PS. 1997. Microhabitats and water relations of epiphytic cacti and ferns in a lowland neotropical forest. *Biotropica* **29:** 261-270.

Angiosperm Phylogeny Group, The. 2003. An update of the Angiosperm Phylogeny Group classification of the orders and families of flowering plants: APG II. *Botanical Journal of the Linnean Society* **141:** 399-436.

Barbour MG. 1973. Desert dogma reexamined: root/shoot productivity and plant spacing. *The American Midland Naturalist* **89:** 41-57.

Beck E. 1994. Cold tolerance in tropical alpine plants. In: Rundel PW, Smith AP, and Meinzer FC (eds.) *Tropical Alpine Environments*. Cambridge University Press, Cambridge.

Bell AD. 1991. *Plant Form: an Illustrated Guide to Flowering Plant Morphology*. Oxford University Press, Oxford.

Benson L. 1982. *The Cacti of the United States and Canada*. Stanford University Press, Stanford, California.

Benson L and Walkington DL. 1965. The southern California prickly-pears—invasion, adulteration, and trial by fire. *Annals of the Missouri Botanical Garden* **52:** 262-273.

Bobich EG and Nobel PS. 2001a. Biomechanics and anatomy of cladode junctions for two Opuntia (Cactaceae) species and their hybrid. *American Journal of Botany 88:* 391-400.

Bobich EG and Nobel PS. 2001b. Vegetative reproduction as related to biomechanics, morphology and anatomy for four cholla cactus species in the Sonoran Desert. *Annals of Botany* **87:** 485-493.

Bobich EG and Nobel PS. 2002. Cladode junction regions and their biomechanics for arborescent platyopuntias. *International Journal of Plant Sciences* **163:** 507-517.

Canadell J, Jackson RB, Ehleringer JR, Mooney HA, Sala OE, and Schulze ED. 1996. Maximum rooting depth of vegetation types at the global scale. *Oecologia* **108:** 583-595.

Cannon WA. 1911. *The Root Habits of Desert Plants*. Publication 131, Carnegie Institution of Washington, Washington, D.C.

Cannon WA. 1916. Distribution of the cacti with especial reference to the role played by the root response to soil temperature and soil moisture. *American Naturalist* **50:** 435-442.

Carlquist S. 1960. Wood anatomy of Asterae (Compositae). *Tropical Woods* **113:** 54-84.

Carlquist S. 1980. Further concepts in ecological wood anatomy, with comments on recent work in wood anatomy and evolution. *Aliso* **9:** 499-553.

Carlquist S. 2001. Wood anatomy of Fouquieriaceae in relation to habit, ecology, and systematics: nature of meristems in wood and bark. *Aliso* **19:** 137-163.

Cody ML. 1984. Branching patterns in columnar cacti. In: Margaris NS, Arianous-tou-Farragitako M, and Oechel WC (eds.) *Being Alive on Land. Tasks for Vegetation Studies, vol. 13.* W. Junk, The Hague. Pp. 201-236.

Cody ML. 1986. Structural niches in plant communities. In: Diamond J and Case TJ (eds.) *Community Ecology.* Harper and Row, Publishers, New York. Pp. 381-405.

Cornejo DO and Simpson BB. 1997. Analysis of form and function in North American columnar cacti (Cactaceae). *American Journal of Botany* **84:** 1482-1501.

Cyr RJ, Lin B-L, and Jernstedt JA. 1988. Root contraction in hyacinth. II. Changes in tubulin levels, microtubule number and orientation associated with differential cell expansion. *Planta* **174:** 446-452.

Dittmer HJ. 1959. A method to determine the length of individual roots. *Bulletin of the Torrey Botanical Club* **86:** 59-61.

Drennan PM and Nobel PS. 1996. Temperature influences on root growth for *Encelia farinosa* (Asteraceae), *Pleuraphis rigida* (Poaceae), and *Agave deserti* (Agavaceae) under current and doubled CO_2 concentrations. *American Journal of Botany* **83:** 133-139.

Dubrovsky JG and North GB. 2002. Root structure and function. In: Nobel PS (ed.) *Cacti: Biology and Uses.* University of California Press, Berkeley, California. Pp. 41-56.

Dubrovsky JG, North GB, and Nobel PS. 1998. Root growth, developmental changes in the apex, and hydraulic conductivity for *Opuntia ficus-indica* during drought. *New Phytologist* **138:** 75-82.

Ehleringer J, Mooney HA, Gulmon SL, and Rundel PW. 1981. Orientation and its consequences for *Copiapoa* (Cactaceae) in the Atacama Desert. *Oecologia* **46:** 63-67.

Esau K. 1977. *Anatomy of Seed Plants.* Wiley, New York.

Evans LS, Imson GJ, Kim JE, and Kahn-Jetter Z. 2004a. Relationships between number of stem segments on longest stems, retention of terminal stem segments and establishment of detached terminal stem segments for 25 species of Cylindropuntia and Opuntia (Cactaceae). *Journal of the Torrey Botanical Society* **131:** 195-203.

Evans LS, Kahn-Jetter Z, and Butwell S. 2004b. Forces necessary for mechanical failure of terminal joints of several species of *Cylindropuntia* and *Opuntia* (Cactaceae). *Journal of the Torrey Botanical Society* **131:** 311-319.

Fahn A. 1990. *Plant Anatomy,* 4th edn. Pergamon Press, Elmsford, New York.

Fisher JB. 1975. Eccentric secondary growth in *Cordyline* and other Agavaceae (Monocotyledonae) and its correlation with auxin distribution. *American Journal of Botany* **62:** 292-302.

Fisher JB and Stevenson JW. 1981. Occurrence of reaction wood in the branches of dicotyledons and its role in tree architecture. *Botanical Gazette* **142:** 82-95.

Fournier M, Bailleres H, and Chanson B. 1994. Tree biomechanics: growth, cumulative prestresses, and reorientations. *Biomimetics* **2:** 229-251.

Franco-Vizcaíno E, Goldstein G, and Ting IP. 1990. Comparative gas exchange of leaves and bark in three stem succulents of Baja California. *American Journal of Botany* **77:** 1272-1278.

Geller GN and Nobel PS. 1984. Cactus ribs: influence on PAR interception and CO_2 uptake. *Photosynthetica* **18:** 482-494.

Geller GN and Nobel PS. 1986. Branching patterns of columnar cacti: influences on PAR interception and CO_2 uptake. *American Journal of Botany* **73:** 1193-1200.

Gentry HS. 1982. *Agaves of Continental North America*. The University of Arizona Press, Tucson, Arizona.

Gibson AC. 1973. Comparative anatomy of secondary xylem in Cactoideae (Cactaceae). *Biotropica* **5:** 29-65.

Gibson AC. 1977a. Vegetative anatomy of *Maihuenia* (Cactaceae) with some theoretical discussions of ontogenetic changes in xylem cell types. *Bulletin of the Torrey Botanical Club* **104:** 35-48.

Gibson AC. 1977b. Wood anatomy of opuntias with cylindrical to globular stems. *Botanical Gazette* **138:** 334-351.

Gibson AC. 1978. Wood anatomy of platyopuntias. *Aliso* **9:** 270-307.

Gibson AC. 1981. Vegetative anatomy of *Pachycormus* (Anacardiaceae). *Botanical Journal of the Linnean Society* **83:** 273-284.

Gibson AC. 1982. The anatomy of succulence. In: Ting IP and Gibbs M (eds.) *Crassulacean acid metabolism: proceedings of the fifth annual symposium in botany, January 14-16th, commemorating the seventy-fifth anniversary of the Agricultural Experiment Station at the University of California Riverside*. American Society of Plant Physiologists, Rockville, Maryland. Pp. 1-15.

Gibson AC. 1996. *Structure-Function Relations of Warm Desert Plants*. Springer-Verlag, Berlin.

Gibson AC and Nobel PS. 1986. *The Cactus Primer*. Harvard University Press, Cambridge, Massachusetts.

Goldstein G, Andrade JL, and Nobel PS. 1991. Differences in water relations parameters for the chlorenchyma and parenchyma of *Opuntia ficus-indica* under wet *versus* dry conditions. *Australian Journal of Plant Physiology* **18:** 95-107.

Hacke UG, Sperry JS, Pockman WT, Davis SD, and McCulloh KA. 2001. Trends in wood density and structure are linked to prevention of xylem implosion by negative pressure. *Oecologia* **126:** 457-461.

Hellmers H, Horton JS, Juhren G and O'Keefe J. 1955. Root systems of some chaparral plants in southern California. *Ecology* **36:** 667-678.

Henrickson J. 1969a. An introduction to the Fouquieriaceae. *Cactus & Succulent Journal* **41:** 97-105.

Henrickson J. 1969b. The succulent fouquierias. *Cactus & Succulent Journal* **41:** 178-184.

Henrickson J. 1969c. Anatomy of periderm and cortex of Fouquieriaceae. *Aliso* **7:** 97-126.

Henrickson J. 1972. A taxonomic revision of the Fouquieriaceae. *Aliso* **7:** 439-537.

Hickman JC (ed.). 1993. *The Jepson Manual: Higher Plants of California*. University of California Press, Berkeley, California.

Kahn-Jetter Z, Evans LS, Liclican E and Pastore M. 2001. Compressive/tensile stresses and lignified cells as resistance components in joints between stem segments *Opuntia fulgida* and *Opuntia versicolor* (Cactaceae). *International Journal of Plant Sciences* **162:** 579-587.

Kausch W. 1965. Beziehungen zwischen Wurzeiwachstum. Transpiration und CO_2-Gaswechsel bei einigen Kakteen. *Planta* **66:** 229-238.

Koller AL and Rost TL. 1998. Leaf anatomy in *Sansevieria* (Agavaceae). *American Journal of Botany* **75:** 615-633.

Landrum JV. 2001. Wide-band tracheids in leaves of genera in Aizoaceae: the systematic occurrence of a novel cell type and its implications for the monophyly of the subfamily Ruschioideae. *Plant Systematics and Evolution* **227:** 49–61.

Lüttge U, Medina E, Cram WJ, Lee HSJ, Popp M, and Smith JAC. 1989. Ecophysiology of xerophytic and halophytic vegetation of a coastal alluvial plain in northern Venezuela. II. Cactaceae. *New Phytologist* **111:** 245-251.

Mabberley DJ. 1974a. Branching in pachycaul senecios: the Durian Theory and the evolution of angiospermous trees and herbs. *New Phytologist* **73:** 967-975.

Mabberley DJ. 1974b. Pachycauly, vessel-elements, islands and the evolution of arborescence in 'herbaceous' families. *New Phytologist* **73:** 977-984.

Mauseth JD. 1993. Water-storing and cavitation-preventing adaptations in wood of cacti. *Annals of Botany* **72:** 81-89.

Mauseth JD. 2004a. The structure of photosynthetic succulent stems in plants other than cacti. *International Journal of Plant Sciences* **165:** 1-9.

Mauseth JD. 2004b. Wide-band tracheids are present in almost all species of Cactaceae. *Journal of Plant Research* **117:** 69-76.

Mauseth JD and Plemons BJ. 1995. Developmentally variable polymorphic woods in cacti. *American Journal of Botany* **82:** 1199-1205.

Mauseth JD, Uozumi Y, Plemons BJ, and Landrum JV. 1995. Structural and systematic study of an unusual tracheid type in cacti. *Journal of Plant Research* **108:** 517-526.

Molina-Freaner F, C Tinoco-Ojanguren and KJ Niklas. 1998. Stem biomechanics of three columnar cacti from the Sonoran Desert. *American Journal of Botany* **85:** 1082-1090.

Niklas KJ. 1988. Dependency of the tensile modulus on transverse dimensions, water potential and cell number of pith parenchyma. *American Journal of Botany* **75:** 1286-1292.

Niklas KJ. 1992. *Plant Biomechanics: an Engineering Approach to Form and Function.* University of Chicago Press, Chicago, Illinois.

Niklas KJ and Buchman SL. 1994. The allometry of saguaro height. *American Journal of Botany* **81:** 1161-1168.

Niklas KJ, Molina-Freaner F, and Tinoco-Ojanguren C. 1999. Biomechanics of the columnar cactus *Pachycereus pringlei. American Journal of Botany* **78:** 1252-1259.

Niklas KJ, Molina-Freaner F, Tinoco-Ojanguren C, and Paolillo Jr DJ. 2000. Wood biomechanics and anatomy of *Pachycereus pringlei. American Journal of Botany* **87:** 469-481.

Niklas KJ, Molina-Freaner F, Tinoco-Ojanguren C, and Paolillo Jr DJ. 2002. The biomechanics of *Pachycereus pringlei* root systems. *American Journal of Botany* **89:** 17-21.

Niklas KJ, Molina-Freaner F, Tinoco-Ojanguren C, Hogan CJ, and Paolillo Jr DJ. 2003. On the mechanical properties of the rare endemic cactus *Stenocereus eruca* and the related species *S. gummosus. American Journal of Botany* **90:** 663-674.

Niklas KJ, Cobb ED, and Marler T. 2006. A comparison between the record height to stem diameter allometries of pachycaulis and leptocaulis species. *American Journal of Botany* **97:** 79-83.

Nilsen ET, Sharifi MR, Rundel PW, Forseth IN, and Ehleringer JR. 1990. Water relations of stem succulent trees in north-central Baja California. *Oecologia* **82:** 299-303.

Nobel PS. 1988. *Environmenal Biology of Agaves and Cacti.* Cambridge University Press, New York.

Nobel PS and Meyer RW. 1991. Biomechanics of cladodes and cladode-cladode junctions for *Opuntia ficus-indica* (Cactaceae). *American Journal of Botany* **78:** 1252-1259.

Nobel PS. 2002. Ecophysiology of roots of desert plants, with special emphasis on agaves and cacti. In: Waisel Y, Eshel A, and Kafkafi U (eds.) *Plant Roots: The Hidden Half, 3rd edn.* Marcel Dekker, New York. Pp. 961-973.

Nobel PS. 2005. *Physicochemical and Environmental Plant Physiology, 3rd edn.* Elsevier Academic Press, Burlington, Massachusetts.

Nobel PS, Cavelier J and Andrade JL. 1992. Mucilage in cacti: its apoplastic capacitance, associated solutes, and influence on tissue water relations. *Journal of Experimental Botany* **43**: 641-648.

Nobel PS, Miller PM, and Graham EA. 1992. Influence of rocks on soil temperature, soil water potential, and rooting patterns for desert succulents. *Oecologia* **92**: 90-96.

North GB and Nobel PS. 1991. Changes in hydraulic conductivity and anatomy caused by drying and rewetting roots of *Agave deserti* (Agavaceae). *American Journal of Botany* **78**: 906-915.

North GB and Nobel PS. 1992. Drought-induced changes in hydraulic conductivity and structure in roots of *Ferocactus acanthodes* and *Opuntia ficus-indica*. *New Phytologist* **120**: 9-19.

North GB and Nobel PS. 1994. Changes in root hydraulic conductivity for two tropical epiphytic cacti as soil moisture varies. *American Journal of Botany* **81**: 46-53.

North GB, Huang B, and Nobel PS. 1993. Changes in structure and hydraulic conductivity for root junctions of desert succulents as soil water status varies. *Botanica Acta* **106**: 126-135.

North GB, Martre P, and Nobel PS. 2004. Aquaporins account for variations in hydraulic conductance for metabolically active root regions of *Agave deserti* in wet, dry, and rewtted soil. *Plant, Cell and Environment* **27**: 219-228.

Olson ME. 2003. Stem and leaf anatomy of the arborescent Cucurbitaceae *Dendrosicyos socotrana* with comments on the evolution of pachycauls from lianas. *Plant Systematics and Evolution* **239**: 199-214.

Ortega-Rubio A, Romero-Schmidt H, Arguelles-Méndez C, Castellanos-Vera A. 1995. Effect of wind on a *Ferocactus fordii* var. *fordii* population on Piedra Island, Baja California Sur, Mexico. *Journal of Arid Environments* **31**: 15-19.

Peinado M, Delgadillo J, Aguirre JL. 2005. Plant associations of El Vizcaino Biosphere Reserve, Baja California Sur, Mexico. *The Southwestern Naturalist* **50**: 129-149.

Pierson EA, Turner RM. 1998. An 85-year study of saguaro (*Carnegiea gigantea*) demography. *Ecology* **79**: 2676-2693.

Preston C.E. 1900. Observations on the root system of certain Cactaceae. *Botanical Gazette* **30**: 348-351.

Price CA and Enquist BJ. 2006. Scaling of mass and morphology in plants with minimal branching: an extension of the WBE model. *Functional Ecology* **20**: 11-20.

Pütz N. 2002. Contractile roots. In: Waisel Y, Eshel A, and Kafkafi U (eds.) *Plant Roots: The Hidden Half, 3rd edn.* Marcel Dekker, New York. Pp. 975-987.

Racine CH and Downhower JF. 1974. Vegetative and reproductive strategies of *Opuntia* (Cactaceae) in the Galapagos Islands. *Biotropica* **6**: 175-186.

Rimbach A. 1922. Die Wurzelverkürzung bei den groоen Monokotylenformen. *Bericht der Deutschen botanischen Gesellschaft* **40**: 196-202.

Rowley GD. 1987. *Caudiciform and Pachycaul Succulents: Pachycauls, Bottle-, Barrel- and Elephant Trees and Their Kin.* Strawberry Press, Mill Valley, California.

Rundel PW and Nobel PS. 1991. Structure and function in desert root systems. In: Atkinson D (ed.) *Plant Root Growth. An Ecological Perspective.* Blackwell Scientific, Oxford, pp. 349-378.

Schenk HJ and Jackson RB. 2002. Rooting depths, lateral roots spreads and below-ground/above-ground allometries of plants in water-limited ecosystems. *Journal of Ecology* **90:** 480-494.

Scott FM. 1932. Some features of the anatomy of *Fouquieria splendens. American Journal of Botany* **19:** 673-678.

Smith JAC and Nobel PS. 1986. Water movement and storage in a desert succulent: anatomy and rehydration kinetics for leaves of *Agave deserti. Journal of Experimental Botany* **37:** 1044-1053.

Sisson WB. 1983. Carbon balance of *Yucca elata* Engelm. during a hot and cool period in situ. *Oecologia* **57:** 352-360.

Stokes A. 2002. Biomechanics of tree root anchorage. In: Waisel Y, Eshel A, and Kafkafi U (eds.) *Plant Roots: The Hidden Half, 3rd edn.* Marcel Dekker, New York. Pp. 175-186.

Stevenson DW. 1980. Radial growth in *Beaucarnea recurvata. American Journal of Botany* **67:** 476-489.

Stone-Palmquist ME and Mauseth JD. 2002. The structure of enlarged storage roots in cacti. *International Journal of Plant Sciences* **163:** 89-98.

Talley S and Mungal G. 2002. Flow around cactus-shaped cylinders. *Center for Turbulence Research Annual Research Briefs.* Pp. 363-376.

Talley S, Iaccarino G, Mungal G, and Mansour N. 2001. An experimental and computational investigation of flow past cacti. *Center for Turbulence Research Annual Research Briefs.* Pp. 51-63.

Turner RM, Bowers JE, and Burgess TL. 1995. *Sonoran Desert Plants: An Ecological Atlas.* The University of Arizona Press, Tucson Arizona.

von Willert DJ. 1992. *Life Strategies of Succulents in Deserts: with Special Reference to the Namib Desert.* Cambridge University Press, New York.

Zimmermann MH, Wardrop AB, and PB Tomlinson. 1968. Tension wood in aerial roots of *Ficus benjamina* L. *Wood Science and Technology* **2:** 95-104.

Chapter 2

WATER TRANSPORT PROCESSES IN DESERT SUCCULENT PLANTS

Paul J. Schulte

Perspectives in Biophysical Plant Ecophysiology: A Tribute to Park S. Nobel, pp. 39-55
Edited by: E. De la Barrera and W.K. Smith
© 2009 by The Author
Book Compilation © 2009 Universidad Nacional Autónoma de México

Introduction

As transpiring land plants, desert succulents depend on the process of water transport for acquiring and distributing water within the various organs and tissues of the plant. Although not reaching the heights of tall trees with subsequent problems associated with transport path length, succulent species often occur in arid environments and must survive extended periods of drought with extremely high evaporative demand and very low soil water potentials. Their photosynthetic tissues are not particularly desiccation tolerant and do not survive drought by the tolerance mechanisms occurring in many desert shrubs[1], such as *Larrea tridentata* (creosote bush) common to the Mojave Desert of the southwestern United States (Smith *et al.* 1997). As we will see, succulent plants depend on a number of anatomical structures, morphological characteristics and physiological processes such as large volumes of tissues specialized for water storage and it is these characteristics that allow them to survive extended drought by maintaining relatively high tissue water potentials.

A number of recent reviews present general aspects of plant water transport—the basis for the cohesion-tension theory as concerned with transport mechanisms and driving forces for flow (Sperry *et al.* 2002, 2003), the concepts of cavitation and embolism of conduits in the xylem (Davis *et al.* 1999; Tyree *et al.* 1999), and the evolution of transport structures (Sperry 2003). Therefore except for a brief review in this chapter, I will focus on aspects of these processes with special relevance for desert succulent plant species. It should be recognized, of course, that the evolution of cacti would have depended on many other physiological characteristics in addition to water storage such as mechanical support (Cornejo and Simpson 1997). Mechanical needs would vary tremendously between shorter barrel cacti (like *Ferocactus* spp.) and columnar cacti like the saguaro (*Carnegiea gigantea*) of the Sonoran desert, but will not be considered in this review (see Molina-Freaner *et al.* 1998 and Niklas *et al.* 2000).

Basic principles

The cohesion-tension theory

The cohesion-tension theory states that water transport from roots to shoots of terrestrial plants occurs because of tensions generated in the transpiration stream as a result of water loss from evaporative surfaces such as leaves (for a thorough review, see Steudle 2001). This tension places the liquid water in a metastable state in that should a gas bubble develop or be drawn into a tracheid or vessel from an adjacent cell, the bubble can expand and block that conduit as an embolism. Despite con-

---

[1] Given their relatively high tissue osmotic potentials (low solute concentrations), succulent desert species were not considered as true xerophytes by Walter and Stadelmann (1974) in the sense of having tissues that tolerate low water potential.

siderable recent debate over the cohesion-tension theory (Tyree 1997), most plant physiologists appear to support the theory although research is still needed regarding a number of mechanistic aspects such as conduit refilling.

Water flowing through specialized conduits in the xylem of plants encounters resistance because of shear forces generated where the water is in contact with cell walls and various other structures such as pit canals and pit membranes. Far from being fixed resistances, the network of resistances in the xylem has a highly dynamic nature because of (1) cavitation and subsequent embolism of conduits, (2) refilling of embolized conduits, and (3) variation in properties of individual cell structures due to changes in the ionic composition of the xylem sap (Zwieniecki *et al.* 2001; Holbrook *et al.* 2002).

Water transport in the plant also encounters resistance in flow through living cells of roots, leaves, and water storage tissues. Dominating this pathway are the cell's plasma membrane and tonoplast, structures where aquaporins may play an important role and may also confer a variable nature to this resistance (Steudle 2001). In roots, the development of an endodermis during root maturation has important consequences for nutrient uptake as well as radial flow resistance by blocking the apoplast pathway and forcing flow through living cell membranes (Steudle 2001).

Water storage in living cells of root, stem, or leaf organs is also an important characteristic of the transport pathway, particularly for succulent plant species. The term capacitance is often applied to this property by analogy with charge storage in electrical circuits. Although perhaps less significant for small plants, stored water may contribute up to 25% of daily transpiration for large trees (Phillips *et al.* 2003). For succulent desert plants, over half of the water transpired during the period of stomatal opening at night may be contributed by water from storage, this storage being recharged when the stomata are closed during the day (Schulte and Nobel 1989). Stored water can therefore act as a buffer against daily water potential changes. But as will be seen later, water storage can also provide an important source capable of maintaining turgor in photosynthetic leaf or stem tissues during long-term drought for succulent plants in arid environments.

Cavitation and embolism

According to the cohesion-tension theory, water in many places within the plant exists in a metastable state. Once a water column under tension breaks due to the formation of an air bubble or the entry of air from an air-filled space outside of the cell or in an adjacent cell, that water column can no longer transmit the tension required to transport water in the xylem. Thus a cavitation, or bubble forming event, may immediately lead to an embolism, or blockage of a vessel. Importantly, tremendous variation exists between plant species, plant organ (root, stem, leaf), along with xylem conduit size and structure with respect to the likelihood of embolism at a given water potential, and these issues have been thoroughly reviewed elsewhere (see Sperry *et al.* 2003). One important factor, conduit diameter, shows an interesting relationship with the vulnerability of that conduit to cavitation. For cavitation induced by tension in the xylem, within a species and organ of that plant, wider conduits cavitate first with increasing tension but the relationship does not appear to be true across plant species, where the nature of the pit membrane pores may be more important than conduit size (Hacke and Sperry 2001). On the other hand, conduit diameter does appear to be significant in general for freezing-induced embolism (Pitterman and Sperry 2003).

Mechanisms responsible for cavitation have been described as either homogeneous nucleation, where a gas bubble forms spontaneously within the liquid water under tension in the xylem tracheids or vessels, or heterogeneous nucleation, where an air bubble is drawn in from an adjacent cell (also called air-seeding). The occurrence of embolism resulting from freezing appears to depend on the volume (and hence diameter) of the conduits, and therefore narrower tracheids or vessels may be less vulnerable to embolism due to freezing. Recently, it has been suggested that although these diameter relationships are valid within a species, tracheids are not inherently less susceptible to freezing-induced embolism than vessels (Pitterman and Sperry 2003).

A less well understood process concerns the refilling of embolized xylem conduits. Historically, it was believed that embolized conduits could only be refilled through the generation of positive pressures in the xylem, such as through root pressure (active uptake of solutes that can generate a positive pressure through osmosis). More recently, mechanisms for the collapse of air bubbles while small tensions still exist have been discussed. However, it is evident experimentally that the recovery of hydraulic conductance appears to occur while substantial tensions exist in the xylem, and so further efforts are underway to consider refilling mechanisms. There is evidence that the positive pressures of living parenchyma cells adjacent to cavitated tracheids might play a role in the refilling process (see Holbrook *et al.* 2002 for further discussion). Finally, one might suggest that a current appreciation of the dynamic nature of the cavitation-refilling process is a positive legacy of the debate over the validity of the cohesion-tension theory, inspired in part by the arguments of Martin Canny and others in the literature (Canny 1995, 2001; Comstock 1999; Stiller and Sperry 1999).

Water relations and transport in succulents

The Crassulacean acid metabolism (CAM) pathway, with its pattern of stomatal activity allowing for reduced water loss for a given CO_2 uptake, and the presence of succulent tissues acting as a storage or buffer for water are perhaps the most obvious characteristics of succulent species enabling their survival in arid environments (Kluge and Ting 1978; von Willert *et al.* 1992). As will be seen later, these same characteristics are also important among many epiphytic species which can experience episodic drought.

In considering the various anatomical and physiological features of succulent plants that enable them to survive drought, we typically do not think of succulents as desiccation tolerant species having tissues that tolerate low water potential or water content. Tissue water potentials generally do not reach low values common to desert shrubs like *Larrea tridentata* of −6 to −8 MPa (Smith *et al.* 1997). Correspondingly, the osmotic potentials of cells in succulent tissues are typically expected to be relatively high and this is the case for many species. However, as noted by von Willert *et al.* (1992), some succulent species reach −3 to −5.5 MPa water potentials after long drought. Exceptions in the family Mesembryanthemaceae also described by von Willert *et al.* (1992) can have tissue osmotic potentials as low as −6.5 MPa during drought. Nonetheless, most of the characteristics of succulent plants that enable them to survive drought are involved with acquiring and maintaining large quantities of stored water and reducing the loss of that water to the atmosphere during extended drought. One of the amazing abilities of succulent species is their ability to maintain high water potentials after months of drought; for example a CAM epiphyte *Tillandsia utriculata* whose water potential only decreased from about −0.75 to −1.25 MPa after two months of drought (Stiles and Martin 1996). As a necessary consequence of large amounts of stored water and high tissue water potentials, together with the low water potentials found in dry soil, these plants must also prevent the loss of that stored tissue water back into the soil.

Anatomical and morphological adaptations to drought

A number of the adaptations of desert succulent plants to periodic drought involve anatomical or morphological characteristics, the most obvious perhaps being the presence of large volumes of water storage tissues in roots, stems or leaves (see also Chapter 1 for a review of anatomy). Water storage tissues associated with succulence in leaves of monocots like *Agave deserti* are typically located internal to photosynthetic tissues (Fig. 2.1) or in the pith of stems for some cacti (Barcikowski and Nobel 1984). Important characteristics of these cells aside from their volume include having a low modulus of elasticity, a mechanical property of the cell wall which determines how changes in cell volume correspond to a given change in cell water potential (Nobel 1983). Thus for a tissue capable of providing large quantities of water with only a small change in water potential, cells must have highly elastic walls (low elastic modulus, expressed as dP/dV times cell volume, where P and V are pressure and volume, respectively). One measurement of this property gave a value of 3.3 MPa for succulent tissues (Stiles and Martin 1996). This value for the elastic modulus is at the low end or lower (more elastic) than values commonly found among various woody plants (3–12 MPa, Borghetti *et al.* 2004; 3–13 MPa, Sack *et al.* 2003; 8–21 MPa, Burghardt and Riederer 2003; 8–24 MPa, Tognetti *et al.* 2002). On the other hand, Youngman and Heckathorn (1992) have measured much higher values with the succulent *Suaeda cal-*

Figure 2.1. *Ferocactus acanthodes* from the Mojave Desert in Nevada. The plant on the left was about 1.3 m tall and 0.45 m in diameter. A cross-section through one rib (right) shows the thin chlorenchyma outer tissue layer (scale bar has 1 mm subdivisions). The remainder of the stem interior to the chlorenchyma is made up of primarily water storage and vascular tissues.

ceoliformis (8-19 MPa), but this plant is a halophyte and as noted later also has much lower osmotic potentials than most succulent plants. The elasticity of plant cells may vary seasonally or with drought (see above papers) but, in general, this parameter has not been characterized well among succulent plant species.

Succulent stems having large quantities of elastic cells whose volume can change considerably (high capacitance) will necessarily undergo a large change in diameter with seasonal changes in water content. Nevertheless, a physical disruption of the epidermis does not necessarily follow from geometrical considerations alone, assuming all the cells lose water volume equally. As discussed elsewhere, however, succulent organs tend to lose water preferentially from the inner tissues and less so from the outermost, photosynthetic tissues. Therefore, large changes in diameter that are primarily due to changes in the water volume of inner cells will lead to tensions or compressions of the outer portions of the stem. The succulent stems of species in genera such as *Ferocactus* and *Carnegiea* are not smoothly circular in cross-section, but convoluted with a pattern of ribs. It was suggested quite early (Spaulding 1905), that such stems can accomplish these volume changes without physical disruption because the ribbing allows the stem to increase in volume without a proportional increase in surface area. The ribbing habit of course also increases the surface area of a stem for a given volume in comparison with a smooth stem of equal volume. As noted above, the presence of the ribs would allow changes in total stem volume without requiring shrinkage or expansion of the outermost cells. Mauseth (2000) analyzed the tradeoffs between enhanced shrinkage ability and increased surface area for hypothetical stems with various numbers of ribs. The analysis suggested that different scenarios of numbers of ribs and the height of ribs could be important for succulent species exposed to various habitats. A dry environment with small amounts of rain throughout the year (hence relatively modest changes in stem volume) might have stems with shallow ribs as opposed to a dry environment with highly variable rainfall and concomitant large changes in stem volume, where stems might have a greater number of deeper ribs.

Aside from water storage, succulent species may also have adaptations to xeric environments involving the anatomical characteristics of xylem conduits. Nonperforate conducting cells (tracheids) in the xylem of many succulent species have secondary annular or helical thickenings that project markedly into the tracheid lumen. These cells have been referred to as vascular tracheids by some authors (Gibson 1973, 1977; Bobich and Nobel 2001) and wide-band tracheids by others (Mauseth et al. 1995; Mauseth and Stone-Palmquist 2001; Landrum 2002). Such cells are suggested to allow water diffusion readily in axial as well as lateral directions and also allow the expansion and contraction that occurs in succulent tissues without disrupting the conducting xylem (Gibson 1996). The annular to helical thickenings do not resist dimensional changes in the axial direction, and this cell type appears to be common in root or shoot organs that show dramatic contraction with dehydration, particularly when rigid fibers are lacking (Mauseth and Stone-Palmquist 2001). These authors found that the occurrence of wide-band tracheids in shoots or roots was related to the size and shape of cacti. Succulent species with tall stems where support would be important had fibrous wood, whereas shorter or globose succulents often had prominent wide-band tracheids because these organs would be more dependent on turgor for support. Wide-band tracheids also occur in many leaf-succulent genera of the Aizoaceae family (commonly called ice plants), being especially dominant in the secondary or higher order veins of leaves (Landrum 2001, 2002). Although the functioning of these cells for water transport has not yet been studied, perhaps there is an important interaction in succulent tissues for resistance to collapse of the xylem conduits and yet flexibility in the face of tissue dimensional changes due to loss or gain of stored water.

Water channels and membrane permeability

An important discovery for our understanding of water flow through living cells was the presence of aquaporins or specialized water channels in plant membranes. The general structure, function, and important roles of aquaporins in plants have been reviewed (Tyerman et al. 1999). Among succulent species, Martre et al. (2001) suggest that for Opuntia acanthocarpa, aquaporin activity (changes in opening of channels) accounts for roughly half of the changes in root hydraulic conductance during drought and rewetting of the soil for distal regions of the root. As noted later, such changes in aquaporin activity may be similarly important for roots of *Agave deserti* during drought (North and Nobel 2000).

Concerns have been raised over the use of mercury compounds in studies of aquaporins because these compounds may have other toxic effects on plant cells (Barrowclough *et al.* 2000). In many experiments, the application of $HgCl_2$ and measurement of hydraulic conductivity is often followed by an application of 2-mercaptoethanol to reverse the inhibitory effects of the former compound on aquaporins. Mercury containing compounds like $HgCl_2$ appear to interact with sulphydryl groups in the aquaporin protein, while 2-mercaptoethanol acts as a scavenging agent because of its own sulphydryl (Tyerman *et al.* 1999). Much further work will be necessary before the role of aquaporins in succulent plant species can be understood. For example, it is noted (Steudle 1994; Steudle and Peterson 1998) that water flow through living tissues is a combination of flow through the apoplast outside of the plasma membrane and also one or more cell-to-cell pathways involving flow across membranes and through the cytoplasm. Thus the potential response of conductivity to changes in aquaporins will depend on the degree to which flow is dependent upon the pathways which include the crossing of a membrane. Further, it is likely that water can also cross membranes independent of aquaporins (through other channels or directly through the lipid bilayer).

Root system isolation and regrowth

There are at least two situations in which root hydraulic conductivity may be important for desert succulents: conductivity as a limitation to water uptake ability when the soil is wet and conductivity changes during the onset of drought that may serve to limit the loss of water from root systems to the drying soil. It became apparent from earlier studies of extended drought for *Agave deserti* (Schulte and Nobel 1989) that modifications to the hydraulic pathway between the root and soil must occur to prevent water loss from the succulent stem and leaves back into the soil through the roots. At the time, it was not clear what mechanisms might exist to prevent this occurrence, but subsequent work has suggested a number of possibilities. A series of modifications to roots and to the contact between roots and the soil are very important for isolating succulent plants from the drying soil as drought progresses. These modifications include embolism of root xylem conduits, the formation of air spaces within the root cortex, increased suberization of cortical cells, and the formation of air gaps between the root and soil (North and Nobel 1991, 1992, 1997, 2000). See Chapter 1 in the present text for a detailed discussion of root systems.

Studies of the hydraulic conductivity of xylem in roots and leaves of the succulent monocots *Agave deserti* and *A. tequilana* demonstrate that these species have a relatively vulnerable xylem with respect to cavitation, corresponding to their desiccation intolerant or avoiding habit (Linton and Nobel 2001). In that study, leaves had higher axial conductivity than roots when the soil was wet, but after extended drought, conductivity was lower in roots than leaves, leading to the further suggestion that such conductivity changes due to embolism were an important component to isolating the succulent plant tissues from water loss to the dry soil. Roots in particular may be highly vulnerable to cavitation in the xylem and loss of conducting ability.

Water storage – capacitance and cell elasticity

It has long been recognized that succulence plays an important role in the water relations of many arid environment plants by providing a water source during periods when water is not available from the soil. Water storage in the pith and cortex of large columnar cacti like *Carnegiea gigantea* and *Ferocactus acanthodes* allows these plants to survive extended drought leading to loss of over 80% of their total stem water (Barcikowski and Nobel 1984). Similarly, a study of the leaf succulent monocot *Agave deserti*, showed that stored water allowed this plant to maintain turgor in the chlorencyma after 8 months of drought leading to the loss of about half of the total leaf water (Schulte and Nobel 1989). Over a daily basis, stored water in succulent organs can act as a buffer for leaves as shown in a study of water loss by transpiration and water uptake by roots of *Agave deserti*, where loss occurred primarily at night but uptake was nearly constant over the day (Graham and Nobel 1999). Thus although stomata are closed during the day, water uptake from the soil may continue because of the recharge of storage tissues, which may have provided approximately 40% of the daily water loss. In a comparison of epiphytic cacti and ferns in a tropical forest, Andrade and Nobel (1997) found that a relatively high water storage capacitance among the epiphytic cacti accounted for their greater occurrence at the driest sites that were studied. The epiphytic cacti were able to continue photosynthetic activity longer into drought periods as a result of their stored water and CAM-mediated nighttime stomatal opening cycle. Thus the presence of water storage tissue is an important adaptation in a variety of environments.

Water transport between storage and photosynthetic tissues and the forces driving such redistribution are important components in the ability of succulence to support plant survival during extended drought. Early work with cacti suggested that the nonphotosynthetic water storage tissues located in the pith or inner cortex regions interior to the chlorenchyma would preferentially lose water to the chlorenchyma during drought such that more water was lost from storage tissues than from the outer photosynthetic tissue (Barcikowski and Nobel 1984). In their studies of *Carnegiea gigantea*, *Ferocactus acanthodes*, and *Opuntia basilaris*, it was found that 80, 82, and 65 percent, respectively, of water loss during desiccation came from water storage tissues. It was also suggested that solute loss in pith storage tissues and a lower osmotic potential (greater solute concentration) in the chlorenchyma were important components in this process. Nerd and Nobel (1991) also indicate an important role for osmotic potential changes in the storage tissues allowing for water flow into photosynthetic tissues. Although the osmotic potential in storage tissues decreased, predictions based on the van't Hoff relation suggest that a 3-fold greater decrease than observed should have occurred because of water content changes alone, thus suggesting that solute quantities (moles per cell) decreased, perhaps partly due to the polymerization of sugars into starch or mucilage polysaccharides. Mucilages in extracellular (apoplast) regions of succulent tissues also act in water storage, perhaps accounting for nearly one-third of the water in storage tissues of *Opuntia ficus-indica* cladodes (Goldstein *et al.* 1991).

In several succulents of the Cactacaea, Mauseth (1995) has observed that during drought, outer photosynthetic portions of the stem cortex retain turgor as well as the innermost portions of the cortex, but a middle region loses turgor, suggesting the presence of highly elastic cell walls that allow them to collapse as water is released from this water-storage tissue. Succulents of *Peperomia carnevalii* lost more water during drought from hydrenchyma (water storage tissue) than from chlorenchyma, perhaps due the presence of an osmotic gradient between these tissues (Herrera *et al.* 2000). Although leaf thickness declined by 35%, these changes were fully due to hydrenchyma water loss and the chlorenchyma tissue thickness remained constant. For *Peperomia magnoliaefolia*, drought leading to the loss of 50% of the leaf water produced relative water content decreases in the hydrenchyma of 75-85%, but only 15-25% for chlorenchyma. A study of a CAM epiphyte also indicated that water is lost preferentially from the hydrenchyma in support of chlorenchyma tissue (Nowak and Martin 1997)

Another process important in the role of stored water during drought involves the transport of water between different organs within the plant. In studies of *Senecio medley-woodii* during a two month drought, Donatz and Eller (1993) observed a gradual dehydration of leaves progressing from the oldest leaves to younger leaves; the older leaves were eventually shed while the young leaves were able to maintain or even increase transpiration. After 10 days of drought, water was no longer acquired from the soil and the younger leaves were maintaining transpiration with water translocated from the older leaves. As discussed earlier for water translocation between storage and photosynthetic tissues within the same organ, it has been suggested that such translocation between leaves is also dependent on an osmotic gradient between these leaves along a shoot. In a study of two succulent species from South Africa, Tuffers *et al.* (1995) also considered the possibility of water transport from old to young leaves during drought. For both species during drought, the older leaves are progressively shed. The experimental removal of older leaves as the drought progressed caused a reduction in physiological activity (including photosystem activity) of young leaves for *Prenia sladeniana* but not for *Delosperma tradescantioides*, suggesting for the former species that water flow from old to young leaves is an important component in drought avoidance. *Peperomia columella* is a succulent species with shoots having numerous succulent leaves composed of water-storing window tissue (a transparent tissue extending from the epidermis down into the leaf) which allows light penetration to the interior photosynthetic tissue. Christensen-Dean and Moore (1993) have shown that the relative volumes of these two tissues varies from young to old leaves along a shoot, with old leaves having a greater proportion of window tissue. During drought, the window tissues of older leaves decrease in volume while increasing in volume in young leaves, suggesting that water is translocated from the older to younger leaves. On the other hand, Rabas and Martin (2003) suggest that water does not flow from old to young leaves during drought for the three succulent species, *Carpobrotus edulis*, *Kalanchoe tubiflora*, and *Sedum spectabile*. Photosynthesis and leaf water contents were measured with and without removal of older leaves. Photosynthesis of young leaves was not enhanced by the presence of old leaves. Also, the removal of older leaves during drought lead to either no change or actually an enhancement of water status of the young leaves, suggesting that the older leaves may not have been water suppliers but water competitors.

Tissue osmotic potential

Osmotic adjustment, or an active increase in solute concentration, is often described as an important plant response to drought and a number of studies indicate that this occurs in succulent species as well. Osmotic potentials for several succulent species are shown in Table 2.1 for wet and drought conditions. Solutes are typically more concentrated in the photosynthetic tissues as compared to water storage tissues and osmotic potential usually, though not always, appears to decline with drought. The two species in the genus *Suaeda* are succulent halophytes and the low osmotic potentials are undoubtedly related to the high salinity habitat. In studies of *Suaeda fruticosa*, a leaf succulent halophyte in Pakistan, Gulzar and Khan (1998) found significant changes in tissue osmotic potential over the course of the day, but it is not clear if the measured changes were due to changes in tissue water content (as opposed to an active increase or decrease in solutes).

For the all-cell succulent[2] *Aptenia cordifolia*, Herppich and Peckmann (1997) noted that although leaf water potential declined with drought, leaf turgor was maintained constant because of active osmotic adjustment. Total solutes (expressed as molar quantity per unit leaf area) increased by 50%, which combined with a reduction in tissue water during drought decreased osmotic potential by 100%. Approximately half of the observed solute accumulation was due to citrate, which nearly doubled in concentration.

Model simulations of water flow through *Ferocactus acanthodes* (Schulte *et al.* 1989) have suggested that the 24-hour osmotic potential changes associated with the CAM cycle are of sufficient magnitude to drive a back and forth flow of water between storage and photosynthetic tissues. Therefore the osmotic potential cycle may be important for the internal redistribution of water within succulents, although this flow result was not

---

[2] a species where succulence and water storage is derived from all parenchyma cells, thus lacking clearly separable storage and photosynthetic tissues (see von Willert *et al.* 1992).

Table 2.1. Osmotic potential (MPa) of the cell sap for tissues of various succulent species. For species with definable water storage and photosynthetic tissues, data are shown separately for the storage and chlorenchyma. The terms wet and dry refer to well-watered and drought conditions, except in the case of *Suaeda fruticosa*, where they refer to coastal and inland sites.

Species	Storage		Chlorenchyma		Whole leaf	
	wet	dry	wet	dry	wet	dry
Agave deserti[a]	−0.95	−0.86	−1.02	−1.52		
			−1.33	−1.52		
Ferocactus acanthodes[b]	−0.48		−0.6[j]			
			−1.0[k]			
Opuntia ficus-indica[c]	−0.53	−0.68	−0.65	−0.75		
O. ficus-indica[d]	−0.50	−0.75	−0.65	−0.75		
Peperomia carnevalli[e]	−0.57	−0.72	−1.22	−0.76		
Aptenia cordifolia[f]					−0.80	−1.40
Tillandsia ionata[g]					−0.46	−0.74
Suaeda fruticosa[h]					−3.3	−4.4[l]
					−5.2	−6.1[m]
S. calceoliformis[i]					−1.9	−2.2

[a] – Smith *et al.* 1987; [b] – Schulte *et al.* 1989; [c] – Goldstein et al. 1991; [d] – Nerd and Nobel 1991; [e] – Herrera *et al.* 2000; [f] – Herppich and Peckmann 1997; [g] – Nowak and Martin 1997; [h] – Gulzar and Khan 1998; [i] – Youngmann and Heckathorn 1992; [j] – beginning of the night; [k] – end of the night; [l] – early morning; [m] – noon.

confirmed with measurements on actual plants. Such osmotically-driven acquisition of water from storage tissues is also important for maintaining turgor in *Peperomia magnoliaefolia* (Schmidt and Kaiser 1987) and *Peperomia carnevalii* (Herrera *et al.* 2000).

Cavitation and embolism

Studies considering embolism and cavitation vulnerability among succulent plants are not particularly numerous. Linton and Nobel (1999) found that root xylem was highly vulnerable to cavitation, with the more mesic *Opuntia ficus-indica* more vulnerable than *Agave deserti* and suggested that this vulnerability may be important to reduce water loss from roots during drought. For roots, embolism of xylem conduits would help to hydraulically isolate the highly succulent aboveground stem from the dry soil. Succulents typically maintain high water potential and must therefore isolate from the soil and hence the xylem vulnerability may be important. Refilling after drought would also be important for establishing xylem connections for newly produced roots.   Drought tolerant desert shrubs like *Larrea tridentata* maintain contact with soil water and can continue water uptake despite water potentials of −10 MPa and have a correspondingly invulnerable xylem (Pockman and Sperry 2000).   Cavitation vulnerability and the effects of cavitation on xylem hydraulic conductance were considered for two species of *Agave* by Linton and Nobel (2001).   After 100 days of drought, leaf water potentials of *Agave deserti* and *Agave tequilana* reached −2.0 and −3.4 MPa respectively.   Predictions of the effect of these water potentials on hydraulic conductance suggested that conductance would decline by 41 and 80% for *Agave deserti* and *Agave tequilana*, respectively.   The species difference in these reductions did not reflect differences in the vulnerability of the xylem, but differences in the water potentials reached by the two species after drought.   For these two *Agave* species, the leaf water potentials leading to 50% loss of hydraulic conductivity were between −2.0 and −2.5 MPa (Linton and Nobel 2001).   Therefore succulent plants are relatively vulnerable to cavitation in comparison with many upland (non-riparian) desert shrubs, but of course typically do not reach the low tissue water potentials found in the non-succulent species (Pockman and Sperry 2000; Linton and Nobel 2001).

Conclusions and future work

Succulent plants combine a number of fascinating characteristics that contribute to their survival in deserts and other environments (such as the epiphytic habit) where they may be subject to drought.   The ability to store water and retranslocate water between tissues or even between organs allows photosynthetic and presumably meristematic tissue to retain turgor for periods of sometimes many months when the air and soil are extremely dry.   Retranslocation between tissues appears to be driven by changes in tissue osmotic potentials.   Were the active decrease in solutes in water storage cells and increase in photosynthetic cells not to occur, the large decrease in water content of storage cells would passively concentrate solutes, ultimately opposing the driving force for flow into photosynthetic tissues.   In addition to gradual solute changes with drought, it appears that the osmotic potential changes occurring at night in the photosynthetic tissues of CAM succulents are important for establishing the driving forces for water retranslocation, perhaps even during the course of a single day.

A number of important physiological processes as discussed in this review involve structural features of cells.   Water storage tissues are composed of large volume cells with highly elastic walls – a combination that allows them to supply water while minimizing the change in cell water potential.   For some succulent species, many of the water conducting cells in the xylem are imperforate vascular tracheids with banded thickenings around their walls that may provide strength to resist cell collapse while readily allowing exchange of water with surrounding cells.   It would be useful to have further work assessing the elastic modulus of cells in succulent tissues as compared to photosynthetic tissues and how this parameter changes with cell turgor and over the course of a drought. In addition, further study is merited on the role of vascular tracheids with banded thickenings, both their occurrence as well as the mechanisms by which they may affect water transport.

While the study of aquaporins has been expanding among plant physiologists, there is relatively less work considering succulent species. Given the critical role of water storage and redistribution, it would be interesting and important to know if aquaporins play a role in regulating flow between storage and photosynthetic tissues in aboveground stems and leaves. In addition to the control of water redistribution by conductance changes, the driving forces for flow appear to be controlled by changes in tissue osmotic potential. The physiological mechanisms underlying the control of such processes at the cellular and molecular level are also important for future study. Although we have long known that water storage tissues enhance survival for succulent plants, it is not just the presence of stored water that is important, but how it is used – how and where it is retranslocated within the plant during drought.

Literature cited

Andrade JL and Nobel PS. 1997. Microhabitats and water relations of epiphytic cacti and ferns in a lowland neotropical forest. *Biotropica* **29**: 261-270.

Barcikowski W and Nobel PS. 1984. Water relations of cacti during desiccation. *Botanical Gazette* **145**: 110-115.

Barrowclough DE, Peterson CA, and Steudle E. 2000. Radial hydraulic conductivity along developing onion roots. *Journal of Experimental Botany* **51**: 547-557.

Bobich EG and Nobel PS. 2001. Biomechanics and anatomy of cladode junctions for two *Opuntia* (Cactaceae) species and their hybrid. *American Journal of Botany* **88**: 391-400.

Borghetti M, Magnani F, Fabrizio A, and Saracino A. 2004. Facing drought in a Mediterranean post-fire community: tissue water relations in species with different life traits. *Acta Oecologica* **25**: 67-72.

Burghardt M and Riederer M. 2003. Ecophysiological relevance of cuticular transpiration of deciduous and evergreen plants in relation to stomatal closure and leaf water potential. *Journal of Experimental Botany* **54**: 1941-1949.

Canny MJ. 1995. A new theory for the ascent of sap. Cohesion supported by tissue pressure. *American Journal of Botany* **75**: 343-357.

Canny MJ. 2001. Contributions to the debate on water transport. *American Journal of Botany* **88**: 43-46.

Christensen-Dean GA and Moore R. 1993. Development of chlorenchyma and window tissue in leaves of *Peperomia columella*. *Annals of Botany* **71**: 141-146.

Comstock JP. 1999. Why Canny's theory doesn't hold water. *American Journal of Botany* **86**: 1077-1081.

Cornejo DO and Simpson BB. 1997. Analysis of form and function in North American columnar cacti (tribe Pachycereeae). *American Journal of Botany* **84**: 1482-1501.

Davis SD, Sperry JS, and Hacke UG. 1999. The relationship between xylem conduit diameter and cavitation caused by freezing. *American Journal of Botany* **86**: 1367-1372.

Donatz M and Eller BM. 1993. Plant water status and water translocation in the drought deciduous CAM-succulent *Senecio medley-woodii*. *Journal of Plant Physiology* **141:** 750-756.

Gibson AC. 1973. Comparative anatomy of secondary xylem in Cactoideae (Cactaceae). *Biotropica* **5:** 29-65.

Gibson AC. 1977. Wood anatomy of opuntias with cylindrical to globular stems. *Botanical Gazette* **138:** 334-351.

Gibson AC. 1996. *Structure-Function Relations of Warm Desert Plants*. Springer-Verlag, Berlin, New York.

Goldstein G, Andrade JL, and Nobel PS. 1991. Differences in water relations parameters for the chlorenchyma and the parenchyma of *Opuntia ficus-indica* under wet versus dry conditions. *Australian Journal of Plant Physiology* **18:** 95-107.

Graham EA and Nobel PS. 1999. Root water uptake, leaf water storage and gas exchange of a desert succulent: implications for root system redundancy. *Annals of Botany* **84:** 213-223.

Gulzar S and Khan MA. 1998. Diurnal water relations of inland and coastal halophytic populations from Pakistan. *Journal of Arid Environments* **40:** 295-305.

Hacke U and Sperry JS. 2001. Functional and ecological xylem anatomy. *Perspectives in Plant Ecology, Evolution, and Systematics* **4:** 97-115.

Herppich WB and Peckmann K. 1997. Responses of gas exchange, photosynthesis, nocturnal acid accumulation and water relations of *Aptenia cordifolia* to short-term drought and rewatering. *Journal of Plant Physiology* **150:** 467-474.

Herrera A, Fernández MD, and Taisma MA. 2000. Effects of drought on CAM and water relations in plants of *Peperomia carnevalii*. *Annals of Botany* **86:** 511-517.

Holbrook NM, Zwieniecki MA, and Melcher PJ. 2002. The dynamics of "dead wood": maintenance of water transport through plants stems. *Integrative and Comparative Biology* **42:** 492-496.

Kluge M and Ting IP. 1978. *Crassulacean acid metabolism. Analysis of an ecological adaptation*. Springer-Verlag, Berlin, New York.

Landrum JV. 2001. Wide-band tracheids in leaves of genera in Aizoaceae: the systematic occurrence of a novel cell type and its implications for the monophyly of the subfamily Ruschioideae. *Plant Systematics and Evolution* **227:** 49-61.

Landrum JV. 2002. Four succulent families and 40 million years of evolution and adaptation to xeric environments: What can stem and leaf anatomical characters tell us about their phylogeny? *Taxon* **51:** 463-473.

Linton MJ and Nobel PS. 1999. Loss of water transport capacity due to xylem cavitation in roots of two CAM succulents. *American Journal of Botany* **86:** 1538-1543.

Linton MJ and Nobel PS. 2001. Hydraulic conductivity, xylem cavitation, and water potential for succulent leaves of *Agave deserti* and *Agave tequilana*. *International Journal of Plant Sciences* **162:** 747-754.

Martre P, North GB, and Nobel PS. 2001. Hydraulic conductance and mercury-sensitive water transport for roots of *Opuntia acanthocarpa* in relation to soil drying and rewetting. *Plant Physiology* **126:** 352-362.

Mauseth JD. 1995. Collapsible water-storage cells in cacti. *Bulletin of the Torrey Botanical Club* **122:** 145-151.

Mauseth JD. 2000. Theoretical aspects of surface-to-volume ratios and water-storage capacities of succulent shoots. *American Journal of Botany* **87:** 1107-1115.

Mauseth JD and Stone-Palmquist ME. 2001. Root wood differs strongly from shoot wood within individual plants of many Cactaceae. *International Journal of Plant Sciences* **162**: 767-776.

Mauseth JD, Uozumi Y, Plemons BJ, and Landrum JV. 1995. Structural and systematic study of an unusual tracheid type in cacti. *Journal of Plant Research* **108**: 517-526.

Molina-Freaner F, Tinoco-Ojanguren C, and Niklas KJ. 1998. Stem biomechanics of three columnar cacti from the Sonoran Desert. *American Journal of Botany* **85**: 1082-1090.

Nerd A and Nobel PS. 1991. Effects of drought on water relations and nonstructural carbohydrates in cladodes of *Opuntia ficus-indica*. *Physiologia plantarum* **81**: 495-500.

Niklas KJ, Molina-Freaner F, Tinoco-Ojanguren C, and Paolillo DJ. 2000. Wood biomechanics and anatomy of *Pachycereus pringlei*. *American Journal of Botany* **87**: 469-481.

Nobel PS. 1983. *Biophysical Plant Physiology and Ecology*. WH Freeman, San Francisco.

North GB and Nobel PS. 1991. Changes in hydraulic conductivity and anatomy caused by drying and rewetting roots of *Agave deserti* (Agavaceae). *American Journal of Botany* **78**: 906-915.

North GB and Nobel PS. 1992. Drought-induced changes in hydraulic conductivity and structure in roots of *Ferocactus acanthodes* and *Opuntia ficus-indica*. *New Phytologist* **120**: 9-19.

North GB and Nobel PS. 1997. Root-soil contact for the desert succulent *Agave deserti* in wet and drying soil. *New Phytologist* **135**: 21-29.

North GB and Nobel PS. 2000. Heterogeneity in water availability alters cellular development and hydraulic conductivity along roots of a desert succulent. *Annals of Botany* **85**: 247-255.

Nowak EJ and Martin CE. 1997. Physiological and anatomical responses to water deficit in the CAM epiphyte *Tillandsia ionantha* (Bromeliaceae). *International Journal of Plant Sciences* **158**: 818-826.

Phillips NG, Ryan MG, Bond BJ, McDowell NG, Hinckley TM, and Čermák J. 2003. Reliance on stored water increases with tree size in three species in the Pacific Northwest. *Tree Physiology* **23**: 237-245.

Pittermann J and Sperry JS. 2003. Tracheid diameter is the key trait determining the extent of freezing-induced embolism in conifers. *Tree Physiology* **23**: 907-914.

Pockman WT and Sperry JS. 2000. Vulnerability to xylem cavitation and the distribution of Sonoran Desert vegetation. *American Journal of Botany* **87**: 1287-1299.

Rabas AR and Martin CE. 2003. Movement of water from old to young leaves in three species of succulents. *Annals of Botany* **92**: 529-536.

Sack L, Cowan PD, Jaikumar N, and Holbrook NM. 2003. The 'hydrology' of leaves: co-ordination of structure and function in temperate woody species. *Plant, Cell and Environment* **26**: 1343-1356.

Schmidt JE and Kaiser WM. 1987. Response of the succulent leaves of *Peperomia magnoliaefolia* to dehydration. *Plant Physiology* **83**: 190-194.

Schulte PJ and Nobel PS. 1989. Responses of a CAM plant to drought and rainfall: Capacitance and osmotic pressure influences on water movement. *Journal of Experimental Botany* **40**: 61-70.

Schulte PJ, Smith JAC, and Nobel PS. 1989. Water storage and osmotic pressure influences on the water relations of a dicotyledonous desert succulent. *Plant, Cell and Environment* **12**: 831-842.

Smith SD, Monson RK, and Anderson JE. 1997. *Physiological Ecology of North American Desert Plants.* Springer-Verlag, Berlin, New York.

Smith JAC, Schulte PJ, and Nobel PS. 1987. Water flow and water storage in *Agave deserti*: osmotic implications of crassulacean acid metabolism. *Plant, Cell and Environment* **10:** 639-648.

Spaulding ES. 1905. Mechanical adjustment of the sahuaro (*Cereus giganteus*) to varying quantities of stored water. *Bulletin of the Torrey Botanical Club* **32:** 57-68.

Sperry JS. 2003. Evolution of water transport and xylem structure. *International Journal of Plant Science* **164:** s115-s127.

Sperry JS, Stiller V, and Hacke UG. 2003. Xylem hydraulics and the soil-plant-atmosphere continuum: opportunities and unresolved issues. *Agronomy Journal* **95:** 1362-1370.

Sperry JS, Hacke UG, Oren R, and Comstock JP. 2002. Water deficits and hydraulic limits to leaf water supply. *Plant, Cell and Environment* **25:** 251-263.

Steudle E. 1994. Water transport across roots. *Plant and Soil* **167:** 79-90.

Steudle E. 2001. The cohesion-tension mechanism and the acquisition of water by plant roots. *Annual Review of Plant Physiology and Plant Molecular Biology* **52:** 847-875.

Steudle E and Peterson CA. 1998. How does water get through roots? *Journal of Experimental Botany* **49:** 775-788.

Stiller V and Sperry JS. 1999. Canny's compensating pressure theory fails a test. *American Journal of Botany* **86:** 1082-1086.

Stiles KC and Martin CE. 1996. Effects of drought stress on CO_2 exchange and water relations in the CAM epiphyte *Tillandsia utriculata* (Bromeliaceae). *Journal of Plant Physiology* **149:** 721-728.

Tognetti R, Raschi A, and Jones MB. 2002. Seasonal changes in tissue elasticity and water transport efficiency in three co-occurring Mediterranean shrubs under natural long-term CO_2 enrichment. *Functional Plant Biology* **29:** 1097-1106.

Tüffers AV, Martin CE, and Vonwillert DJ. 1995. Possible water movement from older to younger leaves and photosynthesis during drought stress in two succulent species from South Africa, *Delosperma tradescantoides* Bgr. and *Preniasladeniana* L. Bol. (Mesembryanthemaceae). *Journal of Plant Physiology* **146:** 177-182.

Tyerman SD, Bohnert HJ, Maurel C, Steudle E, and Smith JAC. 1999. Plant aquaporins: their molecular biology, biophysics and significance for plant water relations. *Journal of Experimental Botany* **50S:** 1055-1071.

Tyree MT. 1997. The cohesion-tension theory of sap ascent: current controversies. *Journal of Experimental Botany* **48:** 1753-1765.

Tyree MT, Salleo S, Nardini A, Lo Gullo MA, and Mosca R. 1999. Refilling of embolized vessels in young stems of laurel. Do we need a new paradigm? *Plant Physiology* **120:** 11-21.

Walter H and Stadelmann E. 1974. A new approach to the water relations of desert plants. In: Brown R (ed.) *Desert Biology vol. II.* Academic Press, New York. Pp. 213-310.

von Willert DJ, Eller BM, Werger MJA, Brinckmann E, and Ihlenfeldt H-D. 1992. *Life Strategies of Succulents in Deserts: with Special Reference to the Namib desert.* Cambridge University Press, Cambridge.

Youngman AL and Heckathorn SA. 1992. Effect of salinity on water relations of two growth forms of *Suaeda calceoliformis*. *Functional Ecology* **6:** 686-692.

Zwieniecki MA, Melcher PJ, and Holbrook NM. 2001. Hydrogel control of xylem hydraulic resistance in plants. *Science* **291:** 1059-1062.

Chapter 3

TEMPERATURE INFLUENCES ON PLANT SPECIES OF ARID AND SEMI-ARID REGIONS WITH EMPHASIS ON CAM SUCCULENTS

Philippa M. Drennan

Perspectives in Biophysical Plant Ecophysiology: A Tribute to Park S. Nobel, pp. 57-94
Edited by: E. De la Barrera and W.K. Smith
© 2009 by The Author
Book Compilation © 2009 Universidad Nacional Autónoma de México

Introduction

Temperature exerts an important, often limiting influence on plants. Although deserts are defined climatically by low rainfall and vegetationally by a discontinuous cover of usually xerophytic species, high soil and atmospheric temperatures that increase evaporative demand from both abiotic and biotic components of the system exacerbate the effects of low rainfall. While stomatal closure may reduce water loss it also decreases the potential for evaporative cooling of plants. Desert species at higher altitudes and latitudes may also be subjected to limiting low temperatures. Furthermore, the open vegetation and low relative humidity of the desert environment allow intense heating during the day and rapid re-radiation of heat at night resulting in large diurnal fluctuations in temperature. Thus spatial and temporal heterogeneity in the temperature of the environment is common for deserts. The effects of temperature may be especially pronounced for succulent plant species, a strategy selected to survive periods of water stress, but which runs into problems with low amounts of transpiration. Reduced transpiration is especially pronounced for those succulents that have the Crassulacean acid metabolism (CAM) pathway of photosynthesis. Daytime stomatal closure decreases water loss from the plant, but potentially allows high temperatures to develop during the day in organs of large volume, despite the low rates of temperature increase associated with succulence (Ansari and Loomis 1959). In this chapter the influence of temperature on the survival and distribution of especially Crassulacean acid metabolism succulents in arid and semi arid areas will be considered, particularly for those of American deserts for which most information is available (*e.g.*, Nobel 1988).

Crassulacean acid metabolism

All three photosynthetic pathways, C_3, C_4, and CAM, are represented in the species of arid communities. For CAM species, as for C_4 species, the enzyme catalyzing CO_2 fixation is phosphoenol pyruvate carboxylase (PEP-case). The malate formed at night by the PEPcase-catalyzed reaction is stored in the vacuoles of the mesophyll tissue until daytime when it is decarboxylated and assimilated into carbohydrates by 1,5-ribulose bisphosphate/oxygenase (Rubisco) of the C_3 photosynthetic carbon reduction cycle. In concentrating CO_2 in the vicinity of Rubisco, the oxygenase activity of Rubisco is suppressed along with the consequent photorespiration, a process that generally increases with increasing temperature. Thus CAM species and C_4 species typically have higher optimum temperatures for photosynthetic carbon reduction than C_3 species. For CAM species, daytime stomatal closure decouples the atmospheric CO_2 concentration from the internal CO_2 concentration which increases as the malate is decarboxylated. This leads to a water use efficiency (WUE) for CAM species in arid areas that is three- to five-fold higher than for C_3 and C_4 species (Nobel 2005) as the saturation water vapor content of air increases almost exponentially with temperature and the nighttime temperature at the water evaporation sites may average 10-12 °C lower.

For CAM species in arid and semi-arid areas, the nocturnal CO_2 uptake is optimized by low night temperatures that favor stabilization of the active form of PEPcase and carboxylation, while higher temperatures favor decarboxylation (Lüttge 2004). Extrapolation of temperature optima from *in vitro* studies of PEPcase and Rubisco does not identify CO_2 uptake and growth optima (Israel and Nobel 1995), but gas exchange studies show that most agaves and cacti have nighttime optima between 12 °C and 20 °C (Nobel 1988). Increasing nighttime temperatures results in decreased stomatal conductance (as measured by water vapor conductance; Nobel 1988). This is consistent with CAM as a water-conserving strategy, but significantly decreases CO_2 uptake at higher temperatures, *e.g.*, maximum rates of nocturnal CO_2 uptake for *Hylocereus undatus* at day/night temperatures of 40/30 °C are 10 % of those at 30/20 °C after 10 weeks and eliminated by 30 weeks (Nobel and De la Barrera 2002). Nocturnal temperatures much below 5 °C reduce the metabolic activity that is required for both carboxylation and compartmentation of the malate (Lüttge 2004). Thus nighttime temperatures below approximately 10 °C do not result in greater CO_2 uptake, although stomatal/water vapor conductance is high (Nobel 1988). Nonetheless, some species that are native to cold environments, take up CO_2 and accumulate malate at night at environmental and tissue temperatures below freezing, *e.g.*, the high-elevation cactus *Tephrocactus floccosus* (Keeley and Keeley 1989) and the platyopuntia *Opuntia humifusa* (Nobel and Loik 1990) although the amounts are less than at higher temperatures. Optimal temperatures for nocturnal CO_2 uptake acclimate to shifts in the nighttime temperatures, *e.g.*, the cactus *Coryphantha vivipara*, which is distributed from northern Mexico to southern Canada, shows an increase of 8 °C in optimal nighttime temperature for CO_2 uptake concomitant with a temperature shift of 20 °C (Nobel 1988). Daytime temperatures are less critical for CO_2 uptake (Nobel 1988); the optimum temperature for the light reactions is possibly in the range 40-45 °C (Gerwick *et al.* 1978). For those organs dependent on carbon allocation from photosynthesis, optimal temperatures for growth are higher than those for nocturnal CO_2 uptake and closer to the daytime temperatures in the environment, *e.g.*, root growth rates for *Agave deserti* and *Opuntia ficus-indica* are maximal at 30 °C and 27-30 °C, respectively (Drennan and Nobel 1996, 1998).

Crassulacean acid metabolism plants show plasticity and variation in response to both environmental and developmental factors (Lüttge 2004). Obligate CAM plants fix CO_2 exclusively at night, *e.g.*, *Opuntia ficus-indica* (Nobel 1988) or extending into early morning and also in the late afternoon, when uptake of atmospheric CO_2 involves binding with Rubisco *e.g.*, *Agave deserti*, (Nobel 1988). Day/night temperature differences entrain the periodicity of the CAM phases in individual species and may synchronize them at the level of the community (Rascher *et al.* 2006). Additionally, facultative CAM plants can switch between C_3 photosynthesis and CAM.

This is commonly induced by drought (Sayed 2001; Lüttge 2004), but in the C_3/CAM intermediate *Clusia minor*, a shift to CAM is enhanced by increased day/night temperature differences, not absolute temperatures (Haag-Kerwer *et al.* 1992). For the obligate CAM succulent *Plectranthus marrubioides*, eliminating day/night temperature differences results in CAM-cycling (Herppich *et al.* 1998) a weak CAM in which nocturnal respiratory CO_2 is refixed and subsequently decarboxylated during the day when stomata are open and most CO_2 uptake is occurring. Low temperatures induce a switch from CAM to C_3 photosynthesis for *Kalanchoë daigremontiana* (Lüttge 2004). CAM idling, the strongest form of CAM in which stomata are permanently closed and CAM recycles respiratory CO_2 (Lüttge 2004) is usually induced by drought. Thus while temperature modifies CAM, water and WUE are important considerations, especially in arid and semi-arid environments.

Functional morphology

Succulence and cell water content

As both succulence and CAM evolved in response to water stress, water storage may be optimized over the carbon gain and the energy balance of the plant in an evolutionary context (Von Willert *et al.* 1992). Transpiration rates of C_3, C_3-succulent, and CAM-succulent species measured under field conditions in arid parts of southern Africa (Von Willert *et al.* 1992) give a ratio of 10.9:3.3:1. The transpiration of 1 mmole m^{-2} s^{-1} results in an energy dissipation of approximately 45 W m^{-2} at an air temperature of 25 °C (Nobel 2005). Thus, transpirational cooling may dissipate some of the net radiation balance in C_3 species but contributes much less to the thermal balance of C_3-succulents and CAM-succulents. The lack of transpiration may increase stem temperature by 4 °C in cacti (Lewis and Nobel 1977). Succulents are sensitive to freezing low temperatures as their large cell size and high water content (90-94%; Nobel 2005) increase the potential for intracellular freezing and cell rupture.

Growth forms

Both succulence and CAM are associated with diverse growth forms which can be broadly divided into stem versus leaf-succulents (Lüttge 2004). Additionally there are a wide range of architectures and morphologies, even within some more commonly recognized growth form categories. Stem succulents, *e.g.*, most Cactaceae, are leafless and range in size from the large columnar cacti, *e.g.*, *Carnegiea gigantea*, through shorter barrel cacti, *e.g.*, *Ferocactus acanthodes*, to small globose and dwarf cacti, *e.g.*, *Coryphantha vivipara*. The degree of branching varies, some have single (*e.g.*, *Cephalocereus columna-trajani*) or multiple unbranched stems (*e.g.*, *Lophocereus schottii*), while others may have branching to various degrees from a main trunk, *e.g.*, *Carnegiea gigantea*. The opuntiod cacti have jointed cylindrical (*e.g.*, *Opuntia acanthocarpa*) or flattened stems (clad-odes; *Opuntia ficus-indica*). Rosette succulents have short stems and a helical phyllotaxy of leaf like tubercles in the Cactaceae (*e.g.*, *Ariocarpus retusus*) or more commonly leaves, *e.g.*, agaves, yuccas, and some of the Crassulaceae, *e.g.*, *Dudleya saxosa*.   Other Crassulaceae, *e.g.*, *Crassula argenta*, and most of the Aizoaceae are leaf succulents which differ consid-erably in both height and the size and shape of their leaves. Larger species are shrub like, *e.g.*, *Ruschia spp.*, while dwarf leaf-succulents, *e.g.*, *Lithops*, may consist of a single pair of succulent leaves on a very short stem, thus approximating in size and shape a dwarf stem-succulent.   For all growth forms, the height above the ground, organ size, shape, orientation, epider-mal coverings, and position in the environment are important for thermal relations.   Few succulents are annual and deciduousness of the photosyn-thetic organs is not typically associated with most succulent growth forms which depend on the water stored in the succulent organs to overcome long-term drought. Those succulents that are deciduous are primarily cau-diciforms that store water in enlarged, woody, non-photosynthetic tissues and often have well-defined phenologies, *e.g.*, *Pachypodium namaquanum* of the Succulent Karoo which grows and flowers during the winter months (Rundel *et al.* 1995). The ability of some succulents to shed plant parts that are damaged may enable the plant to survive temperature extremes that are lethal to individual organs, thus extending the potential range of the species, albeit at the expense of productivity, *e.g.*, some crassulas (Fig. 3.1).

Volume and surface area to volume ratios

The relatively larger mass of succulent plants and organs is associated with slower heating, but to a potentially higher temperature as the volu-metric heat capacity of succulents is high because of their high water con-tent.   Increased size can act as a thermal buffer with slow rates of change decreasing midday maximum plant temperature and increasing the mini-mum nighttime temperature.   For the columnar cactus *Carnegiea gigan-tea*, the increased midheight stem diameter over latitudinal gradients pos-sibly extends the distribution of this species northwards (Smith *et al.* 1984), especially as increased diameter is predicted to increase minimum apical temperatures for this cold-sensitive species (Nobel 1980a). However, for *Lophocereus schottii*, which also increases in diameter over a similar latitudinal range, a slight reduction in minimum apical temperature with increasing diameter is predicted. Thus, similar morphological trends pos-sibly have different thermal outcomes for different species (Nobel 1984). Nonetheless, for North American columnar cacti, larger stem girths are associated with cooler winter temperatures (which may co-occur with lower rainfalls) both within species and also within a taxon (Cornejo and Simpson 1997). *Carnegiea gigantea* is the most northerly distributed of the columnar cacti and also has the largest diameter stem. The low surface area to volume ratios for succulents, which decrease as the size of a succu-lent organ increases, decrease the surface area across which water loss can occur and hence evaporative cooling.   This is of little impact to the thermal balance for CAM species with their low daytime transpiration rates. However, organs with low surface area:volume ratios have thicker boundary layers than small leaves decreasing heat exchange (loss) by conduction and convection (Nobel 2005). Ribs, spines, and irregular surface morphologies increase turbulent air flow decreasing the thickness of the boundary layer and potentially increasing convective heat loss (Nobel 1988).

Spines, trichomes, and surface micromorphology

Spines and trichomes confer both heat tolerance and frost tolerance on a species (Gibson 1996). In still air, spines and trichomes can increase the boundary layer thickness; however, they contribute to turbulence and possible increased heat loss in air currents (see above), as may waxy microstructures and the papillose surface of some species, *e.g.*, *Lithops* (Gibson 1996). Spines and trichomes also increase reflectance and shading, decreasing daytime surface temperatures by up to 4 °C for *Ferocactus acanthodes* (Nobel 1988) and 6 °C for *Mammillaria dioica* (Nobel 1978). However, spine removal does not result in increased surface temperatures for *Coryphantha vivipara*. This is possibly because convective cooling is enhanced by the underlying surface micromorphology of this species which has pronounced protruberances (nipples: Norman and Martin 1986). Dense mats of trichomes in the apical regions of columnar and barrel cacti insulate the meristem raising nighttime minimum temperatures, *e.g.*, in simulations of the effect of a 10 mm layer of apical pubescence for *Carnegiea gigantea* nighttime minimum temperatures are increased 7 °C (Nobel 1980a, 1982). Thick surface wax possibly also increases reflectance of especially infra-red (Gibson 1996).

Shape and orientation

The orientation of leaves and stems influences the amount of solar radiation absorbed and hence the heat loading of the plant. Plant and organ architecture resulting in predominantly vertical surfaces, *e.g.*, the stems of *Carnegiea gigantea* and the cladodes of platyopuntias, avoids heating from the interception of direct radiation in the middle of the day; the photosynthetic surfaces receive most incident radiation in the cooler early morning and late afternoon (Nobel 1988; Sortibrán *et al.* 2005). *Opuntia puberula*, which has horizontal cladodes, is apparently restricted to shaded microhabitats where cladode temperatures do not differ markedly from air temperatures in either a vertical or horizontal position (Sortibrán *et al.* 2005). By contrast, horizontal cladodes exposed to direct solar radiation have temperatures approximately 13 °C above the air temperature and 7 °C higher than vertically placed cladodes. Similarly, the experimental horizontal placement of the usually vertical cladodes of *Opuntia pilifera* and the columnar cactus *Cephalocereus columna-trajani* result in temperature increases of 12 °C and 20 °C to tissue temperatures of 47 °C and 55 °C, respectively (Zavala-Hurtado *et al.* 1998). Shape as a strategy for reducing

Figure 3.1. *Crassula argenta* two weeks after a night in which temperatures dropped below freezing for approximately 4 hours. The outer leaves of each branch were killed by the freezing low temperatures and abscised within three weeks. The younger inner leaves and bud were thermally insulated by the outer leaves and survived.

heating potentially compromises carbon gain. For platyopuntias, the non-random orientation of the flattened surface of the terminal cladodes of platyopuntias maximizes the interception of photosynthetically active radiation and photosynthesis, *e.g.*, during a winter growing season at a latitude of 34° N most newly initiated cladodes of *Opuntia ficus-indica* face south-north (Nobel 1988). Similarly, branching occurs disproportionately to the south for *C. gigantea* across its geographic range and independent of local temperature and precipitation, thus maximizing interception of photosynthetically active radiation (Geller and Nobel 1986; Drezner 2003). Stem tilting towards the equator in barrel cacti growing at latitudes above 23.5° is associated with increased apical temperatures that maximize reproduction without increasing the temperature of the whole plant. By contrast, the columnar cactus *Cephalocereus columna-trajani*, which has a narrow intertropical distribution north of the equator, tilts away from the equator with a declination angle that is greater than that of the sun during its summer solstice. By comparison to a vertical stem, this reduces the interception of radiant energy in the middle of the day and in the afternoon during the hotter summer months, but increases it during the September equinox when rain is more frequent and most plant growth occurs (Zavala-Hurtado *et al.* 1998). The helical phyllotaxis of agaves in which the leaves in the rosettes are progressively more horizontal towards the base of the plant also maximizes the interception of photosynthetically active radiation while minimizing temperature increases (Nobel 1988). The meristematic region is surrounded by vertical young leaves which may provide thermal buffering (Woodhouse *et al.* 1983). These leaves will intercept less direct radiation near the middle of the day and possibly also insulate the meristematic region at night, thus protecting it from thermal extremes. Vertical organs may also increase convective heat loss.

Position in the environment

The surface of the soil shows greatest variation in temperature, thus low-growing succulents, dwarf or prostrate, are subject to extremes of temperature. Embedded succulents such as *Lithops spp.* in southern Africa experience generally higher temperatures than succulents of different growth form under similar conditions, *e.g.*, *Lithops gracilidelineata*, 41 °C, versus *Zygophyllum simplex*, a bushy plant with very small succulent leaves, −38 °C (Turner and Picker 1993). The leaf temperature of *Lithops* is tightly coupled with that of the soil which has a large thermal capacity by comparison to the small-volume succulent. Thus the photosynthetic tissues a few centimeters below the soil surface experience less variation in temperature than the non-photosynthetic tissues at the surface which can range from 12-46 °C in a single day (Eller and Nipkow 1983; Turner and Picker 1993). Light reaching the photosynthetic tissues through translucent epidermal 'windows' at the exposed surface of the leaf, heats the succulent leaf interior. The greater the transmittance through the windows, the greater the heating (Eller and Grobbelar 1986; Turner and Picker 1993).

Species with clear windows are distributed at higher latitudes in the relatively cool and lower radiation environments (Turner and Picker 1993).

Temperature tolerances

Temperature tolerances and acclimation

Thermal tolerances are difficult to asses in the field where the damage associated with extreme temperature events may take years to become apparent. Furthermore, the stress induced by a given temperature is dependent on the length of exposure to that temperature making comparisons between species difficult. Thus laboratory assays are a valuable tool for quantifying temperature tolerances. In particular the uptake of a vital stain, *e.g.*, neutral red, by chlorenchyma cells in tissues maintained at known temperatures for 1 h can be used to establish a T_{50}, *i.e.*, the temperature at which 50% of the cells are killed (Nobel 1988). Rankings of species with respect to temperature sensitivity as determined by vital stain uptake correlate with the relative amount of damage incurred in the field when these species are exposed to similar conditions (Nobel 1990). Another approach is to approximate cell death from conductivity measurements of electrolyte leakage. Conductivities are standardized against visual assessments of the extent of leaf damage to yield T_{50} values (Van Coller and Stock 1994).

The shoots of succulents thus investigated show wide thermal tolerances from a low of −48 °C for *Opuntia fragilis* (Loik and Nobel 1993) to a high of 67 °C for *Opuntia ficus-indica* (Nobel 1988; Nobel and De la Barrera 2003; Table 3.1). With acclimation to extremes, individual species in the Agavaceae and Cactaceae can tolerate a range of temperatures in excess of 70 °C, however, the few species investigated in the Aizoaceae and the Crassulaceae have a narrower tolerance, for example, *Dudleya saxosa*, with a range of approximately 60 °C (Nobel and Zutta 2007; Table 3.1). Comparisons of trends in temperature tolerances associated with families and growth forms are generally tentative due to the paucity of data for most groups other than cacti and agaves. However, families apparently also differ in the absolute values for high and low temperature tolerances. The Cactaceae and the Agavaceae tolerate temperatures above 60 °C. Some platyopuntias, which extend the latitudinal range of the Cactaceaea beyond that of the Agavaceae, also tolerate extremely low temperatures, *e.g.*, −48 °C for *Opuntia fragilis* (Loik and Nobel 1993). Within the Cactaceae the low-growing, mat-forming platyopuntias are the most cold tolerant species (Table 3.1); the columnar and barrel cacti and the cylindropuntias have low-temperature T_{50} values similar to the Aizoaceae and the Crassulaceae. At high temperature extremes, the cacti are on average the most tolerant with the dwarf Aizoaceae and the Crassulaceae having tolerances of less than 60 °C (Table 3.1). Thermal tolerances of agaves and cacti are also correlated with latitudinal and altitudinal distribution of the species, the more cold tolerant species occurring further north and at higher elevations .

Table 3.1. Thermal tolerances and acclimation for succulents of arid and semi-arid areas compared on the basis of T_{50} values, *i.e.*, the temperature that kills 50 % of the cells in the tissue, and the change in T_{50} for each 10 °C decrease or increase in day/night temperatures.

Family	Species	Organ	Low-temperature T_{50} (°C)	Cold acclimation (°C)	High-temperature T_{50} (°C)	Heat acclimation (°C)	Reference
Agavaceae							
	Agave americana	leaf	−7.4	1.8	63.8	3.3	Nobel and Smith 1983
	Agave deserti	leaf	−16.3	3.9	62.8	3.1	Nobel and Smith 1983
		root	−6.7	1.7	54	2.8	Jordan and Nobel 1984
	Agave parryi	leaf	−19.6	4.8	60.4	2.4	Nobel and Smith 1983
	Agave sisalana	leaf	−6.4	0			Nobel and McDaniel 1988
	Agave tequilana	leaf	−8.0	1.0			Nobel et al 1998
	Agave utahensis	leaf	−17.5	3.9	57.2	1.4	Nobel and Smith 1983
	Agave vilmoriana	leaf	−8.0	1.2			Nobel and McDaniel 1988
Aizoaceae							
	Haworthia retusa	leaf	−7.4	1.4	58.4	1.9	Nobel 1989a
	Haworthia turgida	leaf	−8.8	1.2	58.8	2.0	Nobel 1989a
	Lithops leslei	leaf	−7.6	1.5	58.1	2.1	Nobel 1989a
	Lithops turbiniformis	leaf	−7.0	1.0	59.0	2.0	Nobel 1989a

continues on following page

Table 3.1 (continued)

							Reference
Cactaceae							
Columnar and barrel cacti	*Carnegiea gigantea*	shoot	−8.6	0.5	62.7	2.8	[a]Nobel 1982; [b]Smith et al. 1984
	Ferocactus acanthodes	shoot	−8.7	0.3	66	4	Smith et al 1984
		root	−7.8	1.4	57	2.5	Jordan and Nobel 1984
	Ferocactus viridescens	shoot	−6.1	0.3	65.4	6.1	[a]Nobel 1982; [b]Smith et al. 1984
	Ferocactus wislizenii	shoot	−8.4	0.3	67.0	6.6	[a]Nobel 1982; [b]Smith et al. 1984
	Lophocereus schottii	shoot	−6.8	0.5	64.5	5.0	[a]Nobel 1982; [b]Smith et al. 1984
	Stenocereus thurberi	shoot	−9.0	0.3	64.2	5.4	[a]Nobel 1982; [b]Smith et al. 1984
	Trichocereus candicans	shoot	−7.4	1.0			Nobel 1982
	Trichocereus chilensis	shoot	−7.8	0.9			Nobel 1982
Cylindropuntias	*Opuntia acanthocarpa*	shoot	−4.9	2.8	60.0 to 69.5	6.1	[a,b]Nobel and Bobich 2002; [b]Smith et al. 1984
	Opuntia bigelovii	root	−1.5	0.6	60.8	9.1	Nobel and Bobich 2002
		shoot	−7.3	0.8	60.6	2.8	[a]Nobel 1982; [b]Didden-Zopfy and Nobel 1982
	Opuntia ramosissima	shoot	−4.4		62.5	7.9	[a]Nobel 1988; [b]Smith et al. 1984

continues on following page

***Table 3.1* (continued)**

Group	Species	Organ					Reference
Platyopuntias	Opuntia basilaris	shoot			62.6	5.3	Smith et al. 1984
	Opuntia chlorotica	shoot			63.9	7.2	Smith et al. 1984
	Opuntia fragilis	shoot	−48.9	9.9	57		[a]Loik and Nobel 1993 [b]Ishikawa and Giusta 1996
	Opuntia humifusa	shoot	−25	3.3			Nobel and Loik 1990
	Opuntia ficus-indica	shoot	−8.1	1.1	66.6	4.2	[a]Nobel et al. 1995 [a]Nobel and De la Barrera 2003 [b]Nobel 1988 Nobel and De la Barrera 2003
		root	−6.5				
	Opuntia strepta-cantha	shoot	−10.8	2.7			Nobel 1990
Globose, short, and dwarf cacti	Ariocarpus fissuratus	shoot			66.3	6.2	Nobel et al. 1986
		root			61.1	6.5	
	Coryphantha vivipara	shoot	−20.3	1.7			Nobel 1982
	Epithelantha bokei	shoot			60.5	3.1	Nobel et al. 1986
		root			60.0	5.5	Nobel et al. 1986
	Mammillaria lasiacantha	shoot			62.8	4.0	Nobel et al. 1986
		root			60.1	3.0	Nobel et al. 1986

continues on following page

Table 3.1 (continued)

Family	Species						Reference
Crassulaceae	*Cotyledon orbiculata*	leaf	−6.1 to −10.4	0 to 2			Van Coller and Stock 1994
	Dudleya saxosa	leaf	−4.3	0.3	55.2	2.3	Nobel and Zutta 2007
		root	−4.5	0.5	59.1	2.2	Nobel and Zutta 2007
Fouquieriaceae	*Fouquieria splendens**	leaf	−5.2		58.8		Nobel and Zutta 2005
		stem	−5.5		66.6		Nobel and Zutta 2005
		root	−4.1		55.9		Nobel and Zutta 2005

* a semi-succulent, C_3 species

[a] reference for low temperature data; [b] reference for high temperature data

(Nobel 1982; Nobel and Mc Daniel 1988; Nobel 1990). For *Opuntia fragilis*, low-temperature tolerances of different populations within the species are also correlated with latitude and the minimum temperatures of the environment (Loik and Nobel 1993).

In individual plants thermal tolerances differ with organ type, age, and position within the organ. Across all succulent families and growth forms, roots generally have higher low-temperature limits and lower high-temperature limits than shoots in the same plant (Table 3.1). Sensitivity to especially high temperatures ($T_{50} < 60$ °C; Table 3.1) probably accounts for the lack of roots in the upper approximately 2-5 cm of the soil which is typical even though these succulents are generally very shallow-rooted species (Jordan and Nobel 1984; Nobel and Bobich 2002; Nobel and De la Barrera 2003; Nobel and Zutta 2005). For ocotillo (*Fouquiera splendens*) in which deciduous leaf loss is a survival strategy for drought, the stems have a greater thermal tolerance than either the roots or the leaves, especially for high temperatures (Nobel and Zutta 2005; Table 3.1). Younger organs are more sensitive to extremes of temperature than older organs. First-year cladodes of *Opuntia fragilis* succumb to nighttime temperatures of −3. 5 °C during the growing season and are about 5 °C less tolerant of high temperatures than the cladodes of previous years (Ishikawa and Gusta 1996). Cladodes of *Opuntia ficus-indica* increase in high-temperature tolerance by 6.5 °C and low-temperature tolerance by 2 °C with age up to 10 years (Nobel and De la Barrera 2003). For the leaf succulent *Cotyledon orbiculata*, the younger leaves and especially the bud are extremely sensitive to low temperatures by comparison to the older leaves (Van Coller and Stock 1994; Fig. 3.2), which possibly by their position reduce heat loss from the younger leaves at night. Tolerance differences within organs reflect differences in the thermal environment of the organ. The high temperature tolerance for a cladode of *Opuntia ficus-indica* is maximal ($T_{50} = 70$ °C; Nobel *et al.* 1986) where the cladode enters the soil, and decreases with increasing cladode distance from the soil surface (Nobel *et al.* 1986). Similarly, *Opuntia acanthocarpa* has a greater heat tolerance at the stem base of the plant than at a midheight position (Smith *et al.* 1984). West-facing cladodes and succulent stems also show greater thermal tolerance than the cooler east-facing tissues in the same plant for both *Opuntia basilaris* and *Ferocactus acanthodes* (Smith *et al.* 1984). Thus thermal tolerances are extended even within organs by acclimation.

The ability to acclimate is positively correlated with the thermal tolerance of a species (Fig. 3.3; Nobel 1988) and thus potentially extends the species range while the inability to acclimate can limit distribution. For example, *Agave sisalana* does not cold acclimate and shows freezing damage at higher temperatures than species which are less cold tolerant at mesic day/night temperatures, but which acclimate in response to decreasing temperatures (Nobel and McDaniel 1988). Thus less tolerant taxa are likely more narrowly distributed and in habitats with less temperature variation. Differences in thermal tolerances of species over relatively small geographical but significant thermal gradients such as altitude, *e.g.*, 2 °C

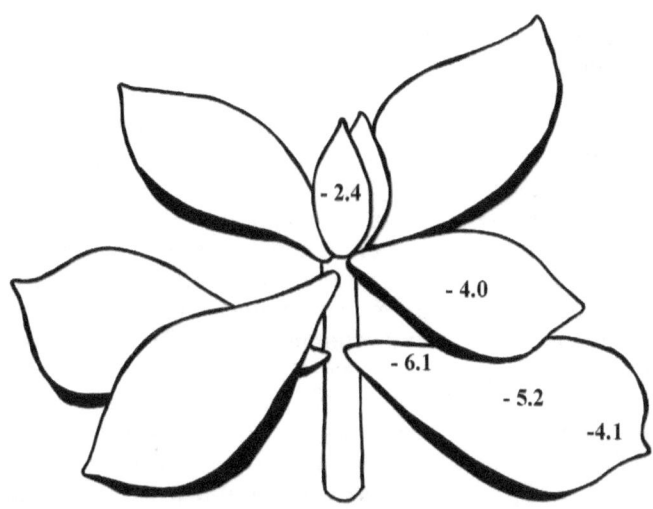

Figure 3.2. Low-temperature T_{50} in °C for the leaves of various ages/position of *Cotyledon orbiculata* var. *orbiculata* (redrawn from Van Coller and Stock 1994).

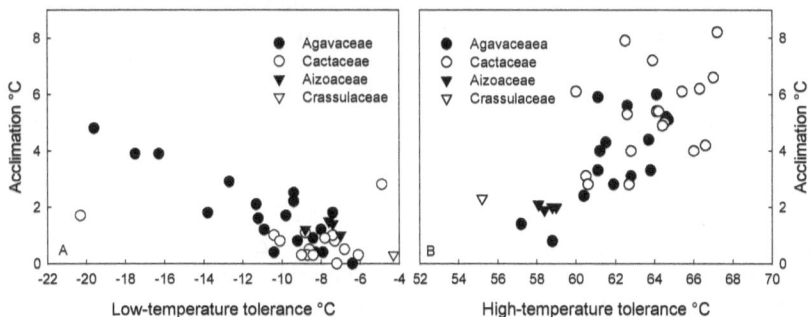

Figure 3.3. Low-temperature (A) and high-temperature acclimation of succulent species as function of their low- and high-temperature tolerance, respectively (data from Nobel 1988 and references cited in Table 3.1).

for *Dudleya saxosa* over 1,000 m (Nobel and Zutta 2007) and seasonal hardening, *e.g.*, 5 °C for *Opuntia bigelovii* over four months (Smith *et al.* 1984) emphasize the importance of acclimation for distribution. Half times for both high- and low-temperature acclimation of 1-4 days indicate a relatively flexible component to acclimation (Smith *et al.* 1984) that would accommodate short-term temperature variations in the environment. However, an age-dependent ability to acclimate for cladodes and fruits of *Opuntia ficus-indica* (Nobel and De la Barrera 2003) suggests that developmental and long-term, possibly cumulative, changes such as increases in solute concentration are important. Furthermore, the fruit of *O. ficus-indica* show cold but not heat acclimation at 4 weeks of age emphasizing that these are different processes. Roots acclimate only to high temperatures which have greater variation and are likely closer to the thermal limits in the upper soil layers than low temperatures (Nobel *et al.* 1986; Nobel and De la Barrera 2003; Nobel and Zutta 2007). Acclimation of the platyopuntias, *e.g.*, *Opuntia ficus-indica*, *Opuntia humifusa*, *Opuntia polyacantha*, and *Opuntia streptacantha*, to low temperatures is accompanied by dehydration of the tissues and the accumulation of solutes which increase the osmotic pressure of the tissue and depress the freezing point, although this alone is insufficient to account for acclimation (Cui and Nobel 1994). An increase in extracellular mucilage and the relative apoplastic fraction of water decrease the likelihood of damaging intracellular ice formation and possibly further avoid damage by slowing the rate of water loss from cells during extracellular freezing (Goldstein and Nobel 1991; Loik and Nobel 1991). The solutes accumulated during acclimation include fructose, glucose, sucrose, mannitol, sorbitol, amino acids, and proline (Goldstein and Nobel 1994; Nobel *et al.* 1995). However, the injection of cladodes of *O. ficus-indica* and *O. humifusa* with nonmetabolizable methylglucose increases cold acclimation for both species without significant change in the levels of these solutes suggesting that the ability to withstand especially freezing low-temperatures is dependent on altered water relations (Nobel *et al.* 1995). High-temperature acclimation is less well characterized but is associated with protein accumulation (Nobel and De la Barrera 2003) and decreases in the lipid/protein ratio of the tonoplast which correlates with increased membrane rigidity (Behzadipour *et al.* 1998). The fatty acid composition of membranes also changes with increasing temperatures, but the role of these changes, especially with respect to fluidity, are not clear (Nobel 1988; Behzadipour *et al.* 1998). For *Opuntia megacantha*, the proline content of the cladodes increases in response to heat stress, especially for mature cladodes (Flores-Hernández *et al.* 2001).

Seed germination, nurse plants, and vegetative reproduction

Optimum temperatures for seed germination are similar to the temperatures that are ideal for the establishment of the species thus positioning the seedlings favorably spatially and temporally in the environment. Additionally, temperature treatments of dormant seeds, *e.g.*, stratification, may act as a cue for germination. For cacti, the optimum temperatures for germination range from 15-30 °C, with germination for many species centered around 25 °C (Nobel 1988; Rojas-Aréchiga and Vázquez-Yanes 2000). Little germination occurs below 12 °C consistent with the low-temperature sensitivity of cacti. In particular, germination of frost-sensitive species with larger growth forms, e.g., *Carnegiea gigantea* and *Stenocereus thurberi*, significantly reduces below 20 °C (McDonough 1964). High temperatures above 35 °C also inhibit germination (Nobel 1988) although, for *Opuntia tomentosa*, dry heat treatment (180 h at 60 °C) which possibly increases the permeability of the seed coat in hard-seeded species, increases germination (Olvera-Carrillo *et al.* 2003). Given the diurnal fluctuations in temperature in arid environments, especially at the soil surface, it is suggested that alternating temperatures may enhance the germination of cacti (Rojas-Aréchiga and Vázquez-Yanes 2000). However, most experiments investigating alternating temperatures have not selected temperatures that bracket the constant temperature and thus do not provide a valid comparison. Germination of both *Mamillaria gaumeri* (Cervera *et al.* 2006) and *Opuntia stricta* (Reinhardt 2000) tested over a range of alternating temperatures, which differed from each other by 10 °C, is optimal at day/night temperatures of 30/20 °C consistent with optimal germination for the Cactaceae at approximately 25 °C. Stratification increases germination for some cacti species which occur in regions with defined moderate to cold winters, e.g., *Opuntia compressa*, *Opuntia polyacantha*, and *Coryphantha vivipara* (Baskin and Baskin 1977; Smreciu *et al.* 1988).

The germination responses to temperature of agaves, yuccas, and ocotillos are similar to those of cacti. Agaves have an average optimum of 25 °C with high- and low-temperature limits of 40 °C and 10 °C, respectively (Freeman 1973a, 1975; Nobel 1988; Peña-Valdivia *et al.* 2006). *Fouquieria splendens* has similar high and low-temperature limits and an optimum germination temperature of 20-25 °C (Freeman, 1973b). However, there are apparent differences in sensitivity to prolonged exposure to 40 °C with *F. splendens* being less sensitive than the sympatric *Agave lechuguilla*, consistent with the observation that it is more likely to germinate in response to summer rains (Freeman *et al.* 1977). High temperature limits to the germination of yuccas have not been established (although the seeds of some species do survive very brief exposure to temperatures of 100 °C as might be experienced in a fire; Keeley and Meyers 1985) but as for cacti and agaves, temperatures lower than approximately 12 °C inhibit germination, which is optimal at 25-26 °C (McClearly and Wagner 1973; Flores and Briones 2001). By contrast, the temperatures for germination of the succulents of the Karoo region in southern Africa are lower than those for succulents and semi-succulents of the New World (Esler 1999). Laboratory studies for the Aizoaceae suggest that germination at day/night temperatures of 20/10 °C is greater than at temperatures of 30/15 °C (Esler 1993). For the leaf succulents *Stoeberia* sp. and *Cephalophyllum spongiosum*, the optimal constant temperatures for germination are 12-17 °C and 17-22 °C, respectively (De Villiers *et al.* 2002). Field observations in this region show that most of the growth forms emerge in response to autumn and winter rains (Van Rooyen *et al.* 1979; Esler 1993; Milton 1995). Indeed irrigation of the Richtersveld in the spring does not result in the germination of species, although there is significant germination the following autumn of both succulents and grasses. Although the season of the rainfall is important for germination to occur, the densities of seeds germinating is dependent on the amount of fall rainfall (Esler 1993; Milton 1995). Seeds of many of the Aizoaceae are maintained in their capsules until such time as rain allows the opening of the capsule and the raindrops disperse the seeds.

Seedlings of succulents in arid and semi-arid regions, which for some species, *e.g.*, *Agave deserti*, have a narrower temperature tolerance range than the adults (Table 3.2) are especially vulnerable to temperature stress. At the end of the first year of growth, seedlings of *Carnegiea gigantea* may be only 3 to 5 mm in height (Steenbergh and Lowe 1976; Drezner 2006). The reduced water storage capacity and higher surface area/volume ratio associated with such small size results in seedlings gaining and loosing heat more rapidly than adult plants (Nobel 1980b). Simulation models for the barrel cactus *Ferocactus acanthodes* predict that maximum surface temperature is 8 °C higher and minimum surface temperature is 3 °C lower in a seedling that is 1 mm in height compared with a 50 mm-tall plant (Nobel 1984). In addition to size, the bulk of a seedling occurs close to the soil surface, which in the open may show large diurnal fluctuations in temperature. Both field measurements and computer modeling show that maximum temperatures of small spherical cacti approach soil surface temperatures that can reach 71 °C in deserts (Nobel *et al.* 1986; Franco and Nobel 1989).

Thus seedlings frequently occur in less extreme micro-environments, for example associated with rocks or nurse plants, the canopy of which may reduce environmental temperature fluctuations. Meta-analysis of published data on nurse-plant interactions shows that most, approximately 53%, are recorded for arid and semi-arid regions (Flores and Jurado, 2003). Furthermore, the large number of species of Cactaceae that are recorded under nurse plants by comparison to cactus species that act as nurse plants (Table 3.3; Flores and Jurado 2003) indicates that the establishment of these succulents in arid and semi-arid areas is facilitated by nurse plants. Indeed, with the exception of the Aizoaceae, for all plant families that have a number of succulent or semi-succulent species and are distributed in arid areas, *viz.*, Agavaceae, Euphorbiaceae, and Fouquieriaceae, more species are recorded as establishing under a nurse plant as opposed to acting as nurse plants (Table 3.3; Flores and Jurado 2003).

The shaded micro-environment under the canopy of a nurse plant reduces temperature extremes of the soil and the atmosphere. Shaded soils are 10-15 °C cooler in summer and 3-4 °C warmer in winter than open soils (Franco and Nobel 1989; Nobel 1989b; Suzán *et al.* 1996). Differences in the size and structure of the canopy of the nurse plant affect the amount of shading and hence soil surface temperature which decreases approximately 2 °C for every 10 % increase in shading (Nobel and Geller 1987). Also gradients in air temperature occur with the outer sub-canopy showing greater

Table 3.2. Temperature tolerances, as determined by vital stain uptake, for seedling versus adult plants of *Agave deserti* and *Ferocactus acanthodes* grown at day/night temperatures of 10/0 °C and 50/40 °C for low-temperature and high-temperature tolerances, respectively.

Day/night temperature (°C)	Species	Growth stage	T_{50}* (°C)	Reference
Low-temperature tolerance 10/0 °C				
	Agave desertii	seedling	−10.4	Nobel 1984
		adult	−16.3	Nobel and Smith 1983
	Ferocactus acanthodes	seedling	−8.2	Nobel 1984
		adult	−8.7	Smith *et al.* 1984
High-temperature tolerance 50/40 °C				
	Agave desertii	seedling	60.7	Nobel 1984
		adult	62.8	Nobel and Smith 1983
	Ferocactus acanthodes	seedling	64.8	Nobel 1984
		adult	66.0	Smith *et al.* 1984

*The temperature that kills 50% of the cells in the tissue.

Table 3.3. The phylogenetic distribution of nurse plant interactions in arid and semi-arid ecosystems. Data, from Flores and Jurado (2003), are the number of species per family recorded as nurse plants or as protégés (*i.e.*, established in association with a nurse plant) for the 10 families with the most species in each category. The succulent/semi-succulent families Agavaceae and Fouquieriaceae are also included.

Family	Number of nurse species	Number of protégé species
Agavaceae	–	5
Aizoaceae	8	8
Asteraceae	15	46
Brassicaceae	–	15
Cactaceae	6	66
Chenopodiaceae	3	7
Euphorbiaceae	6	18
Fabaceae*	49	36
Fagaceae	5	–
Fouquieriaceae	–	3
Poaceae	6	47
Rosaceae	4	4
Rhamnaceae	3	7
Solanaceae	4	11

* Species recorded as Mimosaceae (Flores and Jurado 2003), are included in the Fabaceae.

variation in temperature than the interior (Drezner 2007). Thus smaller nurse plants (*e.g.*, *Ambrosia deltoidea* versus *Cercidium microphyllum*) and those with more open or deciduous canopies (*e.g.*, *Prosopis velutina* versus *Olneya tesota*) are less effective in buffering the thermal environment (Franco and Nobel 1989; Suzán *et al.* 1996). *Carnegiea gigantea* seedlings cluster at the base of the nurse plants (Franco and Nobel 1989; Drezner 2006) where reduction in PAR is greatest, and cluster more tightly if the canopy is open such as under *Larrea tridentata* where approximately 80% of the seedlings occur in the innermost 10% of the canopy by comparison to 50% of the seedlings under the smaller but more dense canopies of *Ambrosia deltoidea* and *Ambrosia dumosa* (Drezner 2006).

The relatively few field measures of the temperature differences for succulents under nurse plants by comparison to those in the open confirm the expected buffering of extremes predicted on the basis of the seasonal and/or diurnal temperature trends in this micro-environment. For the cactus *Peniocereus striatus* and the desert agave *Agave deserti* occurring in the shade of *Olneya tesota* and *Pleuraphis rigida*, respectively, the maximum summer surface temperatures are decreased by 6-11 ºC , and the minimum winter temperatures are increased by 0.5-3 ºC (Nobel 1984; Suzán *et al.* 1996). These differences may be sufficient to maintain the seedlings within their thermal tolerance limits especially at the extremes of distribution. Shading also decreases evapotranspiration and separating the positive effects of thermal buffering from those attributable to higher water contents (Valiente-Banuet and Ezcurra 1991) is not often reported. Higher winter and diurnal minimum temperatures under canopies may decrease the relative humidity of the under canopy atmosphere for which, in the case of *Cercidium microphyllum*, lower dew point temperatures are measured (Drezner 2007). Thus, in the absence of recent rainfall, the under canopy may be drier (Shreve 1931). Additionally, competition for water can reduce the growth of the nurse and clustered seedlings (Franco and Nobel 1989).

The relative trade-offs between a milder thermal environment and other micro environmental factors possibly differ depending on the location of the nurse and seedling within the distribution range of a species. For example, the saguaro cactus shows a southerly bias in distribution under nurse plants at northern, colder sites in Arizona (Drezner and Garrity 2003) where it is close to the limits of its distribution. Available microclimate data suggest that the minimum temperatures on the southern side are higher than for the north, but that direction has less effect on maximum temperatures under the preferred nurse species *Cercidium microphyllum*. At warmer sites in Arizona, at similar latitudes, there is no significant difference between saguaro numbers on the southern versus the northern sides of the nurse plants. In the Vizcaíno Desert and the Gran Desierto de Altar, where solar radiation has a southern azimuth as for the Arizona desert but minimum temperatures are higher, there are approximately 10 × more seedlings of *Carnegiea gigantea* on the northern than the southern side of the nurse plants (Valiente-Banuet and Ezcurra 1991). Northern distributions of cacti under nurse canopies are also observed for several spe cies in Argentina (*Cereus aethiops, Denmoza rhodacantha, Echinopsis leucantha, Gymnocalycium gibbosum, Lobivia formosa*, and *Opuntia longispina*; Méndez *et al.* 2004). Here the northern azimuth of the sun and resulting low PAR and temperatures on the southern side of the nurse plants probably limit seedling growth. The distribution of seedlings in a particular cardinal direction under nurse plants is less apparent in tropical arid areas where the directional effects of solar radiation and hence microclimate differences at the soil surface under the nurse plant are less pronounced (Valiente-Banuet *et al.* 1991; López and Valdivia 2007).

The dependence on a nurse plant for establishment of succulent seedlings is different within and between families (Flores *et al.* 2004), but a lack of studies across the full geographic range for most species makes it difficult to determine whether there are obligate *versus* facultative nurse plant requirements. Freezing sensitive cacti, *e.g, Carnegiea gigantea*, are especially noted for occurring with nurse plants. The freezing sensitive, arborescent cactus *Myrtillocactus geometrizans* does not establish in the open: seedlings are always found beneath a nurse plant, including individuals of *O. streptacantha* which are more freezing tolerant than *M. geometrizans*, but may be out competed by the latter as it grows to maturity (Flores-Flores and Yeaton 2003). Larger seedling size may increase survival in the absence of a nurse plant through increased water storage capacity, better thermal buffering, and possibly decreased heat sensitivity. In *Ferocactus wisilezeni*, larger seed size results in seedlings with a 4 times greater volume than that for *Carnegiea gigantea* seedlings (Bowers 2001). These larger seedlings have a greater survival rate than those of the *C. gigantea* and unlike those of *C. gigantea*, can become established without a nurse plant. There is little evidence that the succulent species of the winter rainfall deserts in southern Africa make use of nurse plants as do the succulents of North America, even though soil surface temperatures during the recruitment of young plants may reach 61 °C (Dean 1992). For many species the seedling survival, which is strongly influenced by competition from neighboring established plants, is greater in open sites than in sites with cover (Milton 1995). The high-temperature tolerances for the seedlings of these species are not known, although adult stone plants (*Lithops*, Aizoaceae) have lower tolerances than cacti, but not exceeded by the environment (Table 3.1; Nobel 1989a). Presumably, since germination is synchronized with winter rainfall and temperatures are not as extreme as in the North American deserts; the seedlings are not thermally stressed. Furthermore, a number of species are apparently associated with rocks which, like nurse plants, ameliorate the microclimate but without competing with the seedling for water or nutrients.

Many succulents also reproduce by cloning. It is common for the terminal joints or cladodes of cacti, whose production is favored by high temperatures for *Opuntia ficus-indica* (Nobel and Castañeda 1998), to fall to the ground where they grow into new individuals. Because of the intense heating of the soil surface, these tissue pieces must show a degree of heat resistance, although their larger volume makes them less susceptible to extremes than the seedlings. The rooted joints of *Opuntia bigelovi* can acclimate by up to 8 °C to a maximum of 60 °C, which would contribute to survival of high soil temperatures (Didden-Zopfy and Nobel 1982). Furthermore, the spines that occur on many of the clonally reproducing species, *e.g.*, *Opuntia ramosissima*, increase reflectance, provide shading, and possibly increase convective heat dissipation by allowing air to flow between the tissue and the soil surface.

Temperature influences on distribution

Geographic distribution

Two regions that show a remarkable abundance of succulent species in the flora are the Sonoran and Mojave Deserts of California (Schmida 1981; Cody 1989) and semi-arid karoo biomes of southern Africa (Werger and Ellis 1981; Van Jaarsveld 1987; Stock *et al.* 1997) where succulent cover can be up to 70% (Cowling *et al.* 1994). At both the global and regional scale, this abundance of succulents is correlated with the predictability of rainfall (Lüttge 2004), *e.g.*, species from the succulent families Aizoaceae and Crassulaceae are most abundant in those areas arid areas of southern Africa (Succulent Karoo) that have the lowest coefficient of variation in their rainfall, albeit that the total rainfall in some areas is less (Cowling and Hilton-Taylor 1999). Deserts with less predictable rainfall (*i.e.*, high variability over a number of years) *e.g.*, the Australian deserts, lack a succulent physiognomy (Lüttge 2004). Succulents also do not occur in very low rainfall deserts where annuals are the dominant component of the vegetation. With the exception of *Caralluma* spp., which occupy shaded, rocky microenvironments in which nocturnal dew formation likely contributes to their water balance (Lüttge 2004); the few annual succulent species that do exist occur in more mesic environments, *e.g.*, *Mesembryanthemum crystallinum*. For those arid regions with CAM-obligate families, *e.g.*, Agavaceae and Cactaceae, the distribution of CAM and succulence are essentially the same as CAM expression and succulence are closely correlated (Kluge *et al.* 2001). Where there are C_3-succulents or CAM-inducible species in a family, *e.g.*, Aizoaceae, the data are not always available to determine the extent of CAM functioning. Succulents with similar growth forms but different photosynthetic pathways may achieve similar WUEs, *e.g.*, *Cotyledon orbiculata* (CAM) and *Othonna opima* (C_3) (Eller and Ferrari 1997), however, the relative contribution of CAM *versus* succulence to temperature limits is not quantified. Of the succulent shrubs in the Succulent Karoo, approximately 25% and 7% are C_3 and CAM-inducible, respectively (Midgley and van der Heyden 1999). For the purposes of discussion, most succulents are assumed CAM-succulents, unless otherwise stated. Within arid and semi-arid areas temperature, especially the limitations imposed by low temperatures, may be an important modifier of the distribution of CAM-succulent species. (Luttge 2004).

In both North and South America, species diversity and growth form diversity of succulents, in particular cacti, decrease with increasing latitude (Fig. 3.4; Cody 1989; Mourelle and Ezcurra 1996, 1997). The number of species declines markedly above 34-36° N and S (Fig. 3.4; Mourelle and Ezcurra 1996). Latitude is an indirect measure of several climatic variables including total rainfall, average annual temperature, and seasonality. It is suggested that for North American cacti the less extreme temperatures at lower latitudes provide greater opportunities for growth at different times of the year allowing the co-existence of more species and growth forms (Cody 1989). Latitude effects are most pronounced for columnar cacti (Fig. 3.4; Mourelle and Ezcurra 1996) whose species richness in Argentina is best predicted by the number of frost-free days, consistent with the low-temperature sensitivity of this growth form (Nobel 1988). For opuntioid and globose cacti in the same geographic region, species richness is best predicted by the percentage of summer rainfall, although for both groups, mean annual temperature explains some of the variation in species richness (Mourelle and Ezcurra 1996). Furthermore, the platyopuntias are the species with the most extreme latitudinal distribution reaching the Patagonian steppe in South America (Mourelle and Ezcurra 1996) and the Canadian prairies in the north (Loik and Nobel 1993). In addition to their tolerance of temperature extremes and the ability to acclimate, the modular growth of the opuntioids allows for a greater plasticity in response to the environment including cladodes becoming prostrate on the surface of the ground where snow cover would insulate them from more extreme conditions (Loik and Nobel 1993). Patterns of species richness are less well quantified for other succulent families in the arid regions of the Americas. In southwestern Africa, succulent plant richness (a total of the genera and subgenera of the Aizoacea and species of *Crassula* and *Zygophyllum*), is strongly associated with low (< 500 mm per annum) but predictable rainfall (< two months per annum without rain; Stock *et al.* 1997). Freezing minimum daily temperatures and maximum daily temperatures > 30 °C correlate with decreased succulent species richness in this area. The direct effects of temperature are not separated from seasonal rainfall patterns for the succulents (Jürgens 1991), but mean temperature of the coldest month is important in explaining the plant diversity in this succulent biome (Thuiller *et al.* 2006).

Species turnover of columnar cactus species along gradients of increasing latitude in Argentina is associated with variation in mean temperature and probably specifically with frost-free days (Mourelle and Ezcurra 1997). In the deserts of North America, species of cacti with cylindrical stems are replaced by similar species of greater cross-sectional stem diameter (Cody 1989). Similarly, the distribution of three arborescent cacti in the southern Chihuahuan Desert in Mexico along a north-west to south-east climatic gradient is strongly influenced by freezing tolerance (Flores-Flores and Yeaton 2003). For the three species, viz., *Myrtillocactus geometrizans*, *Opuntia streptacantha*, and *Opuntia leucotricha*, a relative competitive hierarchy, established by determining canopy interference between adja-

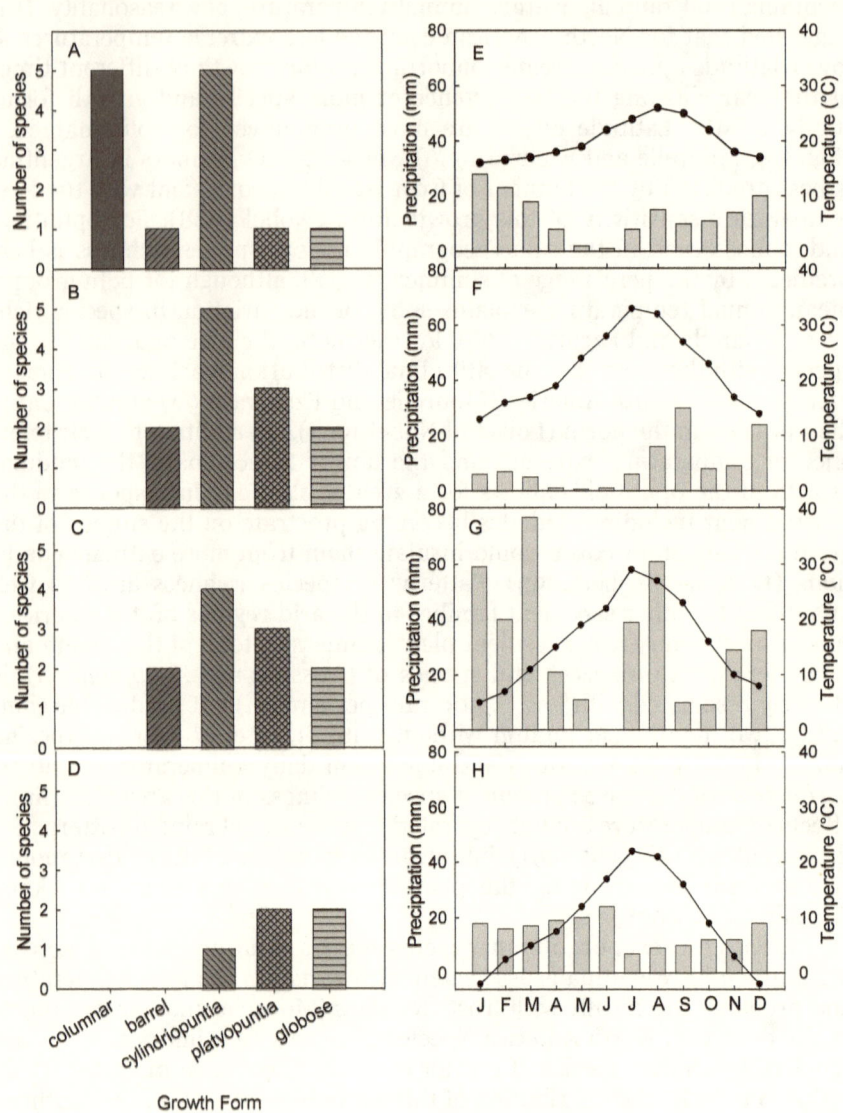

Figure 3.4. The number of species of cacti of different growth forms (A-D) and climate diagrams (E-H) for desert sites which range from Baja California, Mexico (A, E) in the south, northwards through Anza-Borrego, California (B, F) and Red Rock Canyon, Nevada (C, G) to the Greta Basin Desert, Nevada (D, H) (data from Cody 1989).

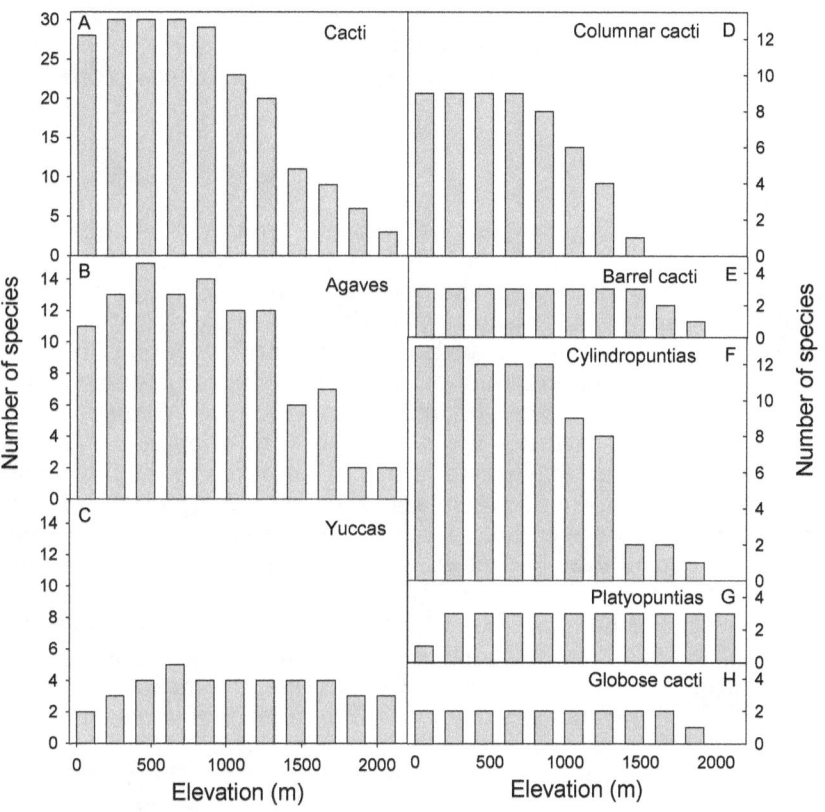

Figure 3.5. The number of species occurring at different elevations in the Sonoran Desert for cacti (A), agaves (B), and yuccas (C), as well as for the different growth forms of cacti: columnar (D), barrel (E), cylindropuntias (F), platyopuntias (G), and globose cacti (H) (Species distributions were obtained from Turner *et al.* 1995).

cent individuals, suggests a potential replacement sequence of *O. leucotri-cha* by *O. streptacantha* and *O. streptacantha* by *M. geometrizans*. Yet *M. geometrizans* is most abundant in the warmer south-east region and absent from the colder north-west where *O. leucotricha* is common. *Opuntia streptacantha* is abundant in the center of the gradient. This distribution is consistent with the freezing tolerance of these species (*O. leucotricha* > *O. streptacantha* > *M. geometrizans*) that was quantified in the field based on visible damage to cladodes following a freezing low temperature event (−12 °C). Thus it is suggested that freezing low temperatures not only limit the distribution of the cacti, but also allow the regional co-occurrence of different species.

Local scale distribution

ALTITUDE—Decreasing temperature possibly limits the distribution of CAM succulent species at higher altitude (Lüttge 2004), although orographical effects on water availability are also important, *e.g.*, the greater abundance of agaves and tillandsias relative to stem succulents is associated with moisture from fog in the cloud belts of arid mountains in Mexico (Martorell and Ezcurra 2002, 2007). However, different altitudinal limits for very similar growth forms with different temperature tolerances suggest that elevational distributions are associated with temperature as does freezing damage which is observed at higher elevations in the field (Nobel and Bobich 2002). For example, two subspecies of *Coryphantha vivipara* with a 700 m difference in their elevational limits have a 3.6 °C difference in their low-temperature T_{50} which is in the range approximated using a lapse rate of −0.6 ° in air temperature per 100 m increase in elevation (Nobel 1982). Furthermore, elevational decreases in the distribution of some CAM succulents, *e.g. Agave utahensis*, approximately 12,000 years ago are suggested to be associated with decreased temperatures during glacial periods (Cole and Arundel 2005). The number of succulent species decreases with increasing elevation in the Sonoran Desert (Fig. 3.5) in particular species richness drops markedly above 1,500m. The abundance of the different growth forms of the cacti with elevation over the latitudes for which data are available (22-35° N; Fig. 3.5) show trends similar to those for latitude. Columnar cacti are not recorded above 1,600 m, (Turner *et al.* 1995). By contrast, in the Zapotitlán Valley in Mexico (18° 20' N) columnar cacti occur to approximately 1,900 m (Pavón *et al.* 2000) presumably associated with generally higher temperatures for a given altitude at a lower latitude. The abundance of cylindryopuntias also declines markedly above approximately 1,500 m (Fig. 3.5). The cacti recorded at elevations of greater than 1,500 m in the Sonoran desert are primarily platyopuntias, in particular, *Opuntia basilaris*, *O. chlorotica*, and *O. phaecantha*. Species richness for both *Agave* spp. and *Yucca* spp. declines with increasing elevation above 400-600 m (Fig. 3.5). However, for the yucca species of the Sonoran Desert there are more species at an elevation of 2,000 m than sea level. Some *Yucca* species, *e.g.*, *Yucca baccata* exhibit CAM (Sayed 2001), however, species with a more northerly distribution, *e.g.*, *Yucca brevifolia*

(T_{50} of -10.4 °C; Loik *et al.* 2000) in the Mojave Desert, are C_3 and not as succulent. Agaves and yuccas of higher latitudes and elevations also have smaller leaves than those at lower latitudes and elevations. Agaves with a small rosette size and thinner leaves are predicted to have slightly higher minimum temperatures than larger species at the same temperature (Nobel and Smith 1983). In the succulent Karoo biome of southern Africa succulent species richness decreases with altitude possibly associated decreased frost-free days (Werger and Morris 1991).

Within an elevational distribution band, the individuals of a species are not necessarily most abundant in a midzone that would possibly represent the average thermal optimum. Indeed frequencies may be greatest where temperatures allow maximum uptake of water during the growing season. Although sensitive to low temperatures (Table 3.1), plant frequency for *Opuntia acanthocarpa* in the Sonoran Desert increases from 1.5 plants per 100 m² at 230 m to 20 at 1,050 m, which is close to its elevational limit (< 1,300 m; Hickman 1993). At this higher elevation, the optimal soil temperatures for root growth occur during the late, but wetter summer, while at the lower elevation optimal soil temperatures occur during late spring and autumn which are typically drier (Nobel and Bobich 2002).

ASPECT AND SLOPE—Slopes of different aspect differ in cover and species composition of the vegetation (Drennan and Nobel 1997). The effect of water availability, radiation, and temperature are difficult to separate, particularly if the moisture supply is strongly directional, *e.g.*, fog deserts such as the Namib Desert where the greater vegetation cover on the west slopes by comparison to the east slopes may be due to increased interception of fog (Jürgens 1986). It has also been suggested that for the more poorly vegetated east slopes, the topographical shading of the slope reduces incident radiation at the plant canopy to well below saturation levels while air temperatures remain high, thus resulting in photosynthesis occurring below the light compensation point (Von Willert *et al.* 1992). Indeed many of the distributions associated with aspect are attributed to limitations in photosynthetically active radiation, *e.g.*, *Opuntia* spp. which show different cladode orientations to maximize interception for a given aspect (Nobel 1988), or to higher soil water contents on slopes receiving less direct incident radiation, *e.g.*, *Agave deserti* is more abundant on north slopes at Agave Hill (where it is the dominant species; Drennan and Nobel 1997; Nobel and Linton 1997). As temperature differences are greater for the soils than the air of different aspects, temperature effects of aspect may be most pronounced for root growth (Nobel and Linton 1997) and root hydraulic conductivity, which decreases with decreasing temperature (Cui and Nobel 1994), as well as for low growing or dwarf species, for which little information on distribution at this scale is available.

The effects of aspect are more pronounced the greater the slope, which affects the angle of light interception. Similar to diurnal heating of vertical *versus* horizontal plant organs, a plateau which receives direct incident irradiation at noon is significantly warmer than a vertically oriented cliff. Thus the saxicolous (cliff-dwelling) habit will likely result in lower plant temperatures for especially low-growing succulents which are significantly influenced by soil temperature (Martorell and Patiño 2006). For species of *Mammillaria* distributed in southern Mexico, plants on plateaus experience midday temperatures approximately 20 % higher than those on cliffs. At low, and presumably warmer, elevations *Mammillaria* populations occur on steeper slopes than at higher altitudes (Fig. 3.6; Martorell and Patiño 2006). Species of cylindropuntias with larger diameter stems and greater low-temperature tolerances (Table 3.1) occur on steeper and possibly cooler slopes in the Mojave and Sonoran Deserts than those with thinner stems. In the Sonoran Desert maximum photosynthetic surface area for *Opuntia bigelovii* (stem diameter 4-6 cm; Hickman 1993) occurs at a slope angle of 24°, for *Opuntia acanthocarpa* (2-2.5 cm) at 12°, and for *Opuntia leptocaulis* (1-1.5 cm) at 6° (Yeaton and Cody 1979).

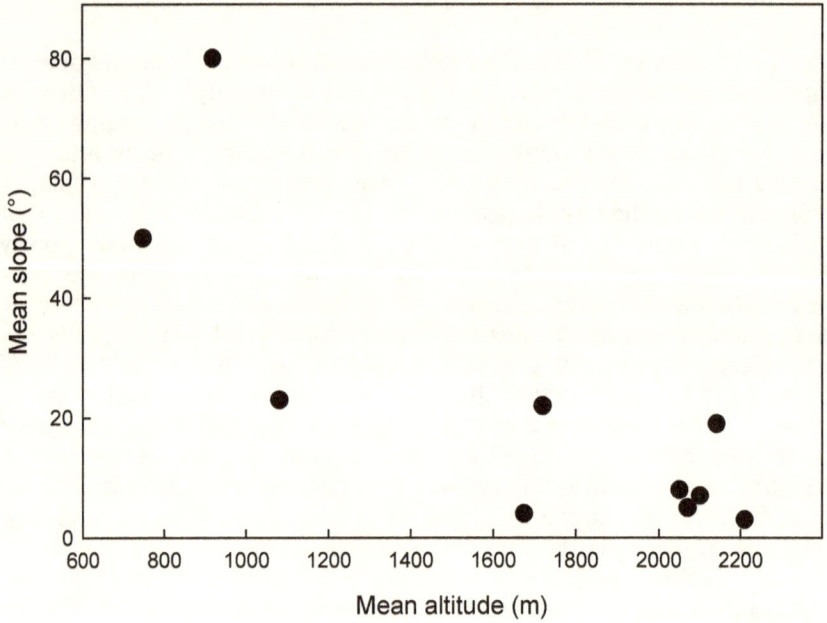

Figure 3.6. Average habitat altitude and slope for 10 species of *Mammillaria* growing in the Tehuacán Valley, Mexico (redrawn from Martorell and Patiño 2006).

ROCKY SUBSTRATES—Thermal properties of rocky substrates differ with rock type. In particular, shortwave absorptance of quartz is low (1:1.3:3.2:3.0 for quartz:dolomite:granite:dry soil at Agave Hill in the Sonoran Desert; Nobel and Zutta 2007). Furthermore, quartz has a high thermal conductivity coefficient (7.2-8.8 W m^{-1} $^{\circ}$C^{-1}) by comparison to granite (1.6-3.4 W m^{-1} $^{\circ}$C^{-1}), dolomite/limestone (1.3-1.7 W m^{-1} $^{\circ}$C^{-1}) and soil (up to approximately 1.0 W m^{-1} $^{\circ}$C^{-1} when wet) and dissipates absorbed shortwave irradiation to the underlying layers (Nobel and Geller 1987; Nobel and Zutta 2007). Thus quartz outcrops are significantly cooler than surrounding soils; temperature reductions exceeding 10 $^{\circ}$C occur for rocky quartz substrates in both the Sonoran Desert of North America and the Namib Desert of southern Africa (von Willert et al. 1992; Schmiedel and Jürgens 2004; Nobel and Zutta 2007). At Agave Hill in the Sonoran Desert, the relatively heat-sensitive Dudleya saxosa (Table 3.1) occurs with a frequency of 3 m^{-2} at sites with > 90 % quartz cover and is seldom present where quartz cover is less than 60% (Nobel and Zutta 2007). The dwarf succulent species which characterize the vegetation of the quartz fields of the Succulent Karoo in southern Africa, e.g., Argyroderma pearsonii and Gibbaeum cryptopodium, can exhibit leaf temperatures 3-5 $^{\circ}$C lower by comparison to individuals occurring outside the field (Schmiedel and Jürgens 2004). Experimental warming of quartz-field vegetation of approximately 5.5 $^{\circ}$C results in substantial mortality of these dwarf succulents, suggesting they are distributed close to their thermal limit (Musil et al. 2005). Rocky substrates may also affect the thermal environment of the plant through shading by individual rocks and larger boulders with outcomes similar to those of nurse plants (Schmiedel and Jürgens 2004).

Global climate change

Increasing environmental temperatures associated with global climate change will increase the potential latitudinal and altitudinal limits for the distribution of many CAM succulents that are currently limited by low temperatures. In particular, the potential areas for the cultivation of commercial species, e.g., Agave salmiana, which are less tolerant of low temperatures and do not acclimate to the same extent as the species of colder regions (Nobel 1996; Nobel et al. 2002), would possibly be increased. However, if nighttime temperatures are too warm (>15 $^{\circ}$C) nocturnal CO_2 uptake may be reduced as occurs for Agave tequilana cultivated in plantations outside of its naturally occurring distribution range (Pimienta-Barrios et al. 2001). This would suggest that distributions would more likely shift than expand, although decreased nocturnal CO_2 uptake can to some extent be compensated for by increased daytime assimilation (Pimienta-Barrios et al. 2001). For naturally occurring populations, the effects of global climate change on future distributions are not easily predicted. In addition to temperature changes, the amount of precipitation, the season of precipitation, and the predictability of the rainfall will likely also change. For American cacti and agaves of semi-arid and arid areas, it is often assumed that low temperatures have a greater influence on the distribution of species than high temperatures which seldom exceed the thermal tolerances of the plants. But, as can be seen from examining some of the local scale distributions, interplay of water availability and temperature are important, *e.g.*, *Opuntia acanthocarpa*. The Succulent Karoo, which has a cool growing season and predictable rainfall (Fig. 3.7, *c.f.* Fig. 3.4E-H) is susceptible to warming. In an experimental approximation of the possible temperature increase in the next century, the dwarf and shrubby succulents of the quartz fields showed up to five times the canopy mortalities of control groups (Musil *et al.* 2005). The tree succulent *Aloe dichotoma* is experiencing increased mortality in the northern (warmer) part of its range and a southerly shift in recruitment consistent with optimum growing conditions now occurring further south (Midgley and Thuiller 2007). Modelling of future habitat suitability for Karoo succulents under potential climate change scenarios suggests that habitat elimination is more likely than range extension, leading to a decrease in the succulent species richness of this area (Midgley and Thuiller 2007).

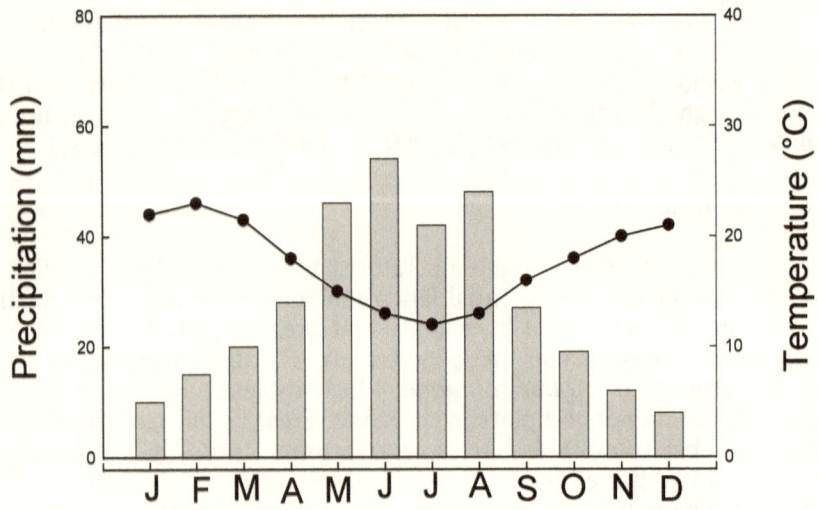

Figure 3.7. A climate diagram for Okiep in the Succulent Karoo (redrawn from Milton *et al.* 1997).

Conclusions and future prospects

Temperature exerts important effects on CAM metabolism including the performance of individual enzymes and the composition of the membranes (Lüttge 2004). For CAM succulents of arid regions, CO_2 uptake is maximized at low nighttime temperatures, with increasing temperatures resulting in stomatal closure in response to decreased stomatal conductance (Nobel 1988). Daytime transpiration is very reduced, especially for obligate CAM species, which reduces the ability of the plant to dissipate some of the net radiation balance through evaporation. The various growth forms of succulents modify the net radiation balance through adaptations such as vertical stems, which decrease direct radiation during the warmest part of the day, and spines, which both shade and increase turbulence of the boundary layer resulting in greater heat loss. The importance of these growth forms as adaptations to the thermal environment is seen in their distributions which are often strongly correlated with differences in the temperature of the environment such as the decrease in columnar cacti with latitude. Morphological plasticity such as stem tilting and cladode orientation maximizes the interception of photosynthetically active radiation and growth while minimizing temperature extremes. The morphology of small CAM succulents differs proportionally from that of larger adult plants and young CAM plants, which are thus more susceptible to temperature extremes, frequently establish in the microenvironment of nurse plants. This contributes to patterns of species richness and distribution in the deserts, especially of North America. Where the growing season is thermally less extreme, or the species is not as sensitive, establishment can occur without interactions with other species. Low temperatures in particular seem important in limiting CAM succulents both physiologically and ecologically. Physiologically some CAM species may shift to a C_3-type metabolism in response to lower temperatures, e.g., Agave vilmoriana (Nobel and McDaniel 1988) and Kalanchoë daigremontiana (Lüttge 2004) which possibly extends their ability to fix CO_2 at otherwise limiting temperatures. Studies on growth forms have focused primarily on the agaves and the larger cacti, but more studies are required of a greater number of growth forms, in particular, those growth forms that are distributed into areas that experience temperatures lower than their apparent thermal tolerance, e.g., Cotyledon orbiculata (Van Coller and Stock 1994), or are distributed more widely globally, e.g., Crassula species. It is possible that smaller species are adapted to and distributed in microenvironments that differ in thermal characteristics from the larger geographic region, e.g., quartz rock outcroppings. There is a paucity of information on the influence of temperature extremes on succulent richness, distribution, and abundance especially at the geographic and global scale. Determining the interaction of especially water availability and temperature in influencing these patterns, will establish the relative importance of these factors and provide better predictions of the effects of global warming and associated climate change.

Literature cited

Ansari AQ and Loomis WE. 1959. Leaf temperatures. *American Journal of Botany* **46:** 713-717.

Baskin JM and Baskin CC. 1977. Seed and seedling ecology of *Opuntia compressa* in Tennessee cedar glades. *Journal of the Tennessee Academy of Science* **52:** 118-122.

Behzadipour M, Ratajczak R, Faist K, Pawlitschek P, Trémolières A, and Kluge M. 1998. Phenotypic adaptation of tonoplast fluidity to growth temperature in the CAM plant *Kalanchoë daigremontiana* Ham. Et Per. is accompanied by changes in the membrane phospholipids and protein composition. *Journal of Membrane Biology* **166:** 61-70.

Bowers JE. 2001. Implications of seed size for seedling survival in *Carnegiea gigantea* and *Ferocactus wislizeni* (Cactaceae). *Southwestern Naturalist* **46:** 272-281.

Cervera JC, Andrade JL, Simá JL, and Graham EA. 2006. Microhabitats, germination, and establishment of *Mamillaria gaumeri* (Cactaceae), a rare species from Yucatan. *International Journal of Plant Sciences* **167:** 311-319.

Cody ML. 1989. Growth-form diversity and community structure in desert plants. *Journal of Arid Environments* **17:** 199-209.

Cole KL and Arundel ST. 2005. Carbon isotopes from fossil packrat pellets and elevational movements of Utah agave plants reveal the Younger Dryas cold period in Grand Canyon, Arizona. Geology 33:713-716.

Cornejo DO and Simpson BB. 1997. Analysis of form and function in North American columnar cacti (Tribe Pachycereeae). *American Journal of Botany* **84:** 1482-1501.

Cowling RM, Esler KJ, Midgley GF, and MA Honig. 1994. Plant functional diversity, species diversity and climate in arid and semi-arid southern Africa. *Journal of Arid Environments* **27:** 141-158.

Cowling RM and Hilton-Taylor C. 1999. Plant biogeography, endemism, and diversity. In: Dean WRJ and Milton SJ (eds.) *The Karoo: ecological patterns and processes*. Cambridge University Press, Cambridge. Pp. 42-56.

Cui M and Nobel PS. 1994. Water budgets and root hydraulic conductivity of opuntias shifted to low temperatures. *International Journal of Plant Sciences* **155:** 167-172.

Dean WRJ. 1992. Temperatures determining activity patterns of some ant species in the southern Karoo, South Africa. *Journal of the Entomological Society of southern Africa* **55:** 149-156.

De Villiers AJ, Van Rooyen MW, and Theron GK. 2002. Germination strategies of Strandveld Succulent Karoo plant species for revegetation purposes: I. Temperature and light requirements. *Seed Science and Technology* **30:** 17-33.

Didden-Zopfy B and Nobel PS. 1982. High temperature tolerance and heat acclimation of *Opuntia bigelovii*. *Oecologia* **52:** 176-180.

Drennan PM and Nobel PS. 1996. Temperature influences on root growth for *Encelia farinosa* (Asteraceae), *Pleuraphis rigida* (Poaceae), and *Agave deserti* (Agavaceae) under current and double CO_2 concentrations. *American Journal of Botany* **83:** 133-139.

Drennan PM and Nobel PS. 1997. Frequencies of major C_3, C_4, and CAM perennials on different slopes in the northwestern Sonoran Desert. *Flora* **192:** 297-304.

Drennan PM and Nobel PS. 1998. Root growth dependence on soil temperature for *Opuntia ficus-indica*: influences of air temperature and a doubled CO_2 concentration. *Functional Ecology* **12:** 959-964.

Drezner TD. 2003. Branch direction in *Carnegiea gigantea* (Cactaceae): Regional patterns and the effect of nurse plants. *Journal of Vegetation Science* **14:** 907-910.

Drezner TD. 2006. Plant facilitation in extreme environments: The non-random distribution of saguaro cacti (*Carnegiea gigantea*) under their nurse associates and the relationship to nurse architecture. *Journal of Arid Environments* **65:** 46-61.

Drezner TD. 2007. An analysis of winter temperature and dew point under the canopy of a common Sonoran Desert nurse and the implications for positive plant interactions. *Journal of Arid Environments* **69:** 554-568.

Drezner TD and Garrity CM. 2003. Saguaro distribution under nurse plants in Arizona's Sonoran Desert: directional and microclimate influences. *The Professional Geographer* **55:** 505-512.

Eller BM and Nipkow A. 1983. Diurnal course of the temperature in a *Lithops sp.* (Mesembryanthemaceae Fenzl) and its surrounding soil. *Plant, Cell and Environment* **6:** 559-565.

Eller BM and Ferrari S. 1997. Water use efficiency of two succulents with contrasting CO_2 fixation pathways. *Plant, Cell and Environment* **20:** 93-100.

Eller BM and Grobbelaar N. 1986. Diurnal temperature variation in and around a *Lithops lesliei* plant growing in its natural habitat on a clear day. *South African Journal of Botany* **52:** 403-407.

Esler KJ. 1993. *Vegetation Patterns and Plant Reproductive Processes in the Succulent Karoo.* Ph.D. Thesis, University of Cape Town, Cape Town.

Esler KJ. 1999. Plant reproductive ecology. In: Dean WJS and Milton SJ (eds) *The Karoo: ecological patterns and processes.* Cambridge University Press, Cambridge. Pp. 123-144.

Flores J and Briones O. 2001. Plant life-form and germination in a Mexican intertropical desert: effects of soil water potential and temperature. *Journal of Arid Environments* **47:** 485-497.

Flores J, Briones O, Flores A, and Sánchez-Colón S. 2004. Effect of predation and solar exposure on the emergence and survival of desert seedlings of contrasting life-forms. *Journal of Arid Environments* **58:** 1-18.

Flores J and Jurado E. 2003. Are nurse-protégé interactions more common among plants from arid environments? *Journal of Vegetation Science* **14:** 911-916.

Flores-Flores JL and Yeaton RI. 2003. The replacement of arborescent cactus species along a climatic gradient in the southern Chihuahuan Desert: competitive hierarchies and response to freezing temperatures. *Journal of Arid Environments* **55:** 583-594.

Flores-Hernández A, Murillo-Amador B, Gracia-Hernandez JL, and Fraga-Palomino HC. 2001. Concentracion de prolina en brotes de cultivares de nopal (*Opuntia megacantha*) sometidos a estrés por calor. *Phyton* (Buenos Aires) **2001:** 15-24.

Franco AC and Nobel PS. 1989. Effect of nurse plants on the microhabitat and growth of cacti. *Journal of Ecology* **77:** 870-886.

Freeman CE. 1973a. Some germination responses of lechuguilla (*Agave lechuguilla* Torr.). *Southwestern Naturalist* **18:** 125-134.

Freeman CE. 1973b. Germination responses of a Texas population of ocotillo (*Fouquieria splendens* Engelm.) to constant temperature, water stress, pH and salinity. *American Midland Naturalist* **89:** 252-256.

Freeman CE. 1975. Germination responses of a New Mexico population of Parry agave (*Agave parryi* Englem. var. *parryi.*). *Southwestern Naturalist* **20:** 69-74.

Freeman CE, Tiffany RS, and Reid WH. 1977. Germination responses of *Agave lecheguilla, A. parryi* and *Fouquieria splendens. Southwestern Naturalist* **22**: 195-204.

Geller GN and Nobel PS. 1986. Branching patterns of columnar cacti: influences on PAR interception and CO_2 uptake. *American Journal of Botany* **73**: 1193-1200.

Gerwick BC, Williams III GJ, Spalding MH, and Edwards GE. 1978. Temperature responses of CO_2 fixation in isolated *Opuntia* cells. *Plant Science Letters* **13**: 389-396.

Gibson AC. 1996. *Structure-function relations of warm desert plants.* Springer-Verlag, Berlin.

Goldstein G and Nobel PS. 1991. The changes in osmotic pressure and mucilage during low-temperature acclimation of *Opuntia ficus-indica. Plant Physiology* **97**: 954-961.

Goldstein G and Nobel PS. 1994. Water relations and low-temperature acclimation for cactus species varying in freezing tolerance. *Plant Physiology* **104**: 675-681.

Haag-Kerwer A, Franco AC, and Lüttge U. 1992. The effect of temperature and light on gas exchange and acid accumulation in the C_3-CAM plant *Clusia minor* L. *Journal of Experimental Botany* 43:345-352.

Herppich WB, Herppich M, and Von Willert DJ. 1998. Ecophysiological investigations on plants of the genus *Plectranthus* (Lamiaceae). Influence of environment and leaf age on CAM, gas exchange and leaf water relations in *Plectranthus marrubioides* Benth. *Flora* **193**: 99-109.

Hickman JC. 1993. *The Jepson Mmanual: Higher Plants of California.* University of California Press, Berkeley.

Ishikawa M and Gusta LV. 1996. Freezing and heat tolerance of *Opuntia* cacti native to the Canadian prairie provinces. *Canadian Journal of Botany* **74**: 1890-1895.

Israel AA and Nobel PS. 1995. Growth temperature versus CO_2 uptake, Rubisco and PEPcase activities, and enzyme high-temperature sensitivities for a CAM plant. *Plant Physiology and Biochemistry* **33**: 345-351.

Jordan PW and Nobel PS. 1984. Thermal and water relations of roots of desert succulents. *Annals of Botany* **54**: 705-717.

Jürgens N. 1986. Untersuchungen zur Ökologie sukkulenter Pflanzen des südlichen Afrika. *Mitteilungen aus dem Institut für Allegemeine Botanik* (Hamburg) **21**: 129-136.

Jürgens N. 1991. A new approach to the Namib Region. I: Phytogeographic subdivision. *Vegetatio* **97**: 21-38.

Keeley JE and Keeley SC. 1989. Crassulacean acid metabolism (CAM) in high elevation tropical cactus. *Plant, Cell and Environment* **12**: 331-336.

Keeley JE and Meyers A. 1985. Effect of heat on seed germination of southwestern *Yucca* species. *Southwestern Naturalist* **30**: 303-305.

Kluge M, Razanoelisoa B, and Brulfert J. 2001. Implications of genotypic diversity and phenotypic plasticity in the ecophysiological success of CAM plants, examined by studies on the vegetation of Madagascar. *Plant Biology* **3**: 214-222.

Lewis DA and Nobel PS. 1977. Thermal energy exchange model and water loss of a barrel cactus, *Ferocactus acanthodes. Plant Physiology* **60**: 609-616.

Loik ME, Huxman TE, Hamerlynck EP, and Smith SD. 2000. Low temperature tolerance and cold acclimation for seedlings of three Mojave Desert Yucca species exposed to elevated CO_2. *Journal of Arid Environments* **46**: 43-56.

Loik ME and Nobel PS. 1991. Water relations and mucopolysaccharide increases for a winter hardy cactus during acclimation to subzero temperatures. *Oecologia* **88**: 340-346.

Loik ME and Nobel PS. 1993. Freezing tolerance and water relations of *Opuntia fragilis* from Canada and the United States. *Ecology* **74:** 1722-1732.

López RP and Valdivia S. 2007. The importance of shrub cover for four cactus species differeing in growth form in an Andean semi-desert. *Journal of Vegetation Science* **18:** 263-270.

Lüttge U. 2004. Ecophysiology of Crassulacean acid metabolism (CAM). *Annals of Botany* **93:** 629-652

Martorell C and Ezcurra E. 2002. Rosette scrub occurrence and fog availability in arid mountains of Mexico. *Journal of Vegetation Science* **13:** 651-662.

Martorell C and Ezcurra E. 2007. The narrow-leaf syndrome: a functional and evolutionary approach to the form of fog-harvesting rosette plants. *Oecologia* **151:** 561-573.

Martorell C and Patiño P. 2006. Globose cacti (*Mammilaria*) living on cliffs avoid high temperatures in a hot dryland of Southern Mexico. *Journal of Arid Environments* **67:** 541-552.

McClearly JA and Wagner KA. 1973. Comparative germination and early growth studies of six species of the genus *Yucca*. *American Midland Naturalist* **90:** 503-508.

McDonough W. 1964. Germination responses of *Carnegiea gigantea* and *Lemairocereus thurberi*. *Ecology* **45:** 155-159.

Méndez E, Guevara JC, and Estévez OR. 2004. Distribution of cacti in *Larrea* spp. shrublands in Mendoza, Argentina. *Journal of Arid Environments* **58:** 451-462.

Midgley GF and Thuiller W. 2007. Potential vulnerability of Namaqualand plant diversity to anthropogenic climate change. *Journal of Arid Environments* **70:** 615-628.

Midgley GF and van der Heyden F. 1999. Form and function in perennial plants. In: Dean WJS and Milton SJ (eds) *The Karoo: ecological patterns and processes.* Cambridge University Press, Cambridge. Pp. 123-144.

Milton SJ. 1995. Spatial and temporal patterns in the emergence and survival of seedlings in arid Karoo shrubland. *Journal of Applied Ecology* **32:** 145-156.

Milton SJ, Yeaton RI, Dean WRJ, and Vlok JHJ. 1997. Succulent Karoo. In: Cowling RM, Richardson DM, and Pierce SM (eds.) *Vegetation of Southern Africa.* Cambridge University Press, Cambridge. Pp. 131-166.

Mourelle C and Ezcurra E. 1996. Species richness of Argentine cacti: A test of biogeographic hypotheses. *Journal of Vegetation Science* **7:** 667-680.

Mourelle C and Ezcurra E. 1997. Differentiation diversity of Argentine cacti and its relationship to environmental factors. *Journal of Vegetation Science* **8:** 547-558.

Musil CF, Schmiedel U, and Midgley GF. 2005. Lethal effects of experimental warming approximating a future climate scenario on southern African quartzfield succulents: a pilot study. *New Phytologist* **165:** 539-547.

Nobel PS. 1978. Surface temperatures of cacti – influences of environmental and morphological factors. *Ecology* **59:** 986-996.

Nobel PS. 1980a. Morphology, surface temperatures, and northern limits of columnar cacti in the Sonoran Desert. *Ecology* **61:** 1-7.

Nobel PS. 1980b. Morphology, nurse plants, and minimum apical temperatures for young *Carnegeia gigantea*. *Botanical Gazette* **141:** 188-191.

Nobel PS. 1982. Low-temperature tolerance and cold hardening of cacti. *Ecology* **63:** 1650-1656.

Nobel PS. 1984. Extreme temperatures and the thermal tolerance for seedlings of desert succulents. *Oecologia* **62:** 310-317.

Nobel PS. 1988. *Environmental Biology of Agaves and Cacti.* Cambridge University Press, Cambridge, UK.

Nobel PS. 1989a. Shoot temperatures and thermal tolerances for succulent species of *Haworthia* and *Lithops*. *Plant, Cell and Environment* **12:** 643-651.

Nobel PS. 1989b. Temperature, water availability, and nutrient levels at various soil depths – consequences for shallow rooted deset succulents, including nurse plant effects. *American Journal of Botany* **76:** 1486-1492.

Nobel PS. 1990. Low temperature tolerance and CO_2 uptake for platyopuntias – a laboratory assessment. *Journal of Arid Environments* **18:** 313-324.

Nobel PS. 1996. Responses of some North American CAM plants to freezing temperatures and doubled CO_2 concentrations: implications of global climate change for extending cultivation. *Journal of Arid Environments* **34:** 187-196.

Nobel PS. 2005. *Physicochemical and Environmental Plant Physiology, 3rd edn.* Elsevier/Academic Press, Burlington, MA.

Nobel PS and Bobich EG. 2002. Plant frequency, stem and root characteristics, and CO_2 uptake for *Opuntia acanthocarpa*: elevational correlates in the northwestern Sonoran Desert. *Oecologia* **130:** 165-172.

Nobel PS and Castañeda M. 1998. Seasonal, light, and temperature influences on organ initiation for unrooted cladodes of the prickly pear cactus *Opuntia ficus-indica*. *Journal of the American Society for Horticultural Science* **123:** 47-51.

Nobel PS, Castañeda M, North G, Pimienta-Barrios E, and Ruiz A. 1998. Temperature influences on leaf CO_2 exchange, cell viability and cultivation range for *Agave tequilana*. *Journal of Arid Environments* **39:** 1-9.

Nobel PS and De la Barrera E. 2002. High temperature and net CO_2 uptake, growth, and stem damage for the hemiepiphytic cactus *Hylocereus undatus*. *Biotropica* **34:** 225-231.

Nobel PS and De la Barrera E. 2003. Tolerances and acclimation to low and high temperatures for cladodes, fruits and roots of a widely cultivated cactus, *Opuntia ficus-indica*. *New Phytologist* **157:** 271-279.

Nobel PS, De la Barrera E, Beilman DW, Doherty JH, and Zutta BR. 2002. The temperature limitations for cultivation of edible cacti in California. *Madroño* **49:** 228-236.

Nobel PS and Geller GN. 1987. Temperature modeling of wet and dry desert soils. *Journal of Ecology* **75:** 247-258.

Nobel PS, Geller GN, Kee SC, and Zimmerman AD. 1986. Temperatures and thermal tolerances for cacti exposed to high temperatures near the soil surface. *Plant, Cell, and Environment* **9:** 279-288.

Nobel PS and Linton MJ. 1997. Frequencies, microclimate and root properties for three codominant perennials in the northwestern Sonoran Desert on north- versus south-facing slopes. *Annals of Botany* **80:** 731-739.

Nobel PS and Loik ME. 1990. Thermal analysis, cell viability and carbon dioxide uptake of a widely distributed north Amercian cactus, *Opuntia humifusa*, at sub-zero temperatures. *Plant Physiology and Biochemistry* **28:** 429-436.

Nobel PS and McDaniel RG. 1988. Low temperature tolerances, nocturnal acid accumulation, and biomass increases for seven species of *Agave*. *Journal of Arid Environments* **15:** 147-155.

Nobel PS and Smith SD. 1983. High and low temperature tolerances and their relationships to the distribution of agaves. *Plant, Cell and Environment* **6:** 711-719.

Nobel PS, Wang N, Balsamo RA, Loik ME, and Hawke ME. 1995. Low-temperature tolerance and acclimation of *Opuntia* spp. after injecting glucose or methylglucose. *International Journal of Plant Sciences* **156:** 496-504.

Nobel PS and Zutta BR. 2005. Morphology, ecophysiology, and seedling establishment for *Fouquiera splendens* in the northwestern Sonoran Desert. *Journal of Arid Environments* **62:** 251-265.

Nobel PS and Zutta BR. 2007. Rock associations, root depth, and temperature tolerances for the "rock live-forever," *Dudleya saxosa*, at three elevations in the north-western Sonoran Desert. *Journal of Arid Environments* **69:** 15-28.

Norman F and Martin CE. 1986. Effects of spine removal on *Coryphantha vivipara* in Central Kansas. *The American Midland Naturalist* **116:** 118-124.

Olvera- Carrillo Y, Márquez-Guzmán J, Barradas VL, Sánchez-Coronado EM, and Orozco-Segovia A. 2003. Germination of the hard seed coated *Opuntia tomentosa* S.D., a cacti from the México valley. *Journal of Arid Environments* **55:** 29-42.

Pavón NP, Hernández-Trejo H, and Rico-Gray V. 2000. Distribution of plant life forms along an altitudinal gradient in the semi-arid valley of Zapotitlan, Mexico. *Journal of Vegetation Science* **11:** 39-42.

Peña-Valdivia CB, Sánchez-Urdaneta AB, Aguirre R, Trejo C, Cárdenas E, and Villegas MA. 2006. Temperature and mechanical scarification on seed germination of 'maguey' (*Agave salmiana* Otto ex Salm-Dyck). *Seed Science and Technology* **34:** 47-56.

Pimienta-Barrios E, Robles-Murguia C, and Nobel PS. 2001. Net CO_2 uptake for *Agave tequilana* in a warm and a temperate environment. *Biotropica* **33:** 321-318.

Rascher U, Bobich EG, and Osmond CB. 2006. The "Kluge-Lüttge Kammer": A preliminary evaluation of an enclosed, Crassulacean acid metabolism (CAM) mesocosm that allows separation of synchronized and desynchronized contributions of pants to whole system gas exchange. *Plant Biology* **8:** 167-174.

Reinhardt CF. 2000. Ecological adaptation of an alien invader plant (*Opuntia stricta*) determines management strategies in the Kruger National Park. *Zeitschrift fuer Pflanzenkrankheiten und Pflanzenschutz Special Issue* **17:** 77-84.

Rojas-Aréchiga M and Vázquez-Yanes C. 2000. Cactus seed germination: a review. *Journal of Arid Environments* **44:** 85-104.

Rundel PW, Cowling RM, Esler KJ, Mustart PM, van Jaarsveld E, and Bezuidenhout H. 1995. Winter growth phenology and leaf orientation in *Pachypodium namaquanum* (Apocynaceae) in the succulent karoo of the Richtersveld, South Africa. *Oecologia* **101:** 472-477.

Sayed OH. 2001. Crassulacean Acid Metabolism 1975-2000, a checklist. *Photosynthetica* **39:** 339-352.

Schmida A. 1981. Mediterranean vegetation of Israel and California: similarities and differences. *Israel Journal of Botany* **30:** 105-123.

Schmiedel U and Jürgens N. 2004. Habitat ecology of southern African quartz fields: Studies on the thermal properties near the ground. *Plant Ecology* **170:** 153-166.

Shreve F. 1931. Physical conditions in sun and shade. *Ecology* **12:** 96-104.

Smith SD, Didden-Zopfy B, and Nobel PS. 1984. High-temperature responses of North American cacti. *Ecology* **65:** 643-651.

Smreciu EA, Currah RS, and Toop E. 1988. Viability and germination of herbaceous perennial species native to southern Alberta Canada grasslands. *Canadian Field Naturalist* **102:** 31-38.

Sortibrán L, Tinoco-Ojanguren C, Terrazas T, and Valiente-Banuet A. 2005. Does cladode inclination restrict microhabitat distribution for *Opuntia puberula* (Cactaceae)? *American Journal of Botany* **92:** 700-708.

Steenbergh WF and Lowe CH. 1976. Ecology of the saguaro I: the role of freezing weather in a warm desert plant population. In: *Research in the Parks. National Park Service Symposium Series no. 1.* National Park Service, Washington, DC., Pp. 49-92.

Stock WD, Allsopp N, van der Heyden F, and Witkowski ETF. 1997. Plant form and function. In: Cowling RM, Richardson DM, and Pierce SM (eds.) *Vegetation of Southern Africa*. Cambridge University Press, Cambridge. Pp. 376-396.

Suzán H, Nabhan GP, and Patten DT. 1996. The importance of *Olneya tesota* as a nurse plant in the Sonoran Desert. *Journal of Vegetation Science* **7**: 635-644.

Thuiller W, Midgley GF, Rouget M, and Cowling RM. 2006. Predicting patterns of plant species richness in megadiverse South Africa. *Ecography* **29**: 733-744.

Turner JS and Picker MD. 1993. Thermal ecology of an embedded dwarf succulent from southern Africa (*Lithops* spp: Mesembryanthemaceae). *Journal of Arid Environments* **24**: 361-385.

Turner RM, Bowers JE, and Burgess TL. 1995. *Sonoran Desert plants. An ecological atlas*. The University of Arizona Press, Tucson.

Valiente-Banuet A and Ezcurra E. 1991. Shade as a cause of the association between the cactus *Neobuxbaumia tetetzo* and the nurse plant *Mimosa luisana* in the Tehuacan Valley, Mexico. *Journal of Ecology* **79**: 961-971.

Valiente-Banuet A, Bolongaro CA, Briones O, Ezcurra E, Rosas M, Nuñez H, Barnard G, and Vázquez E. 1991. Spatial relationships between cacti and nurse shrubs in a semi-arid environment in Central Mexico. *Journal of Vegetation Science* **2**: 15-20.

Van Coller A and Stock WD. 1994. Cold tolerance of the southern African succulent, *Cotyledon orbiculata* L. across its geographical range. *Flora* **189**: 89-94.

Van Rooyen MW, Theron GK, and Grobelaar N. 1979. Phenology of the vegetation in the Hester Malan Nature Reserve in the Namaqualand Broken Veld: 1. General observations. *Journal of South African Botany* **45**: 279-293.

Van Jaarsveld E. 1987. The succulent riches of South Africa and Namibia. *Aloe* **24**: 45-92.

Von Willert DJ, Eller BM, Werger MJA, Brinkmann E, and Ihlenfeldt H-D. 1992. *Life Strategies of Succulents in Deserts with Special Reference to the Namib Desert*. Cambridge University Press, Cambridge, UK.

Werger MJA and Ellis RP. 1981. Photosynthetic pathways in the arid regions of South Africa. *Flora* **171**: 64-75.

Werger MJA and Morris JW. 1991. Climatic control of vegetation structure and leaf characteristics along an aridity gradient. *Annali di Botanica* **49**: 203-215.

Woodhouse RM, Williams JG, and Nobel PS. 1983. Simulation of plant temperature and water loss by the desert succulent, *Agave deserti*. *Oecologia* **57**: 291-297.

Yeaton RL and Cody ML. 1979. The distribution of cacti along environmental gradients in the Sonoron and Mohave Deserts. *Journal of Ecology* **67**: 529-541.

Zavala-Hurtado JA, Vite F, and Ezcurra E. 1998. Stem tilting and pseudocephalium orientation in *Cephalocereus columna-trajani* (Cactaceae): A functional interpretation. *Ecology* **79**: 340-348.

Chapter 4

MICROENVIRONMENTS, WATER RELATIONS, AND PRODUCTIVITY OF CAM PLANTS

José Luis Andrade, J. Carlos Cervera, and Eric A. Graham

Introduction
Water relations of tissues and microhabitats
Water acquisition and microsites
Light microenvironments
Water and CO_2 uptake
Productivity of CAM plants
Water, seed germination, and seedling survival
Conclusions and future prospects

Perspectives in Biophysical Plant Ecophysiology: A Tribute to Park S. Nobel, pp. 95-120
Edited by: E. De la Barrera and W.K. Smith
© 2009 by The Authors
Book Compilation © 2009 Universidad Nacional Autónoma de México

Introduction

Plants that inhabit environments with extreme seasonal water availability show several adaptations to minimize water loss: succulence, rosette-like shoots that capture and retain water (phytotelmata), reduced stomatal size and frequency, cuticles impermeable to water, specialized leaf and stem structures for water absorption, Crassulacean acid metabolism (CAM), special root features, and water-binding mucilage in tissues (Nobel 1988; Lüttge 1989; Benzing 1990; Nobel et al. 1992a; Nobel 1996; Helbsing et al. 2000). The nocturnal CO_2 assimilation of CAM permits great water use efficiency (WUE), because it occurs when air temperature is lower and relative humidity is higher than during the day, which allows less water loss by transpiration. In fact, CAM plants show a much higher WUE that C_3 and C_4 plants under similar conditions (Drennan and Nobel 2000; Winter et al. 2005). Accordingly, CAM plants are typical from arid and semiarid regions and the canopies of tropical forests, representing about 6% of the vascular plant species (Winter and Smith 1996), a greater percentage than that for C_4 species. Additionally, CAM is also found in aquatic plants that inhabit places where CO_2 becomes scarce during the day (Keeley 1996; Keeley and Rundel 2003).

Environmental factors influence the biochemical and physiological attributes of CAM, and CAM plants can be mainly affected by temperature, light level and water status, showing a vast array of plasticity (Cushman 2001; Kluge et al. 2001; Dodd et al. 2002; Lüttge 2004; Winter and Holtum 2005). In fact, there is a continuum between C_3 and CAM plants and some species show low degree of nocturnal CO_2 uptake or internal CO_2 recycling (Pierce et al. 2002a). Furthermore, for many species the contribution of CAM to daily carbon gain becomes proportionally more important as drought progresses (Pierce et al. 2002a; Winter and Holtum 2002; Graham and Andrade 2004). Net CO_2 uptake for CAM plants has been traditionally characterized by a four-phase framework (Osmond 1978) that allows describing the CAM cycle in a comparative manner. During drought, the pattern of the CAM phases is adjusted and some phases can be reduced or lost as drought progresses (Nobel 1988; Dodd et al. 2002; Lüttge 2004). Similarly, environmental conditions of the natural ecosystems where CAM plants inhabit affect the CAM cycle, especially because drought is associated to high temperatures and high photosynthetic photon flux (PPF). Nevertheless, some CAM plants can achieve high productivities compared with C_3 and C_4 plants in similar conditions (Nobel 1991, 1996) and greater attention has been paid to the effects of particular environmental factors on CO_2 uptake for many CAM species, particularly agaves and cacti (Nobel 1988, 1994).

The focus of this chapter is on the environmental factors that affect CAM plants in the microhabitats where they occur. We will emphasize the water relations of epiphytic and terrestrial CAM plants but we will also consider the plant responses to light and temperature. For many CAM plants, little change in net CO_2 uptake occurs during the first days of drought because of the water stored in their tissues and the high WUE of the CAM cycle (Nobel 1988; Lüttge 2004). Succulent CAM plants store appreciable amounts of water in specialized tissues of their leaves and stems and water is transported from the storage tissue to the photosynthetic tissue (chlorenchyma) to allow continuous CO_2 uptake. For instance, some cactus species lose four times more water from the water-storage parenchyma than from the chlorenchyma during drought (Barcikowski and Nobel 1984; Goldstein et al. 1991). With their high WUE, CAM plants are capable to inhabit arid and semiarid habitats, but they are far more numerous in tropical forests, primarily in the epiphytic habitat (Winter and Smith 1996; Lüttge 2004). In tropical forests, deciduousness, rainfall seasonality, and tree host architecture and height generate multiple microhabitats, which are occupied by C_3 and CAM epiphytes. Among these, CAM epiphytes are more abundant in drier forests and tend to occur in the more exposed microhabitats, within the same forest, than C_3 epiphytes (Winter et al. 1983; Smith et al. 1986; Benzing 1990). In fact, it is estimated that about 57% of all epiphytes are CAM plants (Lüttge 2004).

Water relations of tissues and microhabitats

Water storage in CAM plants can be either external, internal or both. Some bromeliads form external water reservoirs or phytotelmata that allows water obtained from rainfall to be stored for several days (Benzing 1990). However, special non-photosynthetical water-storage tissues permit long-term water use in many CAM taxa (Lüttge 2004). Generally, water moves from the water-storage tissues to the photosynthetic tissues for many days or weeks during drought. In fact, this internal water redistribution reflects differences in water relations between tissues. For some cacti, solutes are lost (Barcikowsky and Nobel 1984) or sugars are polymerized (Goldstein et al. 1991) in the water-storage parenchyma during drought to maintain a water potential difference between this tissue and the chlorenchyma, facilitating water movement. Also, the cell-wall elastic modulus of the water-storage parenchyma is about 40% of that of the chlorenchyma, which means that the water-storage parenchyma maintains turgor over a large range of water contents (Goldstein et al. 1991). Additionally, mucilage, with a very high water-binding capacity, occurs in greater amounts in the water-storage parenchyma than in the chlorenchyma (Goldstein et al. 1991) and the amount of mucilage in the water-storage parenchyma is positively correlated to the tissue relative capacitance in several cactus species (Nobel et al. 1992a).

Uprooted cactus plants can survive for several years without water (Szarek and Ting 1975; Nobel 1981; Barcikowski and Nobel 1984), but when some plant tissues lose more than 50% of their total water they can die (Lüttge 2004). However, for *Opuntia ficus-indica*, the chlorenchyma can reversible lose 70% of its water content at full turgor and the water-storage parenchyma can lose 82% (Goldstein et al. 1991). Similarly, for epiphytic bromeliads, tissues of *Tillandsia fasciculata* can recover after losing 60% of the water present at full turgor whereas the tissues of

Guzmania monostachya can lose about 90% in a lowland seasonal forest (Zotz and Andrade 1998); in a deciduous dry forest, tissues of *T. elongata* can lose over 62% of the water present at full turgor and *T. brachycaulos* can lose 48% (Andrade 2003). It is surprising that both massive cacti and epiphytic bromeliads show similar magnitudes of tissue recovery. Furthermore, *T. ionantha* also shows high tolerance to change in tissue hydration (Benzing and Dahle 1971) and the C_3 epiphytic ferns *Polypodium crassifolium* (but Holtum and Winter (1999) found it as a weak CAM) and *P. phyllitidis* can even lose 98% of the leaf succulence and recover 2 days after rewetting (Andrade and Nobel 1997). Rapid succulence recovery is typical of epiphytes and occurs within a few hours or a few days after rewetting (Sinclair 1983; Andrade and Nobel 1997); CAM epiphytes also resume tissue acidity rhythms within 3 days after rewetting (Andrade and Nobel 1996). Consequently, tissue recovery upon rewetting does not explain the occurrence of a CAM plant in a particular microhabitat; it seems that tissue relative capacitance can explain better why a specific CAM plant occurs in certain microhabitats.

Relative capacitance is the capacity of the tissues to maintaining their water potentials when the water content decreases and relates to succulence and to the tissue volume to surface area ratio (Andrade and Nobel 1997; Zotz and Andrade 1998). Relative capacitance for epiphytic CAM plants averages about half as much as terrestrial CAM species (Table 4.1). Also, within epiphytes, relative capacitance is about 2-fold higher for species that occur in more exposed microhabitats than those that occur in shaded microhabitats (Table 4.1); for instance, the highest relative capacitance values are for the epiphytic cacti *Epiphyllum phyllanthus*, *Rhipsalis baccifera* and the epiphytic bromeliad *Tillandsia fasciculata*, which occur more frequently in the more exposed sites of deciduous trees (Andrade and Nobel 1996, 1997; Zotz 1997). Desert CAM plants show a greater relative capacitance than CAM epiphytes because they confront longer dry periods, up to several months. However, in seasonally dry forests periods without water are accompanied by an increase in photosynthetic photon flux (PPF) because of the deciduousness of trees and the reduced amount of clouds, further enhancing desiccating conditions. For instance, PPF above the canopy in a Panamanian lowland forest can be 30% higher during the dry season (Windsor 1990); hence, only the most drought tolerant epiphytes with the highest tissue relative capacitance can establish on deciduous trees (Andrade and Nobel 1997). Also, in a Mexican tropical dry deciduous forest, PPF changes dramatically for epiphytes between the rainy and dry seasons, from 3 to 9 times more PPF during the dry season versus the rainy season (Graham and Andrade 2004). Nevertheless, those epiphytes in exposed positions in the canopy can also have more access to dew deposition, reducing drought stress (Andrade 2003; Graham and Andrade 2004).

Table 4.1. Relative capacitance (MPa^{-1}) for some CAM species. Data were either taken from the original sources or calculated from the relative water content and osmotic potential values (at the turgor loss point) of the pressure-volume curves.

Species	Family	Relative capacitance (MPa^{-1})	Type of vegetation and habitat
Epiphytic species			
Pyrrosia adnascens[1]	Polypodiaceae	0.15	Lowland forest – exposed
P. angustata[1]	Polypodiaceae	0.15	Lowland forest – semi exposed
Dendrobium crumenatum[1]	Orchidaceae	0.27	Lowland forest
D. tortile[1]	Orchidaceae	0.16	Montane forest
Eria velutina[1]	Orchidaceae	0.53	Lowland forest
Guzmania monostachya[7]	Bromeliaceae	0.30	Seasonally dry lowland forest – semi exposed
Tillandsia fasciculata[7]	Bromeliaceae	0.70	Seasonally dry lowland forest – exposed
T. utriculata[5]	Bromeliaceae	0.16	*Quercus* forest – semi exposed
Epiphyllum phyllanthus[6]	Cactaceae	0.55	Seasonally dry lowland forest – exposed
Rhipsalis baccifera[6]	Cactaceae	0.45	Seasonally dry lowland forest – exposed
Terrestrial species			
Opuntia ficus-indica[3]	Cactaceae	0.80	Semi desert, cultivated
O. basilaris[4]	Cactaceae	1.04	Desert
O. acanthocarpa[4]	Cactaceae	0.96	Desert
Echinocereus engelmannii[4]	Cactaceae	1.35	Desert
Ferocactus acanthodes[4]	Cactaceae	0.81	Desert
Agave deserti[2]	Agavaceae	0.18	Desert

[1] – Sinclair 1983; [2] – Nobel and Jordan 1983; [3] – Goldstein *et al.* 1991; [4] – Nobel *et al.* 1992a; [5] – Stiles and Martin 1996; [6] – Andrade and Nobel 1997; [7] – Zotz and Andrade 1998.

Water acquisition and microsites

The roots of many succulent CAM plants have evolved rectifier-like properties, which mean that roots take up water fast when the soil is wet, but reduce water loss to a dry soil during drought (Nobel and Cui 1992; Nobel and North 1996). In early stages of a drought, root hydraulic conductivity decreases rapidly but during a prolonged drought, the soil hydraulic conductivity becomes the limiting factor that prevents water loss from the succulent shoots to the soil (Nobel and Cui 1992). When water is available, roots of CAM plants show high water uptake rates due to the initiation of new roots (Nobel and North 1996). Because most CAM plants have low root:shoot ratios, the roots usually explore small soil volumes, such as rock crevices or tree holes. In fact, lateral roots and soil water content is higher under rocks, whose surfaces can intercept upward movement of water and also offer a cooler platform for water condensation (Nobel *et al.* 1992b). Additionally, Dubrovsky and Gómez-Lomelí (2003) found that cactus lateral root formation is a stable developmental process, which accelerates during water stress; features that can be important for seedling establishment. Also, roots of the epiphytic cacti *Epiphyllum phyllanthus* and *Rhipsalis baccifera* develop sheaths composed of soil particles, root hairs and mucilage, which reduce plant water loss to a dry soil but helps the roots to take advantage of episodic rainfalls (North and Nobel 1994).

Root: shoot ratio is close to zero for all the epiphytic species within the Bromeliaceae, which take up water and nutrients by foliar trichomes (Benzing 1990). These epiphytic bromeliads may also rely on alternative sources of water, such as fog and dew. For instance, in cloud forests, the sites with more fog interception support more diverse and abundant epiphytic Bromeliaceae populations (Cavelier and Goldstein 1989). Fog may also be important for desert succulent leaf-rosette plants, whose leaves act as structures that intercept and harvest fog (Martorell and Ezcurra 2002). Additionally, dew can help maintaining a favorable water balance for two co-occurring *Tillandsia* species during the driest months in a tropical dry deciduous forest in Mexico (Andrade 2003); these species show a vertical stratification, where *T. brachycaulos* grows on the trees from very close to the soil up to the small branches in the canopy. In contrast, *T. elongata*, which is more sensitive to drought, inhabits in more exposed locations where individuals can intercept more rainfall and dew (Andrade 2003; Graham and Andrade 2004). In a forest in Panama, however, vertical stratification of two co-occurring epiphytic bromeliads could be explained by the physiological responses of small individuals. Although adult plants of *Guzmania monostachya* and *Tillandsia fasciculata* show similar water losses, small plants of *G. monostachya* show higher water losses than small individuals of *T. fasciculata* (Zotz and Andrade 1998). In a previous study, *T. fasciculata* was consistently found in higher sites of the canopy than *G. monostachya* (Zotz 1997); presumably because small plants of the latter species could not survive prolonged drought in exposed microsites.

A combination of lack of water and high temperatures in microhabitats can also affect plant growth. The epiphytic bromeliad *T. brachycaulos* occurs within the canopy in a tropical dry deciduous forest and shows great shade tolerance (Graham and Andrade 2004), but individuals that grow in the darkest light microhabitats produce less leaves and flowers, and leaves elongate slower than plants in more exposed microhabitats (Cervantes *et al.* 2005). Common garden experiments showed, however, that these responses are because individuals in the most shaded microhabitats receive less water from rainfall and dew; also, because individuals in those microsites are unable to dissipate the excess of heat during midday (Fig. 4.1; Cervantes *et al.* 2005). In this particular tropical dry forest, this epiphytic bromeliad would have more growth and reproduction in partially shaded microsites within the canopy, where individuals can capture more water and dissipate better the heat. Demographic studies should consider including a factor for some individuals, which grow more in certain microhabitats than in others, to analyze population growth, although this may be a difficult task (Jonzén *et al.* 2002).

Light microenvironments

Light microenvironments can vary enormously for CAM plants that live in places with changing vegetation physiognomy, such as forest and the forest canopies of seasonally tropical forest, where plants that are shaded during the rainy season, can be totally exposed during the dry season. Also, even in very humid tropical forests, the clouds and the sunflecks create a light environment extremely variable for understory CAM plants. Additionally, in tropical forests, epiphytes show a vertical stratification and CAM species tend to occupy more exposed microsites than C_3 species (Winter *et al.* 1983; Benzing 1990; Andrade and Nobel 1996, 1997). However, it is difficult to separate the effects of the light environment from that for the lack of water for CAM plants (as we saw in the previous section) because adaptations to the excess of light correlate with those to drought.

CAM photosynthesis is energetically less efficient than C_3 photosynthesis (Nobel 1991) and supposedly is even less efficient in the understory sites. However, although the frequency of CAM plants is positively correlated with exposure and less water availability within the forests and along different geographical regions (Zotz and Hietz 2001), CAM provides a more rapid response to sunflecks in shade adapted plants than for neighboring C_3 plants (Skillman *et al.* 1999); but also confers photoprotection through the maintenance of the electron transport (Griffiths *et al.* 1986; Maxwell *et al.* 1995). The thermal dissipation of excess energy associated with the xanthophyll cycle helps protecting the photosynthetic apparatus from photoinhibitory damage in plants normally subjected to a combination of high light and water stress (Haslam *et al.* 2003; Lu *et al.* 2003). Additionally, Kondo *et al.* (2004) reported diurnal movement of chloroplasts in the leaves of succulent CAM plants under high light and water stress, suggesting that this mechanism is a common photoprotective strategy used by CAM plants subjected to severe water stress.

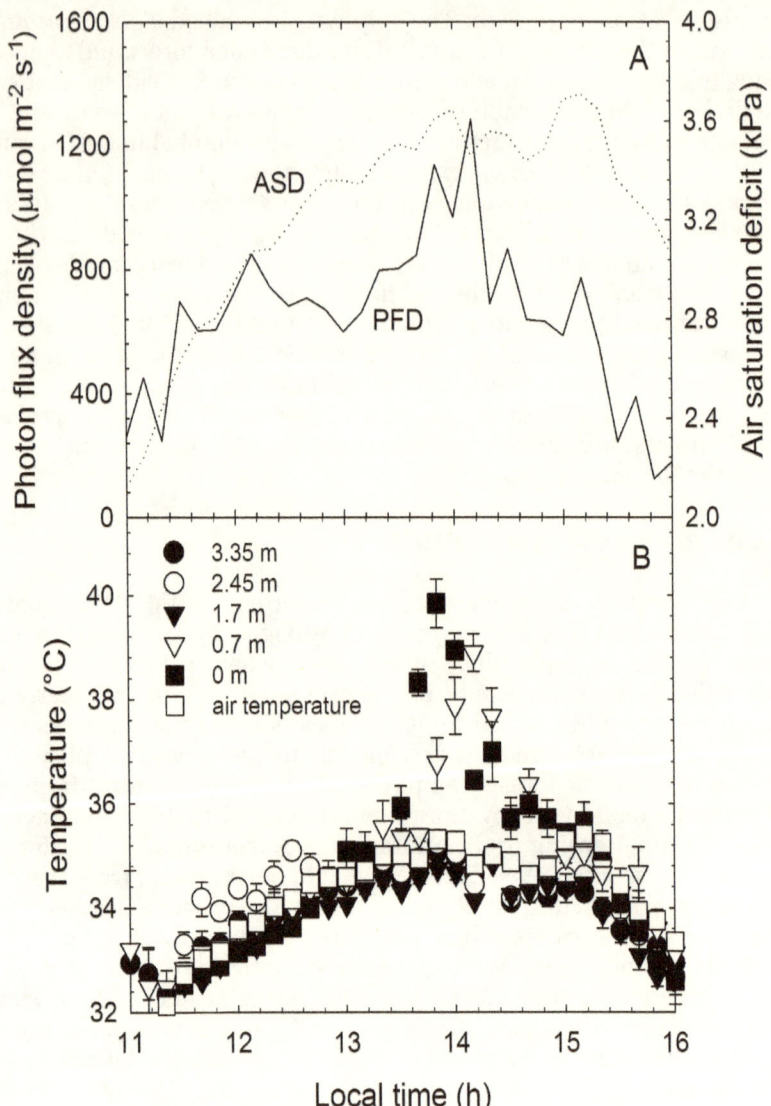

Figure 4.1. Daily course of (A) photosynthetic photon flux (PPF), air saturation deficit (ASD), and (B) leaf and air temperatures during a clear day in the early dry season in a dry forest in Yucatán, Mexico. Data for leaf temperature are means ± s.e. (n = 3 plants) for *Tillandsia brachycaulos* individuals growing at five tree heights (modified from Cervantes *et al.* 2005).

Thanks to inexpensive and portable meters for chlorophyll fluorescence, several photosynthetic parameters can be measured (Maxwell and Johnson 2000). For instance, the quantum yield for the photosystem II (measured as F_v/F_m), which can be a good indicator for the plant photosynthesis performance; the non-photochemical quenching (NPQ), which estimates the photoprotection capacity of the reversible light-induced carotenoid zeaxanthine (NPQ_{fast}) and the long-term damage or photoinhibition (NPQ_{slow}). Sun species exhibit great NPQ_{fast}, independently of the light environment or growth conditions (Johnson *et al.* 1993), and shade species show a reduced NPQ_{fast} and a high photodamage and a high NPQ_{slow} when exposed to sun (Maxwell and Johnson 2000).

Water stress can increase the negative high radiation effects even in succulent CAM plants (Maxwell *et al.* 1995; Zotz and Hietz 2001; Haslam *et al.* 2003; Andrade *et al.* 2006). For instance, for *Tillandsia brachycaulos*, during the rainy season, leaf F_v/F_m was relatively high in the morning, decreases during the day (which indicates both photochemical and non-photochemical quenching), and increases again to the high values in the evening, in three different shade treatments (Fig. 4.2; Cervantes *et al.* 2005). During the dry season, when plants were moderately stressed, the initial and final values for F_v/F_m indicate that photoinhibition occurs, especially in the most exposed individuals.

Water and CO₂ uptake

The uptake of CO_2 by plants, which in turn affects their growth and their productivity, is influenced by air temperature, light (represented as photosynthetic photon flux, PPF), rainfall and soil nutrient levels. In many situations, one environmental factor can have a predominant influence, such as annual rainfall for arid and semiarid-land species (Nobel and Pimienta-Barrios 1995). Since most studies on CO_2 uptake for CAM plants have been done with plants normally found in arid and semi-arid habitats (Ting 1985; Nobel 1988; von Willert *et al.* 1992), we will emphasize the influence of soil water availability on CO_2 uptake of terrestrial CAM plants.

At field capacity and under the usual PPF and temperature conditions of the habitat for CAM plants from arid zones, most of the CO_2 uptake occurs at night. For example, nighttime CO_2 uptake for *Agave fourcroydes*, a cultivated agave from semiarid Yucatán, was nearly 91% of the 325 mmol m^{-2} of the CO_2 fixed in a 24-h period (Nobel 1985). Also, in the Chihuahuan desert, net CO_2 uptake for *A. lechuguilla* over a 24-h period was 172 mmol m^{-2}, of which 85% occurred at night. Lower temperatures and higher PPF favored daytime CO_2 uptake for this species (Nobel and Quero 1986). After 11 days of drought, total daily net CO_2 uptake for *A. fourcroydes* decreased to 71 mmol m^{-2} and to only 2 mmol m^{-2} after 30 days of drought. Three days after watering the plants droughted for 30 days, its maximal net

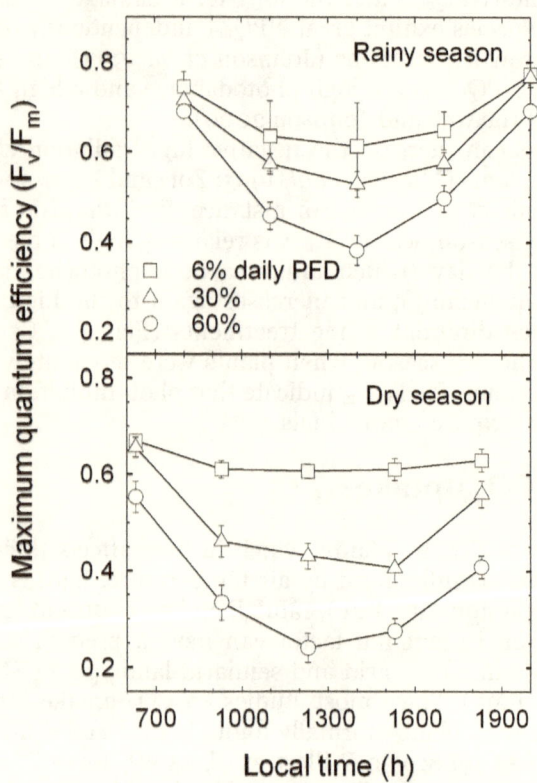

Figure 4.2. Daily course of maximum quantum efficiency (F_v/F_m) for *Tillandsia brachycaulos* during the rainy season (upper panel) and the dry season (lower panel). Plants were acclimated in shade treatments of 60% of total daily photon flux density (PFD; circles), 30% daily PFD (triangles), and 6% daily PFD (squares) for one month before measurements (modified from Cervantes *et al.* 2005).

CO_2 uptake rate was 48% of that under field capacity conditions, and after 7 days it was 91% (Nobel 1985). *Agave lechuguilla*, was more sensitive to drought than *A. fourcroydes*, since after 7 days of drought, net total daily CO_2 was 108 mmol m^{-2} (none of this during daytime) and after 13 days of drought it was only 1.4 mmol m^{-2} (Nobel and Quero 1986). Several epiphytic cacti show diurnal and nocturnal CO_2 uptake (sometimes both diurnal and nocturnal uptakes are of the same magnitude) and short-term droughts caused a shift favoring nocturnal CO_2 uptake only and reducing it to zero after a few days of drought (Nobel and Hartsock 1990; Andrade and Nobel 1997).

Under well-watered conditions, net CO_2 uptake for cacti from subfamily Pereskioideae (characterized by having flattened, photosynthetic leaves; Anderson and Brown 2001) occurred in the leaves, using the C_3 pathway, and total daily net CO_2 uptake ranged from 130 mmol m^{-2} for *Pereskia aculeata*, to 202 mmol m^{-2} for *Pereskia grandifolia*. No CO_2 uptake occurred in the stems. Droughts of 7 and 14 days decreased leaf daytime net CO_2 uptake by an average of 49 and 88%, respectively; at night there is a net loss of CO_2 by leaf (Nobel and Hartsock 1987).

For *Austrocylindropuntia subulata* (a member of subfamily Opuntioideae with fleshy leaves; Anderson and Brown 2001), total daily leaf net CO_2 uptake was 91 mmol m^{-2} under well-watered conditions (95% occurring during the day) and 7 and 14 days of drought reduced leaf daytime net CO_2 uptake by 90 and 100%, respectively. Also, the stems of this species had periods of positive CO_2 uptake (Nobel 1988). However, net CO_2 uptake occurring at night increased from 5% under wet conditions to 71% after 7 days of drought to 99% after 14 days of drought (Nobel and Hartsock 1987). Thus, shifts from predominantly daytime to predominantly nighttime CO_2 uptake can be induced by drought for leafy cacti in subfamily Opuntioideae, indicating a high degree of biochemical versatility. For the largest genus in subfamily Opuntioideae, *Opuntia* (characterized by ephemeral leaves and flattened stems segmented in cladiodes; Anderson and Brown 2001), net CO_2 uptake occurs in the stems during nighttime (Nobel 1988). However, if soil water potential is near field capacity and temperatures are moderate, some net CO_2 uptake occurs at the beginning of the light period (Cui *et al.* 1993; Cui and Nobel 1994). In contrast, for members of subfamily Cactoideae (stems usually not segmented with vestigial or no leaves at all, Anderson and Brown 2001), like columnar and globular cacti, 91 to 99% of the net CO_2 uptake occurs predominantly at night, even when over-watered (Hanscom and Ting 1978; Nobel 1986). Figure 4.3 summarizes gas exchange patterns for agave species and members of the three cacti subfamilies.

As in leafy members of Opuntioideae, some plants have been found to shift the nature of carbon metabolism with changes in their growth environment. In certain C_3 plants, CAM may be induced by drought. Examples are found within families Clusiaceae (Ball *et al.* 1991; De Mattos and Lüttge 2001), Crassulaceae (Smirnoff 1996), Euphorbiaceae (Rundel *et al.* 1999; Ramachandra Reddy *et al.* 2003), Mesembryanthemaceae (Winter and

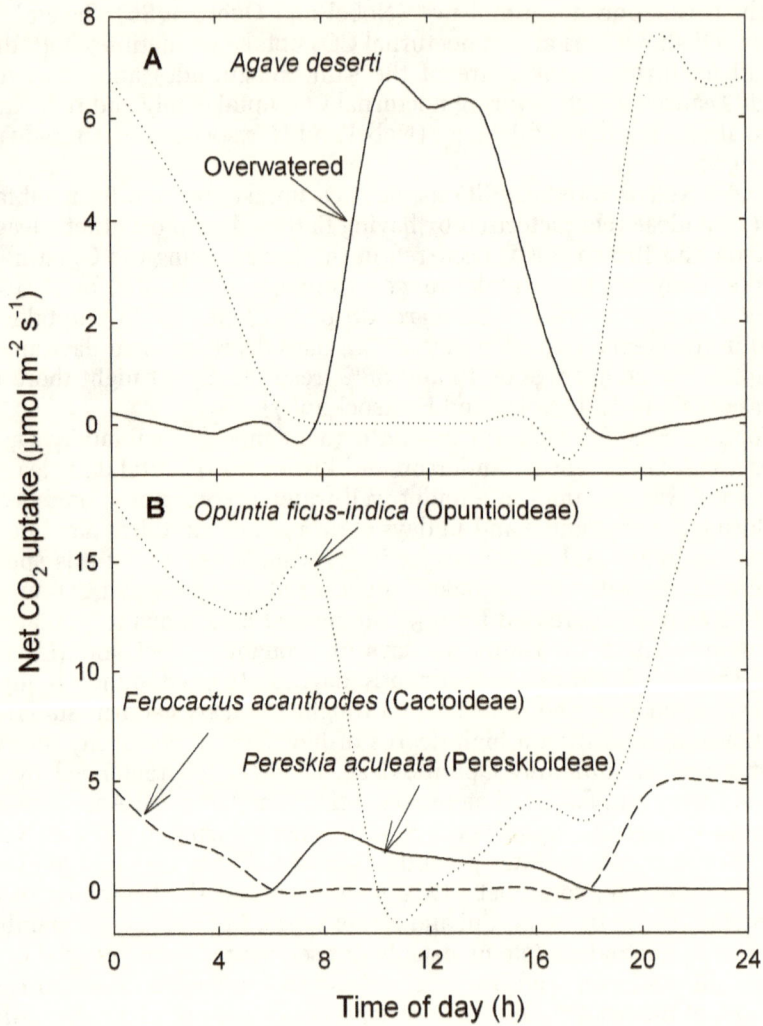

Figure 4.3. Gas exchange patterns for (A) *Agave deserti* and (B) cactus species from different subfamilies (modified from Nobel 1988 and Nobel and Bobich 2002).

Ziegler 1992), Portulacaceae (Martin and Zee 1983; Guralnick and Ting 1988; Herrera *et al.* 1991; Veste *et al.* 2001; Taisma and Herrera 2003), Ruschioideae (Rundel *et al.* 1999) and Vitaceae (Olivares *et al.* 1984). This drought-induced CAM is reversible upon rewatering. There is also a great deal of metabolic flexibility within the Bromeliaceae family. For the CAM bromeliad *Aechmea dactylina* in a high rainfall cloud forest in Panama, *in situ* measurements showed similar rates of nighttime net CO_2 uptake occurred during both the wet and dry seasons; however, the amount of CO_2 assimilated during the early morning and late afternoon was reduced as drought progressed (Pierce *et al.* 2002a). For *Tillandsia* species from a tropical dry forest, gas exchange measurements under controlled conditions showed that for *T. brachycaulos* 12% of the total carbon gain was fixed during daytime, whereas for *T. elongata*, 25% of the carbon was fixed in the late afternoon via C_3 photosynthesis (Graham and Andrade 2004). For both species, after six days of drought, daytime net CO_2 uptake was abolished. Also, nighttime CO_2 uptake was reduced as drought progressed. The shift in photosynthetic pattern is presumed to be an adaptive advantage, besides water conservation, in terms of photosynthetic efficiency, as decarboxylation of nocturnally accumulated malic acid should substantially increase the internal concentration of CO_2 during the day (Martin 1996; Haslam *et al.* 2003). This type of metabolic flexibility should be an adaptation to survive under natural adverse microclimatic scenarios. Other interesting find is the shift from C_4 to CAM that has been found in some members of the genus *Portulaca* (Sayed 1998; Guralnick and Jackson 2001; Lara *et al.* 2003), where enzyme activity and regulation are modified by drought. However, CAM and C_4 photosynthesis operate independently in these plants (Guralnick *et al.* 2002; Sage 2002).

Productivity of CAM plants

Plant productivity represents the cumulative, integrative effects of soil water potential, temperature and photosynthetic photon flux (PPF), as they relate to net CO_2 uptake and carbon storage (Nobel 1988). For instance, appreciable stomatal opening with its accompanying CO_2 uptake for *Agave deserti* only occurs when soil water potential is above −0.5 MPa (Nobel 1976). Also, seasonal changes in temperature had a negative effect on the total daily net CO_2 uptake over a 24-h period made by *Agave tequilana,* since individuals growing at a diurnal/nocturnal temperature of 25/15°C had a total daily net CO_2 uptake of 298 mmol m^{-2} and individuals growing at 35/25°C had a total of 84 mmol m^{-2} (Nobel *et al.* 1998). Additionally, consistent with the shaded habitat of *Hylocereus undatus*, total daily net CO_2 uptake is appreciable at a total daily PPF of only 2 mol m^{-2} d^{-1} and is maximal at 20 mol m^{-2} d^{-1}, above which photoinhibition reduces CO_2 uptake (Nobel and De la Barrera 2004).

All of the physical factors discussed above have an impact on productivity of CAM, which is generally associated with succulent species such as cacti that conserve water and are relatively slow growing (Osmond 1978).

For naturally growing wild CAM plants, CAM is generally a strategy for stress survival and not for high productivity and dominance (Lüttge 2004). This is the case of dwarf cacti, which have an average growth in height of less than 1 mm per year (Nobel 1988). In contrast, C_3 crops such as cassava and alfalfa can have dry weight yields above 30 Mg ha^{-1} year^{-1}. Also C_4 crops, such as sugar cane and corn can have yields between 50 and 70 Mg ha^{-1} year^{-1} (Nobel 1991). However, certain CAM crops such as *Agave four-croydes*, *A. salmiana* and *Opuntia ficus-indica* have similar high productivities under appropriate environmental conditions (Nobel 1991, 1996), relying on the metabolic flexibility of CO_2 uptake that allows them to fix CO_2 via C_3 during the late afternoon (Lüttge 2004).

The responses of plant productivity to soil water potential, air temperature, and photosynthetic photon flux have been characterized for various CAM species (Table 4.2), mainly crops, and these responses have been used to generate an environmental productivity index (EPI; Nobel 1984, 1985, 1986, 1991, 1996). Thus, EPI is a physiological indicator of expected growth and is defined as follows: EPI = Water index × Temperature index × PPF index, and ranges from 1.00 when all of the components are optimal for maximal net CO_2 uptake to 0.00 when at least one of the components is limiting net CO_2 uptake.

Since CO_2 uptake is expressed on leaf area basis for an agave or on a stem area basis for a cactus, and we can measure these areas, the environmental productivity index can be used to calculate CO_2 uptake per plant (Nobel 1988). Assuming that CO_2 is incorporated into carbohydrate and correcting for root respiration, and expressing the data on the basis of the ground area explored by the roots, productivity can be expressed per unit ground area. Plant productivity can be measured as the dry weight gained annually on an area basis, such as development of new leaves. Because EPIs can help predict plant responses in productivity to different environmental scenarios, they can be used to evaluate plant productivity along the distribution range of a given species or to predict plant productivity response to global climate change.

Water, seed germination, and seedling survival

In addition to its influence on net CO_2 uptake and productivity of mature CAM plants, water also influences seed germination and seedling establishment. Unpredictable rainfall, high temperatures and high solar radiation are the main environmental factors limiting seed germination and seedling establishment in arid and semi-arid environments (Franco and Nobel 1989; Rojas-Aréchiga *et al.* 1997; Rojas-Aréchiga and Vázquez-Yanes 2000). Indeed, seedling establishment is a rare and sporadic event since seedlings are often subject to high mortality rates because they may be less tolerant than adults to the extreme environmental fluctuations that occur at the soil surface (Dodd and Donovan 1999).

Table 4.2. Relation of CAM plants to which components of an environmental productivity index have been developed.

Species	Habitat	Reference
Agave deserti	Common agave in north American deserts	Nobel 1984
Agave fourcroydes	CAM crop from Yucatán, México.	Nobel 1985
Ferocactus acanthodes	Common cactus in north American deserts	Nobel 1986
Agave lechuguilla	Common agave in the Chihuahuan desert	Nobel and Quero 1986
Agave tequilana	CAM crop from Jalisco, México	Nobel and Valenzuela 1987
Opuntia ficus-indica	Widely cultivated CAM crop	Nobel 1991
Hylocereus undatus	Widely cultivated CAM crop	Nobel *et al.* 2002

We already know for a large number of cactus species that optimal temperature for germination ranges between 17°C and 34°C, with optimal values at 25°C (Nobel 1988). Also, the effect of light on the induction of seed germination has been studied for several species (Rojas-Aréchiga and Vázquez-Yanes 2000). Exposure to low light levels promotes germination for several globose cactus species (positive photoblastic), and light has no influence on germination of some columnar cacti (Rojas-Aréchiga *et al.* 1997; Flores *et al.* 2006). Germination of photoblastic seeds in arid environments is restricted to occur at the upper fast-drying soil, so the light requirement ensures that germination will take place only after heavy rainfalls that wet the soil for a longer time (Kigel 1995).

Despite this fact, few studies have investigated the effect of soil moisture on cactus seed germination. Germination is reduced approximately 30% for seeds of *Stenocereus thurberi* and *Mammillaria gaumeri* when incubated at −1.0 MPa compared to 0.0 MPa (De la Barrera and Nobel 2003; Cervera *et al.* 2006). Similar reductions occur for seeds of *Pachycereus pringlei* incubated at −1.3 MPa (Nolasco *et al.* 1996); also, germination was highest for three species of *Neobuxbaumia* near field capacity, however there was no germination at −1.0 MPa (Ramírez-Padilla and Valverde 2005). In contrast, seeds of *Pachycereus hollianus* did not germinate when incubated at 0.0 MPa and maximum germination occurred at −0.41 MPa (Flores and Briones 2001). Since rainy periods in deserts are usually of short duration and the top soil layers quickly desiccate and salt up (Kigel 1995), viable seeds have to uptake a sufficient quantity of water in order to germinate. During seed development of several South American cacti, epidermal cells produce a proteinaceous material that improves water uptake and ensures germination with a minimum amount of available water (Bregman and Graven 1997). Another mechanism to ensure seed germination is hydropriming, a memory of the internal changes induced in the seed by an hydration event, which can allow seeds to resist periods of desiccaation and germinate faster after rehydration (Dubrovsky 1996, 1998). Hydropriming also improved seedling survival (Dubrovsky 1996).

Cactus seedlings have higher probabilities of survival under the canopy of perennial shrubs, known as nurse plants (Jordan and Nobel 1981, 1982; Franco and Nobel 1989; Valiente-Banuet and Ezcurra 1991; Godínez-Álvarez *et al.* 1999; Esparza-Olguín *et al.* 2002; Flores *et al.* 2004; Cervera *et al.* 2006). Nurse plants provide protection against direct solar radiation, reduce extreme soil temperatures and most likely reduce soil water loss, meaning more water available, thereby increasing seedling survival and extending its distributional boundaries (Nobel 1980; Franco and Nobel 1989; Flores and Jurado 2003; Cervera *et al.* 2006). Most cacti are succulents that avoid drought by storing water within their tissues (Gibson and Nobel 1986; Ishikawa and Gusta 1996); however, since many cactus species have relatively slow growth rates they have a relatively small water storage capacity during the first growing season (Jordan and Nobel 1979, 1981). Thus, drought tolerance of seedlings is low because the volume to surface area ratio (a relationship of the surface over which water can be transpired to the volume of tissue which over water can be stored) for cactus seedlings is initially low relative to adults (Nobel 1988).

Exposure to high solar radiation increases seedling mortality (Flores *et al.* 2004; Cervera *et al.* 2006; Gallardo-Vásquez and De la Barrera 2007) primarily because of the increase in tissue temperature and in the water loss by transpiration (Gibson and Nobel 1986). The reduction in incoming PPF increased survival and water storage capacity of *Mammillaria gaumeri* seedlings in a coastal sand dune; the volume to surface area ratio was two-fold higher for seedlings receiving 20% of ambient PPF than those receiving 50% of ambient PPF (Cervera *et al.* 2006). At the end of the first growing season, the highest volume to surface area ratio for *M. gaumeri* seedlings was 0.085 cm, half of that volume to surface area ratio for *Ferocactus acanthodes* seedlings (Jordan and Nobel 1981; Cervera *et al.* 2006). Such *F. acanthodes* seedlings can lose 84% of their stored water and still survive, increasing the length of drought that they can tolerate. Even so, the relatively small volume to surface area ratio for seedlings makes this stage the most vulnerable to drought and most of them die at the start of the dry season (Nobel 1988). Data for *Carnegiea gigantea*, *F. acanthodes*, *M. gaumeri* and *Agave deserti* suggest that the establishment of succulents from arid and semi-arid regions may be limited to favourable years with dry seasons that are cooler or relatively wet (Jordan and Nobel 1979, 1981, 1982; Cervera *et al.* 2006).

As we have seen, rain is the most influential environmental variable for seedling survival in arid and semiarid-regions as the length of drought limits cactus seedling establishment to a few suitable years (Jordan and Nobel 1982). Additionally, extreme temperatures restrict seedling establishment and survival since they limit CO_2 uptake and also can damage cacti (Nobel 1988). Freezing temperatures can cause extracellular ice crystal formation in cacti, which draws water out of the cells and can lead to irreversible damage (Nobel 1982). Extremely high temperatures can denature proteins, degrade cell membranes, and disrupt metabolism in general

(Nobel 1988; Srinivasan *et al.* 1996). Acclimation to changes in diurnal/nocturnal temperatures between seasons can result in differences in tolerance to high and low extremes (Nobel and De la Barrera 2003), also potentially increasing survival. Adult individuals of desert dwarf cacti such as *Epithelantha bokei, Mammillaria lasiacantha* and *Ariocarpus fissuratus*, can withstand high temperatures of 64° (Nobel 1988). In contrast, in the tropical dry deciduous forest, seedlings of *Mammillaria gaumeri* and adult individuals of the climbing cactus *Hylocereus undatus* cannot tolerate temperatures above 50°C (Nobel *et al.* 2002; Cervera *et al.* 2006). Figure 4.4 shows the tolerance to extreme temperatures of *M. gaumeri* seedlings acclimated to three different temperature regimes. *M. gaumeri* seedlings are less tolerant to the extreme high temperatures that characterize the dry season, because they are acclimated to the cooler temperatures of the northwinds season (Cervera *et al.* 2006). Position in the canopy can also reduce the high temperature effects for CAM plants. For instance, co-occurring epiphytic bromeliads that grow in exposed locations in the canopy of the same forests, can dissipate better the heat since they are less succulent and can receive more water as precipitation and dew than individuals occurring close to the soil (Graham and Andrade 2004; Cervantes *et al.*, 2005), because soil and air temperatures can be as much as 10 °C higher than air ambient temperatures.

Conclusions and future prospects

Because CAM plants show a vast array of adaptations to their highly variable environment, especially in tropical forests where most CAM plants inhabit (Winter and Smith 1996; Lüttge 2004), we are discussing about multiple microenvironments for numerous life forms. Hence, studies on physiological responses of CAM plants to the environment are far from being complete, since most investigations have been done with adult members of only three families: Agavaceae, Bromeliaceae and Cactaceae, mostly from arid zones (except for the tropical bromeliads; Nobel 1988; Benzing 1990; Winter and Smith 1996; Zotz and Hietz 2001; Andrade *et al.* 2004; Lüttge 2004). Special attention should be put on the environmental physiology of CAM plant seedlings, from their photosynthetic pathway following germination to their survival in different microenvironments (Schmidt *et al.* 2001; Andrade *et al.* 2004; De la Barrera and Andrade 2005; Ayala-Cordero *et al.* 2006; Cervera *et al.* 2006; Gallardo-Vásquez and De la Barrera 2007; Hernández-González and Briones-Villarreal 2007).

Microenvironmental approaches also help to understand several ecological aspects for CAM plants. Many CAM plant species grow and reproduce more under certain microenvironments that allow them to get more water and adequate light (Cervantes *et al.* 2005; Andrade *et al.* 2006; Cervera *et al.* 2006, 2007); those individuals, which grow in other less suitable microenvironments, will get less carbon (less photosynthesis or higher respiration rates), will use photosynthesis products to procure water

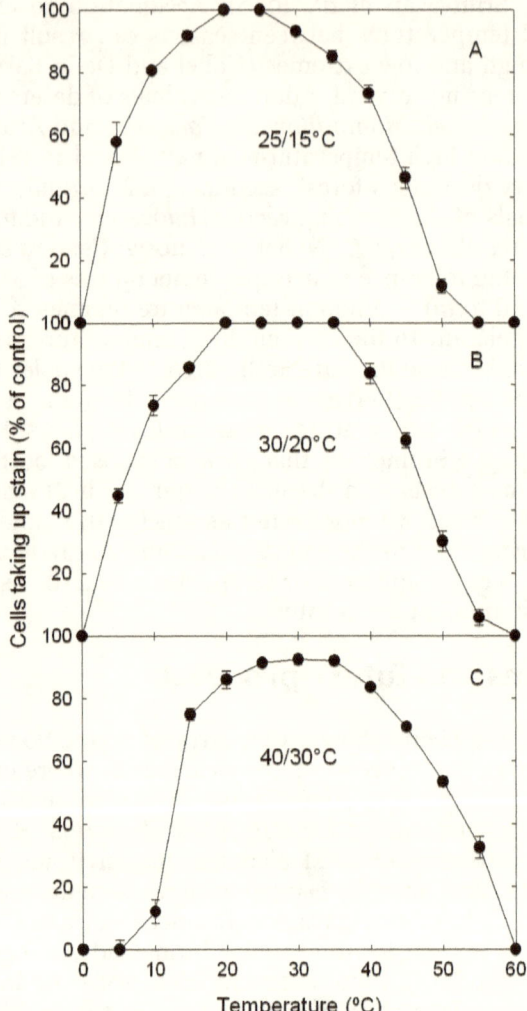

Figure 4.4. Effect of high and low temperature treatments on the uptake of neutral red into chlorenchyma cells of newly germinated *M. gaumeri* seedlings maintained in a growth chamber for 4 weeks at the indicated diurnal/nocturnal temperatures that characterize (A) northwinds season, (B) rainy season, (C) dry season. Samples were held at the indicated temperature for 60 min before the temperature was lowered by 5 °C for the low-temperature treatments or raised by 5 °C for the high-temperature treatments. Data are expressed relative to the control at 25 °C (for which 85% of the chorenchyma cells took up the stain) and are means ± s.e. (*n* = 5 seedlings). Lethal temperature (LT_{50}) was defined as that which halved stain uptake from the maximum occurring for the control (modified from Cervera *et al.* 2006).

and repair reaction centers and would have less defenses against predators and pathogens. So, demographic studies should take into consideration the factor microenvironment within their matrix data, since population growth can be sustained by the individuals in the more favorable microhabitats.

Also, correlations between microhabitats and morphological and physiological features should be taken into consideration. Many CAM plants take up CO_2 during the day according to the more water available and other environmental factors such as low nocturnal temperature and vapor pressure deficit. Because CO_2 uptake via ribulose 1, 5-diphosphate carboxilase/oxygenase (Rubisco) is related with more discrimination against $^{13}CO_2$ (compared to the more abundant $^{12}CO_2$), the comparison between the rates in $^{13}C/^{12}C$ in plant material has become an important tool to measure the relative amount of diurnal CO_2 uptake in CAM plants (Winter and Smith 1996; Winter *et al.* 2005). However, when both isotopic analysis and nocturnal acid accumulation are measured, a good number of species with 'weak' CAM can be detected within the group of C_3-type $\delta^{13}C$ plants (Pierce *et al.* 2002a; Holtum *et al.* 2004; Silvera *et al.* 2005). Further studies on these two groups of plants, strong and weak CAM species, in different microhabitats, are necessary to explore the relationships between microenvironments and CAM. One group of plants that would deserve more attention is the genus *Clusia*, which shows an interesting array of growth forms and physiological plasticity among its CAM species (Winter and Smith 1996; Holtum *et al.* 2004; Winter *et al.* 2005; Lüttge 2006).

Acknowledgements

The authors thank Klaus Winter for rewarding discussions, and the support by the University of California Institute for Mexico and the United States-Consejo Nacional de Ciencia y Tecnología (CONACYT) grant and the Fondo Mixto CONACYT-Gobierno del Estado de Yucatán grant, YUC-2003-C02-042.

Literature cited

Anderson EF and Brown R. 2001. *The cactus family*. Cambridge, UK, Timber Press.

Andrade JL. 2003. Dew deposition on epiphytic bromeliad leaves: an important event in a Mexican tropical dry deciduous forest. *Journal of Tropical Ecology* **19:** 479-488.

Andrade JL, Graham EA, and Zotz G. 2004. Determinantes morfofisiológicos y ambientales de la distribución de epifitas en el dosel de bosques tropicales. In: Marino-Cabrera, H (ed.) *Fisiología Ecológica en Plantas*. Ediciones Universitarias de Valparaíso, Valparaíso, Chile. Pp. 139-156.

Andrade JL and Nobel PS. 1996. Habitat, CO_2 uptake and growth for the CAM epiphytic cactus *Epiphyllum phyllanthus* in a Panamanian tropical forest. *Journal of Tropical Ecology* **12:** 291-306.

Andrade JL, Rengifo E, Ricalde MF, Simá JL, Cervera JC, and Vargas-Soto G. 2006. Microambientes de luz, crecimiento y fotosíntesis de la pitahaya (*Hylocereus undatus*) en un agrosistema de Yucatán, México. *Agrociencia* **40:** 687-697.

Andrade JL and Nobel PS. 1997. Microhabitats and water relations of epiphytic cacti and ferns in a lowland neotropical forest. *Biotropica* **29**: 261-270.

Ayala-Cordero G, Terrazas T, López-Mata L, and Trejo C. 2006. Morpho-anatomical changes and photosynthetic metabolism of *Stenocereus beneckei* seedlings under soil water deficit. *Journal of Experimental Botany* **57**: 3165-3174.

Ball E, Hann J, Kluge M, Lee HSJ, Lüttge U, Orthen B, Popp M, Schmitt A, and Ting IP. 1991. Ecophysiological comportment of the tropical CAM-tree Clusia in the field. II. Modes of Photosynthesis in trees and seedlings. *New Phytologist* **117**: 483-491.

Barcikowski W and Nobel PS. 1984. Water relations of cacti during desiccation: distribution of water in tissues. *Botanical Gazette* **145**: 110-115.

Benzing DH. 1990. *Vascular Epiphytes*. Cambridge University Press, Cambridge, MA.

Benzing DH and Dahle CE. 1971. The vegetative morphology, habitat preference and water balance mechanisms of the bromeliad *Tillandsia ionantha* Planch. *The American Midland Naturalist* **85**: 11-21.

Bregman R and Graven P. 1997. Subcuticular secretion by cactus seeds improves germination by means of rapid uptake and distribution of water. *Annals of Botany* **80**: 525-531.

Cavelier J and Goldstein G. 1989. Mist and fog interception in elfin cloud forests in Colombia and Venezuela. *Journal of Tropical Ecology* **5**: 309-322.

Cervantes SE, Graham EA, and Andrade JL. 2005. Light microhabitats, growth and photosynthesis of an epiphytic bromeliad in a tropical dry forest. *Plant Ecology* **179**: 107-118.

Cervera JC, Andrade JL, Simá JL, and Graham EA. 2006. Microhabitats, germination, and establishment for *Mammillaria gaumeri* (Cactaceae), a rare species from Yucatan. *International Journal of Plant Sciences* **167**: 311-319.

Cervera JC, Andrade JL, Graham EA, Durán R, Jackson PC, and Simá JL. 2007. Photosynthesis and optimal light microhabitats for a rare cactus, *Mammillaria gaumeri*, in two tropical ecosystems. *Biotropica* **39**: 620-627.

Cui M, Miller PM, and Nobel PS. 1993. CO_2 exchange and growth of the Crassulacean acid metabolism plant *Opuntia ficus-indica* under elevated CO_2 in open-top chambers. *Plant Physiology* **103**: 519-524.

Cui M and Nobel PS. 1994. Gas exhange and growth responses to elevated CO_2 and light levels for the CAM species *Opuntia ficus-indica*. *Plant, Cell and Environment* **17**: 935-944.

Cushman JC. 2001. Crassulacean acid metabolism. A plastic photosynthetic adaptation to arid environments. *Plant Physiology* **127**: 1439-1448.

De la Barrera E and Andrade JL. 2005. Challenges to plant megadiversity: how environmental physiology can help. *New Phytologist* **167**: 5-8.

De la Barrera E and Nobel PS. 2003. Physiological ecology of seed germination for the columnar cactus *Stenocereus queretaroensis*. *Journal of Arid Environments* **53**: 297-306.

De Mattos EA and Lüttge U. 2001. Chlorophyll fluorescence and organic acid oscillations during transition from CAM to C_3-photosynthesis in *Clusia minor* L. (Clusiaceae). *Annals of Botany* **88**: 457-463.

Dodd AN, Borland AM, Haslam RP, Griffiths H, and Maxwell K. 2002. Crassulacean acid metabolism: plastic, fantastic. *Journal of Experimental Botany* **53**: 569-580.

Dodd GL and Donovan LA. 1999. Water potential and ionic effects on germination and seedling growth of two cold desert shrubs. *American Journal of Botany* **86**: 1146-1153.

Drennan PM and Nobel PS. 2000. Resposes of CAM species to increasing atmospheric CO_2 concentrations. *Plant, Cell & Environment* **23:** 767-781.

Dubrovsky JG. 1996. Seed hydration memory in Sonoran Desert cacti and its ecological implication. *American Journal of Botany* **83:** 624-632.

Dubrovsky JG. 1998. Discontinuous hydration as a facultative requirement for seed germination in two cactus species of the Sonoran Desert. *Journal of the Torrey Botanical Society* **125:** 33-39.

Dubrovsky JG and Gómez-Lomelí LF. 2003. Water deficit accelerates determinate developmental program of the primary root and does not affect lateral root initiation in a Sonoran Desert cactus (*Pachycererus pringlei*, Cactaceae). *American Journal of Botany* **90:** 823-831.

Esparza-Olguín L, Valverde T, and Vilchis-Anaya E. 2002. Demographic analysis of a rare columnar cactus (*Neobuxbaumia macrocephala*) in the Tehuacán Valley, Mexico. *Conservation Biology* **103:** 349-359.

Flores J and Briones O. 2001. Plant life-form and germination in a Mexican intertropical desert: effects of soil water potential and temperature. *Journal of Arid Environments* **47:** 485-497.

Flores J, Briones O, Flores A, and Sánchez-Colón S. 2004. Effect of predation and solar exposure on the emergence and survival of desert seedlings of contrasting life-forms. *Journal of Arid Environments* **58:** 1-18.

Flores J and Jurado E. 2003. Are nurse-protégé interactions more common among plants from arid environments? *Journal of Vegetation Science* **14:** 911-916.

Flores J, Jurado E, and Arredondo A. 2006. Effect of light on germination of seeds of cactaceae from the Chihuahuan desert, México. *Seed Science Research* **16:** 149-155.

Franco AC and Nobel PS. 1989. Effect of nurse plants on the microhabitat and growth of cacti. *Journal of Ecology* **77:** 870-886.

Gallardo-Vásquez JC and De la Barrera E. 2007. Environmental and ontogenetic influences on growth, photosynthesis, and survival for young pitayo (*Stenocereus queretaroensis*) seedlings. *Journal of the Professional Association for Cactus Development* **9:** 118–135.

Gibson A and Nobel PS. 1986. *The Cactus Primer*. Harvard University Press, Cambridge, Massachussetts, USA.

Godínez-Álvarez H, Valiente-Banuet A, and Banuet LV. 1999. Biotic interactions and the population dynamics of the long-lived columnar cactus *Neobuxbaumia tetetzo* in the Tehuacán Valley, México. *Canadian Journal of Botany* **77:** 203-208.

Goldstein G, Andrade JL, and Nobel PS. 1991. Differences in water relations parameters for the chlorenchyma and the parenchyma of *Opuntia ficus-indica* under wet *versus* dry conditions. *Australian Journal of Plant Physiology* **18:** 95-107.

Graham EA and Andrade JL. 2004. Drought tolerance associated with vertical stratification of two co-occurring epiphytic bromeliads in a tropical dry forest. *American Journal of Botany* **91:** 699-706.

Griffths H, Lüttge U, Stimmel K.-H, Crook CE, Griffths NM, and Smith JAC. 1986. Comparative ecophysiology of CAM and C_3 bromeliads. III.Enviromental influences on CO_2 assimilation and transpiration. *Plant, Cell and Environment* **9:** 385-393.

Guralnick LJ and Ting IP. 1988. Physiological changes in *Portulacaria afra* (L.) Jacq. during a summer drought and rewatering. *Plant Physiology* **85:** 481-486.

Guralnick LJ, Edwards G, Ku MSB, Hockema B, and Franceschi BR. 2002. Photo-synthetic and anatomical characteristics in the C_4-crassulacean acid metabo-lism-cycling plant, *Portulaca grandiflora. Functional Plant Biology* **29**: 763-773.

Guralnick LJ and Jackson MD. 2001. The occurrence and phylogenetics of crassu-lacean acid metabolism in Portulacaceae. *International Journal of Plant Sci-ences* **162**: 257-262.

Hanscom ZI and Ting IP. 1978. Responses of succulents to plant water stress. *Plant Physiology* **61**: 327-330.

Haslam R, Borland A, Maxwell K, and Griffiths H. 2003. Physiological responses of the CAM epiphyte *Tillandsia usneoides* L. (Bromeliaceae) to variations in light and water supply. *Journal of Plant Physiology* **160**: 627-634.

Helbsing S, Riederer M, and Zotz G. 2000. Cuticles of vascular epiphytes: efficient barriers for water loss after stomatal closure? *Annals of Botany* **86**: 765-769.

Hernández-González O and Briones-Villarreal O. 2007. CAM photosynthesis in columnar cactus seedlings during ontogeny: the effect of light on nocturnal acidity accumulation and chlorophyll fluorescence. *American Journal of Bot-any* **94**: 1344-1351.

Herrera A, Delgado J, and Paraguatey I. 1991. Occurrence of inducible Crassulacean acid metabolism in leaves of *Talinum triangulare* (Portulacaceae). *Journal of Experimental Botany* **42**: 493-499.

Holtum JAM and Winter K. 1999. Degrees of Crassulacean acid metabolism in tropical epiphytic and lithophytic ferns. *Australian Journal of Plant Physiology* **26**: 749-757.

Holtum JAM, Aranda J, Virgo A, Gehring HH, and Winter K. 2004. $\delta^{13}C$ values and crassulacean acid metabolism in *Clusia* species from Panama. *Trees* **18**: 658-668.

Ishikawa M and Gusta LV. 1996. Freezing and heat tolerance of *Opuntia* cacti na-tive to the Canadian prairie provinces. *Canadian Journal of Botany* **74**: 1890-1895.

Johnson GN, Young AJ, Scholes JD, and Horton P. 1993. The dissipation of excess excitation-energy in British plant-species. *Plant, Cell and Environment* **16**: 673-679.

Jonzén N, Lundberg P, Ranta E, and Kaitala V. 2002. The irreducible uncertainty of the demography-environment interaction in ecology. *Proceedings of the Royal Society of London Series B-Biological Sciences* **269**: 221-225.

Jordan PW and Nobel PS. 1979. Infrequent establishment of seedlings of *Agave deserti* (Agavaceae) in the northwestern Sonoran Desert. *American Journal of Botany* **66**: 1079-1084.

Jordan PW and Nobel PS. 1981. Seedling establishment of *Ferocactus acanthodes* in relation to drought. *Ecology* **62**: 901-906.

Jordan PW and Nobel PS. 1982. Height distributions of two species of cacti in rela-tion to rainfall, seedling establishment and growth. *Botanical Gazette* **143**: 511-517.

Keeley JE. 1996. Aquatic CAM photosynthesis. In: Winter K and Smith JAC (eds.) *Crassulacean Acid Metabolism. Biochemistry, Ecophysiology and Evolution.* Springer, Berlin. Pp. 281-295.

Keeley JE and Rundel PW. 2003. Evolution of CAM and C_4 carbon-concentrating mechanisms. *International Journal of Plant Sciences* **164**: 55-77.

Kigel J. 1995. Seed germination in arid and semiarid regions. In: Kigel J and Galili G. (eds.) *Seed development and germination.* New York, USA, Marcel Dekker Inc. Pp. 645-699.

Kluge M, Razanoelisoa B, and Brulfert J. 2001. Implications of genotypic diversity and phenotypic plasticity in the ecophysiological success of CAM plants, examined by studies on the vegetation of Madagascar. *Plant Biology* 3: 214-222.

Kondo A, Kaikawa J, Funaguma T, and Ueno O. 2004. Clumping and dispersal of chloroplasts in succulent plants. *Planta* 219: 500-506.

Lara MV, Disante KB, Podestá FE, Andreo CS, and Drincovich MF. 2003. Induction of a Crassulacean acid like metabolism in the C_4 succulent plant, *Portulaca oleraceae* L.: physiological and morphological changes are accompanied by specific modifications in phosphoenolpyruvate carboxylase. *Photosynthesis Research* 77: 241-254.

Lu CM, Qiu NW, Lu QT, Wang BS, and Kuang TY. 2003. PSII photochemistry, thermal energy dissipation, and the xanthophyll cycle in *Kalanchoe daigremontiana* exposed to a combination of water stress and high light. *Physiologia Plantarum* 118: 173-182.

Lüttge U. 1989. *Vascular Plants as Epiphytes*. Springer, Berlin.

Lüttge U. 2004. Ecophysiology of Crassulacean acid metabolism (CAM). *Annals of Botany* 93: 629-652.

Lüttge U. 2006. Photosynthetic flexibility and ecophysiological plasticity: questions and lessons from *Clusia*, the only CAM tree in the neotropics. *New Phytologist* 171: 7-25.

Martorell C and Ezcurra E. 2002. Rosette scrub occurrence and fog availability in arid mountains of Mexico. *Journal of Vegetation Science* 13: 651-662.

Martin CE. 1996. Putative causes and consequences of recycling CO_2 via Crassulacean acid metabolism. In: Winter K and Smith JAC (eds.) *Crassulacean Acid Metabolism. Biochemistry, Ecophysiology and Evolution*. Berlin, Germany, Springer. Ecological Studies. Pp. 192-203.

Martin CE and Zee AK. 1983 C_3 photosynthesis and Crassulacean acid metabolism in a Kansas rock outcrop succulent, *Talinum calycinum* Engelm (Portulacaceae). *Plant Physiology* 86: 562-568.

Maxwell K, Griffiths H, Borland A, Young A, Broadmeadow M, and Fordham C. 1995. Short-term photosynthetic responses of the C_3-CAM epiphyte *Guzmania monostachya* var. *monostachya* to tropical seasonal transitions under field conditions. *Australian Journal of Plant Physiology* 22: 771-781.

Maxwell K and Johnson G. 2000. Chlorophyll fluorescence—a practical guide. *Journal of Experimental Botany* 51: 659-668.

Nobel PS. 1976. Water relations and photosynthesis of a desert CAM plant, *Agave deserti*. *Plant Physiology* 58: 576-582.

Nobel PS. 1980. Morphology, nurse plants, and minimal apical temperatures for young *Carnegeia gigantea*. *Botanical Gazette* 141: 188-191.

Nobel PS. 1981. Influence of freezing temperature on a cactus, *Coryphantha vivipara*. *Oecologia* 48: 194-198.

Nobel PS. 1982. Low-temperature tolerance and cold hardening of cacti. *Ecology* 63: 1650-1656.

Nobel PS. 1984. Productivity of *Agave deserti:* measurement by dry weight and monthly prediction using physiological responses to environmental parameters. *Oecologia* 64: 1-7.

Nobel PS. 1985. PAR, water and temperature limitations on the productivity of cultivated *Agave fourcroydes* (henequen). *Journal of Applied Ecology* 22: 157-173.

Nobel PS. 1986. Relation between monthly growth of *Ferocactus acanthodes* and an envirironmental productivity index. *American Journal of Botany* 73: 541-547.

Nobel PS. 1988. *Environmental Biology of Agaves and Cacti*. Cambridge University Press, Nueva York.

Nobel PS. 1991. Achievable productivities of certain CAM plants: basis for high values compared with C_3 and C_4 plants. *New Phytologist* **119**: 183-205.

Nobel PS. 1994. *Remarkable Agaves and Cacti*. Oxford University Press, New York.

Nobel PS. 1996. High productivity of certain agronomic CAM species. In: Winter K and Smith JAC (eds.) *Crassulacean Acid Metabolism. Biochemistry, Ecophysiology and Evolution*. Springer, Berlin. Pp. 255-265.

Nobel PS and Bobich EG. 2002. Environmental Biology. In: Nobel PS. *Cacti: Biology and Uses*. Berkeley, California, USA, University of California Press. Pp. 57-74.

Nobel PS, Cavelier J, and Andrade JL. 1992a. Mucilage in cacti: its apoplastic capacitance, associated solutes, and influence on tissue water relations. *Journal of Experimental Botany* **43**: 641-648.

Nobel PS, Castañeda M, North GB, Pimienta-Barrios E, and Ruiz A. 1998. Temperature influences on leaf CO_2 exchange, cell viability and cultivation range for *Agave tequilana*. *Journal of Arid Environments* **39**: 1-9.

Nobel PS and Cui M. 1992. Hydraulic conductances of the soil, the root-soil air gap, and the root: changes for desert succulents in drying soil. *Journal of Experimental Botany* **43**: 319-326.

Nobel PS and De la Barrera E. 2003. Tolerances and acclimation to low and high temperatures for cladodes, fruits and roots of a widely cultivated cactus, *Opuntia ficus-indica*. *New Phytologist* **157**: 271-279.

Nobel PS and De la Barrera E. 2004. CO_2 uptake by the cultivated hemiepiphytic cactus, *Hylocereus undatus*. *Annals of Applied Biology* **144**: 1-8.

Nobel PS, De la Barrera E, Beilman DW, Doherty JH, and Zutta BR. 2002. Temperature limitations for cultivation of edible cacti in California. *Madroño* **49**: 228-236.

Nobel PS and Jordan PW. 1983. Transpiration stream of desert species: resistances and capacitances for a C3, a C4 and CAM plant. *Journal of Experimental Botany* **34**: 1379-1391.

Nobel PS and Hartsock TM. 1987. Drought-induced shifts in daily CO_2 uptake patterns from leafy cacti. *Physiologia Plantarum* **70**: 114-118.

Nobel PS and Hartsock TM. 1990. Diel patterns of CO_2 exchange for epiphytic cacti differing in succulence. *Physiologia Plantarum* **78**: 628-634.

Nobel PS, Miller PM, and Graham EA. 1992b. Influence of rocks on soil temperature, soil water potential, and rooting patterns for desert succulents. *Oecologia* **92**: 90-96.

Nobel PS and North GB. 1996. Features of roots of CAM plants. In: Winter K and Smith JAC (eds) *Crassulacean Acid Metabolism. Biochemistry, Ecophysiology and Evolution*. Springer, Berlin. Pp. 266-280.

Nobel PS and Pimienta-Barrios E. 1995. Monthly stem elongation for *Stenocereus queretaroensis*: relationships to environmental conditions, net CO_2 uptake and seasonal variations in sugar content. *Environmental and Experimental Botany* **35**: 17-24.

Nobel PS and Quero E. 1986. Environmental productivity indices for a Chihuahuan desert CAM plant, *Agave lechuguilla*. *Ecology* **67**: 1-11.

Nolasco H, Vega-Villasante F, Romero-Schmidt HL, and Díaz-Rondero A. 1996. The effect of light, salinity, acidity, light and temperature on the germination of seeds of cardón (*Pachycereus pringlei* (S. Wats.) Britton & Rose (Cactaceae). *Journal of Arid Environments* **33**: 87-94.

North GB and Nobel PS. 1994. Changes in root hydraulic conductivity for two tropical epiphytic cacti as soil moisture varies. *American Journal of Botany* **81:** 46-
53.

Olivares E, Urich R, Montes G, Coronel I, and Herrera A. 1984. Occurrence of Crassulacean acid metabolism in *Cissus trifoliate* (L.) Vitaceae). *Oecologia* **61:** 358-
362.

Osmond CB. 1978. Crassulacean acid metabolism: a curiosity in context. *Annual
Review of Plant Physiology* **29:** 379-414.

Patten DT and Dinger BE. 1969. Carbon dioxide exchange patterns of cacti from
different environments. *Ecology* **50:** 686-688.

Pierce S, Winter K, and Griffiths H. 2002a. Carbon isotope ratio and the extent of
daily CAM use by Bromeliaceae. *New Phytologist* **156:** 75-83.

Pierce S, Winter K, and Griffiths H. 2002b. The role of CAM in high rainfall cloud
forests: an *in situ* comparison of photosynthetic pathways in Bromeliaceae.
Plant, Cell and Environment **25:** 1181-1189.

Ramachandra RA, Sundar D, and Gnanam A. 2003. Photosynthetic flexibility in
Pedilanthus tithymaloides Poit, a CAM plant. *Journal of Plant Physiology* **160:**
75-80.

Ramírez-Padilla CA and Valverde T. 2005. Germination response of three con-
generic cactus species (*Neobuxbaumia*) with different degrees of rarity. *Journal
of Arid Environments* **61:** 333-343.

Rojas-Aréchiga M, Orozco-Segovia A, and Vázquez-Yanes C. 1997. Effect of light on
germination of seven species of cacti from the Zapotitlán Valley in Puebla,
México. *Journal of Arid Environments* **36:** 571-578.

Rojas-Aréchiga M and Vázquez-Yanes C. 2000. Cactus seed germination: a review.
Journal of Arid Environments **44:** 85-104.

Rundel PW, Esler KJ, and Cowling RM. 1999. Ecological and phylogenetic patterns
of carbon isotope discrimination in the winter-rainfall flora of the Richtersveld,
South Africa. *Plant Ecology* **142:** 133-148.

Sage R. 2002. Are Crassulacean acid metabolism and C_4 photosynthesis incompatible? *Functional Plant Biology* **29:** 775-785.

Sayed OH. 1998. Phenomorphology and ecophysiology of desert succulents in eastern Arabia. *Journal of Arid Environments* **40:** 177-189.

Schmidt G, Stuntz S, and Zotz G. 2001. Plant size: an ignored parameter in epiphyte
ecophysiology? *Plant Ecology* **153:** 65-72.

Silvera K, Santiago LS, and Winter K. 2005. Distribution of Crassulacean acid metabolism in orchids of Panama: evidence of selection for weak and strong
modes. *Functional Plant Biology* **32:** 397-407.

Sinclair R. 1983. Water relations of tropical epiphytes. II. Performance during
droughting. *Journal of Experimental Botany* **34:** 1664-1675.

Smirnoff N. 1996. Regulation of Crassulacean acid metabolism by water status in
the C_3/CAM intermediate *Sedum telephium*. In: Winter K and SmithJAC (eds.)
Crassulacean Acid Metabolism. Biochemistry, Ecophysiology and Evolution.
Springer, Berlin. Pp. 176-191.

Smith JAC, Griffiths H, and Lüttge U. 1986. Comparative ecophysiology of CAM
and C_3 bromeliads. I. The ecology of the Bromeliaceae in Trinidad. *Plant, Cell
and Environment* **9:** 359-376.

Srinivasan A, Takeda H, and Senboku T. 1996. Heat tolerance in food legumes as
evaluated by cell membrane thermostability and chlorophyll fluorescence techniques. *Euphytica* **88:** 35-45.

Stiles KC and Martin CE. 1996. Effects of drought stress in CO_2 exchange and water
relations in the CAM epiphyte *Tillandsia utriculata* (Bromeliaceae). *Journal of
Plant Physiology* **149:** 721-728.

Szarek SR and Ting IP. 1975. Photosynthetic efficiency of CAM plants in relation to C_3 and C_4 plants. In: Marcelle R (ed.) *Environmental and Biological Control of Photosynthesis*. Dr. W. Junk, The Haghe. Pp. 289-297.

Taisma MA and Herrera A. 2003. Drought under natural conditions affects leaf properties, induces CAM and promotes reproduction in plants of *Talinum triangulare*. *Interciencia* **28**: 292-297.

Ting IP. 1985. Crassulacean acid metabolism. *Annual Review of Plant Physiology* **36**: 595-622.

Valiente-Banuet A and Ezcurra E. 1991. Shade as the cause of association between the cactus *Neobuxbaumia tetetzo* and the nurse plant *Mimosa luisana* in the Tehuacán Valley, Mexico. *Journal of Ecology* **79**: 961-971.

Veste M, Herppich WB, and von Willert DJ. 2001. Variability of CAM in leaf-deciduous succulents from the Succulent Karoo (South Africa). *Basic and Applied Ecology* **2**: 283-288.

Von Willert DJ, Eller BM, Werger MJA, Brinckmann E, and Ihlenfeldt HD. 1992. *Life Strategies of Succulents in Deserts*. Cambridge Studies in Ecology. Cambridge University Press, Cambridge.

Windsor DM. 1990. *Climate and Moisture Variability in a Tropical Forest: Long-Term Records from Barro Colorado Island, Panama*. Smithsonian Contributions to the Earth Sciences, Number 29. Smithsonian Institution Press, Washington, DC.

Winter K, Aranda J, and Holtum JAM. 2005. Carbon isotope composition and water- use efficiency in plants with Crassulacean acid metabolism. *Functional Plant Biology* **32**: 381-388.

Winter K and Holtum JAM. 2002. How closely do the $\delta^{13}C$ values of Crassulacean acid metabolism plant reflect the proportion of CO_2 fixed during day and night? *Plant Physiology* **129**: 1843-1851.

Winter K and Holtum JAM. 2005. The effects of salinity, Crassulacean acid metabolism, and plant age on the carbon isotope composition of *Mesembryanthemum crystalinum* L., a halophytic C_3-CAM species. *Planta* **222**: 201-209.

Winter K and Smith JAC. 1996. An introduction to Crassulacean acid metabolism: biochemical principles and biological diversity. In: Winter K and Smith JAC (eds) *Crassulacean Acid Metabolism. Biochemistry, Ecophysiology and Evolution*. Springer, Berlin. Pp. 1-13.

Winter K, Wallace BJ, Stocker GC, and Roksandic Z. 1983. Crassulacean acid metabolism in Australian vascular epiphytes and some related species. *Oecologia* **57**: 129-141.

Winter K and Ziegler H. 1992. Induction of Crassulacean acid metabolism in *Mesembryanthemum crystallinum* increases reproductive success under conditions of drought and salinity stress. *Oecologia* **92**: 475-479.

Zotz G. 1997. Substrate use of three epiphytic bromeliads. *Ecography* **20**: 264-270.

Zotz G and Andrade JL. 1998. Water relations of two co-occurring epiphytic bromeliads. *Journal of Plant Physiology* **152**: 545-554.

Zotz G and Hietz P. 2001. The physiological and ecology of vascular epiphytes: current knowledge, open questions. *Journal of Experimental Botany* **52**: 2067-2078.

Chapter 5

ECOPHYSIOLOGICAL STUDIES OF PERENNIALS OF THE BROMELIACEAE FAMILY IN A DRY FOREST: STRATEGIES FOR SURVIVAL

Casandra Reyes-García and Howard Griffiths

Introduction
The family Bromeliaceae
Anatomy in the Bromeliaceae and the interaction with
  water and light
  Leaf structure
  Trichomes
  Rooting in bromeliads
Crassulacean acid metabolism
  as a water saving mechanism
High humidity and its role in the survival of perennials
  in a seasonally dry forest
  The effect of transpiration on leaf water
    enrichment in δ¹⁸O
  Oxygen isotopic signatures across the leaf
  High humidity and survival of
    perennial bromeliads

Perspectives in Biophysical Plant Ecophysiology: A Tribute to Park S. Nobel, pp. 121-151
Edited by: E. De la Barrera and W.K. Smith
© 2009 by The Authors
Book Compilation © 2009 Universidad Nacional Autónoma de México

Introduction

Seasonally dry tropical forests constitute environments that can have a shift in the dominant life forms during the year. These forests are defined as being frost free, having an annual rainfall of 250 to 2000 mm and several months of drought (Mooney *et al.* 1995). Seasonal forests have a lower species diversity than moist forests, but a higher structural (plant habit) and physiological diversity (Medina 1995). The forest of Chamela, Mexico, represents one of the most diverse seasonally dry tropical forests (Lott 1993) and is one of the most seasonal regarding precipitation (Bullock 1986). During the 8 months of dry season at Chamela, the landscape appears dominated by families that exhibit Crassulacean acid metabolism (CAM), a metabolism that conserves water by performing gas exchange at night. Only a few C_3 shrubs remain active, while more than 95 % of the trees shed their leaves (Bullock and Solís-Magallanes 1990; Martínez-Yrízar and Sarukhán 1990).

Out of the 950 plant species present at Chamela, 60 exhibit CAM photosynthesis. These perennials are found in four families: Agavaceae (four species), Bromeliaceae (26 species), Cactaceae (19 species) and Orquidaceae (11 species). At the driest upland site, the Bromeliaceae family is represented by three terrestrial and ten epiphytic species; the epiphytes all belonging to the Tillandsioideae sub-family (Reyes-García *et al.* 2008), the most drought resistant (Pittendrigh 1948). In this chapter we will explore strategies of the perennial plants in the family Bromeliaceae to survive and thrive during the prolonged drought by reviewing previous studies using stable isotopes and other ecophysiological tools. Even though the physiology of these plants has been well documented and specific microclimatic requirements for different species have been described, few attempts have been made to develop a niche theory among coexisting species. This leaves the open question on whether the distribution of the species in the habitats responds only to the history and dispersion of the species. Yet, because niche theory has been well developed in dry and seasonal desert communities, which can have great resemblance with the canopy microenvironment; we will compare those case studies and theories of water use strategies to those observed in the Bromeliaceae community of Chamela.

The family Bromeliaceae

The family Bromeliaceae constitutes a monophyletic group of monocots with terrestrial and epiphytic species (Smith *et al.* 1986; Givnish *et al.* 2004) distributed throughout tropical America (Smith and Downs 1974, 1977, 1979). The genus *Tillandsia* is the most diverse of the family. It is represented by epiphytic species, most of which exhibit CAM (Smith *et al.* 1986; Crayn *et al.* 2004). A distinctive character that has allowed this family to successfully exploit the epiphytic habitat is the presence of complex trichomes (Pittendrigh 1948; Benzing 1976; Benzing *et al.* 1978; Adams and Martin 1986b; Nyman *et al.* 1987; Benz and Martin 2006). The characteristics of the bromeliad trichome vary progressively from water repellent trichomes present in the terrestrial species *Pitcairnia recurvata* that still rely on roots for nutrition, to trichomes that absorb water and nutrients in epiphytes such as the species *Tillandsia intermedia* (Nyman *et al.* 1987; Pierce *et al.* 2001). These trichomes allow the epiphytes to take up nutrients and water from precipitation and canopy leachates that can contain amino acids, phosphate and other nutrients (Benzing and Renfrow 1980).

Pittendrigh (1948) classified the Bromeliaceae based on their strategy for resource acquisition dividing them into four categories:

Type I Soil-Root: terrestrial species that take water and nutrients from the soil through the roots.

Type II Tank-Root: species that have a small tank where some water and organic matter can accumulate. These species possess functional roots that invade the base of the tank.

Type III Tank-Absorbing Trichome: epiphytic species with a well developed tank where water and organic matter is stored. The trichomes are the absorptive structures, substituting the roots. They are mainly localized at the base of the leaf.

Type IV Atmosphere-Absorbing Trichome: epiphytic species with poorly developed tanks or lacking them. The absorptive trichomes cover the whole leaf. The roots are only used for attaching the plant to its host.

In these categories there is a gradual detachment from the use of roots as absorptive structures to the use of trichomes, as well as from the use of the soil as a source of nutrients to the use of precipitation. In a more recent review, Benzing (2000) divides the type III-tank into 2 types, one predominantly C_3, all belonging to the Bromelioideae, and a second group mostly CAM, belonging mainly to the Tillandsioideae sub-family. In the present chapter, we will use a simplified classification, and refer to the species as terrestrial , tank, or atmospheric.

Anatomy in the Bromeliaceae and the interaction with water and light

Leaf structure

The family Bromeliaceae has a vegetative form which has an overrepresentation of leaves as opposed to other structures. Leaves carry out many functions, particularly in epiphytes where they absorb water and nutrients. These leaves usually present xerophytic characters to maintain the water absorbed during precipitation events. Among these characters we can find: a thick cuticle, a mesophyll with a differentiated hypodermis and chlorenchyma, a dense trichome cover and low stomatal size and densities (Benzing 1990, 2000; Nowak and Martin 1997).

The leaves have a distinct water storage tissue layer predominantly on the adaxial (upper) side, occupying a larger area in the family Bromelioideae (the terrestrial group used in this study). The abaxial (lower) side is dominated by the chlorenchyma that is distributed around air canals which run longitudinally in terrestrial and tank species in parallel with leaf veins (Benzing 2000). Water movement from water storage tissue to chlorenchyma, in bromeliads, as in other succulent species (Goldstein *et al.* 1991; Tissue *et al.* 1991; Nowak and Martin 1997; Nobel 2006), allows photosynthetic performance to be maintained under low leaf relative water content.

Trichomes

Trichomes have diverse functions. The most conspicuous being herbivore defense, excretion of aromatic compounds, and light reflection (Lüttge *et al.* 1986; Willmer and Ficker 1996; Nowak and Martin 1997; Pérez-Estrada *et al.* 2000; Gassmann and Hare 2005; Benz and Martin 2006). In epiphytic bromeliads, the trichomes have an important role in water uptake, nutrition and light reflection. Lüttge *et al.* (1986) observed that the trichomes of some epiphytic bromeliads can reflect up to 46% of the light that reaches the leaf, constituting an effective mechanism for protection against photoinhibition in excessive light.

Trichomes are usually found associated with stomata, possibly creating a favorable microenvironment. Nevertheless, this protection can also limit gas exchange when the leaf is wet and the trichomes promote the formation of a film of water on the surface that stops gas exchange (Benzing and Renfrow 1971; Lüttge *et al.* 1986; Martin 1994). Nutrient and water absorption is carried out by the trichomes. The external cells of the scale-like trichome are dead and respond by hygroscopic movements which oppress the trichome "wing" close to the surface of the leaf when it is wet, causing the water to move to the base cells by capillarity. The living cells at the base of the trichome lack a cuticle in the transversal cell wall and absorb water and nutrients.

Rooting in bromeliads

Epiphytic bromeliads use roots that are presumed to serve only as holdfasts and lack root hairs (Benzing 2000). Bromeliads have particularly tough roots that usually have a thick walled epidermis and are suberized. This is important in terrestrial species because most of them inhabit hostile environments in which soil can reach high temperatures and damage fine roots (Wan *et al.* 1994). It is possible that terrestrial bromeliads develop "rain roots" which develop following short pulses of rain events, before the main growing season, as has been described for agaves, cacti, and desert shrubs, but this has not been reported to date (Benzing 2000; Nobel 2003; Schwinning and Sala 2004). The development of these fine, fast growing roots allows desert plants to exploit pulse precipitation events effectively.

Crassulacean acid metabolism as a water saving mechanism

The CAM pathway involves carbon intake during the night, when vapor pressure deficit is low and less water is lost by transpiration. The CO_2 taken up is incorporated into malic acid by the enzymes phosphoenolpyruvate carboxylase (PEPc) and malate dehydrogenase and stored in the vacuole. Throughout most of the day, the stomata remain closed, diminishing water loss, and the malate is decarboxylated and assimilated by ribulose biphosphate carboxylase (Rubisco). The pathway is common in xerophytic plants due to its water-saving properties, but also effectively diminishes photorespiration by concentrating the CO_2 inside photosynthetic cells, improving the affinity of Rubisco to this substrate.

The diel cycle of the metabolic pathway is divided into four defined phases (Osmond 1978). Phase one takes place during the night and consists in the assimilation of atmospheric or respiratory CO_2 in oxaloacetate by PEPc. This compound is immediately reduced to malate, which is stored in the vacuole. In phase II, at dawn, both PEPc and Rubisco are active, and there is a peak in gas exchange, before the stomata close. Phase III occurs during the day, when the stomata are usually closed and there is decarboxylation of the malic acid and the released CO_2 is fixed by Rubisco. Phase IV marks the transition between the deactivation of Rubisco, when most of the malic acid has been degraded and the activation of PEPc, coupled to the opening of the stomata (Griffiths *et al.* 1990; Cushman and Bohnert 1999). In the family Bromeliaceae 40% of the species exhibit some level of CAM activity that has evolved independently within the family several times (Crayn *et al.* 2004). The high incidence of CAM in terrestrial bromeliads is associated to hostile substrates like sand and rocky surfaces (Benzing 2000).

High humidity and its role in the survival of perennials in a seasonally dry forest

The effect of transpiration on leaf water enrichment in $\delta^{18}O$

Ecophysiological studies involving stable isotopes as a tool have allowed the visualization of physiological processes that were difficult to measure previously (Reyes-García and Andrade 2007). Stable isotopes are atoms which have the same number of protons but a different number of neutrons and are not radioactive. The different isotopes of an element are found in characteristic proportions in nature, with the heavy isotope generally representing only about 1% of the total abundance. The proportions of the heavy to light isotopes can vary in different biotic and abiotic systems because the heavy isotopes react and diffuse more slowly and form stronger bonds than the lighter isotopes. Isotope fractionation then occurs during physical, chemical, and biological processes that the element undergoes in its natural cycle. This phenomenon has been used to study the interaction of plants with the environment because it is possible to relate the proportion of an isotope present in plant tissue with the source of the element (air or water) and the physiological processes it has been through (Santiago *et al.* 2005).

To determine the isotopic composition of a sample, a mass spectrometer is used to compare the signature of the sample to that of a standard that is arbitrarily given the value of zero. The isotopic signature is by convention expressed in a "delta" notation as follows:

$$\delta^{18}O = [(R_{sample}/R_{standard}) - 1] \times 1000 \qquad \text{Eq. (5.1)}$$

With R being the molar isotope ratio of $^{18}O/^{16}O$. The standard used for determining ^{18}O values is the Standard Mean Ocean Water (SMOW). The resulting δ values are then positive when enriched in the heavy isotope, and negative when depleted with respect to ocean water.

The Craig-Gordon model describes the effect of temperature and humidity on the enrichment in $\delta^{18}O$ of an evaporating lake (Craig and Gordon 1965). Similar to a lake, instantaneous leaf water enrichment takes place when water evaporates from the stomatal cavity favoring the lighter isotopes and leaving behind leaf water enriched in the heavier $\delta^{18}O$ isotope. Under constant temperature and humidity, enrichment continues until an isotopic steady state (ISS) is reached and the transpired $\delta^{18}O$ signal is equal to the leaf input water. The Craig-Gordon model has been modified by Flanagan *et al.* (1991b) to describe leaf water enrichment in $\delta^{18}O$ as transpiration occurs:

$$R_L = \alpha * \left[\alpha^K R_s \left(1 - \frac{w_a}{w_i} \right) + R_a \frac{w_a}{w_i} \right] \qquad \text{Eq. (5.2)}$$

This enrichment involves two fractionation factors, the equilibrium and the kinetic factors. The equilibrium factor (α^*) is temperature dependant and is the result of isotopic fractionation in the phase change of liquid water to vapor. It is expressed as $\alpha^* = R_L/R_V$ and has a value of 1.009 at 25 °C (Majoube 1971). R_L being the ratio of $^{18}O/^{16}O$ for leaf water and R_v the same ratio for vapor. The kinetic factor is a result of the different diffusivities of the heavy and light molecules, and is defined as the ratio of the conductances of the isotopes. It is expressed: $\alpha^k = g/g^{18}$, g being conductance and g^{18} the conductance of the small ^{18}O fraction. α^k has a value of 1.028 (Merlivat 1978), modified for boundary layer conductance giving the value 1.032 (Farquhar et al. 1989). The subscripts s and a correspond to source water and atmospheric water ratios; w_a/w_i is the ratio of ambient to leaf intercellular vapor mole fraction. Using Eq. 5.2, we can predict interactions between microclimatic conditions and the transpiring plant.

Oxygen isotopic signatures across the leaf

Many times in nature, the theoretical steady state of leaf water is never reached (Yakir et al. 1994). This could be due to changing environmental conditions, but also to intrinsic characteristics of the leaf. Leaf veins, which contain non-fractionated water coming from the roots, are constantly diluting the fractionated water generated at substomatal cavities (Farquhar and Lloyd 1993; Farquhar and Gan 2003). Based on this model, there are gradients of enrichment within the leaf from the cells that surround the stomata (highly enriched) to those that are localized near the main veins (non fractionated), creating a string of pools with intermediate $\delta^{18}O$ values. The phenomenon by which the enriched water at the sub-stomatal cavity back-diffuses towards the depleted vein is referred to as the Péclet effect.

Most studies concerning the leaf water $\delta^{18}O$ signal have been developed in Dicotyledonous plants. These studies have found no fractionation differences between C_3 and C_4 plants (Flanagan et al. 1991a). Helliker and Ehleringer (2000) found a different fractionation pattern for Monocotyledonous grasses. In these plants, the signal becomes progressively enriched from the proximal to the distal part of the leaf (Helliker and Ehleringer 2000, 2002; Barnes et al. 2004; Ogee et al. 2007). This is a result of the water in the parallel veins of the long grass leaves becoming progressively enriched by the back diffusion of leaf water from substomatal cavities, previously fractionated by transpiration. Towards the distal areas of the leaf, the water in the vein is more enriched, and the overall $\delta^{18}O$ signal of the leaf water is higher, differences within one leaf can be as high as 60‰ under low humidity (Helliker and Ehleringer 2000, 2002).

High humidity and survival of perennial bromeliads

The influence of atmospheric water vapor on the leaf water $\delta^{18}O$ has recently been investigated (Cernusak et al. 2005; Seibt et al. 2006; Helliker and Griffiths 2007), and allows the visualization of the large quantities of water being exchanged between the plant and atmosphere, that were not previously evident when measuring net water fluxes. When the pineapple (*Ananas comosus*), a terrestrial CAM bromeliad, was manipulated in the greenhouse by changing the night-time relative humidity (75 and 100%) and irrigation (well watered and droughted), changes in the level of enrichment along the leaf were evident (Fig. 5.1a). A progressive enrichment from leaf base to tip occurred under well watered conditions with 75% relative humidity, although it was much lower than that observed on C_3 grasses (Helliker and Ehleringer 2000). After a 3 month drought, when stomata would be closed or under saturating night-time relative humidity, no progressive enrichment was observed in the bromeliad (Fig. 5.1a). Since no water absorbing trichomes are present in the leaves of terrestrial bromeliads (Pittendrigh 1948; Benzing 2000; Pierce *et al.* 2001), and no trichomes can be observed in the mid and tip leaf sections (Table 5.1), the lack of enrichment along the leaf cannot be explained by local water absorption, but by gas exchange at the abundant stomata present at the leaf-tip.

Terrestrial bromeliads that inhabit the dry forest of Chamela experience 8-month-long droughts, when relative humidity at midday can go down to 30% in the dry season, when no canopy cover is present (Reyes-García 2005). Yet, night-time humidity, when gas exchange is performed, is usually not below 75% and often at saturation, as dew and fog events are common (Barradas and Glez-Medellín 1999). *Bromelia karatas*, a terrestrial CAM species, had no progressive enrichment in leaf water $\delta^{18}O$ along the leaf when measured in Chamela during wet, dry or an intermediate dry season with frequent fog and dew events (Fig. 5.1b). This lack of progressive enrichment along the more than 1 m leaves of *B. karatas* can be explained by the constant high humidity during gas exchange and the influence of atmospheric vapor $\delta^{18}O$ in its signal. Further evidence on the constant high humidity conditions at the seasonally dry forest of Chamela, when *B. karatas* performs gas exchange, is the lack of night-time enrichment in leaf water $\delta^{18}O$ when monitored at night vs. midday in the wet and dry seasons (Fig. 5.2). The exchange of water vapor through the stomata does not result in net water gain for the species, but the lack of enrichment along the leaf throughout the seasons evidences a constant high humidity that can be essential for the species to survive the prolonged drought, as it diminishes water loss by transpiration.

Schwinning and Ehleringer (2001) used a computer model to predict the performance of desert plants with contrasting traits. When a plant with a succulent stem was compared to one with a woody stem, the first one was found to conserve a higher capacitance throughout the dry period and maintain leaf conductance. On the other hand, the woody plant would quickly show very negative water potential followed by stomata closure. Nevertheless, as the drought progresses, more water would be conserved in the woody plant as compared to the succulent plant, in which the maintenance of leaf conductance would result in a loss of 90% of its water in 37 days. This is avoided by succulents through stomata closure under desiccating conditions. Succulent perennials at Chamela, can maintain gas exchange for longer periods throughout the year due to the low night-time vapor pressure deficits that allows water to be conserved.

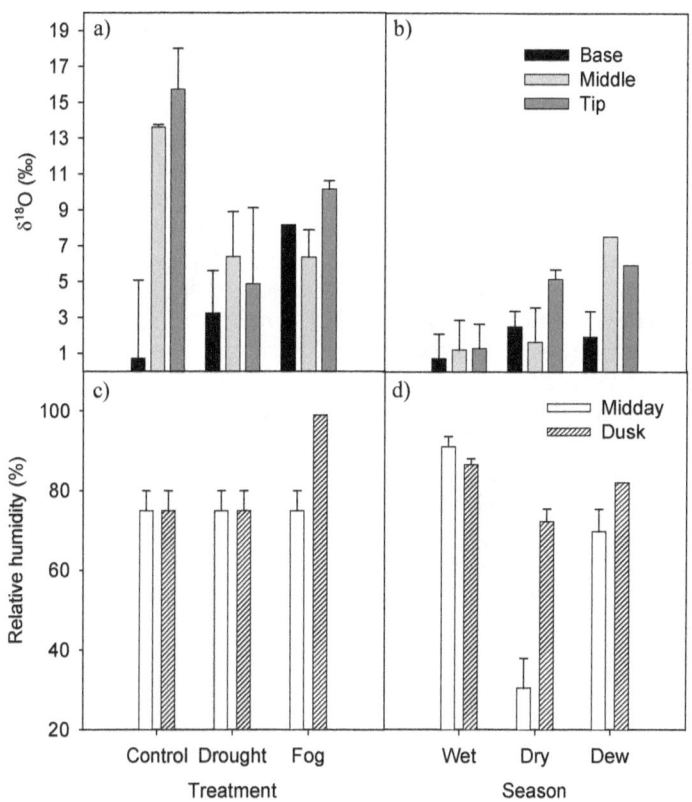

Figure 5.1. Bulk leaf water $\delta^{18}O$ in the leaf sections of two terrestrial bromeliads and relative humidities under greenhouse and field conditions. $\delta^{18}O$ of *A. comosus* under greenhouse conditions, a); *B. karatas* under field conditions, b); and corresponding relative humidities c) and d), respectively. Greenhouse conditions were manipulated to simulate seasonal changes in Chamela. Significant changes between the leaf sections leaf water $\delta^{18}O$ were observed under control conditions and under dew conditions in the field that represents the first three months of drought when dew formation is common. (Reyes-García and Griffiths unpublished data).

The epiphytic habitat as a semi-arid environment: Alternative sources of water

The epiphytic habitat constitutes a semi-arid microclimate in which the lack of water-retaining soil restricts water availability to pulse events from precipitation (Richards 1996; Andrade and Nobel 1997). This inherent characteristic of the habitat makes water the strongest limiting factor for epiphytic growth in vascular plants (Laube and Zotz 2003) even in moist, evergreen forests where rain events are frequent. The different families that have epiphytic representatives show many diverse ways to obtain and store water for survival in the canopy (*i.e.*, tanks and water absorbing trichomes in the Bromeliaceae; the velamen and pseudobulbs in the Orchidaceae; Benzing 1990). Moreover, a higher diversity of epiphytes is found in wetter habitats that exhibit frequent dew or fog events, which can maintain water potentials high between rain events (Sudgen and Robins 1979; Wolf and Flamenco 2003; Kreft *et al.* 2004; Kuper *et al.* 2004).
The semi-arid conditions of the epiphytic habitat allow the comparison between the strategies displayed in epiphytes and desert plants, especially in forests with discontinuous water supply like the dry deciduous forest of Chamela, in which the present study took place. In arid and semi-arid environments, the use of small pulse rain events and alternative water sources like dew can be decisive to survival in between heavier seasonal rains (Goldberg and Novoplansky 1997; Martorell and Ezcurra 2002; Schwinning *et al.* 2003; Chesson *et al.* 2004; Schwinning and Sala, 2004; Sher *et al.* 2004; also see Chapters 4 and 13). Martorell and Ezcurra (2002) found a significant shift towards a dominant rosette morphology at an altitude

Table 5.1. Trichome and stomata density for the terrestrial bromeliad *B. karatas*. Imprints from 3 leaves were taken at Chamela. Density was counted using an optical microscope under 10× magnification. Statistical differences were found in trichome density between base and other leaf section in the abaxial leaf portion and in stomata density among all leaf sections ($p<0.05$; Reyes-García and Griffiths unpublished data).

Leaf section	Trichome density (mm^{-2})		Stomatal density
	Abaxial	Adaxial	(mm^{-2})
Base	32 ± 25	6 ± 4	57 ± 3
Middle	0 ± 0	12 ± 2	31 ± 5
Tip	0 ± 0	8 ± 5	83 ± 11

where cloud formation was frequent. The authors found that the rosette was efficient in funnelling the water to the roots, and hypothesised that the cloud water contributes between 10-100% of the annual water budget of the plant, and thus the advantage of obtaining this additional water source can favor the survival of rosette plants. A reverse direction in xylem sap results from water from fog carried back towards the roots in Californian Redwoods (Dawson 1998; Burgess and Dawson 2004). In Mediterranean habitats, dew contributes to the rehydration of shrubs under water stress in the dry season (Munne-Bosch *et al.* 1999). And in the dry forest of Chamela, dew allows the survival of a bush with inverse phenology that would otherwise dehydrate during the dry season (Barradas and Glez-Medellín 1999).

General strategies found in species from different xerophytic habitats are expected to be present in dry forest epiphytes. A common classification for drought resisting strategies described by Levitt (1980) are the *drought avoiding* and *drought tolerating* strategies (see Chapter 7). Drought avoidance implies that the plant is never under water stress, maintaining a high water content, even when under stressful environmental conditions. Drought avoidance can be achieved through saving water by closing stomata, having water storing tissue, diminished leaf area, thicker cuticle and CAM metabolism. On the other hand, drought tolerance involves highly developed dehydration avoidance and dehydration tolerance. Examples of this is the ability to open stomata at low water potentials due to differential water content in guard cells, uncoupling photosynthesis from transpiration and having low transpiration rates (*i.e.*, by maximizing boundary layers). A continuum of drought avoidance and tolerance strategies are expected to be represented in coexisting xerophytic species.

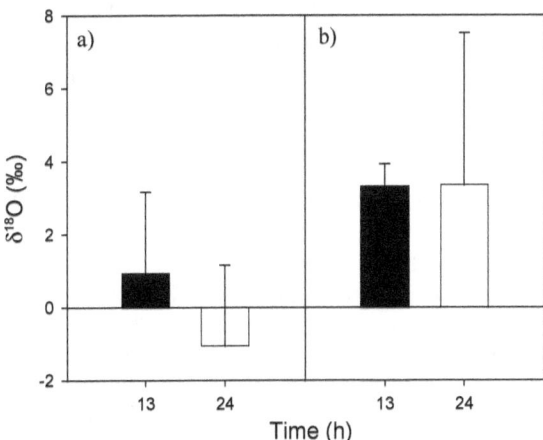

Figure 5.2. Diel changes in bulk leaf water $\delta^{18}O$ in *B. karatas* during a) the wet and b) dry seasons. Three leaves were sampled at 13:00 and 24:00 h in Chamela, Mexico. No significant changes were found during the day; higher values were observed during the dry season compared to the wet season. (Reyes-García and Griffiths unpublished data).

Characterization of a community of bromeliad epiphytes in a seasonally dry forest

Most studies regarding the distribution of epiphytic communities have been performed in moist forests. Under these conditions, all studies find that rough bark promotes high abundance of epiphytic bromeliads in a tree because the wind-dispersed plumed seeds are easily anchored in its bark and also because this bark texture tends to retain water (Hietz and Hietzseifert 1995; Hietz 1997; Wolf and Könings 2001; Winkler *et al.* 2005), allowing the survival of the bromeliads in the most vulnerable part of their lifecycle (Castro-Hernández *et al.* 1999). In the dry forest of Chamela, Reyes-García *et al.* (2008) found that nearly 40% of the bromeliad epiphytes are distributed on only 5% of the trees. When analyzing tree height, canopy cover, bark texture and leaf type, the strongest factor limiting distribution was the presence of compound leaves in the host tree, which probably allow more scattered light to penetrate by leaf fluttering and heliotropic leaf movements. Without this appropriate light environment, the epiphytes would have a lower productivity during the favorable rainy season and lower allocation to reserves and leaf production, which could decrease survival rates in the dry season (Goldberg and Novoplansky 1997).

Different authors have found epiphytic species to distribute in a non-random way among host trees and relate this uncertainty to the yet unknown biology of most species (Flores-Palacios and García-Franco 2006; Laube and Zotz 2006). Species distribution within the trees with high abundance of epiphytes at Chamela is shown in Table 5.2. This distribution was also found to be non-random when a chi-test was performed relating abundance within the tree to total number of individuals counted within the community. No particular association of species was evident, but it was noted that *Tillandsia usneoides*, a species with a wide environmental range and probably the most shade tolerant species of the community, was distributed on trees with single leaves.

Responses of xerophytic plants to pulse precipitation

In the semi-arid habitats the discontinuous and unpredictable water supply favors plant traits that optimize use of small scattered rain events that can constitute a high source of water on an annual basis (Goldberg and Novoplansky 1997; Chesson *et al.* 2004; Schwinning and Sala 2004). Due to the importance of these scattered rain events, the pulse-reserve paradigm as proposed by Noy-Meir (1974) and revised by Reynolds *et al.* (2004) has been one of the most cited paradigms in arid and semi-arid studies.

The model modified by Reynolds *et al.* (2004) describes how a pulse of precipitation can increase soil water content and trigger differential responses on distinct plant functional types, that eventually allocate water to reserves (being seeds, roots, stems, etc.). In this model the author stresses that some pulse events can trigger higher responses (*i.e.*, activation of carbon uptake, new root production) than others due to the effect on soil water content.

The plant functional type is a key determinant of the response to a pulse event as, in dry ecosystems, the segregation of roots in the soil profile determines whether small rain events can be absorbed through shallow roots or larger events are needed for deep roots (*i.e.*, Goldberg and Novoplansky 1997; Schwinning and Ehleringer 2001; Reynolds *et al.* 2004; Schwinning *et al.* 2004). The segregation of roots in the soil profile, concomitant to other genetically based strategies on water use have been described as an axis of niche differentiation. In this axis only species with different functional types (being root dept, drought tolerance, etc.) can coexist without competing and driving each other to extinction, as described in the Lotka-Volterra model (Goldberg and Novoplansky 1997; Reynolds *et al.* 2004; Silvertown 2004).

Schwinning and Sala (2004) developed a model to explain the differential responses to pulse rain events observed in different functional types. In a study performed using several desert grasses and shrubs representative of a variety of functional types, the C_4 grass was seen to have a quicker response than the other types, resuming gas exchange, with a lag of 3 days, after 9 mm of rain were applied (Schwinning *et al.* 2003). Other species

Table 5.2. Frequency table of the epiphytic bromeliads (genus *Tillandsia*) found in the dry forest of Chamela, Mexico. Total numbers of individuals in the 60 trees sampled as well as the distribution within the nine most populated trees are shown. The chi-test shows a distribution different from the expected, given by the total frequency of the species in the whole community ($P \leq 0.01$; Reyes-García and Griffiths unpublished data).

Species	Tree Number									TOTAL
	1	2	3	4	5	6	7	8	9	60 trees
Tillandsia makoyana	3	0	2	10	7	1	6	1	19	64
T. eistetteri	0	2	4	1	1	3	4	2	17	42
T. pseudobaileyi	3	2	0	0	0	8	1	7	3	31
T. intermedia	2	1	0	1	1	0	4	5	0	21
T. recurvata	0	0	6	0	5	2	1	1	5	20
T. usneoides	0	1	0	0	0	0	0	1	0	20
T. ionantha	0	0	0	0	0	0	2	6	2	16
T. rothii	0	0	0	0	0	0	4	1	2	8
T. schiedeana	1	0	0	0	0	0	0	0	0	2
T. caput-medusae	0	0	0	0	0	0	0	0	0	2
Juvenile	2	4	0	2	4	4	0	11	0	30
TOTAL	9	10	12	14	18	18	22	35	48	256

needed much more water to show a response. In the model, the species that respond quickly increasing carbon uptake are called "optimists" and they gain carbon if the pulse is abundant enough to maintain carbon uptake for a substantial period of time (Fig. 5.3). If the pulse event is small, an optimist might lose carbon invested in photosynthetic enzymes or root production that had been activated "expecting" a longer period of rain. A contrary response is the "pessimist," which has a higher threshold of response to pulse events, and does not reactivate metabolic processes that involve carbon investment following small rain events. The pessimist misses the opportunity for higher carbon gain if the pulse turns out to be long.

The lag time between the rain event and the response can be a result of the optimist-pessimist strategy, the efficiency of the plant (*i.e.*, in water use), and of the investment it has to make to up-regulate metabolism. The higher the responses to rain, in terms of carbon investment, the longer the lag time for a response (Fig. 5.3). If the plant's response is to reactivate carbon uptake, carbon first will have to be invested in synthesizing new enzymes or repairing damaged photosystems, during which there will be no positive carbon gain. If the investment involves root production in order to take up the available water, then the lag time before positive carbon gain is obtained will be longer.

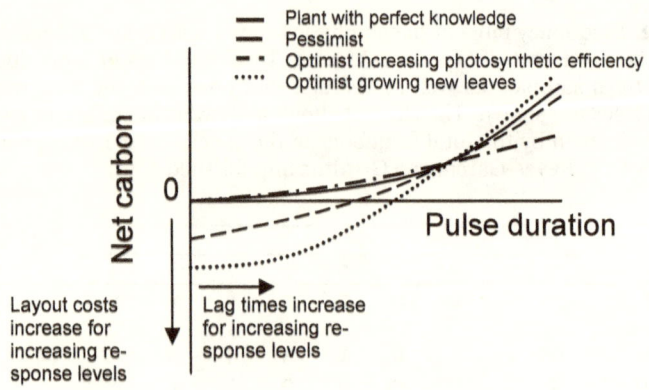

Figure 5.3. Graphical representation of pulse precipitation responses of desert plants. The abscissa represents pulse duration, from a single small rainfall event to large event clusters. The ordinate represents carbon gain derived from use of pulse water or carbon spent in preparation for the use of the pulse. The *solid line* represents the ideal response. The *alternative lines* represent plants adapted to different precipitation patterns. Pessimists invest little or no carbon to respond to pulses. Optimists make larger carbon investments and have higher carbon gain in long pulses, but loose carbon in short pulses. (Modified from Schwinning and Sala 2004).

Water use and pulse precipitation response thresholds in the Bromeliaceae of Chamela

Grime (2001) describes that diversity in the strategies used to take up the limiting resource are characteristic of an environment with a spatial and/or temporal variation in this resource. Rainfall in Chamela is concentrated during five months, from mid-June to mid-October, and occasional rains occur outside these months, with some storms present from January to March (Fig. 5.4). During the first months of the prolonged dry season, there is a high frequency of dew and fog events (Barradas and Glez-Medellín 1999), as well as small pulse events of precipitation generally lower than 1 mm that are not recorded by the local weather station. Thus, water in Chamela is a limiting factor for growth and survival, being highly seasonal and available in different forms, especially for an epiphyte, which has no access to soil water. The high diversity of water use strategies present in the epiphytic bromeliads in this forest supports Grime's theory. This diversity is comparable to that found in the functional types of other semi-arid and arid environments that rely extensively on pulse events (Chesson *et al.* 2004).

Under variable water availability, recovery from drought and metabolic responses to pulse precipitation events (optimist *vs.* pessimist strategies) are important determinants of niche differentiation. We observed differential responses after three short rain events (each < 1 mm) that occurred following three months without recorded precipitation for five coexisting epiphytic bromeliads of the genus *Tillandsia* from the forest of Chamela (Fig. 5.5). After the pulse precipitation events, *Tillandsia pseudobaileyi* had a significant increase in specific leaf weight (SLW; $P \le 0.05$, effect of day on SLW, one way ANOVA), indicative of rehydration for this period of 6 days when no possible leaf growth occurred. Rehydration on this species is concomitant with the higher trichome density it exhibits in comparison to the coexisting species (Table 5.3, $P \le 0.01$, species effect in a 3 way ANOVA combining species, leaf section and exposure in the canopy, Tukey Tests were used to assess differences among the species), which showed only non-significant increases in SLW. Conversely, only *Tillandsia eistetteri* increased carbon uptake by showing a boost in dawn to dusk titratable acidity ($P \le 0.05$, effect of day on acidity, one way ANOVA). This evidences different thresholds of response to pulse events, being *T. eistetteri* the optimist of the group, by reassuming gas exchange under very small changes in water content. All other species showed a more pessimist strategy, even though *T. pseudobaileyi* proved to be very effective at rapidly increasing its water content. The smallest change in SLW is observed in *Tillandsia ionantha*, a species with the lowest trichome cover of the atmospheric species, which lack a water storing tank.

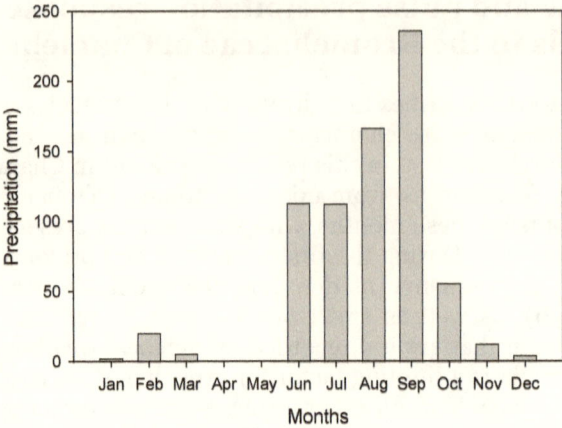

Figure 5.4. Average monthly rainfall in the seasonally dry tropical forest of Chamela, Mexico (1999-2003). Data provided by the meteorological station of Chamela.

Table 5.3. Trichome cover in 6 epiphytic bromeliads (genus *Tillandsia*) of Chamela under contrasting light environments. Field leaf imprints were taken from exposed and shaded bromeliads (n=3) and trichome density was measured using an optical microscope under 10x magnification. *T. makoyana* and *T. pseudobaileyi* showed higher trichome density under exposed conditions, *T. schiedeana* showed the inverse pattern and other species showed no differences. Means and standard errors are shown. (Modified from Reyes-García *et al.* 2008).

Species	Exposed		Shaded	
	abaxial	adaxial	abaxial	adaxial
T. makoyana	37 ± 5	25 ± 2	27 ± 2	19 ± 3
T. eistetteri	49 ± 5	54 ± 5	44 ± 5	57 ± 5
T. pseudobaileyi	55 ± 4	95 ± 6	50 ± 5	66 ± 10
T. ionantha	12 ± 2	37 ± 7	11 ± 2	32 ± 3
T. rothii	66 ± 4	77 ± 8	55 ± 6	55 ± 7
T. schiedeana	36 ± 3	33 ± 5	54 ± 3	43 ± 4

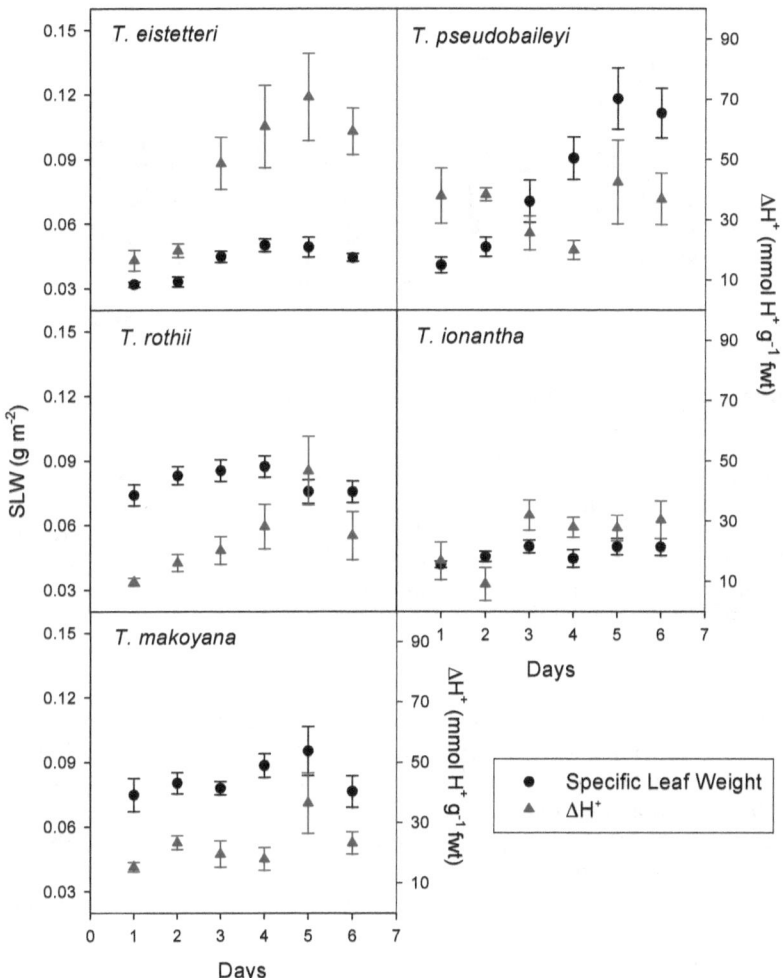

Figure 5.5. Increase in ΔH^+ and in the specific leaf weight (fresh weight to area ratio, SWL) of five epiphytes at Chamela. Data was taken in 6 consecutive days on January 2002 after a series of trace rain events (> 1 mm), following a prolonged 90 day drought. Rain events are marked by arrows. (Reyes-García and Griffiths unpublished data).

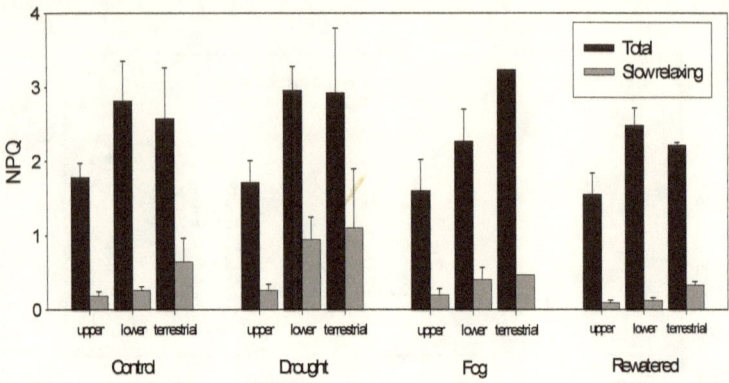

Figure 5.6. Non-photochemical quenching (NPQ) in terrestrial, upper and lower canopy bromeliad species from Chamela. The control, drought, fog, and rewatering treatments performed in the greenhouse in Cambridge. The upper canopy epiphytes are *T. makoyana*, *T. rothii* and *T. eistetteri*, the lower canopy epiphytes are *T. ionantha* and *T. intermedia*, and the terrestrial species *A. cosmosus*. There are significant increases in slow relaxing NPQ in lower canopy epiphytes and the terrestrial during drought ($P \leq 0.01$), sun epiphytes have lower values of total and slow relaxing NPQ compared to the other epiphytes and terrestrial ($P \leq 0.01$). (Reyes-García and Griffiths unpublished data).

Additional laboratory-based studies, that simulated the response to dew and fog events after a prolonged drought, evidenced that before an increase in carbon gain can be appreciated, the epiphytes show other metabolic responses to very small water availability (Reyes-García 2005). In this experiment, five epiphytic bromeliad species (three from the upper canopy, and two from the lower canopy) and one terrestrial bromeliad were droughted (down to 30% relative water content) and then either rewatered or put in a chamber with a humidifier that simulated fog formation during 4 hours around dawn (Fig. 5.6). When non-photochemical quenching (NPQ) was measured, the droughted lower canopy and terrestrial species exhibited an increase in the slow relaxing component of NPQ, which indicates photoinhibition. After 15 days in the fogging chamber, with no other water available, significant decrease in slow relaxing NPQ was observed for these species, values being similar to those measured under control (well watered) conditions.

Field and experimental data suggest that the epiphytic bromeliads of Chamela have differential responses to pulse events, but also, that the species have great efficiency at using this type of precipitation, by showing only a one-day lag time after two trace rain events (*T. eistetteri*), and metabolic repairs to a damaged photosystem after several days of fog/dew formation. Hence, these epiphytic bromeliads seem more effective at using pulse precipitation than previously measured desert plants. In these species with perennial leaves that are responsible for water uptake, no immediate leaf or root production is needed to take advantage of a pulse event. Furthermore, the species in the top canopy showed great capacity for photoprotection, which means less investment in repairing mechanisms when water is available, and a smaller lag time before a response to the pulse. A slower rehydration time is documented for tank species compared to atmospheric species (Benzing and Burt 1970; Benzing 2000), which probably results in longer response times or a pessimist strategy.

By comparing the responses of the epiphytes to models developed for desert species, we can appreciate their high dependence on pulsed rain or dew events that occur during the dry season. Goldberg and Novoplansky (1997) called this season the interpulse, and the favorable season the pulse. They argue that semi-arid species are fine tuned to these events when survival during the interpulse depends on the effective use of such isolated water events. It is likely that this is the case in the epiphytes of the dry forest of Chamela, but the roles of these events remain unclear for the terrestrial species. In the case of *B. karatas*, a study of the seasonal changes in the root system and an analysis of how many interpulse rain events are large enough to infiltrate the soil and activate root uptake would be necessary to define the relevance of these events (Reynolds *et al.* 2004).

Niche differentiation in the epiphytic species of Chamela

In response to root competition for water accumulated in soil, plant life-forms in semi-arid and arid environments appear to specialize by using specific soil layers at certain times of the year (Walter and Stadelmann 1974; Cody 1986; Smith and Nobel 1986; Ehleringer *et al.* 1991; Flanagan *et al.* 1992; Lin *et al.* 1996; Schwinning *et al.* 2003; Chesson *et al.* 2004). This diversification is observed in different locations with unrelated genera (Soriano and Sala 1983; Sala *et al.* 1989; Paruelo *et al.* 1998). Due to the limited canopy space suitable for epiphyte colonization in the seasonal dry forest of Chamela, a similar pattern of specialization in water absorption is evident among the coexisting species.

In addition to the seasonality, rainfall at Chamela is comprised by a high frequency of very small rain events and a few large rain events (Fig. 5.7). When analyzing the pattern of daily rains from 2000 to 2003, it is noted that even in the wet months of July through October, more than 50% of the days had no rain event, and of the days with rains, more than 60% were events that measured less than 10 mm, showing that rain interception could be an important limiting factor, even in the wet months.

This would enhance the relevance of a balanced distribution of the species in the canopy and a niche differentiation related to water interception. During heavy rains, water quickly saturates the capacity of the tanks and the remaining water runs off and can be intercepted by species in the lower canopy. But on smaller rain events, a tank species could block a substantial amount of rain from species growing below.

In Table 5.4 and Fig. 5.8 a synthesis of the water-use strategies exhibited by the epiphytic bromeliads at Chamela is shown (Reyes-García 2005). As has been observed previously in other sites, the species can have a differential distribution based on branch size (Hietz and Hietzseifert 1995; Kernan and Fowler 1995; Hietz 1997; Hietz *et al.* 2002; Merwin *et al.* 2003). Based on photographic record, as well as field and greenhouse observations on distribution and physiological traits, differential use of tree branches in species related to their water and light-use strategies are described.

The species segregate forming guilds according to competition pressure for a suitable space for: light interception during the wet season, water interception irrespective of the magnitude of the event, anchorage place for plumed seeds, high humidity, moderate radiation in dry season, branch with mechanical strength for growth.

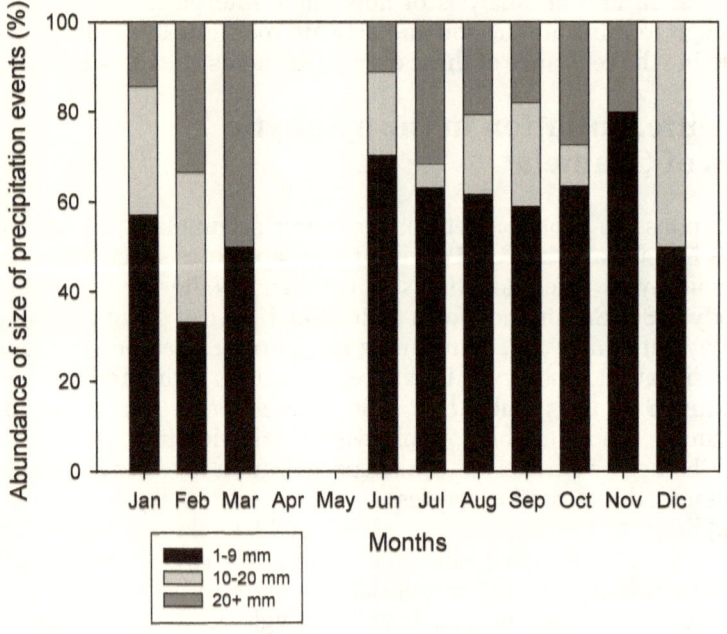

Figure 5.7. Abundance of precipitation with-in size classes per month at Chamela. Percentage of daily precipitation in size classes are presented using data from 2000-2003 provided by the meteorological station at Chamela. Months with no bars had no precipitation events during these years. Most months had more than 50% of the rain events in the size class of 1-9 mm per day, large rain events (20+ mm per day) represented about 20% of monthly rain.

This results in a segregation of species that anchor on different branch sizes and heights along the canopy and experience multiple microenvironmental conditions. This way, the community consists of epiphytes that have five different ways of intercepting water and light: 1. Upper canopy-atmospheric (*Tillandsia eistetteri*); 2. Upper canopy-shallow tank (*T. rothii*); 3. Upper canopy-deep-tank, (*Tillandsia makoyana*); 4. Lower canopy-main trunk-atmospherics (*T. ionantha, T. pseudobaileyi*); 5. Lower canopy-hanging-atmospherics (*Tillandsia intermedia, Tillandsia usneoides*).

The distribution within the vertical canopy profile, with the species at the bottom being of a smaller size; and those higher up being large tank species segregating along different branch types (primary, secondary), suggest that water interception could be a limitation. In this respect, we can observe that the lower canopy species tend to use alternative sources of water by utilizing the stemflow that leaks along the main trunks and branches and can constitute a high proportion of the water that reaches the ground, even during only dew events (Clark *et al.* 1998).

Two species hang in a vine-like fashion, *T. intermedia* and *T. usneoides,* abundant trichomes suggest rapid rehydration. The species *T. usneoides,* was the only species that was found to be more abundant in trees with simple leaves (Reyes-García *et al.* 2008). This is probably the most shade tolerant species of this community, given its distribution range (Martin *et al.* 1985; Martin *et al.* 1986; Haslam *et al.* 2003) and can effectively use the small amount of light that reaches the understory of a tree that would be unsuitable for other epiphytes. Both species lack the photoprotective strategies of the species from the upper canopy and benefit from drops of rainfall that fall from the leaves after a rain event.

A temporal niche differentiation is observed in these species as well as spatial. The tank species can tolerate less drought and become dormant after prolonged periods without rain (Adams and Martin 1986a; Zotz and Andrade 1998; Reyes-García 2005). Yet, these species also experience a prolonged productive season due to the water stored in the tank that can last a few weeks after the rain is gone, when the atmospheric species are already loosing water. A similar phenomenon could be true for the terrestrial species that can access soil water for a period after precipitation has stopped. Conversely, the atmospherics experience drought almost instantly, a few days after the heavy seasonal rains stop, but can maintain their metabolism under these conditions.

Implications of climate change on the Bromeliad community

There is evidence to suggest that plants coexisting in an environment where water is limiting have fine-tuned adaptations to exploit precipitation events (Schwinning *et al.* 2003, 2004; Reynolds *et al.* 2004; Schwinning and Sala 2004; Sher *et al.* 2004). Thus, a change in the pre-

Table 5.4. Species composition and functional groups of the dry forest of Chamela. The species, life form, distribution along the vertical profile of the canopy, rooting, limits to distribution and pulse event response are shown for bromeliads of Chamela. (Modified from Reyes-García 2005).

Species	Life form	Distribution	Rooting	Limits to Distribution	Pulse Event Response
T. eistetteri	Atmospheric	Upper canopy	Thin, external, tertiary branches	High night-time humidity and dew formation, high light for carbon gain all year	Optimist, lower threshold of all species. 3 d rehydration under saturating watering regime
T. rothii	Shallow tank	Upper canopy	Medium sized, external, secondary branches	High water and light interception during the wet season, medium weight of tank	Pessimist. 10 d or less rehydration under saturating watering regime
T. makoyana	Deep tank	Upper canopy	Thick, horizontal, primary branches	High water and light interception during the wet season, heavy weight of tank	Pessimist. 10 d rehydration under saturating watering regime

continues on following page

Table 5-4 (continued)

T. intermedia	Atmospheric	Lower canopy	Hanging from bushes and lower branches	Low light incidence during dry season to prevent photodamage	High trichome cover suggests effective rehydration
T. usneoides	Atmospheric	Lower canopy	Hanging from bushes and lower branches	Low light incidence during dry season to prevent photodamage	—
T. ionantha	Atmospheric	Lower canopy	Main trunk	Low light incidence during dry season to prevent photodamage, throughfall main water source	—
T. pseudobaileyi	Atmospheric	Lower canopy	Main trunk	Low light incidence during dry season to prevent photodamage, throughfall main water source	Optimist. 2 d rehydration under 3 mm watering regime

cipitation abundance and distribution could affect these communities. One possible effect is that a shift from scattered pulse events to a higher frequency of heavier rains, as has been predicted in some models for tropical America (Easterling *et al.* 2000; Groisman *et al.* 2004, 2005), will allow faster growing plants, less adapted to drought, to invade and dominate these environments, causing a long term change in species composition (Sala and Lauenroth 1982; Schwinning *et al.* 2003; Sher *et al.* 2004). A second prediction derives from a population study performed on the epiphyte *Tillandsia brachycaulus* during a 3 year period (Mondragón *et al.* 2004). In this study, it was observed that the population growth index, which indicates whether the number of individuals in a population is growing (>1), stable (1), or decreasing (<1), was only above 1 in the wettest year, suggesting that the species would eventually become extinct if the dry years are more frequent than the wet years. A prediction that higher frequency of dry years over wet years is expected in the dry forests in Mexico (Villeres-Ruiz and Trejo-Vázquez 1997), suggesting the fragile ecological balance of these populations.

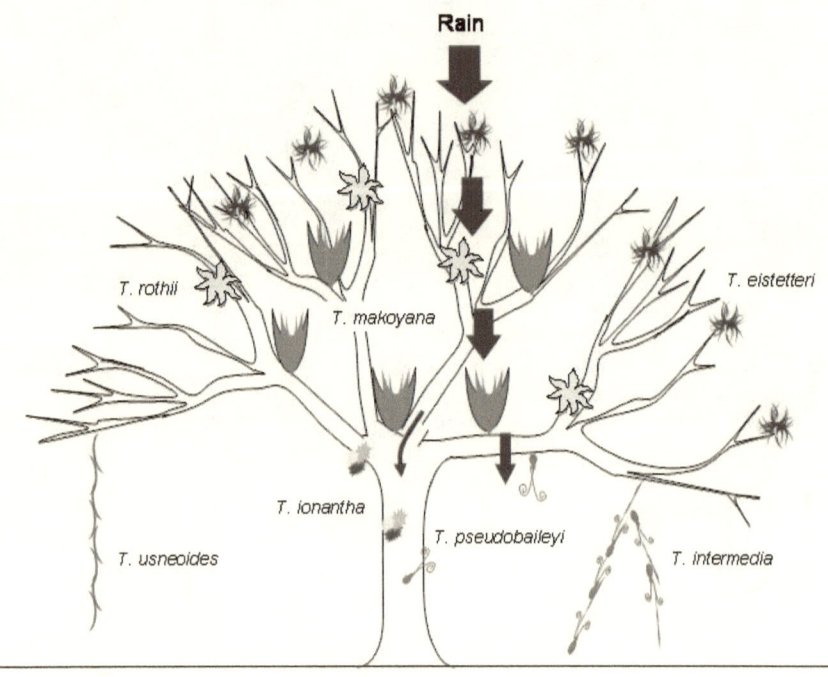

Figure 5.8. Distribution of the epiphytic species of Chamela among the strata and branches. The amount of rain intercepted is signaled by the decrease in the size of the arrow.

Conclusions and future prospects

Perennials inhabiting a seasonally dry tropical forest are subject to many changes in resource availability throughout the year. Strategies for survival during prolonged drought were analyzed in this chapter, and compared to strategies previously described for desert plants using the bromeliad community in the forest of Chamela, Mexico as a case study. The great diversity of life forms within the Bromeliaceae, from terrestrial representatives, to those with water impounding tanks and the atmospherics are represented at Chamela, and a variety of niches are occupied by them. High humidity during nocturnal gas exchange in these CAM plants helps maintain water content high for longer throughout the dry period. Frequent fog and dew formation that condenses on the leaves could also be an important source of water, as it is constant, and can be collected by the tanks. Pulse precipitation is also a frequent source of water and most species showed a high capacity to activate photosynthesis or other metabolic activity after receiving small quantities of water. There was a stratification of epiphytic species within the canopy that could be related to water use strategies. This community, because it is fine tuned to rain frequency, humidity, fog and dew events, could be highly sensitive to climate change.

Even though many studies have dealt with ecophysiological traits of the Bromeliaceae, few have dealt with it in terms of niche differentiation and distribution had assumed to be random within host trees. But recent efforts to understand epiphyte distribution in host trees recognize a high species-dependent uncertainty related to the yet unknown biology of most epiphytic species (Flores-Palacios and García-Franco 2006; Laube and Zotz 2006). Further studies involving the physiology and distribution of Bromeliaceae species are needed to understand niche differentiation and whether ecological exclusion is possible in some environments.

Acknowledgements

We thank the Station of Biology of Chamela (IB-UNAM), for support, facilities, and the climatic data provided, and Glyn D. Jones for assistance while processing isotope samples. CR-G received a Ph. D. scholarship from CONACyT and a complementary bursary from the Cambridge Overseas Trust.

Literature cited

Adams WW and Martin CE. 1986a. Physiological consequences of changes in life form of the Mexican epiphyte *Tillandsia deppeana* (Bromeliaceae). *Oecologia* **70**: 298-304.

Adams WW and Martin CE. 1986b. Morphological-changes accompanying the transition from juvenile (atmospheric) to adult (tank) forms in the Mexican epiphyte *Tillandsia deppeana* (Bromeliaceae). *American Journal of Botany* **73**: 1207-1214.

Andrade JL and Nobel PS. 1997. Microhabitats and water relations of epiphytic cacti and ferns in a lowland neotropical forest. *Biotropica* **29**: 261-270.

Barnes B, Farquhar G, and Gan K. 2004. Modelling the isotope enrichment of leaf water. *Journal of Mathematical Biology* **48**: 672-702.

Barradas VL and Glez-Medellín MG. 1999. Dew and its effect on two heliophile understorey species of a tropical dry deciduous forest in Mexico. *International Journal of Biometeorology* **43**: 1-7.

Benz BW and Martin CE. 2006. Foliar trichomes, boundary layers, and gas exchange in 12 species of epiphytic Tillandsia (Bromeliaceae). *Journal of Plant Physiology* **163**: 648-656.

Benzing DH and Burt KM. 1970. Foliar permeability among 20 species of Bromeliaceae. *Bulletin of the Torrey Botanical Club* **97**: 269-279.

Benzing DH and Renfrow A. 1971. Significance of photosynthetic efficiency to habitat preference and phylogeny among tillandsioid bromeliads. *Botanical Gazette* **132**: 19-30.

Benzing DH. 1976. Bromeliad-trichomes - Structure, function, and ecological significance. *Selbyana* **1**: 330-348.

Benzing DH, Seemann J, and Renfrow A. 1978. Foliar epidermis in Tillandsioideae (Bromeliaceae) and its role in habitat selection. *American Journal of Botany* **65**: 359-365.

Benzing DH and Renfrow A. 1980. The nutritional dynamics of *Tillandsia circinata* in Southern Florida and the origin of the "airplant" strategy. *Botanical Gazette* **14**: 165-172.

Benzing DH. 1990. *Vascular Epiphytes*. Cambridge University Press, New York.

Benzing DH. 2000. *Bromeliaceae: Profile of an Adaptive Radiation*. Cambridge University Press, Cambridge, UK.

Bullock SH. 1986. Climate of Chamela, Jalisco, and trends in the south coastal region of Mexico. *Archives of Meteorolology, Geophysics, and Bioclimatology* **36**: 297-316.

Bullock SH and Solís-Magallanes JA. 1990. Phenology of canopy trees of a tropical deciduous forest in Mexico. *Biotropica* **22**: 22-35.

Burgess SSO and Dawson TE. 2004. The contribution of fog to the water relations of *Sequoia sempervirens* (D. Don): foliar uptake and prevention of dehydration. *Plant Cell and Environment* **27**: 1023-1034.

Castro-Hernández JC, Wolf JHD, García-Franco JG, and González-Espinosa M. 1999. The influence of humidity, nutrients and light on the establishment of the epiphytic bromeliad *Tillandsia guatemalensis* in the highlands of Chiapas, Mexico. *Revista de Biología Tropical* **47**: 763-773.

Cernusak LA, Farquhar GD, and Pate JS. 2005. Environmental and physiological controls over oxygen and carbon isotope composition of Tasmanian blue gum, *Eucalyptus globulus*. *Tree Physiology* **25**: 129-146.

Chesson P, Gebauer RLE, Schwinning S, Huntly N, Wiegand K, Ernest MSK, Sher, A, Novoplansky A, and Weltzin JF. 2004. Resource pulses, species interactions, and diversity maintenance in arid and semi-arid environments. *Oecologia* **141**: 236-253.

Clark KL, Nadkarni NM, Schaefer D, and Gholz HL. 1998. Atmospheric deposition and net retention of ions by the canopy in a tropical montane forest, Monteverde, Costa Rica. *Journal of Tropical Ecology* **14**: 27-45.

Cody ML. 1986. Structural niches in plant communities. In: Diamond J and Case TJ (eds.) *Community Ecology*. Harper & Row, New York. Pp. 381-405.

Craig H and Gordon LI. 1965. Deuterium and oxygen-18 variations in the ocean and the marine atmosphere. In: Tongiorgi E (ed.) *Proceedings of a Conference on Stable Isotopes in Oceanographic Studies and Paleotemperatures.* Spoleto, Italy. Pp. 9-130.

Crayn DM, Winter K, and Smith JAC. 2004. Multiple origins of Crassulacean acid metabolism and the epiphytic habit in the Neotropical family Bromeliaceae. *Proceedings of the National Academy of Sciences of the United States of America* **101**: 3703-3708.

Cushman JC and Bohnert HJ. 1999. Crassulacean acid metabolism: Molecular genetics. *Annual Review of Plant Physiology and Plant Molecular Biology* **50**: 305-332.

Dawson TE. 1998. Fog in the California redwood forest: Ecosystem inputs and use by plants. *Oecologia* **117**: 476-485.

Easterling DR, Meehl GA, Parmesan C, Changnon SA, Karl TR, and Mearns LO. 2000. Climate extremes: Observations, modeling, and impacts. *Science* **289**: 2068-2074.

Ehleringer JR, Phillips SL, Schuster WSF, and Sandquist DR. 1991. Differential utilization of summer rains by desert plants. *Oecologia* **88**: 430-434.

Farquhar G, Hubick KT, Condon AG, and Richards RA. 1989. Carbon isotope discrimination and water-use efficiency. In: Rundel PW, Ehleringer JR, and Nagy KA (eds.) *Stable Isotopes in Ecological Research.* Springer-Verlang, New York.) Pp. 21-46.

Farquhar GD and Lloyd J. 1993. Carbon and oxygen isotope effects in the exchange of carbon dioxide between terrestrial plants and atmosphere. In: Ehleringer JR, Hall AE, and Farquhar GD (eds.) *Stable Isotopes and Plant Carbon-Water Relations.* Academic Press, San Diego. Pp. 47-70.

Farquhar GD and Gan KS. 2003. On the progressive enrichment of the oxygen isotopic composition of water along a leaf. *Plant, Cell and Environment* **26**: 801-819.

Flanagan LB, Bain JF, and Ehleringer JR. 1991a. Stable oxygen and hydrogen isotope composition of leaf water in C_3 and C_4 plant-species under field conditions. *Oecologia* **88**: 394-400.

Flanagan LB, Comstock JP, and Ehleringer JR. 1991b. Comparison of modeled and observed environmental influences on the stable oxygen and hydrogen isotope composition of leaf water in *Phaseolus vulgaris* L. *Plant Physiology* **96**: 588-596.

Flanagan LB, Ehleringer JR, and Marshall JD. 1992. Differential uptake of summer precipitation among co-occurring trees and shrubs in a Pinyon-Juniper Woodland. *Plant, Cell and Environment* **15**: 831-836.

Flores-Palacios A and García-Franco JG. 2006. The relationship between tree size and epiphyte species richness: testing four different hypotheses. *Journal of Biogeography* **33**: 323-330.

Gassmann AJ and Hare JD. 2005. Indirect cost of a defensive trait: variation in trichome type affects the natural enemies of herbivorous insects on *Datura wrightii. Oecologia* **144**: 62-71.

Givnish TJ, Millam KC, Evans TM, Hall JC, Pires JC, Berry PE, and Sytsma KJ. 2004. Ancient vicariance or recent long-distance dispersal? Inferences about phylogeny and South American-African disjunctions in rapateaceae and bromeliaceae based on ndhF sequence data. *International Journal of Plant Sciences* **165**: S35-S54.

Goldberg D and Novoplansky A. 1997. On the relative importance of competition in unproductive environments. *Journal of Ecology* **85**: 409-418.

Goldstein G, Andrade JL, and Nobel PS. 1991. Differences in water relations parameters for the chlorenchyma and the parenchyma of *Opuntia-ficus-indica* under wet versus dry conditions. *Australian Journal of Plant Physiology* **18**: 95-107.

Griffiths H, Broadmeadow MSJ, Borland AM, and Hetherington CS. 1990. Short-term changes in carbon isotope discrimination identify transitions between C_3 and C_4 carboxylation during Crassulacean acid metabolism. *Planta* **181**: 604-610.

Grime JP. 2001. *Plant Strategies, Vegetation Processes, and Ecosystem Properties* John Wiley & Sons Chichester.

Groisman PY, Knight RW, Karl TR, Easterling DR, Sun BM, and Lawrimore JH. 2004. Contemporary changes of the hydrological cycle over the contiguous United States: Trends derived from in situ observations. *Journal of Hydrometeorology* **5**: 64-85.

Groisman PY, Knight RW, Easterling DR, Karl TR, Hegerl GC, and Razuvaev VAN. 2005. Trends in intense precipitation in the climate record. *Journal of Climate* **18**: 1326-1350.

Haslam R, Borland A, Maxwell K, and Griffiths H. 2003. Physiological responses of the CAM epiphyte *Tillandsia usneoides* L. (Bromeliaceae) to variations in light and water supply. *Journal of Plant Physiology* **160**: 627-634.

Helliker BR and Ehleringer JR. 2000. Establishing a grassland signature in veins: O^{18} in the leaf water of C_3 and C_4 grasses. *Proceedings of the National Academy of Sciences of the United States of America* **97**: 7894-7898.

Helliker BR and Ehleringer JR. 2002. Differential O^{18} enrichment of leaf cellulose in C_3 versus C_4 grasses. *Functional Plant Biology* **29**: 435-442.

Helliker BR and Griffiths H. 2007. Towards a plant-based proxy for the ^{18}O atmospheric water vapor. *Global Change Biology* **13**: 723-733.

Hietz P and Hietzseifert U. 1995. Structure and ecology of epiphyte communities of a cloud forest in central Veracruz, Mexico. *Journal of Vegetation Science* **6**: 719-728.

Hietz P. 1997. Population dynamics of epiphytes in a Mexican humid montane forest. *Journal of Ecology* **85**: 767-775.

Hietz P. Ausserer, J., and Schindler, G. (2002). Growth, maturation and survival of epiphytic bromeliads in a Mexican humid montane forest. *Journal of Tropical Ecology* **18**: 177-191.

Kernan C and Fowler N. 1995. Differential substrate use by epiphytes in Corcovado National Park, Costa Rica: A source of guild structure. *Journal of Ecology* **83**: 65-73.

Kreft H, Koster N, Kuper W, Nieder J, and Barthlott W. 2004. Diversity and biogeography of vascular epiphytes in Western Amazonia, Yasuni, Ecuador. *Journal of Biogeography* **31**: 1463-1476.

Kuper W, Kreft H, Nieder J, Koster N, and Barthlott W. 2004. Large-scale diversity patterns of vascular epiphytes in Neotropical montane rain forests. *Journal of Biogeography* **31**: 1477-1487.

Laube S and Zotz G. 2003. Which abiotic factors limit vegetative growth in a vascular epiphyte? *Functional Ecology* **17**: 598-604.

Laube S and Zotz G. 2006. Neither host-specific nor random: Vascular epiphytes on three tree species in a Panamanian lowland forest. *Annals of Botany* **97**: 1103-1114.

Levitt J. 1980. *Responses of Plants to Environmental Stresses. Vol 2.* Academic Press, London.

Lin GH, Phillips SL, and Ehleringer JR. 1996. Monosoonal precipitation responses of shrubs in a cold desert community on the Colorado Plateau. *Oecologia* **106:** 8-17.

Lott E. 1993. Annotated checklist of the vascular flora of the Chamela Bay region, Jalisco, Mexico. *Occasional Papers of the California Academy of Sciences* **148:** 1-60.

Ludwig F, Dawson TE, Prins HHT, Berendse F, and de Kroon H. 2004. Below-ground competition between trees and grasses may overwhelm the facilitative effects of hydraulic lift. *Ecology Letters* **7:** 623-631.

Lüttge U, Klauke B, Griffiths H, Smith JAC, and Stimmel KH. 1986. Comparative ecophysiology of CAM and C_3 bromeliads. V. Gas- exchange and leaf structure of the C_3 bromeliad *P. integrifolia*. *Plant, Cell and Environment* **9:** 411-419.

Majoube M. 1971. Oxygen-18 and deuterium fractionation between water and steam. *Journal de Chimie Physique et de Physico-Chimie Biologique* **68:** 1423-1436.

Martin CE, McLeod KW, Eades CA, and Pitzer AF. 1985. Morphological and physio-logical responses to irradiance in the CAM epiphyte *Tillandsia usneoides* L. (Bromeliaceae). *Botanical Gazette* **146:** 489-494.

Martin CE, Eades CA, and Pitner RA. 1986. Effects of irradiance on Crassulacean acid metabolism in the epiphyte *Tillandsia usneoides* L. (Bromeliaceae). *Plant Physiology* **80:** 23-26.

Martin CE. 1994. Physiological ecology of the Bromeliaceae. *Botanical Review* **60:** 1-82.

Martínez-Yrízar A and Sarukhán J. 1990. Litterfall patterns in a tropical deciduous forest in Mexico over a 5-Year period. *Journal of Tropical Ecology* **6:** 433-444.

Martorell C and Ezcurra E. 2002. Rosette scrub occurrence and fog availability in arid mountains of Mexico. *Journal of Vegetation Science* **13:** 651-662.

Medina E. 1995. Diversity of life forms of higher plants in neotropical dry forests. In: Bullock SH, Mooney HA, and Medina E (eds.) *Seasonally Dry Tropical Forests*. Cambridge University Press, Cambridge, UK. Pp. 221-242.

Merlivat L. 1978. Molecular diffusivities of $(H_2O)-O^{16}$ $Hd^{16}O$, and $(H_2O)-O^{18}$ in Gases. *Journal of Chemical Physics* **69:** 2864-2871.

Merwin MC, Rentmeester SA, and Nadkarni NM. 2003. The influence of host tree species on the distribution of epiphytic bromellads in experimental monospeci-fic plantations, La Selva, Costa Rica. *Biotropica* **35:** 37-47.

Mondragón D, Durán R, Ramírez I, and Valverde T. 2004. Temporal variation in the demography of the clonal epiphyte Tillandsia brachycaulos (Bromeliaceae) in the Yucatan Peninsula, Mexico. *Journal of Tropical Ecology* **20:** 189-200.

Mooney HA, Bullock SH, and Medina E. 1995. Introduction. In: Bullock SH, Mooney HA, and Medina E (eds.) *Seasonally dry tropical forests*. Cambridge University Press, Cambridge, UK. Pp. 1-8.

Munne-Bosch S, Nogues S, and Alegre L. 1999. Diurnal variations of photosynthesis and dew absorption by leaves in two evergreen shrubs growing in Mediterra-nean field conditions. *New Phytologist* **144:** 109-119.

Nobel PS. 2003. *Environmental Biology of Agaves and Cacti*. Cambridge University Press, Cambridge.

Nobel PS. 2006. Parenchyma-chlorenchyma water movement during drought for the hemiepiphytic cactus *Hylocereus undatus*. *Annals of Botany* **97:** 469-474.

Nowak EJ and Martin CE. 1997. Physiological and anatomical responses to water deficits in the CAM epiphyte *Tillandsia ionantha* (Bromeliaceae). *International Journal of Plant Sciences* **158:** 818-826.

Noy-Meir I. 1974. Desert ecosystems: higher trophic levels. *Annual Review of Ecology and Systematics* **5:** 195-214.

Nyman LP, Davis JP, Odell SJ, Arditti J, Stephens GC, and Benzing DH. 1987. Active uptake of amino-acids by leaves of an epiphytic vascular plant, Tillandsia-Paucifolia (Bromeliaceae). *Plant Physiology* **83:** 681-684.

Ogee J, Cuntz M, Peylin P, and Bariac T. 2007. Non-steady-state, non-uniform transpiration rate and leaf anatomy effects on the progressive stable isotope enrichment of leaf water along monocot leaves. *Plant, Cell and Environment* **30:** 367-387.

Osmond CB. 1978. Crassulacean acid metabolism – curiosity in context. *Annual Review of Plant Physiology and Plant Molecular Biology* **29:** 379-414.

Paruelo JM, Jobbagy EG, Sala OE, Lauenroth WK, and Burke IC. 1998. Functional and structural convergence of temperate grassland and shrubland ecosystems. *Ecological Applications* **8:** 194-206.

Pérez-Estrada LB, Cano-Santana Z, and Oyama K. 2000. Variation in leaf trichomes of *Wigandia urens*: environmental factors and physiological consequences. *Tree Physiology* **20:** 629-632.

Pierce S, Maxwell K, Griffiths H, and Winter K. 2001. Hydrophobic trichome layers and epicuticular wax powders in Bromeliaceae. *American Journal of Botany* **88:** 1371-1389.

Pittendrigh CS. 1948. The bromeliad-*Anopheles*-malaria complex in Trinidad. I. The bromeliad flora. *Evolution* **2:** 58-89.

Reyes-García C. 2005. *Niche differentiation in coexisting CAM bromeliads from the seasonally dry forest of Chamela (Mexico).* Ph. D. Thesis, University of Cambridge, Cambridge, UK.

Reyes-García C and Andrade JL. 2007. Los isótopos estables del hidrógeno y el oxígeno en los estudios ecofisiológicos de plantas. *Boletín de la Sociedad Botánica Méxicana* **80:** 19-28.

Reyes-García C, Griffiths H, Rincón E, and Huante P. 2008. Niche differentiation in tank and atmospheric epiphytic bromeliads of a seasonally dry forest. *Biotropica* **40:** 168–175.

Reynolds JF, Kemp PR, Ogle K, and Fernandez RJ. 2004. Modifying the 'pulse-reserve' paradigm for deserts of North America: precipitation pulses, soil water, and plant responses. *Oecologia* **141:** 194-210.

Richards PW. 1996. *The Tropical Rainforest.* Cambridge University Press, N.Y.

Sala OE and Lauenroth WK. 1982. Small rainfall events - an ecological role in semiarid regions. *Oecologia* **53:** 301-304.

Sala OE, Golluscio RA, Lauenroth WK, and Soriano A. 1989. Resource partitioning between shrubs and grasses in the Patagonian steppe. *Oecologia* **81:** 501-505.

Santiago LS, Silvera K, Andrade JL, and Dawson TE. 2005. El uso de isótopos estables en biología tropical. *Interciencia* **30:** 28-35.

Schwinning S and Ehleringer JR. 2001. Water use trade-offs and optimal adaptations to pulse-driven arid ecosystems. *Journal of Ecology* **89:** 464-480.

Schwinning S, Starr BI, and Ehleringer JR. 2003. Dominant cold desert plants do not partition warm season precipitation by event size. *Oecologia* **136:** 252-260.

Schwinning S and Sala OE. 2004. Hierarchy of responses to resource pulses in and and semi-arid ecosystems. *Oecologia* **141:** 211-220.

Schwinning S, Sala OE, Loik ME, and Ehleringer JR. 2004. Thresholds, memory, and seasonality: understanding pulse dynamics in arid/semi-arid ecosystems. *Oecologia* **141:** 191-193.

Seibt U, Wingate L, Berry JA, and Lloyd J. 2006. Non-steady state effects in diurnal [18]O discrimination by *Picea sitchensis* branches in the field. *Plant, Cell and Environment* **29:** 928-939.

Sher AA, Goldberg DE, and Novoplansky A. 2004. The effect of mean and variance in resource supply on survival of annuals from Mediterranean and desert environments. *Oecologia* **141:** 353-362.

Silvertown J. 2004. Plant coexistence and the niche. *Trends in Ecology and Evolution* **19:** 605-611.

Smith JAC, Griffiths H, and Lüttge U. 1986. Comparative Ecophysiology of CAM and C-3 Bromeliads. I. The Ecology of the Bromeliaceae in Trinidad. *Plant, Cell and Environment* **9:** 359-376.

Smith LB and Downs RJ. 1974. *Pitcairnioideae (Bromeliaceae) Flora Neotropica. Monograph no. 14, pt. 1.* Hafner Press New York. Pp. 1-662.

Smith LB and Downs RJ. 1977. *Tillandsioideae (Bromeliaceae) Flora Neotropica. Monograph no. 14, pt. 2.* Hafner Press, New York. Pp. 663-1492.

Smith LB and Downs RJ. 1979. *Bromelioideae (Bromeliaceae) Flora Neotropica Monograph no. 14, pt. 3.* Pp. 1493-2142.

Smith SD and Nobel PS. 1986. Deserts. In: Baker NR and Long SP (eds.) *Photosynthesis in Contrasting Environments.* Elsevier, Amsterdam. Pp. 13-62.

Soriano A and Sala OE. 1983. Ecological strategies in a Patagonian arid steppe. *Vegetatio* **56:** 9-15.

Sudgen AM and Robins RJ. 1979. Aspects of the ecology of vascular epiphytes in Colombian cloud forests, I. The distribution of the epiphytic flora. *Biotropica* **11:** 173-188.

Tissue DT, Yakir D, and Nobel PS. 1991. Diel water-movement between parenchyma and chlorenchyma of two desert CAM plants under dry and wet conditions. *Plant Cell and Environment* **14:** 407-413.

Villeres-Ruiz L and Trejo-Vázquez I. 1997. Assessment of the vulnerability of forest ecosystems to climate change in Mexico. *Climate Research* **9:** 87-93.

Walter H and Stadelmann E. 1974. A new approach to the water relations of desert plants. In: Brown Jr GW (ed.) *Desert Biology. vol II. Special Topics on the Physical and Biological Aspects of Arid Regions.* Academic Press, New York. Pp. 213-310.

Wan C, Sosebee RE, and McMichael BL. 1994. Hydraulic-properties of shallow *vs.* deep lateral roots in a semiarid shrub, *Gutierrezia sarothrae. American Midland Naturalist* **131:** 120-127.

Willmer C and Ficker M. 1996. *Stomata.* Chapman & Hall, London.

Winkler M, Hulber K, and Hietz P. 2005. Effect of canopy position on germination and seedling survival of epiphytic bromeliads in a Mexican humid montane forest. *Annals of Botany* **95:** 1039-1047.

Wolf JHD and Könings CJF. 2001. Toward the sustainable harvesting of epiphytic bromeliads: a pilot study from the highlands of Chiapas, Mexico. *Biological Conservation* **101:** 23-31.

Wolf JHD and Flamenco A. 2003. Patterns in species richness and distribution of vascular epiphytes in Chiapas, Mexico. *Journal of Biogeography* **30:** 1689-1707.

Yakir D, Berry JA, Giles L, and Osmond CB. 1994. Isotopic heterogeneity of water in transpiring leaves – Identification of the component that controls the delta-O^{18} of atmospheric O_2 and CO_2. *Plant, Cell and Environment* **17:** 73-80.

Zotz G and Andrade JL. 1998. Water relations of two co-occurring epiphytic bromeliads. *Journal of Plant Physiology* **152:** 545-554.

Chapter 6

ECOPHYSIOLOGY AND FRUIT PRODUCTION OF CULTIVATED CACTI

Paolo Inglese, Giuseppe Barbera,
Giovanni Gugliuzza, and Giorgia Liguori

Perspectives in Biophysical Plant Ecophysiology: A Tribute to Park S. Nobel, pp. 153-166
Edited by: E. De la Barrera and W.K. Smith
© 2009 by The Authors
Book Compilation © 2009 Universidad Nacional Autónoma de México

Introduction

Cactus pear (*Opuntia ficus-indica* L. Mill.) is cultivated in wide range of environments with the consequence of large differences in crop potential, orchard systems and management. These differences may be related to temperature and rainfall range (water availability) but also to the day/night length and, of course, to soil characteristics.

Cactus pear can be utilized in the subsistence and in the market oriented agricultural systems of semi-arid areas. It is able to supply fruit, forage, fodder, and vegetables in specialized plantations or in multipurpose ones (Le Houérou 2002). Fruits can be harvested from July to November in the Northern hemisphere—Mediterranean Basin, California, and Mexico—and from January to April in the Southern one, depending on genotype and genotype × environment interaction. Natural or induced reflowering may extend the ripening period in winter (January - February) in the Northern hemisphere and autumn (September - October) in the southern one. An almost continuous flow of flowering has been reported in Salinas, California (Bunch 1996), resulting in an extended fruit ripening period.

Fruit are consumed fresh or after a relatively short period of post harvest storage. Recently, the diffusion of minimal processed fruits has become common, also to overcome problems related to the presence of glochids in the peel (Inglese *et al.* 2002).

Geographic distribution area

The production of cactus pear fruit is, of course, common in Central America, Mexico being the first producing country, with a large number of cultivars that more than often result from inbreeding between different Opuntias (Pimienta Barrios 1990). Argentina, Chile and California are among the fruit producing, and consuming, countries in the Americas. According to Wessels (1988) 22 spineless Burbank's selections were imported during 1914 in South Africa, not for fruit but forage and fodder production. Good varieties for fruit production were introduced in various parts of that country and 1,200 ha of commercially grown orchards are cultivated in the Northern Province, Ciskey, and Western Cape (Brutsch and Zimmermann 1993; Potgieter and Smith 2006). Ethiopia (Tigray), also produces cactus fruits in commercial plantations, but to a very minor extent and thanks to FAO efforts (Brutsch 1997). In this area, wild stands of cactus pear occur from 2,300 to 3,400 m a.s.l. always under rainfed conditions and with a low excursion of average monthly temperature.

Cactus pear was introduced to Europe, from Central America to Spain after Colombus' first expedition (Diguet 1928). The distribution of cactus pear in the Mediterranean Basin was furthered by the Spanish expansion in Italy of the XVI and XVII centuries, and by the return of the Moors to North Africa, when they were driven out from the Iberian Peninsula in 1610 (Diguet 1928). *Opuntia ficus-indica* spread, from the XVII to the XIX centuries, along the main Mediterranean countries (Diguet 1928) in form of semi-intensive or naturalized plantations, but it was in Sicily, during the XIX and XX century, that specialized, intensive, plantations for fruit production were first established and appropriate technologies were developed on the basis of applied scientific research (Barbera and Inglese 2001).

Cactus pear is naturalized along the western part of the Mediterranean Sea: southern Spain, southern Portugal, Italy and northern Africa. The main areas are in island of Sicily and in Tunisia. Sicily has approximately 5,000 ha of specialized plantations for fruit production (Barbera and Inglese 1993).

In northern Africa we can estimate 300,000 ha in Tunisa (25,000 ha in the Kasserine Governorate; Le Houérou 2002), 50,000 ha in Morocco (Arba *et al.* 2002), that include a large extent of defensive hedges and non-specialized or semi-natural groves.

In Algeria, cactus pear is localized in the coastal areas of Teniet, El Had, and Annaba, and in the inland areas of Tebessa and Batna where 3,550 ha are cultivated (Inglese 1995). Bastawros (1994) reports 2,000 ha of semi-specialized plantations in the provinces of Kalubia, Giza and El-Fayum in Egypt.

Cactus pear is diffused, but much less popular as fruit crop, in the Middle East, and irrigated fruit orchards are developing fast since the 1990s in Israel (Nerd *et al.* 1991) and Jordan (Abu Zurayk 1994).

Italy and Israel export part of their production to Europe and North America, while the north-African production goes for self-consumption or in the local markets (Inglese 1995).

Responses to temperature

In its native highlands of Mexico, cactus pear is cultivated in semi-arid areas, where the annual rainfall is concentrated in the summer during the fruit development period. In the Mediterranean Basin, Middle East, North and East Africa, the dry season coincides with the long and hot summer, when vegetative and reproductive growth occur.

In the area of greatest biodiversity in the central plateau of Mexico (elevation 1,800 to 2,200 m) annual rainfall is less than 500 mm, the mean annual temperatures range from 16 to 18 °C and the long term daily maximum temperatures for the hottest month do not exceed 35 °C (Pimienta Barrios 1990). Similar conditions exist in Sicily where rainfall is in the range of 500 mm and mean annual temperature ranges from 15 to 18 °C, with peaks of 37 °C in August, during the fruit development period (Inglese 1995). While *O. ficus-indica* currently occurs in extensive areas in north Africa (Monjauze and Le Houerou 1965) and the Republic of South Africa (Wessels 1988) it is conspicuously absent in regions that have conditions of less than 350 mm rainfall and daily summer maximum temperatures greater than 42 °C such as the Sahelian African cities of Niamey and Khartoum, and from the California Mojave Desert in California. One exception to the presence of *O. ficus-indica* where temperatures are greater than 40 °C and the mean annual precipitation less than 350 mm exist are in Adigrat, Ethiopia (Brutsch 1997). Le Houérou (2002) reports plantations in

Aziza (Lybia) where the maximum temperature may exceed 50 °C. *Opuntia ficus-indica* cladodes cannot survive at 70 °C (Nobel 1994).

In any case, the optimal day/night temperature regime for nocturnal CO_2 uptake by *O. ficus-indica* is 25/15 °C. Higher or lower day/night temperatures result in a sharp decrease of carbon assimilation, leading to poor plant growth and production (Nobel 1994) and, eventually, to low crop value. High temperatures (> 30 °C) during the fruit development period affect fruit shape if they occur during the initial stages of bud development or fruit growth, and shorten the third stage of fruit growth, when most of the growth of edible flesh occurs, leading to advanced and early ripening, with small fruits, low firmness and sugar content. High temperatures during fruit development enhance fruit sensitivity to low (< 8 °C) temperatures during post-harvest storage (Inglese *et al.* 2002), reducing fruit post harvest storage and shelf-life period (Inglese *et al.* 2002).

On the other hand, daily temperatures below 15 °C delay fruit ripening time and result in thicker fruit peel and lower soluble solid content and peel colour (Nerd *et al.* 1991; Inglese 1995; Liguori *et al.* 2006).

High temperatures are, indeed, one of the major constraints for a production of high quality fruits in areas with hot and dry summer. As a matter of fact, the optimal daily temperature for CO_2 uptake decreases from 17 °C under wet conditions to 14 °C after seven weeks of drought (Nobel 1994).

"Dry but not too hot" could be the motto for *O. ficus-indica* for fruit production.

The number of days required to reach commercial harvest maturity changes with the time of bloom and with prevailing temperatures during the fruit development period, but the thermal time measured in terms of Growing Degree Hours (GDH) from bloom to harvest does not change to the same extent (40-43×10^3; Inglese *et al.* 1999; Liguori *et al.* 2006). In other words, the duration of the fruit development period is well explained in terms of GDH rather than in number of days, and a different accumulation pattern of the GDH accounts for the variability in fruit ripening time that may occur from year to year and in relation to the location site. GDH could be utilized to estimate the length of the fruit development period and the fruit ripening period for a new area of cactus pear cultivation.

Nerd and Mizrahi (1995) found that detached cladodes that experienced low winter temperatures produce the most fruit buds the following spring. Similar results were reported by Gutterman (1995) who examined 18 light × temperature combinations of detached cladodes and found that detached cladodes produced significantly more fruit with 8 h of light grown outside in cool temperatures than with 8 h of light in a heated greenhouse. Nobel and Castañeda (1988) indicate an increase in fruit production on detached cladodes held at 15/5 °C against cladodes left at 25/15 °C day/night temperature. Potgieter and Smith (2006) report a strong influence of environment on fruit yield, with highest fruit yield being obtained in areas with warm summers and cool winters of South Africa. They also report a strong genotype × environment interaction, indicating a different plasticity of cactus pear cultivars, in terms of temperature requirement for optimal fruit production. However, cactus pear produces fruits in the Valley of Catamarca, north-west Argentina, lower Tigray (Ethiopia) and in the Canary Islands, where no more than 100 Chilling Units accumulate in winter, and it is able to reflower several times in the same season as it naturally occurs in Chile or California, or as it artificially induced in Italy and Israel (Inglese 1995; Liguori *et al.* 2006). Apparently, these out-of-season blooms occur with no relation with endodormancy and the rest period could result from ecodormancy rather than true rest or endodormancy, as it happens for *Vitis vinifera* (Faust 2000).

In *Opuntia ficus-indica*, it is possible to have another flush of fruits taking away the first spring flush of flowers and cladodes (Inglese *et al.* 2002). Temperature affects plant reflowerng aptitude. As a matter of fact, plant responses to the removal of the spring flush (SFR) is highly affected by prevailing temperatures at the removal time (Barbera *et al.* 1991; Brutsch and Scott 1991; Nieddu and Spano 1992). High day/night temperatures (> 30/20 °C) result in a larger proportion of new cladodes rather than fruits, while lower day/night temperatures (< 20/15 °C) may not result in rebudding.

Double flowering occurs naturally in Salinas area in California, where fruits are picked from September to March (Inglese 1995), and in the Central Region in Chile, where fruit harvest lasts from February to April and from July to September (Sudzuki *et al.* 1993).

A winter crop may occur in Israel, on current year cladodes following extensive fertigation applied soon after harvesting the summer crop in July. However, winter fruits are poor in sugars and show a lower flesh percent than summer fruits (Nerd *et al.* 1993). In Italy a return bloom occurs in July following the complete SFR, in May, resulting in a late fruit harvest season in October-November (Inglese *et al.* 2002).

Responses to water availability and quality

The water-use efficiency (WUE) of cactus pear is quite high, since De Kock (1980) and Le Houérou (2002) reported values of 4.0 and 3.3 mg DM g^{-1} H_2O, compared to about 1.3 for wheat and 1.4 for barley. Moreover, *O. ficus-indica* achieves these limits with low water inputs. WUE of the same order of magnitude can be achieved with C_4 plants such as pearl millet, sorghum, and maize, but water consumption is then three to five times higher in absolute figures (Le Houérou 2000).

Opuntia ficus-indica, like most cacti, is very sensitive to anoxia in the root zone and therefore cannot withstand any prolonged water logging.

Because of its high drought resistance and high water-use efficiency, cactus pear is usually cultivated without irrigation. The absolute minimum requisite for rainfed cultivation is *ca.* 200 mm yr^{-1} provided that soils are sandy and deep enough. On silty and loamy soils the minimum requisite is 300 to 400 mm mean annual precipitation.

However, in areas with no summer rains and where annual rainfall is less than 300 mm, the plants require supplementary irrigation to get adequate yields and good fruit quality (Barbera 1984; Van Der Merwe *et al.* 1997; Gugliuzza *et al.* 2002).

In terms of fruit production, the economic value of the crop (fruit size) can be diminished largely before the plant shows any apparent symptom of water stress. It is also difficult to find morphological or physiological parameters useful to schedule irrigation treatments (like stem or leaf water potential, carbon exchange rate, sap flow, trunk diameter, etc.). Indeed, the plant is able to recycle internal water and to supply water to the fruit through the phloem (Nobel 2002).

Water shortage reduces flower bud induction and fruit size, and irrigation can be complementary and it is necessary to get an acceptable fruit size (> 120 g f.w.) if no rainfalls (or less than 30-50 mm) occur during the fruit development period (Gugliuzza *et al.* 2002).

Irrigation alone cannot make up for reduced size when there are a high (> 9) number of fruits per cladode (Gugliuzza *et al.* 2002). No more than 6 fruits can be left on each of the 1-year-old fertile cladodes, since a greater crop load result in a sharp reduction of fruit size, even if plants are irrigated.

Even if irrigated or thinned, fruits of the summer flush do not attain the same size as those of the second flush resulting from SFR (Barbera 1984). Fruit of the first flush have a shorter development period than the second flush ones, and, particularly, a shorter duration of Stage III of the fruit development period. Moreover, fruits of the first flush reach Stage III in August, when the average daily temperature (day/night) is 30/23 °C, with peaks of 35/28 °C, whereas fruits of the second flush reach Stage III in a period (October) characterized by temperatures (25/15 °C) more favourable for cladode CO_2 uptake (Nobel and Hartsock 1984).

The natural crop load may influence the effect of thinning. This can be explained considering that at full bloom the flower fresh weight reaches 35% of harvest fruit fresh weight (Inglese *et al.* 1995). This endorses the opportunity of early thinning to get a significant effect on fruit size.

Furthermore, irrigation cannot counteract the effect of very high temperature, since it affects carbon exchange rate more than transpiration (Inglese *et al. in litteris*). For this particular reason, that is quite common in the Mediterranean Basin, some farmers apply black nets (30%-40% shade) over the rows to reduce day temperature, with low reduction in light availability for nocturnal carbon assimilation.

The shallow root system benefits from microjet irrigation but furrow irrigation can be performed to reduce costs.

Opuntia ficus-indica is seriously affected by salinity stress. Gersani *et al.* (1993) showed that a concentration of 30 mol m^{-3} (equivalent to 1.76 ppt NaCl) reduced growth by 40% as compared to non saline control while a concentration of 100 mol m^{-3} (5.85 ppt NaCl) reduced growth by 93%.

Yields and crop management

There are just few observations on dry matter partitioning for mature, fruiting plants of cactus pear. Considering the dry weight of the fruits and current-year cladodes at harvest, the growth ratio of the 1-year-old clad-odes (17 ± 1.5 g) and a total amount of 300 fruits and 130 current-year cladodes (in the first and in the second flush), the proportion of annual dry matter allocated to the fruits, or Harvest Index (HI) is 35% for the first flush and 45% for the second flush. These HI values are slightly lower than those reported for a very efficient fruit crop like peach. Differences between the first and the second flush depend on the higher fruit dry weight and the lower cladode dry weight measured for the second flush (Inglese *et al.* 1999). However, there is no information related to cultivar and training systems. García de Cortazar and Nobel (1992), showed a marked increase of vegetative *vs.* reproductive growth as a result of high density planting systems.

Inglese *et al.* (1994) and Luo and Nobel (1994) investigated the source to sink relationship on mature fruiting plants, indicating a massive carbon flow of assimilates among cladodes of different age. Young developing cladodes apparently compete with fruits, as indicated by their higher Absolute Growth Rate (AGR). However, they became source of carbohydrates at an early stage of their development (Luo and Nobel 1993) that coincides with the development of flowers or the earliest stage of the FDP. Relative sink strength changes along with the developmental stages of the seasonal growth of fruit and cladodes (Inglese *et al.* 1999). The fruits become the major sink during Stage III as indicated by the sharp reduction of cladode AGR that occur at that stage.

Competition between fruit *vs.* cladode growth, as well as the reduction of the number of new cladodes following SFR can be sources of plant alternate bearing behavior that also changes with genotype (Inglese *et al.* 2002).

Alternate bearing of cactus pear depends on the reduction of the fertile cladodes with results from a scarce vegetative activity the previous year. The number of flower per fertile cladode is more stable year by year but depends on cladode age, being highest in 1-year-old cladode. On the other hand, it is clear that the tree needs a large renewal of the fruiting cladodes every year and this require new strategies of pruning able to develop a large number of new cladodes every year which should not result in proportional increase of plant size. One solution could be the increase of the number of plants per hectare; the other could be topping the plants every two or three years to improve cladode development from the lower part of the canopies (Inglese *et al.* 2002a).

Investigation on planting and training systems that maximize fruit yield and quality completely lack and do represent one of the most outstanding challenges for future research, together with the reduction of field variability in yield and fruit quality. Indeed, fruit that do not attain the prime quality in term of size have 50% to 70% less market value than the top quality ones and their relative occurrence is often higher than 50%.

Understanding the sources of this variability that depends on plant (geno-type) × environment interaction, would greatly enhance orchard profitabil-ity without necessarily increasing yields.

Fruit productivity of *Opuntia ficus-indica* is, in fact, extremely variable. Yields of 20 to 30 tons ha^{-1} are reported in Israel and Italy (Barbera and Inglese 1993; Nerd and Mizrahi 1993) and 10 to 30 tons ha^{-1} in South Af-rica (Wessels 1988; Brutsch and Zimmerman 1993). The wide variability in yield depends on orchard design (plant spacing), cultural practices, envi-ronmental conditions (including soil type), and cultivar fertility. Plants begin to yield 2 to 3 years after planting, reaching their maximum potential 6 to 8 years after planting, which they bear for 25 to 30 years and even longer, depending on pruning and overall orchard management.

For a mature plant, most (80-90%) 1-year-old cladodes bear fruits ac-counting for 90% of the annual yield.

However, they show a wide fertility range, depending on plant age, en-vironmental conditions, and their state of growth. The number of fruiting cladodes left on a plant every year depends on plant spacing, and ranges from 100 to 120 for 350 to 400 plants ha^{-1} to 20 to 30 for 1,000 to 1,200 plants ha^{-1} The closer the plant spacing, the higher the pruning intensity and frequency needed. A mature cactus pear tree produces new fruits and cladodes every year at a ratio of 4:1 (Barbera and Inglese 1993). Most of the flowers occur on one-year-old terminal cladodes, and new cladodes usually develop on two-year-old or even older cladodes (Inglese *et al.* 1994). Vegetative and reproductive buds appear contemporarily in spring or early summer when the spring flush is removed to induce reflowering (Barbera *et al.* 1991).

Fruit growth follows a double sigmoid pattern in terms of fresh weight, and a pronounced gain in dry weight occurs for the peel during Stage I, for the seeds during Stage II and for the core during Stage III of the fruit de-velopment period (Barbera *et al.* 1992; Nerd and Mizrahi 1997). Cladodes are strong sinks during the sigmoidal growth period that occurs during the earlier 4-5 weeks of their development. At this stage they switch to a linear growth in terms of dry weight accumulation and from being sinks for car-bohydrate they become sources (Luo and Nobel 1993). The growth of the fruit and the daughter cladode implies a substantial translocation of stored carbohydrates from basal cladodes (Luo and Nobel 1993; Inglese *et al.* 1994). In fact, when more than five fruits develop on a one-year-old fruit-ing cladode, an extensive import of assimilates occurs, particularly during Stage III of fruit growth (Inglese *et al.* 1994).

Cladode fertility is related to the amount of dry weight accumulated per unit surface area (García de Cortazar and Nobel 1992). Cladodes become productive when their dry weight exceeds (EDW) the minimum dry weight for a particular surface area by at least 33 g. Values of EDW become posi-tive after 60-70 days of cladode development and in November they reach 13.1 ± 2.4 g and 10.6 ± 1.2 g respectively for cladodes of the spring and the second flush of the previous season. In the next season, EDW at spring burst was 19.1 ± 2.4 g and 17.6 ± 1.2 g respectively for cladodes of the spring and the second flush of the previous season. Indeed, 1-year old clad-odes, in November, that means 18 and 16 months after burst of spring and second flush cladodes, reached an EDW of 37.1 ± 2.4 g and 32.6 ± 1.2 g (Inglese *et al.* 1999).

High planting densities lead to an extremely high accumulation of dry matter in the vegetative growth, but deeply affect allocation of resources to the fruit (García de Cortazar and Nobel, 1992). Recent studies on fruiting orchard also indicate an impressive capacity of the cactus pear orchard for carbon stock in perennial structures.

Different cultivars account for fruit production in Mexico, Italy, South Africa, California, Argentina and Chile, the most important areas for cactus fruit commercial production, while North Africa countries such as Tunisia, Egypt, Algeria and Morocco have a large fruit production more than often dedicated to local markets (Inglese *et al.* 2002).

Despite this relative large diffusion, cactus pear marketing is seasonal and, due to the poor post-harvest performances of the fruit, covers no more than two months, in each ripening season of each cultivar.

Chances to increase the time span of fruit market are related to (a) sharing fruits from different growing areas, (b) the use of cultivars with different ripening time, (c) the regulation of the flowering period and, in turn, of the fruit ripening period, and (d) the improvement of post harvest strategies.

The most powerful tool to get a longer fruit marketing season is related to cropping strategies and, eventually to the ability of *O. ficus-indica* to reflower and to get different crops in the same year (Inglese *et al.* 2002).

Liguori *et al.* (2006) demonstrated that the double removal of new fruits and cladodes induced a third flush of flowers and cladodes during late August with a fruit production that ripe the following winter (January-March). However, it is clear that the reflowering ability is halved by the double removal, but it is unclear whether this depends on environmental conditions, *i.e.*, higher temperatures, or specific cladode potential. The reduction of the crop depends more on the lower number of flowers per fertile cladode rather than on the number of fertile cladodes. This means that flower evocation is reduced, but, indeed, it is the whole rebudding capacity that is reduced, since the number of current-year cladodes is also reduced by 60%. The length of the fruit development period increases from 100-120 days to 160-190 days for the out-of-season winter crop, depending on a longer third stage of fruit growth that, for FIII fruits, occurred in winter time when temperatures were under the optimal values for fruit growth. To allow regular fruit ripening in cool winters when temperatures are not high enough to support growth, polyethylene covering, although it reduces PPFD, is essential to allow optimal temperatures for cladode photosynthetic activity and fruit growth and ripening. In fact, quality of out-of-season winter fruits depends on winter temperature. The winter fruit production obtained in the field in Israel (Nerd *et al.* 1993) was very low in terms of percent flesh (42%) although fruit size was higher than for the summer crop, while winter fruit obtained after double SFR and covered under polyethylene in late fall, were regular in size and percent flesh with only a slight reduction of total soluble solid content. Nobel (1988) indicated a day/night temperature of 25/15 °C as optimal temperature for cladode CO_2 uptake, with a substantial reduction of photosynthetic activity at 15/5 °C. This explains why polyethylene covering results in excellent fruit size and flesh percent and only in a slight reduction of total soluble solids content.

The most relevant problems of such new cropping strategies are a) to have adequate environmental conditions during winter to get normal fruit ripening and marketable quality, and b) to avoid alternate bearing that will eventually result from the sharp reduction of current-year cladodes that are responsible for the new crop in the following year. While the first issue can be addressed with the recourse to polyethylene covering, the second one needs further research to understand how the production of new cladodes could be implemented after the double removal.

Light and fruit production (flower evocation, fruit growth, ripening, and quality)

Fruits of *O. ficus-indica* generally do not occur on shaded cladodes (García de Cortazar and Nobel 1992), probably because such conditions do not permit the accumulation of sufficient excess dry weight per cladode. Indeed, net assimilation rate of cladodes of *O. ficus-indica* become negative for a total daily photosynthetic photon flux (PPF) of 5 mol m^{-2} day^{-1} and reaches 90% of maximal at 20 mol m^{-2} day^{-1} (Nobel 1988; Nobel and Hartsock 1983). Shading affects flower evocation as was clearly demonstrated by Cicala *et al.* (1997) and Deidda *et al.* (1992). At least 80% of flowering seems to be inhibited if light is withheld the last two months before bloom, indicating that flower evocation occurs closely to bud sprouting. A sufficient light intensity during the winter rest period is therefore essential for a regular bloom. Shading applied within 5 days after the removal of the spring flush also inhibits return bloom (Barbera *et al.* 1993). Shading also affects fruit quality, depending on the extent of shading and sucrose exchange within the tree. Complete shading of the main photoassimilates source for the fruit from 45 to 75 days after bloom has no influence on fruit weight and quality as well as on fruit ripening time. On the other hand a short period (15 days) of imposed shade during earlier stages of fruit growth significantly affects fruit weight, but not total soluble solid content or fruit firmness and ripening time. The fruits do not fully recover to maximal growth even when the fruiting cladode was shaded for a short period, before bloom. As for other fruit trees, such as peach and apple, fruit size at harvest seems to be related to the fruit growth rate attained early during the fruit development period. At this stage, photoassimilate supply from the fruiting cladode becomes crucial to support fruit growth, probably because of the competitive demand of different and actively growing vegetative and reproductive sinks (Inglese *et al.* 1999). This agrees with the fact that thinning is most effective in increasing fruit size when applied not later than 3 weeks after bloom (Inglese *et al.* 1995). Since net assimilation of heavy shaded cladodes is negligible, fruit growth in cladodes shaded for long periods during the fruit development period, must depend on extensive photosynthate translocation from non-shaded branches (Luo and Nobel 1992; Inglese *et al.* 1994). This mobilization of storage carbohydrates makes fruit growth only partially dependent on reduced PPF at fruit canopy location. Fruit ripening in shaded cladodes occur later than in sunlit ones. Fruit ripening varies within the plant and the fruiting cladode, mainly because of a different time course of flower bud formation, development and flowering (Barbera and Inglese 1993). The pattern of PPF distribution within the canopy enhances such variability, since shade delays fruit ripening, according to the length of the shading period.

In conclusion, the wide within-tree variability of fruit quality and ripening time that occurs in cactus pear (Barbera and Inglese 1993) can be partially explained in terms of PPF distribution within the canopy. Planting and training systems as well as plant canopy architecture and demography should be analyzed in relation to light interception and distribution within the tree, and optimal stem area index (SAI) for fruit production needs to be set. These aspects have been widely studied in terms of PAR interception, growth and assimilation of the single cladode (Nobel 1988), but deserve further studies in terms of fruit quality.

Conclusions

The response of *Opuntia ficus-indica* to environmental factors has been largely studied by P. S. Nobel and his findings are the cornerstone for any scientist devoted to this species. However, in terms of horticulture many questions still arise and require further investigation. How to increase yields per hectare and crop value? There is not much research on orchard design and plant pruning strategies, even if the knowledge on crop distribution within the plant is well known. In Italy, the highest yields come from orchards made of 300-400 plants ha^{-1} with 50-60 kg fruits plant^{-1}, which means 400-500 fruits per plant on 90-100 fertile cladodes. Considering that the major constraint for fruit development, in terms of sink-source competition, is the number of fruits per cladode, the only possible strategy is to increase the number of fertile cladodes and to reduce the number of cladodes with less than six fruits. To achieve these goals, adequate pruning strategies must be developed at plant level while orchard design deserves more investigation. Moreover, basic research on genotype comparative crop ability is still largely underestimated as well as genotype × environment interaction. For instance, the Italian cultivar "Gialla," which is considered highly productive, shows large variations in yield, depending on pruning strategies or environment (Inglese *et al.* 2002).

Crop value much depends in crop variability in terms of fruit size, that is the major factor for fruit price in Europe. As a matter of fact, difference between the highest priced fruits and the regular crop can be over 60%. The increase of crop value very much depends on regular fruit thinning, in terms of timing and number of fruits per cladode, and cladode annual pruning.

If the factors affecting fruit size or crop potential have been largely investigated, it is still impossible to define the best environmental and horticultural conditions that maximise fruit quality in terms of flavour, taste, color, flesh crunchiness and post harvest fruit physiology. Recent investigations (Gugliuzza *et al. in litteris*) give evidence of clear relation between fruit quality and cladode position within the canopy, mostly related to light availability.

Finally, it is important to remember that cactus pear plantations can contribute to carbon stock in perennial structures (García de Cortazar and Nobel 1992). This underscores the multifuntional role of this plant, even if cultivated for the major purpose of fruit production.

Literature cited

Abu Zuryak A. 1994. Cactus pear in Jordan. *Cactusnet Newsletter n. 2*. FAO, Rome.

Arba M, Benismail MC., and Mokhtari M. 2002. The prickly pear cactus (*Opuntia* spp., Cactaceae) in Morocco. Main species and cultivars characterization. *Acta Horticulturae* **581**: 103-109

Barbera G. 1984. Ricerche sull'irrigazione del ficodindia. *Frutticoltura* **46**: 49-55.

Barbera G and Inglese P. 1993. *La coltura del ficodindia*. Calderini Edagricole, Bologna.

Barbera G and Inglese P. 2001. *Ficodindia*. Epos ed., Palermo.

Barbera G, Carimi F, and Inglese P. 1991. The reflowering of prickly pear *Opuntia ficus-indica* (L.) Miller: influence of removal time and cladode load on yield and fruit ripening. *Advances in Horticultural Science* **5**: 77-80.

Barbera G, Carimi F, Inglese P, and Panno M. 1992. Physical, morphological and chemical changes during fruit development and ripening in three cultivars of prickly pear, *Opuntia ficus-indica* (L.) Miller. *Journal of Horticultural Science* **67**: 307-312.

Barbera G, Carimi F, and Inglese P. 1993. Effects of GA3 and shading on return bloom of prickly pear *Opuntia ficus-indica* (L.) Mill. *Journal of the South African Society for Horticultural Science* **3**: 9-10.

Bastawros MB. 1994. Cactus pear in Egypt. *Cactusnet Newsletter n 2*. FAO (Rome).

Bunch R. 1996. Cactus pear products at D'Arrigo Bros. *Journal of the Professional Association for Cactus Development* **1**: 100-102.

Brutsch MO. 1997. The beles or cactus pear (*Opuntia ficus-indica*) in Tigray, Ethiopia. *Journal Professional Association for Cactus Development* **2**: 130-141.

Brutsch MO and Scott MB. 1991. Extending the fruiting season of spineless prickly pear (*Opuntia ficus-indica*). *Journal of the South African Society for Horticultural Science* **1**: 73-76;

Brutsch M and Zimmermann H. 1993. The prickly pear (*Opuntia ficus-indica* Cactaceae) in South Africa: Utilization of the naturalized weed and of the cultivated plants. *Economic Botany* **47**: 154-162.

Cicala A, Fabbri A, Di Grazia A, Tamburino A, and Valenti C. 1997. Plant shading and flower induction in *Opuntia ficus-indica* (L.) Mill. *Acta Horticulturae* **438**: 57-64.

Deidda P, Nieddu G, and Spano D. 1992. Reproductive behaviour of cactus pear *Opuntia ficus-indica* (L.) Mill. in Sardinia. *Actas II Congreso Internacional de Tuna y Cochinilla, Santiago de Chile 22-25/9/1992*. Pp. 19-23.

De Kock GC. 1980. Drought-resistant fodder shrub crops in South Africa. In: Le Houérou HN (ed.) *Browse in Africa: The current state of knowledge.* International Livestock Centre for Africa, Addis-Ababa. Pp. 399-410.

Diguet L. 1928. *Les cactes utiles du Mexique.* Archives d'Histoire Naturelle, Soc. Nat. d'Acclim de France, Paris.

Faust M. 2000. Physiological considerations for growing temperate-zone fruit crops in warm climates. In : Erez A (ed.) *Temperate Fruit Crops in Warm Climates.* Kluwer, Dordecht. Pp. 137-156.

García de Cortazar V and Nobel PS. 1992. Biomass and fruit production for the prickly pear cactus, *Opuntia ficus-indica. Journal of American Society for Horticultural Science* **117:** 558-562.

Gersani M, Graham AE, and Nobel PS. 1993. Growth response of individual roots of *Opuntia ficus-indica* to salinity. *Plant, Cell and Environment* **16:** 827-834.

Gugliuzza G, Inglese P, and Farina V. 2002. Relationship between fruit thinning and irrigation on determining fruit quality of cactus pear (*Opuntia ficus-indica*) fruits. *Acta Horticulturae* **581:** 205-210.

Gutterman Y. 1995. Environmental factors affecting flowering and fruit development of *Opuntia ficus-indica* cuttings during the 3 weeks before planting. *Israel Journal of Plant Sciences* **43:** 151-157.

Inglese P. 1995. Orchard planting and management. In: Barbera G, Inglese P, and Pimienta Barrios E (eds.) *Agro-ecology, Cultivation and Uses of Cactus Pear.* Paper 132. FAO, Rome. Pp. 78-91.

Inglese P, Israel AI, and Nobel PS. 1994. Growth and CO2 uptake for cladodes and fruit of the Crassulacean acid metabolism species *Opuntia ficus-indica* during fruit development. *Physiologia Plantarum* **91:** 708-714.

Inglese P, Barbera G, La Mantia T, and Portolano S. 1995. Crop production, growth and ultimate size of cactus pear following fruit thinning. *Hortscience* **30:** 227-230.

Inglese P, Barbera G, and La Mantia T. 1999. Seasonal reproductive and vegetative growth patterns, and resource allocation during Cactus Pear *Opuntia ficus -indica* (L.) Mill. fruit growth. *HortScience* **34:** 69-72.

Inglese, P., Basile, F. and M. Schirra. 2002. Cactus pear fruit production. In: Nobel PS (ed.) *Cacti: Biology and Uses.* California University Press. Pp. 163-183.

Inglese P, Gugliuzza G, and La Mantia T. 2002. Alternate bearing and summer pruning of cactus pear. *Acta Horticulturae* **58:** 202-204.

Le Houérou HN. 2000. Cacti (*Opuntia* spp.) as a fodder crop for marginal lands in the mediterranean basin. *Acta Horticulturae* **581:** 21-46.

Liguori G, Di Miceli C, Gugliuzza G, and Inglese P. 2006. Physiological and technical aspects of cactus pear (*Opuntia ficus-indica* (L.) Mill.) double reflowering and out-of-season winter fruit cropping. *International Journal of Fruit Science* **6:** 23-34.

Luo Y and Nobel PS. 1993. Growth characteristics of newly initiated cladodes of *Opuntia ficus-indica* as affected by shading, drought, and elevated CO_2. *Physiologia Plantarum* **87:** 467-474.

Monjauze A and LeHouerou HN. 1965. Le role des *Opuntia* dans l'Economie agricole de Nord Africaine. *Extrait du Bulletin de l'Ecole Nationale Superieure d'Agriculture de Tunis* **8-9:** 85-165.

Nerd A and Mizrahi Y. 1993. Modern cultivation of prickly pear in Israel. *Acta Horticulturae* **349:** 235-237.

Nerd A, Karady A, and Mizrahi Y. 1991. Out of season prickly pear: fruit characteristics and effect of fertilization and short droughts on productivity. *HortScience* **26:** 337-342.

Nerd A, Mesika R, and Mizrahi Y. 1993. Effect of N fertilizer of autumn flush and cladode N in prickly pear (*Opuntia ficus-indica* (L.) Mill.). *Journal of Arid Environments* **68:** 337-342.

Nerd A and Mizrahi Y. 1995. Effect of low winter temperatures on bud break in *Opuntia ficus-indica. Advances in Horticultural Science* **9:** 1-4.

Nieddu G and Spano D. 1992. Flowering and fruit growth in *Opuntia ficus-indica. Acta Horticulturae* **296:** 153-159.

Nobel PS. 1988. *Environmental biology of Agaves and Cacti.* Cambridge University Press, N.Y.

Nobel PS. 2002. Cactus physiological ecology, emphasazing gas exchange of *Platyopuntias* fruit. *Acta Horticulturae* **58:** 143-150

Nobel PS and Hartsock TL. 1984. Physiological responses of *Opuntia ficus-indica* to growth temperature. *Physiologia Plantarum* **60:** 98-105.

Nobel PS and Castañeda M. 1998. Seasonal light and temperature influences organ initiation for unrooted cladodes of the prickly pear cactus *Opuntia ficus-indica. Journal of the American Society for Horticultural Science* **123:** 47-51.

Pimienta Barrios E. *El Nopal Tunero.* Universidad de Guadalajara, Mexico.

Potgieter J and Smith M. 2006. Genotype × environment interaction in cactus pear (*Opuntia* spp.), additive main effects and multiplicative interaction analysis of fruit yield. *Acta Horticulturae* **728:** 97-104.

Sudzuki F, Muñoz C, and Berger H. 1993. *El Cultivo de la Tuna (Cactus Pear).* Facultad de Ciencias Ciencias Agrarias y Forestales, Universidad de Chile.

Van Der Merwe LL, Wessels AB, and Ferreira DI. Supplementary irrigation for spineless cactus pear. *Acta Horticulturae* **438:** 77-82.

Wessels AB. 1988. *Spineless Prickly Pears.* Perskor, Johannesburg, South Africa.

II. Plant Ecophysiology:
From Grasslands to Alpine Environments

Chapter 7

PHYSIOLOGICAL ADAPTATIONS
OF PERENNIAL GRASSES
TO DROUGHT STRESS

Michelle DaCosta and Bingru Huang

Introduction
   Perennial grasses and water availability
   Mechanisms of drought resistance
Water relations
   Water uptake associated with root size,
     distribution, and activity
   Osmotic adjustment
Carbohydrate metabolism
   Carbon allocation
   Carbohydrate accumulation
   Carbohydrate partitioning between growth
     and storage
Hormonal regulation
   ABA accumulation and regulation of stomata
     and growth
   Interaction of cytokinins and ABA in relation
     to drought resistance
Concluding remarks

Perspectives in Biophysical Plant Ecophysiology: A Tribute to Park S. Nobel, pp. 169-190
Edited by: E. De la Barrera and W.K. Smith
© 2009 by The Authors
Book Compilation © 2009 Universidad Nacional Autónoma de México

Introduction

Perennial grasses and water availability

The grass family is comprised of more species than any other flowering plant family, and also has the most widespread geographic distribution in environments ranging from hot and arid to cold and humid habitats (Nelson and Moser 1995). Grasses are generally classified into cool- and warm-season groups based on their adaptation to specific ranges in temperature and precipitation, which are mainly governed by latitude and altitude. Optimum growth for cool-season grasses occurs at temperatures between 16 to 24 °C (60 to 75 °F), and their distribution is mainly limited by intensity and duration of high temperatures and water deficit (Turgeon 2002). For warm-season perennial grasses, optimum growth occurs between 27 to 35 °C (80 to 95 °F). Unlike cool-season grasses, these species are more highly limited by intensity and duration of cold temperatures. Some cool- and warm-season species may also extend into the transitional zone, which marks the boundary between temperate and subtropical climates that limit more southern or northern adaptation for most grass species (Beard 1973; Turgeon 2002). In addition to temperature and precipitation, several other biotic and abiotic factors play a role in the adaptation and diversity among perennial grasses (CAST 2004), including herbivores, fungi, bacteria, photoperiod, and soil pH (Casler *et al.* 1996).

The availability of water is a major factor limiting distribution, growth, and productivity of cool-season and warm-season grass species. Water availability to plants depends on many biological and physical factors. Some of these factors bring about spatial variation in soil moisture, including root architecture and viability, soil texture, microbial and invertebrate activity, and soil hydraulic conductivity (Nilsen and Orcutt 1996). For instance, grasses having a majority of root systems in the surface soil may deplete water rapidly within this area of the soil while water may still be abundant lower in the soil profile, leading to heterogeneous soil moisture profiles. In addition to spatial variation, water shortage can also result from seasonal variability in precipitation, as well as daily periods of high atmospheric vapor pressure demand.

Issues of decreased water availability have become more of a problem in recent years as a result of a combination of several factors: increasing demand from agriculture, industry, and domestic use, low precipitation, and contamination of potable water supplies. This has resulted in implementation of more stringent restrictions in water use, and has forced growers, homeowners, and many industry sectors to conserve water. With increased pressure to utilize more ecological and economical strategies in the management of perennial grass commodities, studies on how these grasses adapt to stress conditions will become increasingly important as resources such as water become more limited. In particular, insights into the physiological mechanisms that impart drought resistance to perennial grasses will aid in the identification of important characteristics that can serve as selection criteria for improved stress tolerance, and ultimately lead to better selection of highly adapted species and cultivars to environments with limited water resources. Furthermore, efforts that enhance our understanding of physiological effects of drought stress can provide guidelines for practical management strategies that promote the maintenance of high quality and/or productivity of cultivated perennial grasses under limited resource availability.

Mechanisms of drought resistance

Broad mechanisms of drought resistance are generally divided into three categories relating to whether plants escape, avoid, or tolerate conditions of drought stress (Turner 1986). In species that exhibit drought escape strategy, there is an adjustment of the life cycle that leads to growth and development predominately during periods of water availability. Perennial species escape drought periods by becoming dormant during the dry period, whereas annual species complete their entire life cycle prior to dry periods. This strategy is common in regions with distinctive wet and dry periods, such as those in Mediterranean and subtropical climates (Simpson 1981). Examples of perennial grasses that exhibit drought escape characteristics include cocksfoot (*Dactylis glomerata* L.; Volaire and Thomas 1995) and Mediterranean-type tall fescue (*Festuca arundinacea* Schreb.; Assuero *et al.* 2002).

The maintenance of plant water status under drought either by increasing the capacity for water uptake of roots and/or reducing water loss of leaves is defined as drought avoidance. Drought-avoiding plants may exhibit many different characteristics in response to drought, including enhanced root plasticity and root extension deeper in the soil profile for greater extraction of water, decreased leaf growth and/or leaf area, enhanced leaf pubescence, leaf rolling or folding, and modification in number and aperture of the stomatal complex (Simpson 1981; Nilsen and Orcutt 1996; Duncan and Carrow 1999). Tall fescue, buffalograss (*Buchloe dactyloides* (Nutt.) Engelm.), bermudagrass (*Cynodon dactylon* (L.) Pers.), and zoysiagrass (*Zoysia japonica* Steud.) are perennial grasses that have demonstrated drought avoidance characteristics (Burton *et al.* 1954; Marcum *et al.* 1995; Qian *et al.* 1997; Volaire and Lelievre 2001). Some perennial grass species are able to tolerate low leaf water potentials, and maintain metabolic processes even under decreased bulk cellular water content. This mechanism is defined as drought tolerance. Strategies for drought tolerance include osmotic adjustment, maintenance of root viability and membrane stability under dehydration, and accumulation of proteins and other metabolites that function directly or indirectly in structural stabilization. Drought tolerance has been exhibited in grass species like cocksfoot, Kentucky bluegrass (*Poa pratensis* L.), centipedegrass (*Eremochloa ophiuroides* (Munro) Hack.), and seashore paspalum (*Paspalum vaginatum* Swartz) (Perdomo *et al.* 1996; Huang *et al.* 1997; Volaire and Lelievre 2001).

The mechanisms of drought escape, avoidance, and tolerance are not mutually exclusive, and the same plant may utilize more than one strategy in order to adapt to periods of drought stress (Nilsen and Orcutt 1996). The relative importance of each mechanism may depend on drought duration, drought severity, and grass species. Drought avoidance may allow for grass survival and provide for sustained growth and function during short-term drought until soil stored water is depleted, while drought tolerance allows for plants to endure during prolonged periods of drought. Under mild or moderate drought, perennial grasses may continue growth, but at a limited rate. In this case, the maintenance of biomass production is important in grass breeding or sward management. Growth maintenance could be achieved through various adaptive mechanisms, such as maintenance of leaf elongation and photosynthesis and cell turgor by osmotic adjustment in all tissues, increasing water uptake and water use efficiency, and the limitation of senescence and reduction of leaf area. However, under severe drought, growth may completely cease while the survival of the plants becomes important for regrowth after the stress. The physiological traits associated with plant survival include the cessation of leaf elongation, the senescence of aerial tissues, the maintenance of metabolism in the meristems with maintenance of a threshold of dehydration with osmotic adjustment, efficient water uptake from deeper layers of the soil, and the increase of carbohydrate reserves.

The responses to decreases in soil and cellular water content involve changes in various morphological, physiological, and biochemical characteristics as discussed above. These characteristics determine the survival and persistence of plants in water-limiting environments (Beard 1973; Nilsen and Orcutt 1996; Shinozaki and Yamaguchi-Shinozaki 1997; Chaves *et al.* 2003). This chapter provides a review of literature on major physiological mechanisms for drought resistance in perennial grass species, with specific emphasis on water relations, carbohydrate metabolism, and hormonal regulation.

Water relations

The maintenance of cellular hydration in leaves, roots, and meristematic regions is critical for plant survival in dry environments. It may be accomplished by a combination of different strategies associated with avoidance and/or tolerance to drought stress. Common survival strategies in perennial grasses include changes in rooting characteristics to avoid drought and accumulation of compatible solutes for osmotic adjustment to tolerate drought.

Water uptake associated with root size, distribution, and activity

Roots are the main engines for meeting transpirational demand, and play an important role in controlling plant water status to avoid drought injury. Water uptake in grasses is controlled by root size, spatial distribution, and activity or viability (Beard 1973; Huang and Gao 2000). A large root system is important for a plant to explore large soil volume for water uptake under non-limiting soil moisture conditions or when soil moisture is uniform along the soil profile. However, total root mass may not be necessarily correlated with drought resistance when soil moisture is limited or unevenly distributed spatially or temporally. Perdomo *et al.* (1996) reported no difference in total root mass between drought resistant and sensitive cultivars of Kentucky bluegrass. Root distribution was associated with drought avoidance more consistently than root mass in this species (Keeley and Koski 2002).

It is common that water is available at greater depths while soil drying occurs at the surface. Most roots of perennial grasses are found in the top 20-cm soil layers where soil moisture is often limited. In the northwestern Sonoran Desert, a C_4 bunchgrass *Pleuraphis rigida* had mean rooting depths of only 9-10 cm (Nobel 1997). Increased proportion of root mass at deeper soil depths in response to decreased water availability in surface soil is considered an important trait for drought avoidance in many perennial grasses (Sheffer *et al.* 1987; Torbert *et al.* 1990; Hays *et al.* 1991; Salaiz *et al.* 1991; Marcum *et al.* 1995; Carrow 1996a, b;). Greater root length densities of tall fescue, buffalograss, and bermuda grass below 30-cm soil depths also contributed to superior drought avoidance in these species compared to a shallower rooting cultivar of zoysiagrass (Qian*et al.* 1997). Keeley and Koski (2002) and Bonos and Murphy (1999) reported Kentucky bluegrass cultivars with greater percentages of their root systems distributed in deeper soil layers were more drought-resistant than those with shallower roots. Similar results have been observed in forage grasses under both field and greenhouse drought conditions, where species with greater plant root development at deeper soil depths exhibited greater persistence under conditions of drought stress (Volaire 1995; Guenni *et al.* 2002; Boschma *et al.* 2003).

Deep rooting into the soil where water is available for uptake allows the plant to avoid or delay tissue dehydration and maintain leaf water relations necessary for maintenance of physiological and metabolic functions (Molyneux and Davies 1983; Smucker and Aiken 1992; Carrow 1996b; Qian *et al.* 1997; Volaire and Lelievre 2001). Huang (1999) found that deep roots of buffalograss have hydraulic lift properties, which lift water from the deeper soil to the drying surface at night to sustain growth and nutrient uptake of surface roots and prevent leaves from desiccation. Factors or conditions that limit root growth may decrease the potential for drought avoidance. For example, by constricting the root growth of tall fescue within small containers, the ability of the plants to maintain leaf water relations under drought stress was greatly diminished compared to that observed under field conditions for this species (Volaire and Lelievre

2001). The importance of deep root proliferation could be related to continued maintenance and support of growth by providing water and nutrients to the rest of the plant even under conditions where part of the root system is under drying soil, thereby contributing to increased plant survival. In general, there may be a greater carbon allocated to roots, with a greater proportion of newly fixed carbon for roots in soil layers where water maybe more available (Huang and Gao 2000). Generally, drought resistant species or cultivars of perennial grasses have greater root-to-shoot ratio than drought sensitive species or cultivars (Nicolas *et al.* 1985; Volaire 1995; Huang and Fu 2000; Huang and Gao 2000). When grown in monoculture under water deficit, increased root-to-shoot ratios were associated with the ability of Mediterranean-type tall fescue plants to maintain higher water status compared to temperate tall fescue cultivars (Assuero *et al.* 2002). A more detailed discussion of carbon allocation in relation to drought resistance is provided in the following section.

Root viability or activity is a critical factor controlling water uptake and plant survival of long-term drought stress, considering that drought is one of the major causes for root death (Smucker *et al.* 1991). Maintaining viable roots during prolonged periods of drought is also an important factor for resumption of water uptake when soil is re-wetted after irrigation or rain (Huang 2000). Extensive root mortality occurs as a result of drought, especially of fine roots (Huck *et al.* 1987; Huang *et al.* 1997; Huang and Fry 1998) and roots in the upper soil profile (Hayes and Seastedt 1987; Smucker *et al.* 1991). Most grass species have a majority of their root systems confined in surface soil; therefore it is critical to evaluate the variability in root activity under decreasing soil moisture in combination with root size and distribution patterns in order to relate these factors to drought resistance. In a study with seven warm-season turfgrasses, Huang *et al.* (1997) found that high root viability in drought-tolerant species was more highly correlated with better shoot growth than root mass or root length under drought stress. In some cases, root viability or activity may not correlate with root distribution patterns or root mass in different soil layers (Sheffer *et al.* 1987; Huang 2000). Bonos and Murphy (1999) found that increased water uptake activity at 15 to 30 cm in Kentucky bluegrass cultivars was associated with increased drought resistance, but this was not attributable to differences in root mass within these same layers.

Osmotic adjustment

Upon exposure to environmental stresses that cause decreases in cellular water, such as drought, extremes in temperature, and salinity, plants may accumulate non-toxic, or compatible solutes. This active accumulation of solutes triggered by cellular dehydration is termed osmotic adjustment and differs from accumulation of solutes simply due to concentration effect (Blum 1988). Osmotically active solutes are associated with either movement of water into or reduced water efflux from cells. Therefore, it helps to maintain cellular turgor at a given leaf water potential and thus delays wilting of leaves, and enables tissues to sustain growth and metabolic and physiological functions at lower plant water status (Bohnert and Jensen 1996; Ingram and Bartels 1996; Hare *et al.* 1998). Furthermore, osmotic adjustment protects cellular proteins, various enzymes, cellular organelles, and membranes against desiccation injury, as well as protection against oxidative damage (Arakawa 1991; Crowe *et al.* 1992; Rhodes and Hanson 1993; Hoekstra *et al.* 2001).

The types of osmotically active solutes reported in osmotic adjustment are diverse, and typically include low molecular weight compounds such as amino acids (*e.g.*, proline), ammonium compounds (*e.g.*, glycinebetaine), sugars (*e.g.*, fructans, sucrose), polyols (*e.g.*, mannitol), inorganic ions (*e.g.*, potassium, calcium), organic acids (*e.g.*, malate); and hydrophilic proteins (*e.g.*, late embryogenesis abundant, LEA) (Hsiao 1973; Morgan 1984; Zhang *et al.* 1999; Chaves *et al.* 2003). In annual and perennial grasses, some of the solutes associated with osmotic adjustment during drought and other stresses include soluble sugars (Barker 1991; Richardson *et al.* 1992; Premachandra *et al.* 1995; Jiang and Huang 2001a), inorganic ions (Jones *et al.* 1980; Morgan 1992; Jiang and Huang 2001a), and proline (Bokhari and Trent 1985; Marcum and Murdoch 1994). Drought resistance has been positively correlated with osmotic adjustment in leaves of many species, including perennial grasses (Wilson and Ludlow 1983; White *et al.* 1992; Qian and Fry 1997; Geerts *et al.* 1998) and cereal grasses (Saneoka *et al.* 1995; Dorffling *et al.* 1997; Karakas *et al.* 1997; Romero *et al.* 1997; Sheveleva *et al.* 1997). Osmotic adjustment has also been observed in roots of different species, contributing to maintenance of root turgor and growth during drought (Sharp and Davies 1979; Westgate and Boyer 1985; Sharp *et al.* 1990). The capacity for osmotic adjustment in roots varies with species (Oosterhuis and Wullschleger 1988), and has not been as extensively studied in perennial grass species.

Osmotic adjustment contributes to maintenance of leaf turgor, thus leaf growth and biomass accumulation of perennial grasses may continue throughout some period of water deficit (Wilson *et al.* 1980; Turner and Begg 1981). However, osmotic adjustment is not always associated with sustained leaf development during stress, even under sufficient positive turgor (Thomas 1986). Other factors may also contribute to the sensitivity of leaf expansion during drought that could override the effects of osmotic adjustment, such as low leaf water potential (Hsiao 1973; Turner and Begg 1981; Tardieu and Simonneau 1998) and hormone signals from roots in drying soil (Davies *et al.* 1994; Wilkinson and Davies 2002). Osmotic adjustment is crucial for maintaining meristem viability under conditions of desiccation and also aid in the recovery of meristem function upon rehydration, whereby the various solutes are recycled and metabolized and used as energy resources for recovery (Toft *et al.* 1987; Wilson and Ludlow 1983; Chaves *et al.* 2003). Elmi and West (1995) suggested that the viability of the meristematic and leaf elongation regions due to osmotic adjustment enhanced turgor maintenance was a critical factor for survival and recovery after drought. Qian and Fry (1997) found a positive correlation between osmotic adjustment and recovery from prolonged drought in four turfgrasses, where species with highest osmotic adjustment had highest percentage of green leaves two weeks after re-watering. White *et al.* (1992) and Volaire (1995) also found that differential osmotic adjustment of turf and forage grasses, particularly within leaf bases, resulted in contrasted tiller survival and recovery after drought stress.

Osmotic adjustment is not always correlated with increased stress resistance (Maggio *et al.* 1997). In a comparison between cocksfoot and perennial ryegrass, cocksfoot had the least osmotic adjustment even though it was more drought resistant compared to ryegrass plants (Thomas 1986). Under these circumstances, other morphological and/or physiological responses play a more important role compared to osmotic adjustment to allow the plant to withstand prolonged drought stress. Furthermore, there may be variability among species and within genotypes and cultivars of the same species in the degree of osmotic adjustment and types of solutes active in the maintenance of turgor (Bokhari and Trent 1985; Barker *et al.* 1993; Rhodes and Hanson 1993; Qian and Fry 1997; Zhang *et al.* 1999). The degree of osmotic adjustment in perennial grasses may be affected by several other factors, including organ type and/or age (Chaves 1991; Veneklaas and Van den Boogaard 1994; Kameli and Losel 1995; Bajji *et al.* 2001) and the rate of dehydration and stress development (Thomas 1991). Under conditions of rapid dehydration, two forage grasses had decreased magnitude of osmotic adjustment compared to the same plants under a slower development of drought stress (Thomas 1986). Slower rates of water stress allow for more gradual acclimation of physiological parameters such as photosynthesis and osmotic potential, whereas faster rates of dehydration mostly cause rapid cellular injury and thus result in the inability of plants to adjust osmotically before they are killed (Toft *et al.* 1987). The duration of drought may also affect the benefits of osmotic adjustment. In the tropical perennial grass, *Eustachys paspaloides*, osmotic adjustment was more beneficial for the maintenance of growth under shorter periods of drought stress rather than long-term soil drought (Toft *et al.* 1987). However, under conditions of prolonged soil water deficit, osmotic adjustment was more important for survival of tiller bases rather than continued turgidity and maintenance of leaf growth.

Carbohydrate metabolism

Carbohydrates provide energy and carbon skeletons for plant growth and metabolic processes. Carbohydrate metabolism is important for supporting plant growth under mild or moderate drought and plant survival under severe drought stress conditions. Water availability in particular affects several aspects of carbon metabolism, such as the partitioning of carbohydrates between growth and storage, differential patterns of allocation between roots and shoots, as well as carbon consumption and availability during and following stress (Chaves 1991). Therefore, it is essential to recognize different plant strategies for allocation and utilization of available carbon resources, and also understand how these mechanisms are related to drought adaptation.

Carbon allocation

Increases in root-to-shoot biomass ratio during drought stress may result from increased carbon investment in roots more so than in shoots (Bradford and Hsiao 1982; Hamblin *et al.* 1990; Mooney and Winner 1991). The shift in carbon allocation pattern in favor of roots contributes to drought resistance in perennial (Volaire 1995; Assuero *et al.* 2002) and annual grasses (Nicolas *et al.* 1985; Simane *et al.* 1993; Kalapos *et al.* 1996). Huang and Gao (2000) reported that drought-tolerant cultivars of tall fescue had greater proportion of ^{14}C allocated to roots compared to drought-sensitive cultivars. Roots of perennial grasses are also able to adjust metabolic activities and carbon allocation within the root system to cope with drought stress. In a study investigating carbon metabolic responses under surface soil drying, both Kentucky bluegrass and tall fescue exhibited decreased respiration rates in shoots and roots in the upper drying soil layer; in contrast, there was enhanced carbon allocation to those roots in the lower, wet soil layer (Huang and Fu 2000; Huang and Gao 2000). This study suggested that perennial grasses are able to survive long-term drought stress by reducing carbon consumption of roots with little benefits in terms of water and nutrient uptake in the surface drying soil while increasing carbon investment in functional roots to maximize the capacity of water and nutrient uptake in the soil where water is available.

Carbohydrate accumulation

The amounts of total nonstructural carbohydrates (TNC) and water-soluble carbohydrates (WSC) within plant tissue have been utilized as indicators for physiological status of perennial grasses under drought and other stresses (Sheffer *et al.* 1979; Volenec 1986; Hull 1992; Fulkerson *et al.* 1994). Huang and Fu (2000) and Huang and Gao (2000) reported that TNC levels were either unaffected or increased in tall fescue and Kentucky bluegrass in response to drying conditions. Increases in carbohydrates have also been observed for *Dactylis glomerata* (Brown and Blaser 1970; Volaire 1994) and other cool-season grasses (Busso *et al.* 1990). This enhanced carbohydrate supply may be related to observations that decreases in plant growth occur before photosynthesis is significantly affected (Hsiao 1973; Chaves 1991), therefore causing some accumulation of carbohydrates within plant tissue. In the studies conducted by Huang and Fu (2000) and Huang and Gao (2000), carbon accumulation during drought stress was also attributed to lower consumption rates, which was related to decreases in root and canopy respiration rates. Reduced respiration rates have been associated with drought resistance and increased plant survival during drought stress in other species (Nicolas *et al.* 1985). Furthermore, storage carbohydrates, such as fructans and starch, may also be hydrolyzed and contribute to higher concentrations of simple sugars (glucose, fructose, and sucrose) in drought stressed grasses (Suzuki and Chatterton 1993; Spollen and Nelson 1994).

The accumulation of TNC and WSC has been associated with increased drought resistance and survival in grasses (Volaire 1994, Volaire *et al.* 1998). Increases in soluble sugars, in combination with other solutes, may contribute to osmotic adjustment by lowering leaf water potential at which stomatal closure occurs and sustaining turgor in growing regions of leaves and roots (Chaves 1991; Richardson *et al.* 1992; Godde 1999). Additionally, these reserves may be utilized as the primary source of carbohydrates for continued biomass production throughout drought or following drought stress (Chaves 1991). However, utilization of carbohydrates for continued growth during stress may be more advantageous during short periods of water deficit, since drought duration and intensity are important factors in the depletion of carbohydrate reserves. If plants continue to utilize carbohydrates under conditions of moderate drought, a negative carbon balance may result when respiration eventually exceeds photosynthesis if stress is not alleviated (Boschma *et al.* 2003). Therefore, maintenance of photosynthesis under moderate drought stress is critical for carbohydrate accumulation in support of continued tissue growth. Poor survival and persistence has been associated with the depletion of carbohydrate reserves in ryegrass (Arcioni *et al.* 1985) and *Dactylis glomerata* (Volaire 1995) under prolonged drought stress. Cultivars or species that stored carbohydrates instead of continuously utilizing them for growth during long-term drought stress may better be able for the survival of plants. Additionally, utilization of reserve carbohydrates for re-growth of grasses after relief from drought late in the growing season may be at the expense of increasing and replacing storage carbohydrates that would otherwise be used for winter survival and spring re-growth (Karsten and MacAdam 2001).

Carbohydrate partitioning between growth and storage

In some perennial grasses, survival and recovery are more important than continued biomass production under prolonged drought (Volaire *et al.* 1998). Therefore, instead of maintenance of growth, these species may allocate reserves to stem bases and roots as a survival mechanism under prolonged periods of drought. In *Themeda triandra*, a C_4 perennial forage grass, severe drought stress induced TNC accumulation primarily in stem bases and served as important reserve pools for re-growth following stress (Oosthuizen and Snyman 2001). Summer drought-tolerant populations of cocksfoot also accumulated significant amounts of WSC in leaf bases, while more drought susceptible populations continuously utilized carbohydrate reserves (Volaire 1995). This differential use in carbohydrates was associated with increased plant survival during stress and recovery after stress in summer drought-tolerant plants. In contrast, perennial ryegrass hydrolyzed carbohydrates in leaves and made them available for regrowth rather than storage, which depleted carbohydrate reserves and limited recovery potential (Karsten and MacAdam 2001).

Hormonal regulation

For years, researchers have been studying the regulatory mechanisms of shoot and root responses of perennial grasses to drought stress. Abscisic acid (ABA) and cytokinins are two major groups of plant hormones that play important roles in regulating plant responses to decreases in soil water availability (Pospisilova *et al.* 2000; Wilkinson and Davies 2002). The involvement of these hormones in root-to-shoot signaling, particularly through regulation of stomatal behavior and leaf growth, has been implicated in plant resistance to drought (Quarrie 1989, 1993; Pospisilova *et al.* 2000). Various studies with crop species and limited work in perennial grasses have demonstrated genetic variation in ABA and cytokinin accumulation and/or sensitivity also contributes to genotypic differences in drought resistance.

ABA accumulation and regulation of stomata and growth

Stomatal regulation of water loss is a critical factor controlling plant survival in dry environments. Excessive water loss through transpiration can result in water deficit and desiccation of a plant. Rapid stomatal closure in response to drought stress may protect plants from being desiccated. The traditional belief is that stomatal closure is induced by water deficit. Recent studies have confirmed that a chemical signal moving from roots to shoots in response to soil drying may induce stomatal closure and reduce water loss (Blackman and Davies 1985; Zhang and Davies 1989; Davies *et al.* 2002). In experiments utilizing split-root systems in which part of the root system is exposed to drying soil while the remaining roots are maintained in moist soil, it has been found that restrictions in stomatal conductance and growth occur even though roots in the well-watered soil provide sufficient moisture to maintain leaf water relations (Zhang and Davies 1989; Gowing *et al.* 1990).

Abscisic acid is believed to be the primary chemical signal controlling stomatal response to drying soil (Wilkinson and Davies 2002). It is now well established that ABA maybe synthesized in roots exposed to drying soil and transported to shoots, where the hormone triggers a signal transduction cascade eventually leading to a reduction in guard cell turgor, stomatal closure, and an overall decrease in water loss via transpiration (McAinsh *et al.* 1997; Assmann and Shimazaki 1999; Wilkinson and Davies 2002). ABA may also be involved in plant adaptation to drought stress by inhibiting leaf growth or transpirational area (Bacon *et al.* 1998; Alves and Setter 2000), and induction of antioxidant enzymes (Bueno *et al.* 1998; Bellaire *et al.* 2000; Jiang and Zhang 2002) and genes related to drought resistance (Bray *et al.* 1999; Jin *et al.* 2000; Campbell *et al.* 2001; Tamminen *et al.* 2001).

The importance of ABA as a metabolic factor in the regulation of plant tolerance to stresses has received great attention in recent years in other species (Bray 1999; Pekic *et al.* 1995). However, limited information is available on the association of ABA signaling and drought resistance in perennial grasses. Many studies in annual cereal grasses and some in perennial grasses have shown a negative relationship between ABA accumulation, stomatal conductance, and leaf growth; particularly for sorghum (*Sorghum bicolor* (L.) Moench; Auge *et al.* 1995), maize (*Zea mays* L.; Davies and Zhang 1991; Tardieu *et al.* 1992), rice (*Oryza sativa* L.; Bano *et al.* 1993), wheat (*Triticum aestivum* L.; Blum *et al.* 1991), barley (*Hordeum vulgare* L.; Borel *et al.* 1997), tall fescue (Puliga *et al.* 1996), and Kentucky bluegrass (Wang *et al.* 2003; DaCosta *et al.* 2004). The combination of decreased stomatal conductance and growth rate in many of these species is beneficial for plant survival of soil drying due to restriction of leaf transpirational area, which permitted prolonged maintenance of leaf turgor. Because modifications in leaf physiological parameters occurred without any significant reduction in leaf water status, these studies provided evidence for hormonal or chemical regulation of plant responses to soil drying. Three forage grasses demonstrating differences in drought sensitivity exhibited variation in ABA control on leaf physiological responses to drought (Puliga *et al.* 1996). In tall fescue, increases in ABA concentration were associated with inhibition of leaf growth and maintenance of leaf turgor, therefore delaying the detrimental effects of soil drying. In contrast, decreases in leaf turgor of *Eragrostis curvula* and *Sporobolus stapfianus* occurred before any significant increases in ABA were detected. Furthermore, *E. curvula* exhibited more of a drought tolerance strategy, sustaining leaf growth even at low leaf water potential. This study demonstrates that not all species exhibit chemical regulation under soil drying, and that variability in how plants sense and respond to decreases in water availability may be related to specific strategies for drought avoidance or drought tolerance. In many species modification in leaf growth may be more closely related to changes in leaf water status rather than chemical influences (Tardieu and Simonneau 1998).

The relationship between ABA accumulation and drought resistance varies with species and genotypes (Blum and Sinmena 1995; Cellier *et al.* 1998; Wang and Huang 2003). Some species with high leaf ABA accumulation have been reported to be more drought-tolerant than those with low ABA accumulation (Larqué-Saavedra and Wain 1976; Henson *et al.* 1981). In contrast, some studies suggest that low ABA accumulation is positively related to drought resistance, such as in cocksfoot (Volaire *et al.* 1998), wheat (Innes *et al.* 1984; Quarrie 1989), maize (Ilahi and Dorffling 1982), and sorghum (Durley *et al.* 1983). Drought resistance in Kentucky bluegrass cultivars was associated with slower rates of ABA accumulation in leaves compared to drought-sensitive cultivars in response to short-term drought stress (Wang *et al.* 2003). These cultivars were characterized by greater leaf water potential, net photosynthesis, stomatal conductance, and turfgrass quality during drought stress, suggesting that slower rates of ABA accumulation could be beneficial for the maintenance of photosynthesis and growth during short-term drought stress.

Sensitivity of the stomatal complex to ABA accumulation is also an important factor shown to vary among genotypes and species (Quarrie and

Jones 1979; Cellier *et al.* 1998). Wang and Huang (2003) reported that drought-tolerant cultivars of Kentucky bluegrass exhibited higher stomatal sensitivity to changes in leaf ABA content compared to drought-sensitive cultivars, which led to earlier stomatal closure, less cell membrane damage, and overall delay in the decline of overall turfgrass quality. They concluded that drought tolerance of Kentucky bluegrass is associated with the sensitivity of stomata to increases in ABA production. Exogenous ABA application to tall fescue and Kentucky bluegrass leaves improved drought resistance in both species under controlled environmental conditions by inducing stomatal closure and enhancing osmotic adjustment (Jiang and Huang 2001b; Wang *et al.* 2003).

Interaction of cytokinins and ABA in relation to drought resistance

Cytokinins generally have antagonistic physiological effects with ABA, especially in stomatal control. Cytokinins maintain stomatal opening, and thus increase stomatal conductance and transpiration rates (Jewer and Incoll 1980; Blackman and Davies 1985; Incoll and Jewer 1987; Lechowski 1997). Cytokinins also delay leaf senescence (Catsky *et al.* 1996; Naqvi 1999) and enhance photosynthetic protein synthesis and electron transport (Synkova *et al.* 1997). There is typically decreased cytokinin accumulation in drought-stressed plants, which can amplify responses of shoots to increasing ABA levels (Goicoechea *et al.* 1995, 1997; Naqvi 1995). However, it was recently suggested that the actual ratio of ABA/cytokinin seems to be more important for regulation of gas exchange and water relations than absolute changes in concentration of either hormone (Cheikh and Jones 1994; Goicoechea *et al.* 1997; Moore-Gordon *et al.* 1998).

There is considerably less research on how these hormones may be involved in the recuperative ability of grasses from drought stress. Changes in hormonal balance have been reported as important for relief from drought stress and facilitating plant recovery (Ivanova *et al.* 1998; Yordanov *et al.* 1997, 1999). In Kentucky bluegrass, an increase in cytokinin content upon relief of drought corresponded with recovery of physiological parameters (Wang and Huang unpublished results).

Concluding remarks

Perennial grasses are widely distributed across different climatic regions, and often exposed to drought stress, especially in semi-arid and arid areas. Grasses adapted to dry environments exhibit three major mechanisms: drought avoidance, drought tolerance, and drought escape, which together are considered as drought resistance. Various physiological traits of drought resistance have been identified in perennial grass species, including modification of water relations, carbohydrate metabolism, and hormone synthesis or sensitivity. However, much more needs to be explored in drought stress tolerance mechanisms, as drought will continue to be the leading factor that limits grass growth in many areas of the world.

Increased efforts should be made to better understand the factors controlling root survival in drying soils, proteins/genes associated with the accumulation of osmolytes, and also the mechanisms controlling efficient carbon utilization and carbon allocation/partitioning under drought stress. Such information should provide additional arsenal in the improvement of drought tolerance in perennial grasses by using biotechnology or traditional breeding. Greater insight is needed on the relationships between hormone accumulation and physiological characteristics for perennial grasses under drought, and how these may be related to drought resistance in different species. The capability of exploiting aspects of hormonal signaling, accumulation, and sensitivity in perennial grasses could serve as a screening tool for genetic improvement and benefit the development of drought-resistant grasses.

Literature cited

Alves AAC and Setter TL. 2000. Response of cassava to water deficit: leaf area growth and abscisic acid. *Crop Science* **40:** 131-137.

Arakawa T. 1991. Protein-solvent interactions in pharmaceutical formulations. *Pharmaceutical Research* **8:** 285-291.

Arcioni S, Falcinelli M, and Mariotti D. 1985. Ecological adaptation in *Lolium perenne* L.: Physiological relationships among persistence, carbohydrate reserves and water availability. *Canadian Journal of Plant Science* **65:** 615-624.

Assmann SM and Shimazaki K. 1999. The multisensory guard cell, stomatal responses to blue light and abscisic acid. *Plant Physiology* **119:** 809-816.

Assuero SG, Matthew C, Kemp P, Barker DJ, and Mazzanti A. 2002. Effects of water deficit on Mediterranean and temperate cultivars of tall fescue. *Australian Journal of Agricultural Research* **53:** 29-40.

Auge RM, Stodola AJW, Ebel RC and Duan X. 1995. Leaf elongation and water relations of mycorrhizal sorghum in response to partial soil drying: two *Glomus* species at varying phosphorus fertilization. *Journal of Experimental Botany* **46:** 297-307.

Bacon MA, Wilkinson S, and Davies WJ. 1998. pH-regulated leaf cell expansion in droughted plants is abscisic acid dependent. *Plant Physiology* **118:** 1507-1515.

Bajji M, Lutts S, and Kinet JM. 2001. Water deficit effects on solute contribution to osmotic adjustment as a function of leaf ageing in three durum wheat (*Triticum durum* Desf.). *Plant Science* **160:** 669-681.

Bano A, Dorffling K, Bettin D, and Hahn H. 1993. Abscisic acid and cytokinins as possible root-to-shoot signals in xylem sap of rice plants in drying soil. *Australian Journal of Plant Physiology* **20:** 109-115.

Barker DJ. 1991. Physiological responses of sorghum and six forage grasses to water deficits. *Dissertation Abstracts International B Sciences Engineering* **52:** 1135B-1136B.

Barker DJ, Sullivan CY, and Moser LE. 1993. Water deficit effects on osmotic potential, cell wall elasticity, and proline in five forage grasses. *Agronomy Journal* **85:** 270-275.

Beard JB. 1973. *Turfgrass: Science and Culture.* Prentice-Hall, Inc., Englewood Cliffs.

Bellaire BA, Carmody J, Braud J, Gosset GR, Banks SW, Lucas MC, and Fowler TE. 2000. Involvement of abscisic acid-dependent and independent pathways in the up-regulation of antioxidant enzyme activity during NaCl stress in cotton callus tissue. *Free Radical Research* **33:** 531-545.

Blackman PG and Davies WJ. 1985. Root to shoot communication in maize plants of the effects of soil drying. *Journal of Experimental Botany* **36:** 39-48.

Blum A. 1988. *Plant breeding for stress environments.* CRC Press Inc., Boca Raton.

Blum A. and Sinmena B. 1995. Isolation and characterization of variant wheat cultivars for ABA sensitivity. *Plant, Cell and Environment* **18:** 77-83.

Blum A, Johnson JW, Ramseur EL, and Tollener EW. 1991. The effect of a drying top soil and a possible non-hydraulic root signal on wheat growth and yield. *Journal of Experimental Botany* **42:** 1225-1231.

Bohnert HJ and Jensen RG. 1996. Strategies for engineering water-stress tolerance in plants. *Trends in Biotechnology* **14:** 89-97.

Bokhari UG and Trent JD. 1985. Proline concentrations in water stressed grasses. *Journal of Range Management* **38:** 37-38.

Bonos S and Murphy JA. 1999. Growth responses and performance of Kentucky bluegrass under summer stress. *Crop Science* **39:** 770-774.

Borel C, Simonneau T, This D, and Tardieu F. 1997. Stomatal conductance and ABA concentration in the xylem sap of barley lines of contrasting genetic origins. *Australian Journal of Plant Physiology* **24:** 607-615.

Boschma SP, Hill MJ, Scott JM, and Rapp GG. 2003. The response to moisture and defoliation stresses, and traits for resilience of perennial grasses on the Northern Tablelands of New South Wales, Australia. *Australian Journal of Agricultural Research* **54:** 903-916.

Bradford KJ and Hsiao TC. 1982. Physiological responses to moderate water stress. In: Lange OL, Nobel PS, Osmond CB, and Ziegler H (eds.) *Physiological Plant Ecology. II. Water relations and carbon assimilation.* Springer-Verlag, Berlin. Pp. 263-324.

Bray EA, Shih TY, Moses MS, Cohen A, Imai R, and Plant AL. 1999. Water deficit induction of a tomato H1 histone requires abscisic acid. *Plant Growth Regulation* **29:** 35-46.

Brown RH and Blaser RE. 1970. Soil moisture and temperature effects on growth and soluble carbohydrates of orchard grass (*Dactylis glomerata*). *Crop Science* **10:** 213-216.

Bueno P, Piqueras A, Kurepa J, Savoure A, Verbruggen N,Van Montegu M, and Inze D. 1998. Expression of antioxidant enzymes in response to abscisic acid and high osmoticum in tobacco BY-2 cell cultures. *Plant Science* **138:** 27-34.

Burton GW, De Vane EH, and Carter RL. 1954. Root penetration, distribution, and activity in southern grasses measured by yields, drought symptoms, and ^{32}P uptake. *Agronomy Journal* **46:** 229-233.

Busso CA, Richards JH, and Chatterton NJ. 1990. Nonstructural carbohydrates and spring regrowth of two cool-season grasses: interaction of drought and clipping. *Journal of Range Management* **43:** 336-343.

Campbell JL, Klueva NY, Zheng K, Nieto-Sotelo J, Ho THD, and Nguyen HT. 2001. Cloning of new members of heat shockprotein HSP101 gene family in wheat (*Triticum aestivum* (L.) Moench) inducible by heat, dehydration, and ABA. *Biochimica et Biophysica Acta* **1517:** 270-277.

Carrow RN. 1996a. Drought avoidance characteristics of diverse tall fescue cultivars. *Crop Science* **36:** 371-377.

Carrow RN. 1996b. Drought resistance aspects of turfgrasses in the southeast: Root-shoot responses. *Crop Science* **36:** 687-694.

Casler MD, Pederson JF, Eizenga GC, and Stratton SD. 1996. Germplasm and culti-var development. In: Moser LE, Buxton DR, and Casler MD (eds.) *Cool-season Forage Grasses*. American Society of Agronomy, Madison. Pp. 413-469.

Catsky J, Pospisilova J, Kaminek M, Gaudinova A, Pulkrabek J, and Zahradnicek J. 1996. Seasonal changes in sugar beet photosynthesis as affected by exogenous cytokinin N6-(m-hydroxybenzyl)adenosine. *Biologia Plantarum* **38:** 511-518.

Cellier F, Conejero G, Breitler JC, and Casse F. 1998. Molecular and physiological-responses to water deficit in drought-tolerant and drought-sensitive lines of sunflower. *Plant Physiology* **116:** 319-328.

Chaves MM. 1991. Effects of water deficits on carbon assimilation. *Journal of Ex-perimental Botany* **42:** 1-16.

Chaves MM, Maroco JP and Pereira JS. 2003. Understanding plant responses to drought-from genes to the whole plant. *Functional Plant Biology* **30:** 239-264.

Cheikh N and Jones RJ. 1994. Disruption of maize kernel growth and development by heat stress: Role of cytokinin/abscisic acid balance. *Plant Physiology* **106:** 45-51.

Council for Agricultural Science and Technology (CAST). 2004. *Biotechnology - derived, perennial turf and forage grasses: criteria for evaluation*. Issue Paper 25. Council for Agricultural Science and Technology, Ames, Iowa.

Crowe JH, Hoekstra FA, and Crowe LM. 1992. Anhydrobiosis. *Annual Review of Physiology* **54:** 579-599.

DaCosta M, Wang Z and Huang B. 2004. Physiological adaptation of Kentucky bluegrass to localized soil drying. *Crop Science* **44:** 1307-1314.

Davies WJ and Zhang J. 1991. Root signals and the regulation of growth and devel-opment of plants in drying soil. *Annual Review of Plant Physiology and Plant Molecular Biology* **42:** 55-76.

Davies WJ, Tardieu F, and Trejo CL. 1994. How do chemical signals work in plants that grow in drying soil? *Plant Physiology* **104:** 309-314.

Davies WJ, Wilkinson S, and Loveys B. 2002. Stomatal control by chemical signal-ing and the exploitation of this mechanism to increase water use efficiency in agriculture. *New Phytologist* **153:** 449-460.

Dorffling K, Dorffling H, Lesselich G, Luck E, Zimmermann C, Melz G, and Jurgens HU. 1997. Heritable improvement of frost tolerance in winter wheat by *in vitro*-selection of hydroxy proline-resistant proline overproducing mutants. *Euphytica* **93:** 1-10.

Duncan RR and Carrow RN. 1999. Turfgrass molecular genetic improvement for abiotic/edaphic stress resistance. *Advances in Agronomy* **67:** 233-305.

Durley RC, Kannangara T, Seetharama N, and Simpson GM. 1983. Drought resis-tance of sorghum bicolor. 5. Genotypic differences in the concentrations of free and conjugated abscisic phaseic and indole-3 acetic acids in leaves of field-grown drought-stressed plants. *Canadian Journal of Plant Science* **63:** 131-145.

Elmi AA and West CP. 1995. Endophyte infection effects on stomatal conductance, osmotic adjustment and drought recovery of tall fescue. *New Phytologist* **131:** 61-67.

Fulkerson WJ, Slack K, and Lowe KF. 1994. Variation in the response of *Lolium* genotypes to defoliation. *Australian Journal of Agricultural Research* **45:** 1309-1317.

Geerts P, Buldgen A, Diallo T, and Dieng A. 1998. Drought resistance by six Sene-galese local strains of *Andropogon gayanus* var. *bisquamulatus* through osmo-regulation. *Tropical Grasslands* **32:** 235-242.

Godde D. 1999. Adaptations of the photosynthetic apparatus to stress conditions. In: Lerner HR (ed.) *Plant Responses to Environmental Stresses*. Marcel Dek-ker, New York. Pp. 449-474.

Goicoechea N, Antolín MC, and Sánchez-Díaz M. 1997. Gas exchange is related to the hormonal balance in mycorrhizal or nitrogen-fixing alfalfa subjected to drought. *Physiologia Plantarum* **100:** 989-997.

Goicoechea N, Dolezal K, Antolín MC, Strnad M, and Sánchez-Díaz M. 1995. Influence of mycorrhizae and *Rhizobium* on cytokinin content in drought-stressed alfalfa. *Journal of Experimental Botany* **46:** 15431549.

Gowing DJG, Davies WJ, and Jones HG. 1990. A positive root-sourced signal as an indicator of soil drying in apple, *Malus* × *domestica* Borkh. *Journal of Experimental Botany* **41:** 1535-1540.

Guenni O, Marin D, and Baruch Z. 2002. Responses to drought of five *Brachiaria* species. I. Biomass production, leaf growth, root distribution, water use and forage quality. *Plant and Soil* **243:** 229-241.

Hamblin A, Tennant D, and Perry MW. 1990. The cost of stress: Dry matter partitioning changes with seasonal supply of water and nitrogen to dryland wheat. *Plant and Soil* **122:** 47-58.

Hare PD, Cress WA, and Van Staden J. 1998. Dissecting the roles of osmolyte accumulation during stress. *Plant, Cell and Environment* **21:** 535-553.

Hayes DC and Seastedt TR. 1987. Root dynamics of tall grass prairie in wet and dry years. *Canadian Journal of Botany* **65:** 787-791.

Hays KL, Barber JF, Kenna MP and McCollum TG. 1991. Drought avoidance mechanisms of selected bermuda grass genotypes. *Hort Science* **26:** 180-182.

Henson IE, Mahalakshmi V, Bidinger FR, and Alagarswamy G. 1981. Genotypic variation in pearl millet (*Pennisetum americanum* (L.) Leeke), in the ability to accumulate abscisic acid in response to water. *Journal of Experimental Botany* **32:** 899-910.

Hoekstra FA Golovina EA and Buitink J. 2001. Mechanisms of plant desiccation tolerance. *Trends in Plant Science* **6:** 431-438.

Hsiao TC. 1973. Plant responses to water stress. *Annual Review of Plant Physiology* **24:** 519-570.

Huang B. 1999. Water relations and root activities of *Buchloe dactyloides* and *Zoysia japonica* in response to localized soil drying. *Plant and Soil* **208:** 179-186.

Huang B. 2000. Role of root morphological and physiological characteristics in drought resistance of plants. In: Wilkinson RE (ed.) *Plant-Environment Interactions.* Marcel Dekker, New York. Pp. 39-63.

Huang B and Fry JD. 1998. Root anatomical, physiological, and morphological responses to drought stress for tall fescue cultivars. *Crop Science* **38:** 1017-1022.

Huang B and Fu J. 2000. Photosynthesis, respiration, and carbon allocation of two cool-season perennial grasses in response to surface soil drying. *Plant and Soil* **227:** 17-26.

Huang B and Gao H. 2000. Root physiological characteristics associated with drought resistance in tall fescue cultivars. *Crop Science* **40:** 196-203.

Huang B, Duncan RR, and Carrow RN. 1997. Drought-resistance mechanisms of seven warm-season turfgrasses under surface soil drying: II. Root aspects. *Crop Science* **37:** 1863-1869.

Huck MG, Hoogenboom G, and Peterson CM. 1987. Soybean root senescence under drought stress. In: Taylor HM (ed.) *Minirhizotron Observation Tubes: Methods and Applications for Measuring Rhizosphere Dynamics.* American Society of Agronomy Publishing, Madison. Pp. 109-121.

Hull RJ. 1992. Energy relations and carbohydrate partitioning in turfgrasses. In: Waddington DV, Carrow RN, and Shearman RC (eds.) *Turfgrass.* Vol. 32. American Society of Agronomy, Inc., Crop Science Society of America, Inc., Soil Science Society of America, Inc., Madison. Pp. 175-205.

Ilahi I and Dorffling K. 1982. Changes in abscisic acid and proline levels in maize varieties of different drought resistance. *Physiologia Plantarum* **55:** 129-135.

Incoll LD and Jewer PC. 1987. Cytokinins and stomata. In: Zeiger E, Farquhar GD, and Cowan IR (eds.) *Stomatal Function.* Stanford University Press, Stanford. Pp. 281-292.

Ingram J and Bartels D. 1996. The molecular basis of dehydration tolerance in plants. *Annual Review of Plant Physiology and Plant Molecular Biology* **47:** 377-403.

Innes P, Blackwell RD, and Quarrie SA. 1984. Some effects of genetic variation in drought-induced abscisic acid accumulation on the yield and water use of spring wheat. *Journal of Agricultural Science* **102:** 341-351.

Ivanova AP, Stefanov KL and Yordanov IT. 1998. Effect of cytokinin 4-PU-30 on the lipid composition of water stressed soybeans. *Biologia Plantarum* **41:** 155-159.

Jewer PC and Incoll LD. 1980. Promotion of stomatal opening in the grass *Anthephora pubescens* Neesby a range of natural and synthetic cytokinins. *Planta* **150:** 218-221.

Jiang M and Zhang J. 2002. Water stress-induced abscisic acid accumulation triggers the increased generation of reactive oxygen species and up-regulates the activities of antioxidant enzymes in maize leaves. *Journal of Experimental Botany* **53:** 2401-2410.

Jiang Y and Huang B. 2001a. Osmotic adjustment and root growth associated with drought preconditioning enhanced heat tolerance in Kentucky bluegrass. *Crop Science* **41:** 1168-1173.

Jiang Y and Huang B. 2001b. Protein alteration in tall fescue in response to drought stress and abscisic acid. *Crop Science* **42:** 202-207.

Jin S, Chen CCS, and Plant AL. 2000. Regulation by ABA of osmotic-stress-induced changes in protein synthesis in tomato roots. *Plant, Cell and Environment* **23:** 51-60.

Jones MM, Osmond CB, and Turner NC. 1980. Accumulation of solutes in leaves of sorghum and sunflower in response to water deficit. *Australian Journal of Plant Physiology* **7:** 193-205.

Kalapos T, Van den Boogaard R and Lambers H. 1996. Effect of soil drying on growth, biomass allocation and leaf gas exchange of two annual grass species. *Plant and Soil* **185:** 137-149.

Kameli A and Losel DM. 1995. Contribution of carbohydrates and other solutes to osmotic adjustment in wheat leaves under water stress. *Journal of Plant Physiology* **145:** 363-366.

Karakas B, Ozias-Akins P, Stushnoff C, Suefferheld M, and Rieger M. 1997. Salinity and drought tolerance of mannitol-accumulating transgenic tobacco. *Plant, Cell and Environment* **20:** 609-619.

Karsten HD and MacAdam JW. 2001. Effect of drought on growth, carbohydrates, and soil water use by perennial ryegrass, tall fescue, and white clover. *Crop Science* **41:** 156-166.

Keeley SJ and Koski AJ. 2002. Root distribution and ET as related to drought avoidance in *Poa pratensis.* In: Thain E (ed.) *Science and Golf IV: Proceedings of the World Scientific Congress of Golf.* Routledge, NewYork. Pp. 555-563.

Larqué-Saavedra A and Wain RL. 1976. Studies on plant growth-regulating substances. XLII. Abscisic acid as a genetic character related to drought tolerance. *Annals of Applied Biology* **83:** 291-297.

Lechowski Z. 1997. Stomatal response to exogenous cytokinin treatment of the hemiparasite *Melampyrum arvense* L. before and after attachment to the host. *Biologia Plantarum* **39:** 13-21.

Maggio A, Bressan RA, Hasegawa PM, and Locy RD. 1997. Moderately increased constitutive proline does not alter osmotic stress tolerance. *Physiologia Plantarum* **101:** 240-246.

Marcum KB and Murdoch CL. 1994. Salinity tolerance mechanisms of six C_4 turfgrasses. *Journal of The American Society for Horticultural Science* **119:** 779-784.

Marcum KB, Engelke MC, Morton SJ, and White RH. 1995. Rooting characteristics and associated drought resistance of zoysiagrass. *Agronomy Journal* **87:** 534-538.

McAinsh MR, Brownlee C, and Hetherington AM. 1997. Calcium ions as second messengers in guard cell signal transduction. *Physiologia Plantarum* **100:** 16-29.

Molyneux DE and Davies WJ. 1983. Rooting pattern and water relations of three pasture grasses growing in drying soil. *Oecologia* **58:** 220-224.

Mooney HA and Winner WE. 1991. Partitioning response of plants to stress. In: Mooney HA, Winner WE, and Pell EJ (eds.) *Response of Plants to Multiple Stresses.* Academic Press, San Diego. Pp. 129-141.

Moore-Gordon CS, Cowan AK, Bertling I, Botha CEJ, and Cross RHM. 1998. Symplastic solute transport and avocado fruit development: a decline in cytokinin/ABA ratio is related to appearance of the Hass small fruit variant. *Plant and Cell Physiology* **39:** 1027-1038.

Morgan JM. 1984. Osmoregulation and water stress in higher plants. *Annual Review of Plant Physiology* **35:** 299-319.

Morgan JM. 1992. Osmotic components and properties associated with genotypic differences in osmoregulation in wheat. *Australian Journal of Plant Physiology* **6:** 67-76.

Naqvi SSM. 1995. Plant/crop hormones under stressful conditions. In: Pessarakli M (ed.) *Handbook of Plant and Crop Stress.* Marcel Dekker, NewYork. Pp. 645-660.

Naqvi SSM. 1999. Plant hormones and stress phenomena. In: Pessarakli M (ed.) *Handbook of Plant and Crop Stress.* Marcel Dekker, New York. Pp. 709-730.

Nelson CJ and Moser LE. 1995. Morphology and systematics. In: Barnes RF, Miller DA, and Nelson CJ (eds.) *Forages. Volume 1: An Introduction to Grassland Agriculture.* Iowa State UniversityPress. Pp. 15-30.

Nicolas ME, Lambers H, Simpson RJ, and Dalling MJ. 1985. Effect of drought on metabolism and partitioning of carbon in two wheat varieties differing in drought-tolerance. *Annals of Botany* **55:** 727-742.

Nilsen ET and Orcutt DM. 1996. *Physiology of Plants Under Stress: Abiotic Factors.* John Wiley & Sons, New York.

Nobel PS. 1997. Root distribution and seasonal production in the northwestern Sonoran Desert for a C_3 subshrub, a C_4 bunchgrass, and a CAM leaf succulent. *American Journal of Botany* **84:** 949-955.

Oosterhuis DM and Wullschleger SD. 1988. Drought tolerance and osmotic adjustment of various crops in response to water stress. *Arkansas Farm Research* **37:** 12.

Oosthuizen IB and Snyman HA. 2001. The influence of water stress on non-structural carbohydrate concentration in *Themeda triandra*. *South African Journal of Botany* **67**: 53-57.

Pekic S, Stikic R, Tomljanovic L, Andjelkovic V, Ivanovic M, and Quarrie SA. 1995. Characterization of maize lines differing in leaf abscisic acid content in the field. 1. Abscisic acid physiology. *Annals of Botany* **75**: 67-73.

Perdomo P, Murphy JA, and Berkowitz GA. 1996. Physiological changes associated with performance of Kentucky bluegrass cultivars during summer stress. *Hort Science* **31**: 1182-1186.

Pospisilova J, Synkova H, and Rulcova J. 2000. Cytokinins and water stress. *Biologia Plantarum* **43**: 321-328.

Premachandra GS, Hahn DT, Rhodes D, and Joly RJ. 1995. Leaf water relations and solute accumulation in two grain sorghum lines exhibiting contrasting drought tolerance. *Journal of Experimental Botany* **46**: 1833-1841.

Puliga S, Vazzana C, and Davies WJ. 1996. Control of crops leaf growth by chemical and hydraulic influences. *Journal of Experimental Botany* **47**: 529-537.

Qian Y and Fry JD. 1997. Water relations and drought tolerance of four turfgrasses. *Journal of The American Society for Horticultural Science* **122**: 129-133.

Qian YL, Fry JD, and Upham WS. 1997. Rooting and drought avoidance of warm-season turfgrasses and tall fescue in Kansas. *Crop Science* **37**: 905-910.

Quarrie SA. 1989. Abscisic acid as a factor in modifying drought resistance. In: Cherry JH (ed.) *Environmental Stress in Plants: Biochemical and Physiological Mechanisms*. Springer-Verlag, Berlin. Pp. 27-37.

Quarrie SA. 1993. Understanding plant responses to stress and breeding for improved stress resistance-the generation gap. In: Close TJ and Bray EA (eds.) *Plant Responses to Cellular Dehydration During Environmental Stress, Proceedings of the 16th Annual Riverside Symposium in Plant Physiology*. Vol. 10. American Society of Plant Physiologists, Rockville. Pp.224-245.

Quarrie SA and Jones HG. 1979. Genotypic variation in leaf water potential, stomatal conductance and abscisic acid concentration in spring wheat subjected to artificial drought stress. *Annals of Botany* **44**: 323-332.

Rhodes D and Hanson AD. 1993. Quaternary ammonium and tertiary sulfonium compounds in higher plants. *Annual Review of Plant Physiology and Plant Molecular Biology* **44**: 357-384.

Richardson MD, Chapman GW, Hoveland CS, and Bacon CW. 1992. Sugar alcohols in endophyte-infected tall fescue under drought. *Crop Science* **32**: 1060-1061.

Romero C, Belles JM, Vaya JL, Serrano R, and Culianez-Macia FA. 1997. Expression of they east trehalose-6-phosphate synthase gene in transgenic tobacco plants: pleiotropic phenotypes include drought tolerance. *Planta* **201**: 293-297.

Salaiz TA, Shearman RC, Riordan TP, and Kinbacher EJ. 1991. Creeping bentgrass cultivar water use and rooting responses. *Crop Science* **31**: 1331-1334.

Saneoka H, Nagasaka C, Hahn DT, Yang WJ, Premachandra GS, Joly RJ, and RhodesD. 1995. Salt tolerance of glycinebetaine-deficient and -containing maize lines. *Plant Physiology* **107**: 631-638.

Sharp RE and Davies WJ. 1979. Solute regulation and growth by roots and shoots of water-stressed maize plants. *Planta* **147**: 43-49.

Sharp RE, Hsiao TC and Silk WK. 1990. Growth of the maize primary root at low water potentials. II. Role of growth and deposition of hexose and potassium in osmotic adjustment. *Plant Physiology* **93**: 1337-1346.

Sheffer KM, Watschke TL, and Duich JM. 1979. Carbohydrate sampling in Kentucky bluegrass turf. *Agronomy Journal* **71**: 301-304.

Sheffer KM, Dunn JH, and Minner DD. 1987. Summer drought response and rooting depth of three cool-season turfgrasses. *Hort Science* **22**: 296-297.

Sheveleva E, Chmara W, Bohnert HJ, and Jesnsen RG. 1997. Increased salt and drought tolerance by D-ononitol production in transgenic *Nicotiana tabacum* L. *Plant Physiology* **115**: 1211-1219.

Shinozaki K and Yamaguchi-Shinozaki K. 1997. Gene expression and signal transduction in water stress response. *Plant Physiology* **115**: 327-334.

Simane B, Peacock JM, and Struik PC. 1993. Differences in developmental plasticity and growth rate among drought-resistant and susceptible cultivars of durum wheat (*Triticum turgidum* L. var. *durum*). *Plant and Soil* **157**: 155-166.

Simpson GM. 1981. *Water Stress on Plants*. Praeger Publishers, New York.

Smucker AJM and Aiken RM. 1992. Dynamic root responses to water deficits. *Soil Science* **154**: 281-289.

Smucker AJM, Nuñez-Barrios A, and Ritchie JT. 1991. Root dynamics in drying soil environments. *Belowground Ecology* **1**: 1-5.

Spollen WG and Nelson CJ. 1994. Response of fructan to water deficit in growing leaves of tall fescue. *Plant Physiology* **106**: 329-336.

Suzuki M and Chatterton NJ. 1993. Fructans in crop production and preservation. In: Chatterton NJ (ed.) *The Science and Technology of Fructans*. CRC Press, Boca Raton. Pp. 232-233.

Synkova H, Wilhelmova N, Sestak Z and Pospisilova J. 1997. Photosynthesis in transgenic plants with elevated cytokinin contents. In: Pessarakli M (ed.) *Handbook of Photosynthesis*. Marcel Dekker, New York. Pp. 541-552.

Tamminen I, Makela P, Heino P, and Palva ET. 2001. Ectopic expression of ABI3 gene enhances freezing tolerance in response to abscisic acid and low temperature in *Arabidopsis thaliana*. *The Plant Journal for Cell and Molecular Biology* **25**: 1-8.

Tardieu F and Simonneau T. 1998. Variability among species of stomatal control under fluctuating soil water status and evaporative demand: modeling isohydric and anisohydric behaviors. *Journal of Experimental Botany* **49**: 419-432.

Tardieu F, Zhang J, and Davies WJ. 1992. What information is conveyed by an ABA signal from maize roots in drying field soil? *Plant, Cell and Environment* **15**: 185-191.

Thomas H. 1986. Effect of rate of dehydration on leaf water status and osmotic adjustment in *Dactylis glomerata* L., *Lolium perenne* L. and *L. multiflorum* Lam. *Annals of Botany* **57**: 225-235.

Thomas H. 1991. Accumulation and consumption of solutes in swards of *Lolium perenne* during drought and after rewatering. *New Phytologist* **118**: 35-48.

Toft NL, McNaughton SJ, and Georgiadis NJ. 1987. Effects of water stress and simulated grazing on leaf elongation and water relation of an East African grass, *Eustachys paspaloides*. *Australian Journal of Plant Physiology* **14**: 211-226.

Torbert HA, Edwards JH, and Pederson JF. 1990. Fescues with large roots are drought tolerant. *Applied Agricultural Research* **5**: 181-187.

Turgeon AJ. 2002. *Turfgrass Management*. 6th edn. Pearson Education, Inc., Upper Saddle River.

Turner NC. 1986. Crop water deficits: a decade of progress. *Advances in Agronomy* **39**: 1-51.

Turner NC and Begg JE. 1981. Plant-water relations and adaptation to stress. *Plant and Soil* **58**: 97-131.

Veneklaas E and Van den Boogaard R. 1994. Leaf-age structure effects on plant water use and photosynthesis of two wheat cultivars. *New Phytologist* **128**: 331-337.

Volaire F. 1994. Effects of summer drought and spring defoliation on carbohydrate reserves, persistence and recovery of two populations of cocksfoot (*Dactylis glomerata*) in a Mediterranean environment. *Journal of Agricultural Science* **122:** 207-215.

Volaire F. 1995. Growth, carbohydrate reserves, and drought survival strategies of contrasting *Dactylis glomerata* populations in a Mediterranean environment. *Journal of Applied Ecology* **32:** 56-66.

Volaire F and Thomas H. 1995. Effects of drought on water relations, mineral uptake, water-soluble carbohydrate accumulation and survival of two contrasting populations of cocksfoot (*Dactylis glomerata* L.). *Annals of Botany* **75:** 513-524.

Volaire F and Lelievre F. 2001. Drought survival in *Dactylis glomerata* and *Festuca arundinacea* under similar rooting conditions in tubes. *Plant and Soil* **229:** 225-234.

Volaire F, Thomas H, and Lelievre F. 1998. Survival and recovery of perennial forage grasses under prolonged Mediterranean drought. *New Phytologist* **140:** 439-449.

Volenec JJ. 1986. Nonstructural carbohydrate in stem base components of tall fescue during regrowth. *Crop Science* **26:** 122-127.

Wang Z and Huang B. 2003. Genotypic variation in abscisic acid accumulation, water relations, and gas exchange for Kentucky bluegrass exposed to drought stress. *Journal of The American Society for Horticultural Science* **128:** 349-355.

Wang Z, Huang B, and Xu Q. 2003. Effects of abscisic acid on drought responses of Kentucky bluegrass. *Journal of The American Society for Horticultural Science* **128:** 36-41.

Westgate ME and Boyer JS. 1985. Osmotic adjustment and the inhibition of leaf, root, stem and silk growth at low water potentials in maize. *Planta* **164:** 540-549.

White RH, Engelke MC, Morton SJ, and Ruemmele BA. 1992. Competitive turgor maintenance in tall fescue. *Crop Science* **32:** 251-256.

Wilkinson S and Davies WJ. 2002. ABA-based chemical signaling: the coordination of responses to stress in plants. *Plant, Cell and Environment* **25:** 195-210.

Wilson JR and Ludlow MM. 1983. Time trends for change in osmotic adjustment and water relations of leaves of *Cenchrus ciliaris* during and after water stress. *Australian Journal of Plant Physiology* **10:** 15-24.

Wilson JR, Ludlow MM, Fisher MJ and Schulze E-D. 1980. Adaptation to water stress of the leaf water relations of four tropical forage species. *Australian Journal of Plant Physiology* **7:** 207-220.

Yordanov I, Velikova V, and Tsonev T. 1999. Influence of drought, high temperature, and carbamide cytokinin 4-PU-30 on photosynthetic activity of bean-plants. 1. Changes in chlorophyll fluorescence quenching. *Photosynthetica* **37:** 447-457.

Yordanov I, Tsonev T, Goltsev V, Merakchiiska-Nikolova M, and Georgieva K. 1997. Gas exchange and chlorophyll fluorescence during water and high temperature stresses and recovery: probable protective effect of carbamide cytokinin 4-PU30. *Photosynthetica* **33:** 423-431.

Zhang J and Davies WJ. 1989. Abscisic acid produced in dehydrating roots may enable the plant to measure the water status of the soil. *Plant, Cell and Environment* **12:** 73-81.

Zhang J, Nguyen HT, and Blum A. 1999. Genetic analysis of osmotic adjustment in crop plants. *Journal of Experimental Botany* **50:** 291-302.

Chapter 8

TREE FUNCTIONAL STRATEGIES IN BRAZILIAN SAVANNAS

Augusto C. Franco

Introduction
Soil water extraction, root distribution,
  and hydraulic redistribution of soil water
Seasonality of climate and leaf phenology
Tree water relations
> *Seasonal changes in leaf water potential*
> *Osmotic adjustment and seasonal changes*
> * in elastic modulus (ε)*
> *Wood density, water storage,*
> * and hydraulic architecture*
> *Hydraulic conductivity, vulnerability*
> * curves, and embolism repair*
> *Stomatal conductance and transpiration*
Photosynthesis
> *Photoprotective mechanisms*
> * and shade tolerance*
> *CO_2 assimilation, water use efficiency,*
> * and leaf phenology*
Conclusion and future prospects

Perspectives in Biophysical Plant Ecophysiology: A Tribute to Park S. Nobel, pp. 191-219
Edited by: E. De la Barrera and W.K. Smith
© 2009 by The Author
Book Compilation © 2009 Universidad Nacional Autónoma de México

Introduction

Seasonally dry tropical regions are covered by savannas and dry forests, which occupy a greater area worldwide than tropical rainforests (Olivares and Medina 1992; Murphy and Lugo 1995). Tropical dry forests are tree-dominated ecosystems, while a xeromorphic, fire-tolerant grass layer is an important component of savannas. In the Neotropics these two vegetation types are floristically and ecologically distinct and their component species may react differently to environmental changes (Pennington *et al.* 2000). This chapter addresses the savannas of Central Brazil, known locally as "cerrado"[1]. The cerrados of Central Brazil are the second most extensive plant formation in South America and cover 2.0×10^6 km², an area which is approximately the same as that of Western Europe (Eiten 1972; Oliveira-Filho and Ratter 2002).

Most of the cerrado occurs in regions with a yearly rainfall that ranges from 750-800 mm to 2000 mm (Eiten 1972). Rainfall in isolated savanna areas adjacent to contiguous forests in the Amazon region can reach higher values (Sanaiotti *et al.* 2002). It is subject to regular and predictable annual drought from May to September, which is a major determinant of ecosystem structure and function. High irradiances, high air temperatures and low relative humidities impose a consistently high evaporative demand during the prolonged dry season, when potential evaporation rates greatly exceed rainfall. Water in the upper soil layers is severely depleted as evidenced by their extremely low soil water potential during this period (Franco 1998, 2002), while the deeper soil layers remain moist even after several months without rain (Jackson *et al.* 1999). Because of the large inter-annual variations in monthly rainfall patterns, evaporation may even exceed rainfall in some months during the wet period as well. Average annual temperature ranges from 20 to 26 °C with diurnal temperature ranges of 20 °C being common during the dry season (Eiten 1972). Frost events are uncommon and occur only at the southern limit of the cerrado region. The nutrient-poor, acid soils with high levels of Aluminum represent an additional limiting factor for plant growth. P is particularly limiting in these ecosystems (Haridasan 2000, 2001).

The cerrado contains a remarkably complex community structure rich in endemic woody species. About 800 species of trees and large shrubs are present in the cerrado region (Oliveira-Filho and Ratter 2002). Individual sites may contain 70 or more different species of trees (Felfili and Silva Jr 1993). The most important families, in terms of species numbers are Leguminosae, Myrtaceae, Malpighiaceae, Vochysiaceae, Melastomataceae and Rubiaceae (Oliveira-Filho and Ratter 2002). The upper canopy of cerrado vegetation typically consists of 6- to 8-m-tall, deciduous and evergreen trees that display a wide range of root habits, from shallow- to deep-

---

[1] The cerrado biome was named after the vernacular term for its predominant vegetation type, a woody savanna of sclerophyllous shrubs and small trees (Eiten 1972; Oliveira-Filho and Ratter 2002).

rooted (Table 8.1; Rawitscher 1948; Jackson *et al.* 1999). Most of the roots are contained in the first meter belowground, although a smaller but significant proportion of the root biomass is still found to depths of 6 to 8 meters (Abdala *et al.* 1998) and even deeper (Rawitscher 1948).

The emphasis of this chapter will be on plant traits related to carbon assimilation and water relations. Carbon acquisition and water balance are not only major constraints to plant fitness, but carbon influx and attendant processes are coupled to water availability. Moreover, most species showed a prolonged midday depression of photosynthetic rates during both the wet and dry seasons (Franco 1998; Moraes and Prado 1998; Franco and Lüttge 2002). Under these circumstances, in addition to a constrain in daily carbon gain, the photosynthetic apparatus can be damaged by the absorption of excessive light energy, unless photoprotective mechanisms operate to alleviate photoinhibition. On the other hand, canopy shading by the grass layer can restrict seedling establishment and growth (Nardoto *et al.* 1998; Braz *et al.* 2000). The effects of canopy shading on CO_2 assimilation can become critical for tree seedling growth and survival in dense cerrado woodlands (Kanegae *et al.* 2000). Thus, plant performance in savanna ecosystems can be greatly affected by small changes in plant size and of canopy structure of the overlying vegetation, as it was observed in desert ecosystems (Franco and Nobel 1988, 1989).

Soil water extraction, root distribution, and hydraulic redistribution of soil water

Models explaining the structure and function of savanna ecosystems typically involve water and nutrients as limiting resources and a two-layered soil-water model of tree-grass coexistence (Walker and Noy-Meir 1982; Sarmiento 1984; Medina and Silva 1990; Scholes and Walker 1995). According to this model, the dense system of shallow roots of grasses make them superior competitors for water in the upper part of the soil profile, whereas trees are deeply rooted and therefore have exclusive access to a deeper, more predictable water source. However, cerrado trees are not homogeneous in terms of rooting patterns and they display a wide range of rooting habits from shallow-rooted to deep-rooted species (Table 8.1).

Moreover, neotropical savanna trees can have extended lateral roots at depths of 20 to 50 cm in addition to the deep vertical system (Sarmiento 1984). Thus, *Curatella americana*, *Byrsonima crassifolia*, and *Bowdichia virgiloides* have the bulk of their roots directly below the area of main concentration of the grass roots, and at the same time reach soil layers several times deeper than the herbaceous plants. These tree species are also commonly found in the cerrados of Central Brazil (Ratter *et al.* 1996). The presence of a dimorphic root system was also reported for *Byrsonima crassa* and *Blepharocalyx salicifolius* in a cerrado site in Central Brazil (Moreira *et al.* 2003). These results were extended by Sternberg *et al.* (2005), who used deuterium-enriched water experiments and extensive root excavations to confirm the presence of long range lateral roots in several cerrado trees. They hypothesized that the extensive lateral root development in these nutrient-poor, acid soils may be an adaptation to better exploit the nutrients at the top layer of the soil profile.

The occurrence of hydraulic lift could be expected in species with a dimorphic root system. Hydraulic lift is the passive movement of water from the lower wetter layers to the upper dryer layers of the soil profile via the plant root (Richards and Caldwell 1987; for a review see for instance Jackson *et al.* 2000). The hydraulic redistribution of water downward (from wetter upper layers to dryer lower layers of the soil profile) has also been reported and called inverse hydraulic lift (Schulze *et al.* 1998). Hydraulic lift in cerrado trees was first reported by Scholz *et al.* (2002). Moreira *et al.* (2003) used deuterated water experiments to demonstrate that water moves up through the central taproot, to lateral roots, to the soil and into neighboring small shrubs and trees in a cerrado site. Their results also showed that roots other than the central taproot contribute a major part of the water uptake by the plant. Thus, lateral roots may have sinker root tips that penetrate the soil profile deeply.

The magnitude of this process and its importance for plant water balance in cerrado ecosystems still needs to be thoroughly evaluated. Hydraulic lift could nearly double the loss of water by evapotranspiration in a sugar maple stand (Jackson *et al.* 2000). On the other hand, hydraulic lift in an *Artemisia tridentata* stand could increase total transpiration during a period of 100 days by only 3.5%, although it could be as high as 20% for some days (Ryel *et al.* 2002). It appears that hydraulic lift makes only a relatively small contribution to the water budget of neighboring plants in the cerrado (Moreira *et al.* 2003). Nevertheless, the addition of small amounts of water to the superficial soil layers could be enough to prevent cavitation of the shallow roots (Brooks *et al.* 2002) or damage to mutualistic associations with organisms living inside or close to the root surfaces (Querejeta *et al.* 2003). It could also be important to maintain or enhance nutrient uptake by fine roots (Caldwell *et al.* 1998).

Thus, the large species-specific differences in rooting depth (Table 8.1), the presence of a dimorphic root system in many species and the occurrence of hydraulic lift result in a pattern of soil water exploitation that is much more complex than it is postulated by the two-layered soil water model. Indeed, cerrado trees extract water along the whole soil profile to the water table (Jackson *et al.* 1999). Moreover, the pattern is dynamic and it may shift to lower depths during the dry season. On the other hand, soil water balance measurements suggest that a significant amount of the annual precipitation remains unexploited by the roots (Silva *et al.* 2003).

Seasonality of climate and leaf phenology

Despite the pronounced seasonality in rainfall, drought does not synchronize vegetative phenology in tropical dry forests and savannas (Eamus and Prior 2001; Rivera *et al.* 2002). The cerrados of Central Brazil constitute mosaics of trees of different phenological types (Table 8.1; Franco *et al.* 2005; Lenza and Klink 2006). Leaf fall does peak in the dry season, but there is considerable inter-specific variation in the timing of bud break and

Table 8.8.1. Patterns of leaf phenology and depth of the root system of 12 cerrado trees. Evergreen leaf-exchangers simultaneously shed the leaves and produce new ones, while the briefly deciduous remain leafless for short periods of time during the dry season. Leafless is the period that the trees were without leaves of any type, including young and senescent leaves. Depth of the root system was based on comparisons of stable hydrogen isotope composition of stem xylem water and soil water that was collected at different depths. (Adapted from Franco *et al.* 2005; for *Caryocar brasiliense* from Maia 1999; for *Myrsine guianensis* from Lenza and Klink 2006).

Species	Family	Leafless	Leaf flush	Root depth
Evergreen				
Schefflera macrocarpa	Araliaceae	0	Throughout the year	Shallow
Miconia ferruginata	Melastomata-ceae	0	Wet season	Shallow
Roupala montana	Proteaceae	0	End of dry season	Shallow
Myrsine guianensis	Myrsinaceae	0	Most of the year	Deep[1]
Evergreen leaf-exchanger				
Sclerolobium paniculatum	Leguminosae–Caesalpinioi-deae	0	Wet season	Shallow
Vochysia elliptica	Vochysiaceae	0	End of dry season	Deep
Ouratea hexasperma	Ochnaceae	0	End of dry season	Shallow
Briefly deciduous				
Caryocar brasiliense	Caryocaraceae	Less than 3 weeks	End of dry season	Deep[1]
Dalbergia miscolo-bium	Leguminosae–Papilionoideae	Less than 3 weeks	End of dry season	Deep
Pterodon pubescens	Leguminosae–Papilionoideae	Less than 3 weeks	End of dry season	Shallow
Deciduous				
Kielmeyera coriacea	Clusiaceae	More than 3 weeks	End of dry season	Deep
Qualea grandiflora	Vochysiaceae	More than 3 weeks	End of dry season	Deep
Qualea parviflora	Vochysiaceae	More than 3 weeks	End of dry season	Inter-mediate

[1]Based on unpublished measurements of stable hydrogen isotope composition.

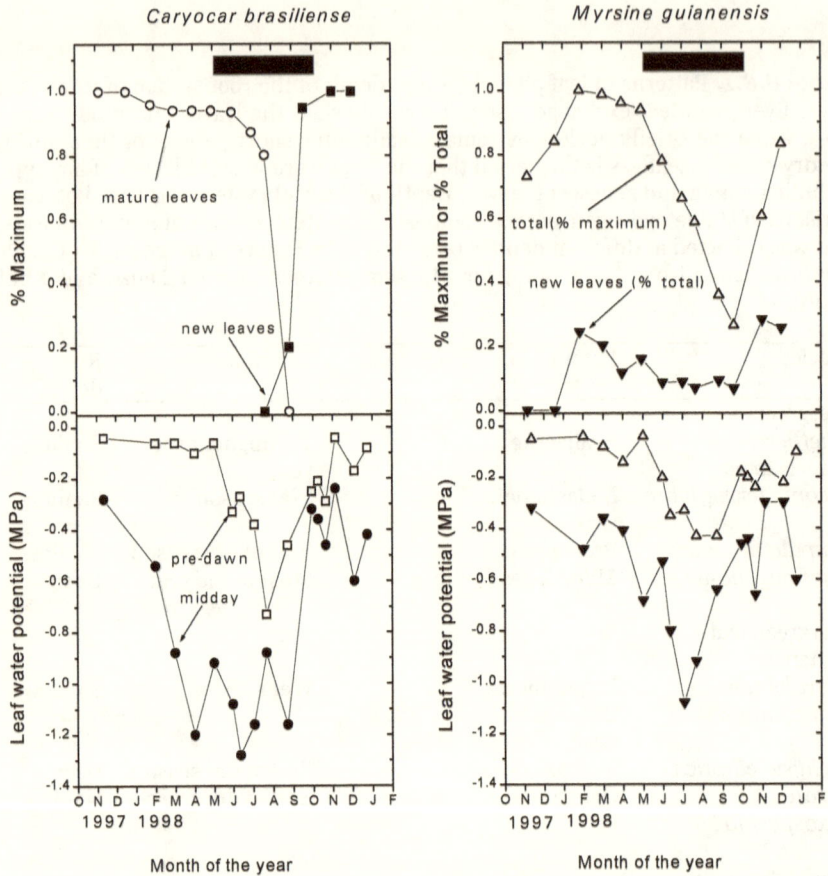

Figure 8.1. Monthly variation in the number of leaves, pre-dawn and midday water potential for *Caryocar brasiliense* and *Myrsine guianensis* in a cerrado site at Reserva Ecológica do IBGE. In *C. brasiliense*, the number of leaves was expressed as % of the number of leaves that were measured in October 1997 (leaf cohort of 1997) or November 1998 (leaf cohort of 1998). In *M. guianensis*, the total number of leaves of each month was expressed as a % of the number of leaves that were measured in February 1998, while the new leaves were expressed as a % of the total number of leaves that were counted at each measurement. Leaf phenology was followed in several branches of 10 different individuals, which included five that were used for the leaf water potential measurements. Maximum values of stomatal conductance (G_{wv}) for *C. brasiliense* decreased from 158-219 mmol m^{-2} s^{-1} in the wet season to 39-59 mmol m^{-2} s^{-1} in the dry season whereas G_{wv} in *M. guianensis* decreased from 165-353 mmol m^{-2} s^{-1} to 40-80 mmol m^{-2} s^{-1}. The dark bars indicate the dry season. Adapted from Maia (1999).

flushing. Many evergreens flush new leaves continuously throughout most of the year (Fig. 8.1). Leaf-exchanging species simultaneously shed the leaves and produce new ones at the end of the dry season. Deciduous species flush by the end of the dry season. However, several species are briefly deciduous in the sense that they remain leafless for short periods of time of less than 3 weeks (Table 8.1; Fig. 8.1; Franco *et al.* 2005; Lenza and Klink 2006).

The cause of the variation in time and intensity of leaf phenophases in seasonal dry forests and savannas has been much debated (Reich 1995; Wright 1996; Eamus and Prior 2001; Rivera *et al.* 2002). Although synchronous bud break in spring commonly observed in several species may be the result of a photoperiodic response, in other species bud break is probably determined mainly by the seasonal variation in tree water status at a given site (Reich 1995; Eamus and Prior 2001; Rivera *et al.* 2002).

Irrespective of the cause, bud break and leaf expansion in the dry season can only occur if the tree is fully hydrated. Because seasonal changes on osmotic pressure are small (Bucci 2001; also, see section on osmotic adjustments) and only the upper soil layers are really depleted of water (Franco 2002), the depth of the root system is critical. Most deciduous trees have a deep-rooted system (Table 8.1), which allows access to permanently wet soil layers, while the leafless period will greatly reduce water loss by the tree. Under these circumstances, the recovery of tree water balance and bud break may occur, provided that reserves of soil water are sufficient to support whole-plant rehydration that precedes leaf flushing in the absence of rain. Moreover, several cerrado species have stems with thick bark and large internal water storage capacity (Coradin 2000; Scholz *et al.* 2007). Stored water can buffer the impact of seasonal drought and enable trees to flower and flush new leaves during the dry season (Borchert 1994).

On the other hand, some shallow-rooted evergreens are also able to flush before the onset of the rainy season. Therefore, we can expect that leaf flushing in cerrado trees is mostly the result of adjustments in the tree-water supply and demand, driven by a complex interplay of partial stomatal closure and partial or total canopy shedding, as well as the depth of the root system and the size of internal water reservoirs. Moreover, it remains to be studied whether redistribution of soil water by hydraulic lift is enough to affect tree rehydration and leaf phenology in cerrado ecosystems.

Tree water relations

Seasonal changes in leaf water potential

Imbalances between water supply and demand result in changes in plant water status, generally assessed by measurements of leaf water potential (Ψ_l). If one assumes negligible or low nocturnal transpiration, plant rehydration could occur and a water balance between the plant and the soil should be reached by the end of the night. Under these circumstances, pre-dawn Ψ_l would be a measure of the water potential of the soil adjacent to the root and therefore, of the maximum water status the plant can achieve. This relationship would depend on the root distribution in the soil and root water extraction may shift to deeper layers as the dry season progresses (Weltzin and McPherson 1997). On the other hand, processes such as nighttime transpiration could prevent equilibration along the soil to leaf continuum. In this case, predawn Ψ_l may be significantly more negative than the water potential of the soil accessed by the roots (Donovan *et al.* 1999, 2001). Thus, seasonal variations in predawn Ψ_l in evergreen cerrado species may partially reflect nocturnal transpiration that would limit plant recharge during night-time (Bucci *et al.* 2004a; 2005). Minimum Ψ_l is generally reached between midday and early afternoon, when the evaporative demand of the atmosphere reaches its peak.

In the cerrados of Central Brazil, seasonal variations in pre-dawn and midday Ψ_l are relatively small (Fig. 8.1), when compared to trees of Venezuelan dry forests (Sobrado 1986), but similar to evergreen species of Australian savannas (Duff *et al.* 1997; Myers *et al.* 1997). Pre-dawn Ψ_l generally remains between −0.1 and −0.3 MPa in the wet season, and −0.3 to −0.8 MPa in the dry season (Mattos *et al.* 1997; Franco 1998; Mattos 1998; Meinzer *et al.* 1999; Bucci *et al.* 2005; Franco *et al.* 2005). These values suggest contact with relatively moist soil at depth. Minimum Ψ_l is in the range of −1 to −3 MPa in the dry season (Perez and Moraes 1991; Franco 1998; Meinzer *et al.* 1999; Franco *et al.* 2005; Bucci *et al.* 2005).

The small seasonal variations in minimum Ψ_l are the result of strong stomatal control of evaporative losses coupled with a decrease in total leaf surface area per tree during the dry season and access to the deep moist soil layers. The leaf surface area per unit of sapwood area (LA/SA), an index of potential architectural constraints on water supply in relation to transpirational demand is about 1.5 to 8 times greater in the wet season compared to the dry season for most of the species (Bucci *et al.* 2005). Strong stomatal limitation of both daily transpiration rates and total daily transpiration was reported to occur during both the wet and the dry seasons (Meinzer *et al.* 1999; Bucci 2001). In the particular case depicted in Fig.1, maximum values of stomatal conductance decreased by about 75% from the wet to the dry season. Thus, the considerable reduction in total leaf area and stomatal conductance were enough to induce a recovery of pre-dawn Ψ_l of adult trees of *Caryocar brasiliense* and *Myrsine guianensis* in the absence of any rainfall at the end of the dry season.

Osmotic adjustment and seasonal changes in elastic modulus (ε)

In many drought-tolerant plants, water stress typically leads to an accumulation of solutes in the cytoplasm and vacuole, thus maintaining turgor pressure despite the decrease in water potentials. Solute accumulation in response to water stress is termed osmotic adjustment and involves an increase in the number of solute molecules per cell rather than a decrease in the amount of water in the cell. A decrease in osmotic potential at full turgor in response to drought is evidence of osmotic adjustment and it is associated with a decrease in osmotic potential at the turgor loss point, as well (Jones and Turner 1978).

Although cerrado trees are exposed to a 5-month long dry period, large drought-induced changes in osmotic potential are not expected, given the small seasonal variation in pre-dawn Ψ_l. Indeed, seasonal changes in osmotic potentials at full turgor and at the turgor loss point of cerrado trees were small (Perez and Moraes 1991; Bucci 2001). Small seasonal changes in osmotic potential or no changes were also measured in the Australian savannas (Eamus and Prior 2001) and in Venezuelan savanna trees (Meinzer *et al.* 1983). In both cases, there was little seasonality in leaf water potentials. Osmotic adjustment is much more common in Venezuelan dry forests, which experienced significantly lower rainfall (900 mm; Sobrado 1986).

It is yet to be determined whether osmotic adjustments is an important mechanism for drought tolerance in cerrado woody seedlings, which can attain pre-dawn Ψ_l as low as −4 MPa (Hoffmann *et al.* 2004). Drought-induced osmotic adjustment was found in saplings of the Australian savanna trees *Eucalyptus tetrodonta* and *E. ferdinandiana*, but not in adult plants (Eamus and Prior 2001).

The elastic properties of plant cell walls, usually measured by the cell's volumetric elastic modulus (ε), must also be considered when analyzing plant response to drought. Because ε affects the rate of decrease in turgor pressure as the plant tissue loses water, adjustments in ε could also affect the turgor maintenance capacity of a plant in response to drought. There are few measurements of ε for cerrado plants, which ranged from 5 to 25 MPa (Bucci 2001). Seasonal changes in ε were relatively small and may be more related to leaf senescence and the production of new leaves in the end of the dry season than as a response to seasonal drought.

Wood density, water storage, and hydraulic architecture

Variation in wood density is correlated with changes in a suite of characteristics related to stem water storage capacity, the efficiency of xylem water transport, regulation of leaf water status, and avoidance of turgor loss (Meinzer 2003). This seems to be the case for cerrado trees as well. In a study of the water relations of six dominant trees in a cerrado site, they all shared the same common negative exponential relationship between sapwood saturated water content and wood density, during the peak of the dry season (Bucci *et al.* 2004b). Thus, sapwood water storage, or capacitance, diminished with increasing wood density, as it was found in other studies (Borchert 1994; Stratton *et al.* 2000; Meinzer 2003).

With respect to hydraulic architecture, specific and leaf-specific hydraulic conductivity[2] decreased, and the leaf area:sapwood area ratio in-

---

[2] Hydraulic conductivity (kg m s^{-1} MPa^{-1}) was calculated as $k_h = J_v \ / \ (\Delta P/\Delta X)$, where J_v was the flow rate through the branch or petiole segment (kg s^{-1}) and

creased more than 5-fold as wood density increased from 0.37 to 0.71 g cm^{-3} across all individuals and species (Bucci *et al.* 2004b). If both sapwood water storage capacity and water transport efficiency are negatively correlated with wood density, it is reasonable to postulate that species with greater wood density would experience larger daily fluctuations in leaf water status (Meinzer 2003). Indeed, wood density was a good predictor of minimum (midday) Ψ_l and total daily transpiration, both of which decreased linearly with increasing wood density in a similar fashion across all individuals and species (Bucci *et al.* 2004b). Moreover, the time of onset of sap flow in the morning and the maximum sap flow tended to occur progressively earlier in the day as wood density increased. On the other hand, stomatal conductance, specific leaf area, and osmotic potential at the turgor loss point, all decreased linearly with increasing wood density. Thus, wood density apparently constrained the evolutionary options related to plant water economy and hydraulic architecture in cerrado woody species (Bucci *et al.* 2004b).

Hydraulic conductivity, vulnerability curves, and embolism repair

Savanna trees should have an efficient water transport system to keep pace with transpiration rates without causing an excessive drop in leaf water potential, given the high temperature and high evaporative demand throughout the year, coupled with the pronounced seasonality of precipitation (Sarmiento *et al.* 1985). In cerrado woody species, leaf-specific hydraulic conductivity (k_l), which is an in-vitro measurement of water transport efficiency obtained with detached sections of terminal stems is strongly correlated with apparent soil-to-leaf hydraulic conductance[3] (G_t), an *in vivo* estimate of whole-plant water transport efficiency (Bucci *et al.* 2004b). Moreover, the diurnal range in leaf water potential was negatively related to k_l for six of the Cerrado tree species, suggesting that higher water transport efficiency resulted in smaller variation in leaf water potential (Bucci *et al.* 2004b). Values of k_l which ranged from 4 to 26 × 10^{-4} kg m^{-1} s^{-1} MPa^{-1} (Bucci *et al.* 2003, 2004b), were relatively high compared to values measured in trees of savannas and seasonal tropical forests, but within

---

$\Delta P/\Delta X$ is the pressure gradient across the segment (MPa m^{-1}). Specific hydraulic conductivity (k_s: kg m^{-1}s^{-1}MPa^{-1}) was obtained as the ratio of k_h and the cross sectional area of the active xylem. Leaf-specific conductivity (k_l: kg m^{-1}s^{-1}MPa^{-1}) was obtained as the ratio of k_h and the leaf area distal to the branch or petiole segment.

[3] The apparent leaf area-specific hydraulic conductance of the soil/root/leaf pathway (G_t) was determined as $G_t = E/\Delta\Psi$, where $\Delta\Psi$ is the difference between the current Ψ_l and the Ψ of the soil, and E is the average transpiration rate per unit leaf area determined from sap flow measurements at the time of Ψ_l measurements. Ideally, this relationship only applies under steady-state conditions, when transpiration and uptake are equal. In transient conditions, there is also a storage component, but is usually considered small compared with water uptake (Tyree et al. 1991; Goldstein et al. 1998). Maximum values of G_t in cerrado trees ranged from about 1 to 16 mmoml m^{-2} s^{-1} (Meinzer et al. 1999; Bucci *et al.* 2005).

the range of values reported to tropical rainforest trees (Eamus and Prior 2001). On the other hand, specific hydraulic conductivity (k_s) ranged from 0. 3 to 1.1 kg m^{-1} s^{-1} MPa^{-1}, which is similar to reported values for savanna and seasonal tropical trees (Eamus and Prior 2001).

Blockage of the xylem conduits by air reduces hydraulic conductivity in plants, and therefore long distance transport (Holbrook and Zwieniecki 1999). Xylem conduits are prone to embolism because the xylem sap is under tension. Cavitation must be prevented if continuity of the water column in the xylem is to be maintained. This can be particularly critical under the high evaporative demand conditions that prevail during the long dry season in the cerrado.

Xylem vulnerability curves describe the relationship between loss of hydraulic conductivity and the xylem water potential that induced the loss of conductivity (Tyree and Ewers 1996). Bucci *et al.* (2003) constructed vulnerability curves for two typical cerrado trees, the shallow-rooted *Schefflera macrocarpa* and the deep-rooted *Caryocar brasiliense*. Percentage loss of conductivity began to increase more rapidly below about −1.0 MPa in both species, reached 50% at about −1.7 MPa, while the 100% conductivity loss point occurred at about −3.5 MPa. Similar values were measured for *Curatella americana* in the llanos of Venezuela, a species that also occurs in the Brazilian cerrados and appears to have access to subsoil water reserves (Sobrado 1996). Values were lower for tropical dry forest trees. For instance, the 50% conductivity loss point ranged from −1.65 to −3.82 MPa in six tree species of Venezuelan dry forests (Sobrado 1997) and from −1.4 to −5.6 MPa in four coexisting tree species from seasonally dry forest of Northen Australia (Choat *et al.* 2003).

The regulation of xylem tension by partial stomatal closure and leaf area adjustments may limit cavitation, embolism and consequent loss of hydraulic conductivity within the plant (Meinzer *et al.* 1999; Naves-Barbiero *et al.* 2000; Bucci 2001). This could be of critical importance in the wet season as well; the large increase in the amount of transpiring leaves in this time of the year does impose a constraint in the water conduit system. Indeed many cerrado trees restrict their transpiration rates both in the wet and the dry season and whole-plant transpiration remains relatively constant throughout the year irrespective of the large changes in total leaf area (Bucci *et al.* 2005).

However, diurnal embolism repair, particularly under high evaporative demand conditions, may be also a pre-requisite for maintaining efficient long-distance movement of water to the transpiring leaves of cerrado trees. Indeed, Bucci *et al.* (2003) provided strong evidence that these trees are capable of repairing embolized vessels while the xylem sap in the non-embolized vessels was still under tension. In a field study, they reported that petiole k_s of *Schefflera macrocarpa* and *Caryocar brasiliense* decreased sharply with increasing transpiration rates and declining leaf water potentials during the morning. Petiole k_s increased during the afternoon while the plants were still transpiring and the water in the non-embolized vessels was still under tension. They used dye experiments to confirm that these diurnal variations in k_s were associated with embolism formation and repair. As a possible mechanism, Bucci *et al.* (2003) suggested that the rate of refilling of the embolized vessels in these two species would be a function of internal pressure imbalances. An increase in osmotically active solutes in cells outside the vascular bundles at around midday would lead to water uptake by these cells. The concurrent increase in tissue volume would be partially constrained by the cortex, resulting in a transient pressure imbalance that would drive radial water movement in the direction of the embolized vessels, thereby refilling them and restoring water flow. Consistent with this, petiole sugar content was highest in earlier afternoon and removal of cortex or longitudinal incisions in the cortex prevented afternoon recovery of k_s and refilling of embolized vessels (Bucci *et al.* 2003). Furthermore, Domec *et al.* (2006) demonstrated not only that daily embolism and refilling was a common occurrence in lateral roots of four co-occurring cerrado tree species but that refilling was probably a function of internal pressure balances in agreement with the results of Bucci *et al.* (2003) for vessels in petioles.

Stomatal conductance and transpiration

The regulation of stomatal opening plays a major role in the control of transpiration, minimizing the effects of increases in the evaporative demand of the atmosphere (increase in vapor pressure deficit) as a force driving transpiration. Studies of regulation of water use at the plant level with sap flow sensors have shown that, despite the potential access to deep soil water, cerrado trees exhibited reduced transpiration due to partial stomatal closure during both the dry and wet seasons (Meinzer *et al.* 1999; Naves-Barbiero *et al.* 2000; Bucci 2001). For most species, sap flow increased sharply in the morning, briefly attained a maximum value by about 09:30 to 12:00, then decreased sharply, despite steadily increasing solar irradiation and evaporative demand. This decrease was particularly strong in the dry season, when high values of vapor pressure deficit prevail during most of the daylight hours. In some cases, transpiration rates briefly recovered in late afternoon (Naves-Barbiero *et al.* 2000).

This rapid decrease in flow rate after an early peak may represent the limits of an internal reservoir that is recharged at night, or may be the result of limited capacity of the root system to absorb water in sufficient quantities to sustain high rates of transpiration over a longer period (Bucci *et al.* 2005). Despite a nearly three-fold increase in mean air saturation deficit between the wet and dry season, total daily sap flow per plant did not differ significantly between seasons for six of the eight study species, but daily transpiration per unit leaf area (E) was significantly higher for most of the species during the dry season. Thus, reductions in stomatal conductance (G_{wv}) and partial leaf loss acted in parallel to stabilize water balance and the decrease in G_{wv} was greater in species exhibiting only a modest decline in leaf area during the dry season. On the other hand, sustained nocturnal transpiration, which ranged from 15 to 22% of daily transpiration during the dry season, prevented the full recharge of internal reservoirs (Bucci *et al.* 2005).

The diurnal and seasonal changes in stomatal conductance are mostly a response to changes in air saturation deficit (D). G_{wv} decreased sharply with increasing D, strongly limiting transpiration during the dry season. A single function relating G_{wv} to D accounted for about 80% of the variation in G_{wv} observed across all species (Bucci *et al.* 2005). However, the response function of stomata to increasing leaf-to-air vapour pressure deficit in many plants varies with plant water status (Franco *et al.* 1994; Prior *et al.* 1997; Thomas and Eamus 1999). Stomatal sensitivity to changes in leaf water potential are also expected in cerrado trees, where maximum stomatal conductances are lower in the dry season, when pre-dawn Ψ_l is also lower (Franco 1998; Franco *et al.* 2005; Bucci *et al.* 2005). Leaf water potential at stomatal closure in potted seedlings of cerrado species exposed to water withholding ranged from −2.4 to −3.9 MPa (Prado *et al.* 1994; Sassaki *et al.* 1997; Moraes and Prado 1998).

There are two possible mechanisms that could couple soil water availability to stomatal conductance. First, there could be an increased synthesis of ABA in roots that are in direct contact with the drying soil. Its subsequent transport in the transpiration stream to leaves would trigger stomatal closure (Gowing *et al.* 1993; Loewenstein and Pallardy 1998; Thomas and Eamus 1999). Alternatively, as soil water availability declines, the water column in the xylem comes under increasing tension and xylem cavitation can occur. Declines in stem hydraulic conductivity as pre-dawn Ψ_l declined have been observed in *Eucalyptus tetrodonta*, an evergreen tree of north Australian savannas (Thomas and Eamus 1999), and in several other tree species (Franks *et al.* 1995; Williams *et al.* 1997). Thus, stomatal closure at earlier stages of leaf desiccation could be an important protective mechanism against xylem cavitation, that could deprive leaves of water supply and potentially cause leaf death (Tyree and Sperry 1989; Sperry and Pockman 1993). Stomatal closure and xylem cavitation were closely linked in a group of tropical dry forest trees, as shown by a strong linear correlation between the water potential at incipient cavitation (20% loss of xylem conductivity) and stomatal closure (50% reduction in stomatal conductance; Brodribb *et al.* 2003). The prevention of cavitation by stomatal closure could be particularly critical for seedlings of evergreen cerrado species, which are particularly exposed to drought during the first years of development. At the seedling stage, the root system is still fairly superficial (Rizzini 1965; Moreira and Klink 2000) and the plant does not have access to the deeper moist soil layers during the dry season (Hoffmann *et al.* 2004).

Photosynthesis

Photoprotective mechanisms and shade tolerance

Plants adjust to changes in the prevailing irradiance levels in an attempt to optimize and preserve the functioning of the photosynthetic apparatus. A precise balance between the use of absorbed light for photosynthesis and the safe dissipation of potentially harmful excess light energy can be critical for cerrado plants. The high atmospheric evaporative demand and hydraulic constraints result in strong stomatal limitation of transpiration and a prolonged midday depression of photosynthetic rates in sunny days in the wet and dry seasons (Franco 1998; Moraes and Prado 1998; Meinzer *et al.* 1999; Naves-Barbiero *et al.* 2000; Franco and Lüttge 2002). In addition to the large effects on daily carbon gain and water loss, this may lead to photodamage, unless compensatory mechanisms to alleviate photoinhibition are operational. Three pigments of the xanthophyll cycle, *i.e.*, violaxanthin, zeaxanthin, and antheraxanthin are thought to play a key role in photoprotective thermal energy dissipation (see for instance Gilmore (1997) and Müller *et al.* (2001), for reviews on this subject). Adjustments in the partitioning of electron flow between assimilative and non-assimilative processes such as photorespiration, could also be important. Thus, CO_2 and phosphoglycerate supplied via photorespiration to the Calvin-Benson cycle could be crucial to maintain electron flow at high irradiance levels when the supply of external CO_2 is limited by stomatal closure (Valentini *et al.* 1995; Kozaki and Takeba 1996; Muraoka *et al.* 2000).

Franco and Lüttge (2002) investigated the implications to photoprotection of photochemical and non-photochemical processes in five cerrado woody species that differed in photosynthetic capacity and in the duration and extension of the midday depression of photosynthesis. Their study highlighted the role of diurnal regulation of non-photochemical energy dissipation in PSII as a photoprotective mechanism during the midday depression period. They also presented some evidence of the importance of high levels of photorespiration to maintain electron flow under high irradiance when the supply of external CO_2 was limited by stomatal closure. However, further studies are needed to assess its contribution as a regulatory photoprotective mechanism.

Mattos (1998) reported diurnal courses of the maximum quantum yield of PSII (*i.e.*, the quantum efficiency if all PSII centres were open) in six cerrado trees at the peak of the dry season. Notwithstanding the differences in the degree of diurnal decline in the maximum efficiency of PSII measured as quenching of chlorophyll fluorescence signals, this decline was completely reversible by the end of the day. However, a photoinactivation slowly reversible component of fluorescence quenching was detected in some species, as evidenced by the low pre-dawn values of the maximum quantum yield of PSII. In a recent study, Franco *et al.* (2007) reported that diurnal adjustments in non-photochemical quenching were tightly correlated with the zeaxanthin levels in two cerrado tree species, underlining the significant role of zeaxanthin in the regulation of the energy dissipation processes under the high irradiance levels and high evaporative demand of the atmosphere that are typical of the end of the dry season. They also reported that although an efficient co-regulation of photochemical and non-photochemical quenching and adjustments in the partitioning of electron flow between assimilative and non-assimilative processes were operating, the photoprotective capacity was momentarily surpassed as irradiance quickly increased after sunrise. However, both species were able to recover and adjust their photoprotective mechanisms to the prevailing irradiances in the course of the day.

In addition to the five-month dry season, rain-free periods are frequent within the wet season (Hoffmann 1996; Kanegae *et al.* 2000), when the trees would have a fully developed canopy. Under these conditions, a strong stomatal regulation of transpiration is necessary to minimize the effects of a combination of high evaporative demand of the atmosphere in rainless days coupled with a higher leaf area that can impose a strong demand on the water transport system of the tree (Mattos *et al.* 2002). There is some evidence that cerrado woody species are capable of rapidly responding to these dry spells, by stomatal regulation of transpiration, coupled with reversible adjustments in non-photochemical quenching to divert the energy in excess to drive photosynthesis to radiationless dissipation (Mattos *et al.* 2002). This process was readily reversible, following a single short rainfall event.

Adult cerrado trees are exposed to high irradiances, but canopy shading can restrict seedling growth in the initial phases of plant development. Leaves of cerrado woody species typically reach 90% of maximum photosynthetic values at photosynthetic photon flux densities (PPFD; the flux of photons between 400 and 700 nm wavelength) of 600 to 1,200 μmol m^{-2} s^{-1}, which is about 30% to 60% of full sunlight (Prado and Moraes 1997). Some species saturate at even higher light levels (Franco and Lüttge 2002). PPFD compensation point ranges from 10 to 50 μmol m^{-2} s^{-1} at leaf temperature in the range of 25 to 30 °C. Open cerrado vegetation types are covered with a grass layer, typically 40 to 50 cm tall. For instance, PPFD measurements suggested that 5-cm-tall *Kielmeyera coriacea* and *Dalbergia miscolobium* under a grass canopy would not receive enough sunlight to reach even 50% of their photosynthetic capacity during the daylight period (Nardoto *et al.* 1998; Braz *et al.* 2000). The effects of canopy shading on CO_2 assimilation would become critical for seedling growth and survival in closed-canopy woodland type savannas (Kanegae *et al.* 2000). Because of shading, species characteristic of open habitats may not be able to grow in closed canopy sites, whereas photoinhibition can be an important stress factor for young plants in fully–exposed habitats. Shading also affects biomass accumulation and partition to roots (Felfili *et al.* 2001; Salgado *et al.* 2001). Root storage can be critical to survive the drought period or perturbations such as fire (Hoffmann and Franco 2003; Hoffmann *et al.* 2004).

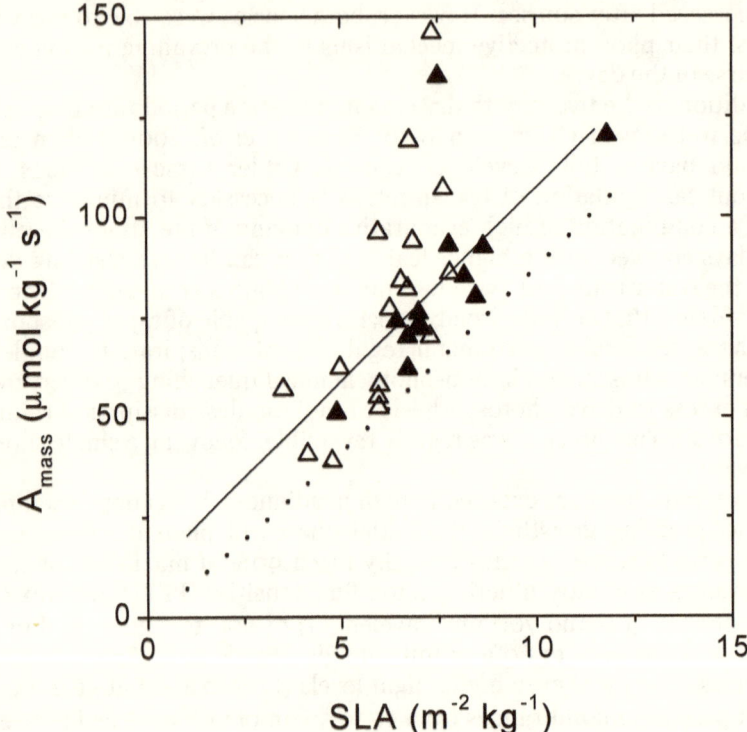

Figure 8.2. Relationship between maximum CO_2 assimilation on a mass basis (A_{mass}) and specific leaf area (SLA). Solid triangles are deciduous trees and open triangles are evergreen trees. Data from Medina and Francisco (1984), Prado and Moraes (1997), and Franco *et al.* (2005). The solid line describes the log-log relationship of A_{mass} and SLA fitted to the data ($\log y = 1.083 + 0.963\log x$; $r=0.67$; $P<0.0001$). The dotted line describes the log-log relationship fitted to data from six biomes ($\log y = -0.22 + 1.08\log x$; Reich *et al.* 1999).

Thus, reported differences in the range of several species along a gradient of increasing tree cover in cerrado ecosystems (Goodland 1971; Goodland and Ferri 1979) may reflect species difference in shade tolerance and acclimation potential. Indeed, integrative studies of spatial distribution coupled with physiological measurements of seedling performance and potential for acclimation of the photosynthetic apparatus are critically needed.

CO_2 assimilation, water use efficiency, and leaf phenology

The coexistence of deciduous and evergreen trees in cerrado environments implies contrasting strategies of resource use and leaf carbon balance, which would be affected by the strong rainfall seasonality. On a leaf area basis, deciduous and evergreen cerrado trees had similar rates of maximum CO_2 assimilation (A_{CO_2}) at the peak of the wet season (Franco et al. 2005). Stomatal conductances at maximum A_{CO_2} were also similar between the two phenological types. Differences in maximum CO_2 assimilation per unit leaf area and stomatal conductance between evergreen and deciduous trees are frequently small in savannas and dry deciduous forests (Goldstein et al. 1989; Sobrado 1991, 1994; Eamus and Cole 1997). However, deciduous species typically have a higher photosynthetic rate per unit leaf mass than evergreen species and higher specific leaf area (Chabot and Hicks 1982; Reich et al. 1992; Prado and Moraes 1997), which is also the case for cerrado trees (Franco et al. 2005).

Photosynthesis is strongly affected by nitrogen availability and it is well known that photosynthetic capacity increases linearly with leaf N concentration (Field and Mooney 1986; Reich et al. 1995, 1998). Indeed, Eamus and Prior (2001) showed that, for a range of tree species from different seasonally dry ecosystems, maximum CO_2 assimilation rates on a mass basis (A_{mass}) increases linearly with increasing leaf N, with no clear distinction between evergreen and deciduous species. This seems to be the case for cerrado trees as well, where deciduous and evergreen trees also shared the same linear relationship between A_{mass} and leaf N (Franco et al. 2005). Moreover, A_{mass} increases with specific leaf area (SLA) for both evergreen and deciduous species (Fig. 8.2). The slope of the relationship describing the log-log plot of A_{mass} and SLA (0.96) is similar to the general one fitted for data across biomes (1.08; Reich et al. 1999). Lower values of SLA generally reflect more investment of N in non-photosynthetic components (Field and Mooney 1986), with thicker and/or denser leaves having both a lower A_{mass} per unit N and a smaller change in A_{mass} per variation in N (Reich and Walters 1994). Because SLA and wood density are also strongly linearly related (Bucci et al. 2004b), variation in physiological and morphological traits appear to be subjected to similar selective pressures. This results in a general functional convergence not only in plant hydraulic architecture and water relations but in leaf structure and photosynthetic capacity, as well.

The trade-off between carbon gain and water use can be critical for evergreen and deciduous species in savanna environments, which are subjected to a prolonged dry season. This trade-off is generally assessed in terms of water use efficiency, here defined as the ratio of A_{CO2}/G_{wv}. There are however large uncertainties involved in attempting to integrate instantaneous measurements to represent a whole season. Measurements of carbon isotopic discrimination (Δ)[4] are typically used to assess WUE on a seasonal basis (Ehleringer and Cooper 1988; Ehleringer 1994). Carbon isotopic discrimination is largely dependent on the ratio of intercellular to ambient CO_2 concentrations (p_i/p_a) prevailing when the leaf carbon is assimilated (Farquhar et al. 1982, 1989). A seasonal decrease in p_i/p_a (lower values of Δ) should reflect diffusional limitations to CO_2 uptake from reductions in stomatal conductance and/or an increase in carboxylation efficiency (higher CO_2 assimilation rates; Ehleringer 1994). Thus, a stronger stomatal control in a shallow-rooted deciduous species resulted in lower foliar values of Δ, in comparison to deep-rooted evergreen species in a Venezuelan seasonally tropical dry forest (Sobrado and Ehleringer 1997). On the other hand, foliar values of Δ decreased toward the end of the dry season in both the evergreen *Curatella americana* and the deciduous *Godmania macrocarpa* in a Venezuelan savanna (Medina and Francisco 1994). The authors concluded that the similarity of WUE in both species were the result of higher photosynthetic capacity in *G. macrocarpa* and lower stomatal conductances in *C. americana*. Eleven of the 13 species studied by Mattos et al. (1997) in a cerrado area in southern Brazil showed a decrease in leaf Δ (higher water use efficiency) that ranged from 0.3 to 4‰ during the dry season. They related their results to a larger decrease in G_{wv} in comparison to A_{CO2}. However, they based this conclusion on gas exchange results for only two of the 13 species. They also did not include information on leaf phenology.

Irrespective of leaf phenoloy, most cerrado woody species showed a decline in the maximum CO_2 assimilation rates and stomatal conductances in the dry season, although the magnitude of the decrease was larger for stomatal conductance (Franco 1998; Moraes and Prado 1998; Franco et al. 2005). However, deciduous cerrado species flush new leaves by the end of the dry period (Table 8.1). The highest leaf concentrations of both N and P were measured in these newly mature leaves of the deciduous trees (Fig. 8.3a,b). Given that the effects of drought on stomatal conductance are apparently similar for deciduous and evergreen cerrado trees (Bucci et al. 2005) and that they share a common linear relationship between A_{mass} and leaf N (Franco et al. 2005), the newly mature leaves of deciduous species with high N and P concentrations could be able to achieve higher water use efficiency, by having higher photosynthetic capacity. Indeed, foliar values of Δ were significantly lower for deciduous species in the dry season (Fig. 8.3c). The differences in terms of leaf N and P and water use efficiency between evergreen and deciduous species decreased during the wet season.

---

[4] Carbon isotopic discrimination (Δ) is calculated as $(\delta^{13}C_{air} - \delta^{13}C_{plant})/(1 + \delta^{13}C_{plant})$, where $\delta^{13}C_{plant}$ is the carbon isotope ratio of the plant material and $\delta^{13}C_{air}$ is that of the air (−8‰), as described by Farquhar and Richards (1984).

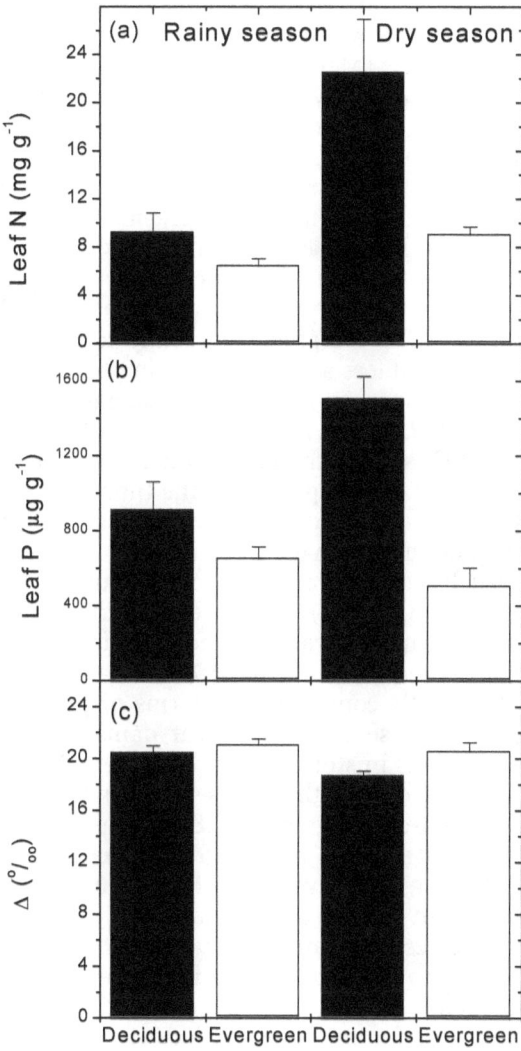

Figure 8.3. Rainy and dry season values of (a) N and (b) P concentrations, and (c) carbon isotopic discrimination (Δ) for leaves of 6 evergreen and 4 deciduous tree species growing in a cerrado site at the IBGE reserve. Bars are means ± 1 SE. Data were adapted from Franco *et al.* (2005). A multivariate analysis of variance was applied to analyze the effect of leaf phenology and season on leaf N, P, and Δ. Leaf phenology (*P*<0.0001), season (*P*=0.004) and the interaction season × phenology (*P*=0.007) had a significant effect by the Wilk's Lambda, Roy's Greatest Root, Hotelling-Lawley Trace, or Pillain Trace tests. The analyses were performed with the statistical software StatView 5 (SAS Institute Inc.).

The effects of drought on WUE at the crown level would be substantial as regulation of total daily water loss is enough to buffer seasonal changes, but not to substantially reduce tree water loss during the dry season. As discussed above, deciduous and evergreen species showed little seasonal variation in total water loss (Meinzer *et al.* 1999; Bucci *et al.* 2005). Thus, one would expect a decrease in WUE at the crown level. This could be critical for an evergreen tree, as it would not reduce water loss significantly in the dry season, but it would fix less carbon.

One could still expect that carbon return per unit dry mass invested nitrogen and carbon should be higher in evergreens because leaves are photosynthetically active for longer periods (Sobrado 1991) and of lower amortized leaf construction costs to replace the tree crown, provided that the leaves are held by more than a year (Givnish 2002). On the other hand, maximum carbon return in evergreens is greatly constrained by the combined effects of partial leaf loss and reductions in photosynthetic rates during the dry season (Franco 1998) and leaf damage by herbivores and pathogens (Marquis *et al.* 2002). Several evergreens flush leaves before the rains begin (Table 8.1), so that the most vulnerable stage of development (the new leaf stage) has already passed by the time the herbivorous insect activity begins (Marquis *et al.* 2002). However, leaf damage by pathogen attack is much higher than damage by herbivores in cerrado ecosystems and fully expanded leaves continued to accrue damage throughout their lives (Marquis *et al.* 2001, 2002). The lower SLA of the evergreen species also suggests larger leaf construction and maintenance costs and this should also be considered (Sobrado 1991; Eamus and Prior 2001). Thus, evergreens may be greatly constrained in terms of producing leaves with long leaf life-span, because of accrued leaf damage by herbivores and pathogens and reductions in stomatal conductance and partial leaf loss to stabilise water balance during the dry season. Indeed, leaf life-span is about one year for leaf-exchangers and 18 months for most other evergreens, while it is about 10 to 11 months for the deciduous species. (Maia 1999; Franco *et al.* 2005). However, leaf life-span in some species such as *Miconia ferruginata* can be two years or more.

Deciduous species compensated for their shorter leaf pay back period by maintaining lower leaf construction costs (higher SLA), a short leafless period and the capacity to resprout by the end of the dry season. They remain completely leafless for only short periods (few weeks to two months) and leaf flushing occurs primarily in the late dry season (Table 8.1; Maia 1999; Naves 2000; Rivera *et al.* 2002). This allows them to quickly achieve full crown development and maximal carbon gain when the rains begin. In contrast, full crown development in many evergreen trees is achieved only in the middle of the rainy season (Table 8.1; Fig. 8.1). Moreover, the higher leaf N and P concentrations of the newly developed crown of deciduous species would potentially allow higher photosynthetic rates than those of evergreen trees at the end of the dry season and during the transition from the dry to the wet period (Franco *et al.* 2005). On the other hand, this apparently requires the maintenance of a well-developed deep-root system that characterizes most deciduous species (Table 8.1). The higher cost of maintaining a deep-root system can be minimized by producing long-lived roots that have low diameter, low tissue density and low nutrient content (Reich *et al.* 2003). These aspects will have to be addressed in comparative studies of root traits in a broad range of cerrado species. Studies on plant WUE and on plant respiration are also needed.

Conclusions and future prospects

Most of the work on water relations of cerrado trees was performed in adult plants. The effects of drought on seedlings need to be better evaluated. Based on the range of wood density from 0.37 to 0.71 (Bucci *et al.* 2004b), their hydraulic architecture was not designed to support very negative pressures (Hacke and Sperry 2001). The root system of cerrado tree seedlings expands slowly and there is evidence that their roots are exposed to the superficial dry soil layers in the first years of life (Hoffmann *et al.* 2004). Thus, studies of the relations between stomatal closure, xylem vulnerability and embolism repair mechanisms in seedlings of a range of species with differing leaf-loss and lead flushing strategies are critically needed. Moreover, the water relations of roots and their impacts on plant survival in savannas remain to be investigated. Roots, particularly small roots, are more susceptible to drought-induced cavitation than branches, which poses them as the weak link in the plant hydraulic continuum under drought conditions (Hacke and Sperry 2001).

Both discrete (*e.g.*, leaf phenology) and continuous (*e.g.*, wood density, SLA) plant functional categories have been used with relatively success in studies of plant adaptations to the cerrado environment. Despite the high diversity of trees in cerrado ecosystems, shared similar functional relationships between leaf- and whole-plant-level traits among species point out that selective pressures impose strong constraints on functional trait variability. However, a limited number of taxa from a restricted number of plant families have been studied. Leaf-life span and related leaf traits needs to be determined in a larger range of cerrado trees to evaluate how differences in leaf-life span will impact resource use. In addition to physiological constraints, the impacts of leaf herbivory and pathogen attack have to be considered in the development of cost-benefit models for leaves of cerrado trees with contrasting phenologies. However, they will still be of limited application if they are not linked to studies of root traits.

Studies on constraints to seedling development are critically needed. In terms of aboveground biomass, cerrado tree seedlings grow slowly in natural conditions (Rizzini 1965; Franco 2002). However, this is probably not the result of inherently low growth rates. In a comparative growth analysis study, Hoffmann and Franco (2003) showed that cerrado and forest wood seedlings had similar growth rates under low and high light or nutrient treatments. These results would imply that the slow growth of cerrado trees is more a consequence of the larger carbon allocation to roots linked to low availability of nutrients in the soil, and light limitation by canopy shading. Thus, one can expect that seedlings of woody plants develop a tree canopy layer in the grass matrix through a slow process. The effects of grass root competition for nutrients need to be evaluated, and the light regimes along the physiognomic gradient need a better characterization. Research is also needed to characterize shade and light acclimatization to increased carbon gain for plants growing in different cerrado physiognomies.

Acknowledgements

This chapter was written during a sabbatical leave at Harvard University, supported by the Rockefeller Foundation. Research funds from Conselho Nacional de Desenvolvimento Cientifico e Tecnologico (CNPq) and the Programa de Apoio a Nucleos de Excelencia-PRONEX are also greatly acknowledged.

Literature cited

Abdala GC, Caldas LS, Haridasan M, and Eiten G. 1998. Above and belowground organic matter and root:shoot ratio in a cerrado in Central Brazil. *Brazilian Journal of Ecology* 2: 11-23.

Brodribb TJ, Holbrook NM, Edwards EJ, and Gutiérrez MV. 2003. Relations between stomatal closure, leaf turgor and xylem vulnerability in eight tropical dry forest trees. *Plant, Cell and Environment* 26: 443-450.

Borchert R. 1994. Soil and stem water storage determine phenology and distribution of tropical dry forest trees. *Ecology* 75: 1437-1449.

Braz VS, Kanegae MF, and Franco AC. 2000. Estabelecimento e desenvolvimento de *Dalbergia miscolobium* Benth. em duas fitofisionomias típicas dos cerrados do Brasil Central. *Acta Botanica Brasilica*, 14: 27-35.

Brooks JR, Meinzer FC, Coulomb R, and Gregg J. 2002. Hydraulic redistribution of soil water during summer drought in two contrasting Pacific Northwest forests. *Tree Physiology* 22: 1107-1117.

Bucci SJ. 2001. *Arquitectura hidráulica y relaciones hídricas de árboles de sabanas neotropicales: efectos de la disponibilidad de agua y nutrientes*. Ph.D. thesis, Universidade de Buenos Aires, Buenos Aires, Argentina.

Bucci SJ, Scholz FG, Goldstein G, Meinzer FC, Sternberg L da SL. 2003. Dynamic changes in hydraulic conductivity of petioles of two savanna species: factors and mechanisms contributing to the refilling of embolized vessels. *Plant, Cell and Environment* 26: 1633-1645.

Bucci SJ, Scholz FG, Goldstein G, Meinzer FC, Hinojosa JA, Hoffmann WA, and Franco AC. 2004a. Processes preventing nocturnal equilibration between leaf and soil water potential in tropical savanna woody species. *Tree Physiology* 24: 1119-1127.

Bucci SJ, Goldstein G, Meinzer FC, Scholz FG, Franco AC, and Bustamente M. 2004b. Functional convergence in hydraulic architecture and water relations of tropical savanna trees: from leaf to whole plant. *Tree Physiology* 24: 891-899.

Bucci SJ, Goldstein G, Meinzer FC, Franco AC, Campanello P, and Scholz FG. 2005. Mechanisms contributing to seasonal homeostasis of minimum leaf water potential and predawn disequilibrium between soil and plant water potential in Neotropical savanna trees. *Trees* 19: 296-304.

Caldwell MM, Dawson TE, and Richards JH. 1998. Hydraulic lift - consequences of water efflux from the roots of plants. *Oecologia* 113: 151-161.

Chabot BF and Hicks DJ. 1982. The ecology of leaf life spans. *Annual Review of Ecology and Systematics* **13:** 229-259.

Choat B, Ball M, Luly J, and Holtum J. 2003. Pit membrane porosity and water stress-induced cavitation in four co-existing dry rainforest tree species. *Plant Physiology* **131:** 41-48.

Coradin VT. 2000. *Formação de anéis de crescimento e sazonalidade da atividade cambial de dez espécies lenhosas do Cerrado.* Ph.D thesis, Universidade de Brasília, Brasília, Brazil.

Domec J-C, Scholz FG, Bucci SJ, Meinzer FC, Goldstein G, and VillaLobos-Vega R. 2006. Diurnal and seasonal variation in root embolism in neotropical savanna woody species: impact on stomatal control of plant water status. *Plant, Cell and Environment* **29:** 26-35.

Donovan LA, Grise DJ, West JB, Pappert RA, Alder AA, Richards JH. 1999. Predawn disequilibrium between plant and soil water potential in two cold desert shrubs. *Oecologia* **120:** 209-217.

Donovan LA, Linton MJ, Richards JH. 2001. Predawn plant water potential does not necessarily equilibrate with soil water potential under well-watered conditions. *Oecologia* **129:** 328-335.

Duff GA, Myers BA, Williams RJ, Eamus D, O'Grady A, and Fordyce IR. 1997. Seasonal patterns in soil moisture, vapour pressure deficit , tree canopy cover and predawn water potential in a northern Australian savanna. *Australian Journal of Botany* **45:** 211-224.

Eamus D and Cole SC. 1997. Diurnal and seasonal comparison of assimilation, phyllode conductance and water potential of three Acacia and one Eucalyptus species in the wet-tropics of Australia. *Australian Journal of Botany* **45:** 275-290.

Eamus D and Prior L. 2001. Ecophysiology of trees of seasonally dry tropics: comparisons among phenologies. *Advances in Ecological Research* **32:**113-197.

Ehleringer JR. 1994. Variation in gas exchange characteristics among desert plants. In: Schulze E-D and Caldwell MM (eds.) *Ecophysiology of Photosynthesis.* Ecological Studies Series. Springer, New York. Pp. 361-392.

Ehleringer JR, Cooper TA. 1988. Correlations between carbon isotope ratio and microhabitat in desert plants. *Oecologia* **76:** 562-566.

Eiten G. 1972. The cerrado vegetation of Brazil. *The Botanical Review* **38:** 201-341.

Farquhar GD, Ehleringer JR, and Hubick KT. 1989. Carbon isotope discrimination during photosynthesis. *Annual Review Plant Physiology and Plant Molecular Biology* **40:** 503-537.

Farquhar GD, O'Leary MH, and Berry JA. 1982. On the relationship between carbon isotope discrimination and the intercellular carbon dioxide concentration in leaves. *Australian Journal of Plant Physiology* **9:** 121-137.

Farquhar GD and Richards RA. 1984. Isotopic composition of plant carbon correlates with water-use efficiency of wheat genotypes. *Australian Journal of Plant Physiology* **11:** 539-552.

Felfili JM, Franco AC, Fagg CW, and Sousa-Silva JC. 2001. Desenvolvimento inicial de espécies de Mata de Galeria. In: Ribeiro JF, Fonseca CEL da, and Sousa-Silva JC. (eds.) *Cerrado Caracterização e Recuperação de Matas de Galeria.* Embrapa Cerrados, Planaltina, Brazil. Pp 779-81.

Felfili JM and Silva Jr. MC. 1993. A comparative study of cerrado (*sensu stricto*) vegetation in Central Brazil. *Journal of Tropical Ecology* **9:** 277-289.

Field C and Mooney H. 1986. The photosynthesis-nitrogen relationship in wild plants. In: Givnish (ed.) *On the Economy of Plant Form and Function.* Cambridge University Press, Cambridge. Pp. 25-55.

Franco AC. 1998. Seasonal patterns of gas exchange, water relations and growth of *Roupala montana*, an evergreen savanna species. *Plant Ecology* **136**: 69-76.

Franco AC. 2002. Ecophysiology of cerrado woody plants. In: Oliveira PS and Marquis RJ (eds.) *The Cerrados of Brazil: Ecology and Natural History of a Neotropical Savanna*. Columbia University Press, New York. Pp. 178-197.

Franco AC, Bustamante M, Caldas LS, Goldstein G, Meinzer FC, Kozovits AR, Rundel P, and Coradin VTR. 2005. Leaf functional traits of Neotropical savanna trees in relation to seasonal water deficit. *Trees* **19**: 326-335.

Franco AC and Lüttge U. 2002. Midday depression in savanna trees: coordinated adjustments in photochemical, efficiency, photorespiration, CO_2 assimilation and water use efficiency. *Oecologia* **131**: 356-365.

Franco AC, Matsubara S, Orthen B. 2007. Photoinhibition, carotenoid composition and the co-regulation of photochemical and non-photochemical quenching in neotropical savanna trees. *Tree Physiology* **27**: 717-725.

Franco AC and Nobel PS. 1988. Interactions between seedlings of *Agave deserti* and the nurse plant *Hilaria rigida*. *Ecology* **69**: 1731-1740.

Franco AC and Nobel PS. 1989. Effect of nurse plants on the microhabitat and growth of cacti. *Journal of Ecology* **77**: 870-886.

Franco AC, Soyza AG de, Virginia RA, Reynolds JF, and Whitford WG. 1994. Effects of plant size and water relations on gas exchange and growth of the desert shrub *Larrea tridentata*. *Oecologia* **97**: 171-178.

Franks PJ, Gibson A, and Bachelard EP. 1995. Xylem permeability and embolism susceptibility in seedlings of *Eucalyptus camaldulensis* Dehnh. from two different climatic zones. *Australian Journal of Plant Physiology* **22**: 15-21.

Gilmore AM. 1997. Mechanistic aspects of xanthophyll cycle-dependent photoprotection in higher plant chloroplasts and leaves. *Physiologia Plantarum* **99**: 197-209.

Givnish TJ. 2002. Adaptive significance of evergreen vs. deciduous leaves: solving the triple paradox. *Silva Fennica* **36**: 703-743.

Goldstein G, Andrade JL, Meinzer FC, Holbrook NM, Cavalier J, Jackson P, and Celis A. 1998. Stem water storage and diurnal patterns of water use in tropical forest canopy trees. *Plant, Cell and Environment* **21**: 397-406.

Goldstein G, Rada F, Rundell P, Azocar A, and Orozco A. 1989. Gas exchange and water relations of evergreen and deciduous tropical savanna trees. *Annales des Science Forestières* **46** (suppl.): 448s-453s.

Goodland R. 1971. A physiognomic analysis of the cerrado vegetation of Central Brazil. *Journal of Ecology* **59**: 411-419.

Goodland R and Ferri MG. 1979. *Ecologia do Cerrado*. Editora da Universidade de São Paulo, São Paulo, Brazil.

Gowing DJG, Jones HG, and Davies WJ. 1993. Xylem-transported abscisic acid: the relative importance of mass and its concentration in the control of stomatal aperture. *Plant, Cell and Environment* **16**: 453-459.

Hacke UG and Sperry JS. 2001. Functional and ecological xylem anatomy. *Perspectives in Plant Ecology, Evolution and Systematics* **4**: 97-115.

Haridasan M. 2000. Nutrição mineral de plantas nativas do cerrado. *Revista Brasileira de Fisiologia Vegetal* **12**: 54-64.

Haridasan M. 2001. Nutrient cycling as a function of landscape and biotic characteristics in the cerrados of Central Brazil. In: McClain ME, Victoria RL, and Richey JE (eds) *The Biogeochemistry of the Amazon Basin*. Oxford University Press, New York. Pp. 68-83.

Hoffmann WA. 1996. The effects of fire and cover on seedling establishment in a neotropical savanna. *Journal of Ecology* **84**: 383-393.

Hoffmann WA and Franco AC. 2003. Comparative growth analysis of tropical forest and savanna woody plants using phylogenetically independent contrasts. *Journal of Ecology* **91:** 475-484.

Hoffmann WA, Orthen B, and Franco AC. 2004. Constraints to seedling success of savanna and forest trees across the savanna-forest boundary. *Oecologia* **140:** 252-260.

Holbrook NM and Zwieniecki MA. 1999. Xylem refilling under tension. Do we need a miracle? *Plant Physiology* **120:** 7-10.

Jackson PC, Meinzer FC, Bustamante M, Goldstein G, Franco A, Rundel PW, Caldas L, Igler E, Causin F. 1999. Partitioning of soil water among tree species in a Brazilian Cerrado ecosystem. *Tree Physiology* **19:** 717-724.

Jackson RB, Sperry JS, and Dawson TE. 2000. Root water uptake and transport: using physiological processes in global predictions. *Trends in Plant Science* **5:** 482-488.

Jones MM and Turner NC. 1978. Osmotic adjustments in leaves of sorghum in response to water deficits. *Plant Physiology* **61:** 122-126.

Kanegae MF, Braz VS, and Franco AC. 2000. Efeitos da seca sazonal e disponibilidade de luz na sobrevivência e crescimento de *Bowdichia virgilioides* em duas fitofisionomias típicas dos cerrados do Brasil Central. *Revista Brasileira de Botânica* **23**: 459-468.

Kozaki A and Takeba G. 1996. Photorespiration protects C_3 plants from photooxidation. *Nature* **384:** 557-560.

Lenza E and Klink CA. 2006. Comportamento fenológico de espécies lenhosas em um cerrado sentido restrito de Brasília, DF. *Revista Brasileira de Botânica* **29:** 627-638.

Loewenstein NJ and Pallardy SG. 1998. Drought tolerance, xylem sap abscisic acid and stomatal conductance during soil drying: a comparison of young plants of four temperate deciduous angiosperms. *Tree Physiology* **18:** 421-430.

Maia JMF. 1999. *Variações Sazonais das Relações Fotossintéticas, Hídricas e Crescimento de Caryocar brasiliense e Rapanea guianensis em um Cerrado Sensu Stricto*. Master's thesis, Universidade de Brasília, Brasília, Brazil.

Marquis RJ, Diniz IR, and Morais HC. 2001. Patterns and correlates of interspecific variation in foliar insect herbivory and pathogen attack in Brazilian cerrado. *Journal of Tropical Ecology* **17:** 1-23.

Marquis RJ, Morais HC, Diniz IR. 2002. Interactions among cerrado plants and their herbivores: unique or typical? In: Oliveira PS and Marquis RJ (eds.) *The Cerrados of Brazil: Ecology and Natural History of a Neotropical Savanna.* Columbia University Press, New York. Pp 306-328.

Mattos EA. 1998. Perspectives in comprative ecophysiology of some Brazilian vegetation types: leaf CO_2 and H_2O exchange, chlorophyll *a* fluorescence and carbon isotope discrimination. In: Scarano FR and Franco AC (eds.) *Ecophysiological Strategies of Xerophytic and Amphibious Plants in the Neotropics.* Series Oecologia Brasiliensis, vol. 4. Universidade Federal do Rio de Janeiro, Rio de Janeiro, Brazil. Pp. 1-22.

Mattos EA de, Reinert F, and Moraes JAPV de. 1997. Comparison of carbon isotope discrimination and CO_2 and H_2O exchange between the dry and the wet season in leaves of several cerrado woody species. *Revista Brasileira de Fisiologia Vegetal* **9:** 77-82.

Mattos EA de, Lobo PC, and Joly CA. 2002. Overnight rainfall inducing rapid changes in photosynthetic behaviour in a cerrado woody species during a dry spell amidst the rainy season. *Australian Journal of Botany* **50:** 241-246.

Medina E and Francisco M. 1994. Photosynthesis and water relations of savanna tree species differing in leaf phenology. *Tree Physiology* **14**: 1367-1381.

Medina E and Silva JF. 1990. Savannas of northern South America: a steady state regulated by water-fire interaction on a background of low nutrient availability. *Journal of Biogeography* **17**: 403-413.

Meinzer FC. 2003. Functional convergence in plants responses to the environment. *Oecologia* **134**: 1-11.

Meinzer FC, Goldstein G, Franco AC, Bustamante M, Igler E, Jackson P, Caldas L, and Rundel PW. 1999. Atmospheric and hydraulic limitations on transpiration in Brazilian cerrado woody species. *Functional Ecology* **13**: 273-282.

Meinzer FC, Seymour V, and Goldstein G. 1983. Water balance in developing leaves of four tropical savanna woody species. *Oecologia* **60**: 237-243.

Moraes JAPV, Prado CHBA. 1998. Photosynthesis and water relations in cerrado vegetation. In: Scarano FR and Franco AC (eds.) *Ecophysiological Strategies of Xerophytic and Amphibious Plants in the Neotropics*. Series Oecologia Brasiliensis, vol. 4. Universidade Federal do Rio de Janeiro, Rio de Janeiro, Brazil. Pp. 45-63.

Moreira AG and Klink CA. 2000. Biomass allocation and growth of tree seedlings from two contrasting Brazilian savannas. *Ecotropics* **13**: 43-51.

Moreira MZ, Scholz FG, Bucci SJ, Sternberg LS, Goldstein G, Meinzer FC, and Franco A.C. 2003. Hydraulic lift in a neotropical savanna. *Functional Ecology* **17**: 573-581.

Müller P, Li X, and Niyogi KK. 2001. Non-photochemical quenching. A response to excess light energy. *Plant Physiology* **125**: 1558-1566.

Muraoka H, Tang Y, Terashima I, Koizumi H, and Washitani I. 2000. Contributions of diffusional limitation, photoinhibition and photorespiration to midday depression of photosynthesis in *Arisaema heterophyllum* in natural high light. *Plant, Cell and Environment* **23**: 235-250.

Murphy PG and Lugo AE. 1995. Dry forests of Central America and the Caribbean. In: Bullock SH, Mooney HA, and Medina E (eds) *Seasonally Dry Tropical Forests*. Cambridge University Press, Cambridge. Pp. 64-92.

Myers BA, Duff GA, Eamus D, Fordyce IR, O'Grady A, Williams RJ. 1997. Seasonal variations in water relations of trees of differing leaf phenology in a wet-dry tropical savanna near Darwin, Northern Australia. *Australian Journal of Botany* **45**: 225-240.

Nardoto GB, Souza MP, and Franco AC. 1998. Estabelecimento e padrões sazonais de produtividade de *Kielmeyera coriacea* (Spr) Mart. nos cerrados do Planalto Central: efeitos do estresse hídrico e sombreamento. *Revista Brasileira de Botânica* **21**: 313-319.

Naves CC. 2000. *Relações Hídricas e Fotossíntese de Duas Espécies Frutíferas do Cerrado*. Master's thesis, Universidade de Brasília, Brasília, Brazil.

Naves-Barbiero CC, Franco AC, Bucci SJ, and Goldstein G. 2000. Fluxo de seiva e condutância estomática de duas espécies lenhosas sempre-verdes no campo sujo e cerradão. *Revista Brasileira de Fisiologia Vegetal* **12**:119-134.

Olivares E and Medina E. 1992. Water and nutrient relations of woody perennials from tropical dry forest in Mexico. *Journal of Vegetation Science* **3**: 383-392.

Oliveira-Filho AT and Ratter JA. 2002. Vegetation physiognomies and woody flora of the Cerrado biome. In: Oliveira PS and Marquis RJ (eds.) *The Cerrados of Brazil: Ecology and Natural History of a Neotropical Savanna*. Columbia University Press, New York. Pp. 91-120.

Pennington RT, Prado DE, and Pendry CA. 2000. Neotropical seasonally dry forests and Quaternary vegetation changes. *Journal of Biogeography* **27**: 261-273.

Perez SCJGA and Moraes JAPV de. 1991. Determinação do potencial hídrico, condutância estomática e potencial osmótico em espécies do estratos arbóreo, arbustivo e herbáceo de um cerradão. *Revista Brasileira de Fisiologia Vegetal* **3:** 27-37.

Prado CHBA, Moraes JAPV de. 1997. Photosynthetic capacity and specific leaf mass in twenty woody species of Cerrado vegetation under field conditions. *Photosynthetica* **33:** 103-112.

Prado CHBA, Moraes JAPV de, and Mattos EA de. 1994. Gas exchange and leaf water status in potted plants Of *Copaifera langsdorffii*. 1. Responses to water stress. *Photosynthetica* **30:** 207-213.

Prior LD, Eamus D, and Duff GA. 1997. Seasonal and diurnal patterns of carbon assimilation, stomatal conductance and leaf water potential in *Eucalyptus tetrodonta* saplings in a wet-dry savanna in northern Australia. *Australian Journal of Botany* **45:** 241-258.

Querejeta, JI, Egerton-Warburton LM, and Allen MF. 2003. Direct nocturnal water transfer from oaks to their mycorrhizal symbionts during severe soil drying. *Oecologia* **134:** 55-64.

Ratter JA, Bridgewater S, Atkinson R, Ribeiro JF. 1996. Analysis of the floristic composition of the Brazilian cerrado vegetation. II. Comparison of the woody vegetation of 98 areas. *Edinburgh Journal of Botany* **53:** 153-180

Rawitscher F. 1948. The water economy of the vegetation of the "campos cerrados" in southern Brazil. *Journal of Ecology* **36:** 237-268.

Reich PB. 1995. Phenology of tropical forests: patterns, causes and consequences. *Canadian Journal of Botany* **73:** 164-174.

Reich PB, Ellsworth DS, and Walters MB. 1998. Leaf structure (specific leaf area) modulates photosynthesis-nitrogen relations: evidence from within and across species and functional groups. *Functional Ecology* **12:** 948-958.

Reich PB, Ellsworth DS, Walters MB, Vose JM, Gresham C, Volin JC and Bowman WD. 1999. Generality of leaf traits relationships: a test across six biomes. *Ecology* **80:** 1955-1969.

Reich PB, Kloeppel BD, Ellsworth DS, and Walters MB. 1995. Different photosynthesis-nitrogen relations in deciduous hardwood and evergreen coniferous tree species. *Oecologia* **104:** 24-30.

Reich PB and Walters MB. 1994. Photosynthesis-nitrogen relations in Amazonian tree species. II. Variation in nitrogen vis-à-vis specific leaf area influences mass and area-based expressions. *Oecologia* **97:** 73-81.

Reich PB, Walters MB, and Ellsworth DS. 1992. Leaf life-span in relation to leaf, plant, and stand characteristics among diverse ecosystems. *Ecological Monographs* **62:** 365-392.

Reich PB, Wright IJ, Cavender-Bares J, Craine JM, Oleksyn J, Westoby M, and Walters MB. 2003. The evolution of plant functional variation: traits, spectra, and strategies. *International Journal of Plant Sciences* **164:** S143-S164.

Richards JH and Caldwell MM. 1987. Hydraulic lift: substantial nocturnal water transport between soil layers by *Artemisia tridentata* roots. *Oecologia* **130:** 173-184.

Rivera G, Elliott S, Caldas LS, Nicolossi G, Coradin VTR, and Borchert R. 2002. Increasing day-length induces spring flushing of tropical dry forest trees in the absence of rain. *Trees* **16:** 445-456.

Rizzini CT. 1965. Experimental studies on seedling development of cerrado woody plants. *Annals of the Missouri Botanical Garden* **52:** 410-426.

Ryel RJ, Caldwell MM, Yoder DK, Or D, and Leffler AJ. 2002. Hydraulic redistribution in a stand of *Artemisia tridentata*: evaluation of benefits to transpiration assessed with a simulation model. *Oecologia* **130:** 173-184.

Salgado MAS, Rezende AV, Felfili JM, Franco AC, and Sousa-Silva JC. 2001. Crescimento e repartição de biomassa em plântulas de *Copaifera langsdorffii* Desf. submetidas a diferentes níveis de sombreamento em viveiro. *Brasil Florestal* **70:** 13-21.

Sanaiotti TM, Martinelli LA, Victoria RL, Trumbore SE, and Camargo PB. 2002. Past vegetation changes in Amazon savannas determined using carbon isotopes of soil organic matter. *Biotropica* **34:** 2-16.

Sarmiento G. 1984. *The Ecology of Neotropical Savannas*. Harvard University Press, Cambridge.

Sarmiento G, Goldstein G, and Meinzer FC. 1985. Adaptive strategies of woody species in neotropical savannas. *Biological Reviews of the Cambridge Philosophical Society* **60:** 315-355.

Sassaki RM, Machado EC, Lagôa AMMA, and Felippe GM. 1997. Effect of water deficiency on photosynthesis of *Dalbergia miscolobium* Benth., a cerrado tree species. *Revista Brasileira de Fisiologia Vegetal* **9:** 83-87.

Scholes RJ and Walker BH. 1995. *An African Savanna. Synthesis of the Nylsvey Study*. Cambridge University Press, Cambridge.

Scholz FG, Bucci SJ, Goldstein G, Meinzer FC, and Franco AC. 2002. Hydraulic redistribution of soil water by savanna trees. *Tree Physiology* **22:** 603-612.

Scholz FG, Bucci SJ, Goldstein G, Meinzer FC, Franco AC, and Miralles-Wilhelm. 2007. Biophysical properties and functional significance of stem water storage tissues in Neotropical savanna trees. *Plant, Cell and Environment* **30:** 236-248.

Schulze ED, Caldwell MM, Canadell J, Mooney HA, Jackson RB, Parson D, Scholes R, Sala OE, and Trimborn P. 1998. Downward flux of water through roots (*i.e.*, inverse hydraulic lift) in dry Kalahari sands. *Oecologia* **115:** 460-462.

Silva LPB, Klink CA, and Silva EM da. 2003. Evapotranspiração em um ecossistema do Cerrado e uma pastagem plantada. In: Claudino-Sales V de, Tonini IM, and Dantas EWC (eds.). *VI Congresso de Ecologia do Brasil. Anais de Trabalhos Completos. Simpósios de Biodiversidade, Unidades de Conservação, Indicadores Ambientais, Caatinga, Cerrado*. Editora da Universidade Federal do Ceará, Fortaleza, Brazil. Pp 517-518.

Sobrado MA. 1986. Aspects of tissue water relations and seasonal changes in leaf water potential components in evergreen and deciduous species coexisting in tropical dry forests. *Oecologia* **68:** 413-416.

Sobrado MA. 1991. Cost-benefit relationships in deciduous and evergreen leaves of tropical dry forest species. *Functional Ecology* **5:** 608-616.

Sobrado MA. 1994. Leaf age effect on photosynthetic rate, transpiration rate and nitrogen content in a tropical dry forest. *Oecologia* **96:** 19-23.

Sobrado MA. 1996. Embolism vulnerability of an evergreen tree. *Biologia Plantarum* **38:** 297-301.

Sobrado MA. 1997. Embolism vulnerability in drought-deciduous and evergreen species of a tropical dry forest. *Acta Oecologia* **18:** 383-391.

Sobrado MA and Ehleringer JR. 1997. Leaf carbon isotope ratios from a tropical dry forest in Venezuela. *Flora* **192:** 121-124.

Sperry JS and Pockman WT. 1993. Limitation of transpiration by hydraulic conductance and xylem cavitation in *Betula occidentalis*. *Plant, Cell and Environment* **16:** 279-287.

Sternberg L da SL, Bucci S, Franco A, Goldstein G, Hoffmann WA, Meinzer FC, Moreira MZ, and Scholz F. 2005. Long range lateral root activity by neo-tropical savanna trees. Plant and Soil **270:** 169-178.

Stratton L, Goldstein G, and Meinzer FC. 2000. Stem water storage and efficiency of water transport: their functional significance in a hawaiian dry forest. *Plant, Cell and Environment* **23:** 99-106.

Thomas DS and Eamus D. 1999. The influence of predawn leaf water potential on stem hydraulic conductivity and foliar ABA concentrations and on stomatal responses to atmospheric water content at constant C_i. *Journal of Experimental Botany* **50:** 243-251.

Tyree MT and Ewers FW. 1996. Hydraulic architecture of woody tropical plants. In: Mulkey SS, Chazdon RL, and Smith AP (eds.) *Tropical Forest Plant Ecophysiology.* Chapman and Hall, New York. Pp. 217-243.

Tyree MT and Sperry JS. 1989. Vulnerability of xylem to cavitation and embolism. *Annual Review of Plant Physiology and Plant Molecular Biology* **40:** 19-48.

Tyree MT, Snyderman DA, Wilmot TR, and Machado J-L. 1991. Water relations and hydraulic architecture of a tropical tree (*Schefflera morototonii*). Data, models and a comparison with two temperate tree species (*Acer saccharum* and *Thuja occidentalis*). *Plant Physiology* **96:** 1105-1113.

Valentini R, Epron D, Angelis P de, Matteucci G, and Dreyer E. 1995. *In situ* estimation of net CO_2 assimilation, photosynthetic electron flow and photorespiration in Turkey oak (*Q. cerris* L.) leaves: diurnal cycles under different levels of water supply. *Plant, Cell and Environment* **18:** 631-640.

Walker BH and Noy-Meir I. 1982. Aspects of the stability and resilience of savanna ecosystems. In: Huntley BJ and Walker BH (eds.) *Ecology of Neotropical Savannas.* Ecological Studies vol. 42, Springer, Berlin. Pp. 556-590.

Weltzin JF and McPherson GR. 1997. Spatial and temporal soil moisture resource partitioning by trees and grasses in a temperate savanna, Arizona, USA. *Oecologia* **112:** 156-164.

Williams JE, Davis SD, and Portwood K. 1997. Xylem embolism in seedlings and resprouts of *Adenostoma fasciculatum* after fire. *Australian Journal of Botany* **45:** 291-300.

Wright SJ. 1996. Phenological responses to seasonality in tropical forest plants. In: Mulkey SS, Chazdon RL, and Smith AP (eds.) *Tropical Forest Plant Ecophysiology.* Chapman and Hall, New York. Pp. 440-460.

Chapter 9

UNUSUAL PHYSIOLOGICAL PROPERTIES OF THE ARID ADAPTED TREE LEGUME *PROSOPIS* AND THEIR APPLICATIONS IN DEVELOPING COUNTRIES

Peter Felker

Introduction
Photosynthesis and water relations
Roots and soil relations
    Depth to water table
    N fixation and cross inoculation in Prosopis
    Water stress experienced when fixing N
    Adaptation to low soil P levels in the field
    N fixation in natural stands
    Salt tolerance
    High pH tolerance
Economically useful traits
Potential opportunities and liabilities with *Prosopis* genetic improvement programs
Genetics, new clones, and asexual reproduction
Potential for combining economic development and management of the weediness
Conclusions

Perspectives in Biophysical Plant Ecophysiology: A Tribute to Park S. Nobel, pp. 221-255
Edited by: E. De la Barrera and W.K. Smith
© 2009 by The Author
Book Compilation © 2009 Universidad Nacional Autónoma de México

Introduction

Despite their traditional outward appearance, the nitrogen fixing trees and shrubs of the genus *Prosopis* possess drought and heat tolerance equivalent to more common "desert type" plants such as the cacti. These trees currently occur in some of the most impoverished and harshest arid ecosystems of the world, *i.e.*, in the African Sahel from Senegal to Somalia, in the Middle Eastern deserts of Yemen, Saudi Arabia, and the Rajasthan desert of India/Pakistan (Pasiecznik *et al.* 2001). Examples of the various forms in Death Valley, Texas and Haiti are shown in Fig. 9.1A-C.

There are three subfamilies in the family Leguminosae. The Papilionoideae contains the common economically important annual legumes such as beans (*Phaseolus*), soybeans (*Glycine max*), alfalfa (*Medicago sativa*), and a few tree legumes such as black locust (*Robinia*). The most primitive subfamily the Caesalpinioideae contains trees such as redbud (*Cercis*), honey locust (*Gleditsia*), and paloverde (*Parkinsonia* and *Cercidium*; Doyle and Luckow 2003). The subfamily Mimosoideae contains many arid trees such as the genera *Acacia* known for its photos of the African savannahs and *Prosopis* which contains the mesquites of North America and the algarrobos of South America. From an economic utility standpoint, the major difference between the *Prosopis* and the *Acacias* is that there are many species of *Prosopis* that produce highly edible pods (with up to 40% sucrose) while almost none of the *Acacias* produce pods edible by humans.

Burkart (1976) described 44 species of *Prosopis* native to four continents (North and South America, Africa, and Asia) and suggested that the species evolved from the primitive *P. africana* species in Gondwanaland some 125 million years ago. Ramírez *et al.* (1999) examined RAPD profiles of species native to all 4 continents and were able to distinguish the species native to the individual continents (in the 5 sections of the genus) and revealed a molecular marker common to old and new world species. A more recent analysis of evolution in the legume family has suggested that the mimosoids and papilonoids evolved later in the Eocene when there was a land bridge connecting North America and Europe about 54 million years ago (Doyle and Luckow 2003). The hypothesis of Doyle and Luckow (2003) notwithstanding, the presence of well adapted native *Prosopis* in harsh deserts of North South America, Africa, the Middle East and India/Pakistan, and the absence of any Papilionoid genera on all 4 continents, suggests to this author that *Prosopis* is an older and more distinct genus than the others. The small size of the *Prosopis* genome (392 to 490 Mbp) also suggests that it is a primitive species.

One important asset of *Prosopis* that sets it apart from many potential economic plants of arid regions, is the presence of a rich interbreeding gene pool. All of the species have 2n = 28 except for some *P. juliflora* which are 4n (Harris *et al.* 2003). With the exception of the atypical shrubby *P. strombulifera*, with low gene heterozygosity that spreads by rhizomes

Figure 9.1. Growth-forms of *Prosopis*. *Prosopis glandulosa* var *torreyana* in Death Valley, California that is the hottest location in the Western hemisphere (A). Dense weedy stand of immature *Prosopis glandulosa* var *glandulosa* in Texas, see man for scale (B). Mature specimen of *Prosopis juliflora* in Haiti with the author's daughters (C).

(Hunziker *et al.* 1986) all the species studied to date are self-incompatible (Simpson 1977; Keys and Smith 1994). Bessega *et al.* (2000), using isozyme data, calculated a 15% selfing rate within most economically important North and South American species. Based on cytological grounds, Hunziker *et al.* (1975) proposed natural hybrids for the Argentine species *P. alba* by *P. nigra*, *P. hassleri* by *P. ruscifolia*, and *P. ruscifolia* by *P. alba*. Saidman and Vilardi (1987) using isozymes on the 7 most economically important arboreal species of Argentina stated that they were so close genetically that each of them should be considered a semi-or sub-species and that as a whole these 7 species would be a super-species or syngameon. We also have observed in replicated field trials (Felker *et al.* 1983) that progeny of the thornless South American species *P. alba* growing in proximity to North American species *P. glandulosa* var. *torreyana* were thorny and thornless and had many traits of the North American species. As described below, this rich genetic pool, in combination with the potential for interspecific hybridization, techniques for rooting trees of elite trees in some species, and grafting in almost all the economically important species, makes it possible to clonally propagate superior hybrids.

Probably due to their economic utility, the vast majority of the physiological and genetic studies have been done on members of the Papilionoideae subfamily with the implicit assumption that the physiology of the other genera would not be radically different. However the evolution of common annual legumes such as soybeans, beans, and alfalfa occurred in temperate and/or regions with rainfalls considerably greater than semiarid regions. In contrast, the evolution of *Acacia* and *Prosopis* occurred in the world's hottest (Death Valley, California) and lowest rainfall regions (Chilean Atacama Desert) and has lead to vast differences in physiological characters such as heat, drought, and salinity tolerance. For example as will be discussed in detail later, common legume genera such as *Pisum*, *Phaseolus* and *Glycine* very poorly tolerate salinities of 2 to 3 dS m^{-1} while both *Prosopis* and *Acacia* have been shown to grow in salinities of 45 and 80 dS m^{-1} respectively. Due to the great energetic cost, N fixation is very sensitive to water stress with nodules abscising or N fixation being greatly reduced even at moderate stress levels. Thus while N fixation in soybeans is reduced to zero at leaf xylem water potentials of -2.8 MPa (Huang *et al.* 1975), N fixation in *Prosopis* proceeds at high rates with xylem water potentials of -3.5 MPa and air temperatures of 45 °C.

Photosynthesis and water relations

In the 1970s, Stanford University scientists O. Björkman and H. Mooney setup a mobile laboratory in Death Valley National Monument in California to measure photosynthesis and water relations in phreatophytic plants like *Prosopis* and non phreatophytes such as *Larrea* (Mooney 1977). *Prosopis* is native to Death Valley and has rather extensive distribution in the valley as noted by Mesquite flats on Death Valley maps. This location was chosen as Death Valley is the hottest location in the western hemisphere and is close to being the hottest location on earth. The Greenland

Ranch meteorological station in Death Valley has a mean daily July maximum temperature of 46.6 °C (115.8 °F) and an absolute maximum temperature of 56.7 °C (134 °F) (47 years of record; U.S. Department of Commerce 1964). Here these workers measured photosynthesis in *Prosopis* at air temperatures of 45 °C and leaf water potentials of −4.5 MPa (Mooney *et al.* 1977). They measured a maximum light saturated photosynthesis rate of 30 mg CO_2/dm²/hr (18 μmol m^{-2} s^{-1}) that they stated was among the highest photosynthetic rates for woody plants.

De Soyza *et al.* (1996) measured water stress and net photosynthesis for the short shrubby *Prosopis glandulosa* on 4 sampling dates on the northern edge of the Chihuahuan desert in New Mexico. Here the mean annual precipitation was 233 mm but unlike Death Valley, the trees did not tap into groundwater. While there were considerable differences between dates and for small and large shrubs, the daily maximum net photosynthesis was greater than 10 μmol m^{-2} s^{-1} and the maximum photosynthesis rate of about 22 μmol m^{-2} s^{-1} occurred for large shrubs at the end of the summer.

On a non-phreatophyte site in Texas with 322 mm annual rainfall, Hansen and Dye (1980) measured water stress and photosynthesis on *Prosopis*. These authors reported that maximum light saturated photosynthesis rates of 19 μmol m^{-2} s^{-1} were associated with xylem water potentials of −3.3 MPa at the beginning of the season and rates of 7.3 μmol m^{-2} s^{-1} were associated with leaf water potentials of −3.7 MPa at the end of the season. These authors stated that *Prosopis* net photosynthesis rate was high compared to reported maximum photosynthetic values for most woody deciduous species of 4.3 to 16 μmol m^{-2} s^{-1}. This is particularly impressive given the low annual precipitation and the low xylem water potentials.

Although photosynthesis was not measured, Nilsen *et al.* (1981) measured leaf water potential, stomatal conductance, and leaf osmotic potential in *Prosopis* in a phreatophytic site in the Sonoran Desert of California where the mean annual rainfall was about 70 mm yr^{-1} and the absolute July maximum temperatures were about 49 °C (Sharifi *et al.* 1983). Using daily comparisons of leaf conductance and water potential, they observed that the leaf stomata stayed open until a leaf water potential of about −4.8 MPa was reached.

In a more mesic (662 mm annual rainfall) site in northern Texas, Ansley *et al.* (1992) compared *Prosopis* water relations under rainout and irrigated conditions. The root systems were containerized to a 2.5 m depth with sheet metal and a plastic vertical barrier. These authors found that, in the treatment without rain or irrigation for 2 months in the summer, the leaf water potential did not go lower than −3.2 MPa. In spite of this lack of water, in the early morning hours the stomatal conductance in rainout treatment was still about 70% of that of the surplus water irrigated treatment.

Flowering and pod production in *Prosopis* is apparently stimulated by stress. Lee and Felker (1992) examined *Prosopis* flower production, pod production, nectar secretion, and xylem water potential in a 450 to 700 mm rainfall gradient over 5 sites in a normal and a dry year. In the wet year, an average of 70 mm rainfall per month occurred for the first 6 months of the year which was the period of flowering and fruit production, but in the dry year only an average of 19 mm per month occurred as a mean for the same sites. Averaged over all 5 sites, the dry year produced a lower pre-dawn xylem water potential of −1.3 MPa than the wet year of −1.0 MPa and had greater inflorescence and 3 times greater pod production. In contrast, in drought tolerant strains of faba bean (*Vicia faba*) being grown under semi-arid conditions (350 mm yr⁻¹) in Australia, the early podding stage was very susceptible to drought stress resulting in 50% reduction in seed yields and a drop in the photosynthetic rate to about 6 μmol m⁻² s⁻¹. The well watered control plants had mid-day leaf water potentials less between −1.0 and −0.6 MPa while the stressed plants were in the −2.0 to −1.5 MPa range (Mwanamwenge *et al.* 1999). These results are not directly comparable since the −1.3 MPa water potentials measured by Lee and Felker (1992) were predawn and would have been much lower at mid-day. Thus we see that *Prosopis* has the interesting physiological trait of having its partitioning into reproductive organs, stimulated, rather than inhibited by drought stress.

In summary, in spite of the fact that many of the measurements on photosynthesis as a function of leaf water potential are 20 years old, the data is impressive. Photosynthesis rates on the order of 10 to 20 μmol m⁻² s⁻¹ measured for *Prosopis* are high for mesic woody trees. The photosynthesis and stomatal conductance of *Prosopis* continued until leaf xylem water potentials reached about −4.8 MPa which is unusually low. Even though the *Prosopis* in Death Valley are tapping into permanent underground water sources, it is probable that even if the soil around the root systems of Papilionoid legumes were maintained at field capacity, most of them would not survive under daily maximum temperatures of 46 °C.

Roots and soil relations

Depth of water table

Prosopis roots can reach water at extraordinary depths of 53 m (Philips 1963). However, it is probable that these roots do not contribute substantially to the growth of the plants since in a study of *Prosopis* height as a function of ground water in the California desert, Meinzer (1927) found that *Prosopis glandulosa* was 3.6 to 6 m tall when the ground water was 3 m deep but was only 0.6 to 0.9 cm tall when the depth to the groundwater was 14 meters.

N fixation and cross inoculation in Prosopis

Nodulation was first reported for *Prosopis* by Khudairi (1957) for *P. stephaniana* in Iraq, then by Basak and Goyal (1972) in India, and by the Canadian, Bailey (1976), in Texas. Felker and Clark (1980) examined nodulation and nitrogen fixation (acetylene reduction) with a rhizobia strain isolated from a *Prosopis* growing in the greenhouse on soil taken from under *Prosopis* in the California desert. These authors examined a wide range of *Prosopis* species including the old world *P. africana*, Peruvian *P. pallida*, *P. tamarugo* from the Chilean Atacama salt deserts, *P. kuntzei* which only has photosynthetic stems (no leaves), the commercial *P. alba* and *P. nigra* from Argentina, and various *Prosopis* species from south-western U.S.A. All of these species were demonstrated to nodulate, reduce acetylene to ethylene, and grow on N free media. However the primitive *P. africana* had the smallest plants, and the lowest N fixation rate. Thus it seemed as if one rhizobia strain probably could cross inoculate all *Prosopis* species. This rhizobia strain was later demonstrated to have exceptional salt tolerance with significant growth in 0.5 M NaCl (Hua *et al.* 1982).

Water stress experienced when fixing N

Due to the high energy cost of breaking the triple bond of atmospheric nitrogen gas to make ammonia, plants in the nitrogen fixing mode have a very high photosynthate requirement. Thus it makes sense that with any kind of stress, the N fixation process is halted to prevent the drain on pho-tosynthate utilization.

In the common bean (*Phaseolus vulgaris*) when water was withheld for 5 or 8 days in greenhouse studies with 24/16 °C day/night temperatures there was a 60% decrease in specific nitrogenase activity (Ramos *et al.* 2003) and a change in leaf xylem water potential from about −0.9 to −0.5 MPa. In a study on the effect of drought stress on soybeans Huang *et al.* (1975) found that the acetylene reduction was completely inhibited a ta leaf water potential of −2.8 MPa. For cow peas (*Vigna unguiculata*), which is probably the most drought tolerant of the annual legumes, Figueiredo *et al.* (1998) found that when the plants were grown in a greenhouse with mini-mum and maximum temperatures of 27 and 35 °C, when the leaf water potential reached −1.1 MPa, the nitrogenase activity declined from 10 μmol per plant per hour to 1 μmol per plant per hour. In contrast, when Felker and Clark (1982) measured *Prosopis* nitrogen fixation rates of 68 μmol per plant per hour in 3-m long soil columns whose tops were close to the roof of the greenhouse in California, the air temperatures ranged from 43 to 47 °C and the leaf water potential from −3.8 to −2.9 MPa. The *Prosopis* N fixation rates in this study compared favorably to values of 40 μmol per plant per hour for soybeans (Thibodeau and Jaworski 1975) and 60 μmol per plant per hour for alder seedlings (Huss-Daniel 1978). Thus not only were the N fixation rates comparable to other plants, but the N fixation occurred at leaf air temperatures and xylem water potentials which would totally inhibit N fixation in other Papilionoid legumes.

Adaptation to low soil P levels in the field

It is generally acknowledged that phosphorus is the most important macronutrient for N fixation in legumes and that addition of P fertilizer often stimulates N fixation and protein in the leaves. For arid soils, which are often alkaline, the sodium bicarbonate extraction technique of Olsen and Sommers (1982) is most often used. For the forage legumes, alfalfa, alsike clover, birdsfoot trefoil, red clover and a grass-legume combination, Olsen soil test values of 0-3, 4-7, 8-11, 12-15, and 16+ mg kg^{-1} were very limiting, limiting, medium, high and very high, respectively (Dahnke *et al.* 1992).

In a survey of seven sites in South Texas where *Prosopis* naturally occurred, Geesing *et al.* (2000) found that the mean Olsen extractable P level was 4.6 mg kg^{-1} with a 95% confidence interval of 2.1. In spite of these low soil P levels, the mean N for all sites was 3.34 % with a 95% confidence interval of 0.13. This mean 3.34% N for all sites corresponds to a protein concentration of 20.8% which is substantial for this low level of soil P. Wightman and Felker (1990) examined various soil and leaf nutrients for the same *Prosopis alba* clone grown on experimental sites with contrasting productivity in south Texas. In spite of the fact that the maximum Olsen bicarbonate extractable P on the high productivity site was only 1.27 mg kg^{-1} (*vs.* 2.12 mg kg^{-1} on the low productivity site), the leaf N on the low productivity site ranged from 2.9 to 3.2% on the low productivity site *vs.* 3.0 to 4.2% N on the high productivity site (However, 60 kg P ha^{-1} was side dressed at the time of planting). The old work of Drake and Streckel (1955) may be relevant in understanding how *Prosopis* can function under such low native soil P levels. These authors found that due to their high cation exchange capacity the root systems of legumes were able to extract calcium from insoluble calcium-phosphorus complexes in the soil, thus effectively solubilizing P. Whatever the mechanism by which *Prosopis* is able to utilize low P containing soils while maintaining leaf protein concentrations of 20%, it would be valuable to exploit this potential in the field and to conduct further studies to determine the mechanism by which *Prosopis* functions on such low P concentrations.

N fixation in natural stands

In spite of the fact that *Prosopis* had been demonstrated to nodulate and fix N in greenhouse studies, there was doubt that *Prosopis* actually fixed N in field settings since nodules could not be found on the root systems (Martin 1948). An intensive U.S. National Science Foundation study where *Prosopis* obtained its water from a 3.5- to 5-m deep perched water table on a harsh desert site in California with a mean July maximum temperature of 47 °C and 70 mm annual rainfall (Sharifi *et al.* 1983), was initiated to resolve this issue of N fixation in natural stands. This study used N balances and the natural abundance of ^{15}N/^{14}N to unequivocally demonstrate that *Prosopis* fixed a minimum of about 30 kg N ha^{-1} yr^{-1} (Rundel *et al.* 1982; Shearer *et al.* 1983; Virginia *et al.* 1984). Rundel *et al.* (1982) suggested that *Prosopis* plantations with virtually complete canopy cover (*vs.* 30% on this site) might be able to achieve N fixation rates of 150 kg N ha^{-1} yr^{-1}. Virginia and Jarrell (1983) found that in the top 30 cm of soil there were 4,400 kg ha^{-1} more total N and 790 kg ha^{-1} more nitrates under the canopy than outside the canopy (assuming a bulk density of 1.2kg l^{-1}).

The high N fixation rates in the harsh California site notwithstanding, there was considerable reticence within the scientific community on whether *Prosopis* fixed N in native range settings since nodules could not be located on *Prosopis* in the field. Given the presence of *Prosopis* on approximately 45 million ha of low N and C containing soils in southwestern U.S.A. and northwestern Mexico (Johnson and Mayeux 1990), and the controversy concerning the encroachment of mesquite (*Prosopis glandulosa* var. *glandulosa* and *P. velutina* principally) onto overgrazed ecosystems, resolution of the importance of *Prosopis* to N cycling in these ecosystems was of considerable importance.

Johnson and Mayeux (1990) provided a fundamental breakthrough on this issue when they examined cores from 65 to 400 cm deep from under the canopies of *Prosopis* and found nodules on 19 trees at five locations in the eastern portion of the range of *Prosopis*. While they were unable to find nodules from some individuals on the drier western portion of the range, *Prosopis* seedlings nodulated when grown in soils from almost all of these sites. These authors concluded that:

> honey mesquite must be a significant contributor to the nitrogen budget of range ecosystems in south-western United States and Mexico and the genus *Prosopis* is likely to play an important role on a global scale.

While Johnson and Mayeux (1990) were able to find nodules on most of the trees on their sites, they did not measure N fixation on these rangeland sites. Due to the deep and extensive root systems, neither acetylene reduction nor ^{15}N enrichment is possible, leaving the only realistic method of assessment being natural abundance $^{15}N/^{14}N$ methods pioneered by Shearer and Kohl (Shearer *et al.* 1983). Villagra-López and Felker (1997) attempted to measure N fixation in natural stands as function of P fertilization and silvicultural management practices, and while they did not observe any treatment effect on N fixation in field settings, they made the serendipitous discovery that the percentage of N derived from biological fixation was inversely proportional to tree size ($r^2 = 0.90$, $p = 0.0001$). This implied that as the trees grew, they built up N under their canopy and then this increased soil N repressed biological N fixation. Geesing *et al.* (2000) followed up on this hypothesis by measuring nitrogen fixation using natural abundance methods on seven native stands of *Prosopis* each of which contained small, medium, and large trees. These workers found highly significant correlations between tree diameter and increase in soil C, N and P under *vs.* outside the canopies with maximum increases of 17.7 Mg ha^{-1} C, 4.4 Mg ha^{-1} N, and 13 kg ha^{-1} P. Moreover they reported highly significant negative regressions between percent of N fixed by the trees and the soil nitrate under the canopies. This physiological response confirmed that the trees were in fact fixing N. These authors reemphasized the classic work of Jenny (1940) on the negative relations between mean annual soil temperature and soil N and C. This was important since arid soils, that contain low water and thus have low heat capacities, have the highest soil temperatures and lowest soil N and C of the world's ecosystems. Possibly these low soil C and N contents are what has stimulated the world's arid ecosystems to have evolved such a high frequency of leguminous trees and shrubs of *Acacia* and *Prosopis*.

Salt tolerance

The negative impact of salinity on growth of plants in irrigated and non-irrigated areas of the world's arid regions continues to be a major problem. Within the family Leguminosae, virtually all of the important annual legumes that belong to the subfamily Papilonoideae such as soybeans, beans, peas, and cowpeas are highly salt sensitive and suffer yield reductions from salinities with conductivities as low as 2 to 3 dS m^{-1} (Richards *et al.* 1954; Ayers and Westcott 1985). Alfalfa is the most salt tolerant of the commercial legume species showing yield reductions of 50% at salinities of 9.6 dS m^{-1} (Ayers and Westcott 1985). In contrast, our work (Felker *et al.* 1981; Rhodes and Felker 1987; Velarde *et al.* 2003), as well as others (Ahmad *et al.* 1994; Baker *et al.* 1995), has shown that some *Prosopis*, especially *P. pallida*, *P. juliflora*, *P. tamarugo*, and *P. alba* have individual plants with rapid growth at seawater salinity or 45 dS m^{-1} which is nearly 20 times greater than salinities that can be tolerated by annual temperate legumes. Elite *Prosopis* trees with superior growth in these trials have been cloned by rooting of cuttings (Velarde *et al.* 2003) and established in seed orchards at the Universidad Católica de Santiago del Estero, Argentina.

It is to be noted that in the same subfamily as *Prosopis*, the Mimosoideae, some *Acacias* had 100% survival at 95 dS m^{-1} which is more than double the salinity of ocean water (Craig *et al.* 1991). The nitrogen-fixing woody salt tolerant genus *Casuarina* also has individual trees and species that can grow at salinities of 0.55 M NaCl which is equivalent to seawater (El-Lakany and Luard 1982; Ng 1987). Among these woody salt tolerant species, only *Prosopis* has portions that are highly edible and timber suitable for high quality furniture.

Identification and understanding of the mechanism of salt tolerance in *Prosopis* may have relevance to current commercially important legumes. The majority of the molecular work characterizing salt tolerance has been conducted on *Arabidopsis*, salt-water algae, and yeast mutants (Bohnert *et al.* 1999; Hasegawa *et al.* 2000). Although Winicov (1998) has reported a transcriptional regulator for gene expression in salt tolerant alfalfa, *Prosopis* is not a halophyte in that it is not able to grow with facility at 40 to 60 dS m^{-1} or to absorb and then secrete sodium salts on the leaves. As clonal propagation of *Prosopis pallida* by stem cuttings is moderately easy, this could be a useful model system for salt tolerance in the legume family. It would be interesting to search for cDNAs in highly salt tolerant *Prosopis* clones that were not present in the low salt tolerant clones in hopes that this information would be relevant to common annual legumes.

The tantalizing scenario of growing *Prosopis* in coastal deserts with seawater irrigation as originally suggested by Epstein *et al.* (1979) for other plants appears to be in the realm of possibility since (1) a few *P. alba* and *P. pallida* grew at seawater salinity, (2) only very limited selections from the germplasm base have been examined, and (3) other field management techniques (*e.g.*, provision of divalent cations Ca and Mg, P, and critical micronutrients) may alleviate the stress. Due to the pressing human needs in coastal deserts where *P. pallida* is adapted, *i.e.*, Mauritania, Somalia, Ethiopia, Yemen, India, *etc.*, it seems reasonable to extensively collect *P. pallida* with the objective of finding highly salt tolerant clones also possessing good growth rates and palatable sweet pods as reported by Alban *et al.* (2001).

High pH tolerance

Moderately alkaline soils, from pH 7.5 to 8.2 are common on rain fed semi-arid regions of western North America where there is a calcareous parent material. Traditional semi-arid crops such as sorghum and cotton can be grown on these soils taking care to correct macronutrient deficiencies such as phosphorus and trace elements such as Fe, Zn and Cu that have limited availability due to insoluble oxides or carbonates. In areas where irrigation mismanagement or where natural lack of drainage is combined with high evaporation rates, soil pH levels may reach 9.0 to 10.4. Virtually none of the commercial crops and only very few highly adapted plant species can survive on soils of these latter pH values.

Vast areas of the highly alkaline types of soils occur in Argentina (Ragonese 1951) and on about 2.5 million ha in India (Singh *et al.* 1989a, b). Installation of tile drainage systems combined with leaching and treatment with gypsum can be used to reclaim these soils, but this is an expensive option. In India the Central Soil Salinity Research Institute (Grewal and Abrol 1986; Singh *et al.* 1988, 1989a, b; Singh 1995, 1996) has been the world leader in using combinations of trees and grasses to reclaim these high-pH soils. *Prosopis juliflora* was able to grow satisfactorily without amendments up to pH 9.0, but these authors found that when the soil pH was 10.4 it was necessary to plant the trees in augerholes with amendments of 3 kg of gypsum and 8 kg of farmyard manure per hole (Singh 1996). Twenty years after such treatments, the initial soil pH of 10.4 decreased to 9.18 under *Eucalyptus tereticornis*, 9.03 under *Acacia nilotica*, 8.67 under *Albizia lebbek*, 8.15 under *Terminalia arjuna*, and 8.03 under *Prosopis juliflora*. In Argentina, where *P. alba* is being used to reclaim these high pH soils, surface applications of elemental sulfur, micronutrients and K increased growth 42% over the untreated control (Velarde *et al.* 2005).

This author is unaware of trials with Papilionoid legumes on soils with pH in the 8.5 to 10.0 range. It seems likely that analogous to the much higher salinity tolerance of *Prosopis* than annual legumes, this genus also possesses greater tolerance to high pH soils.

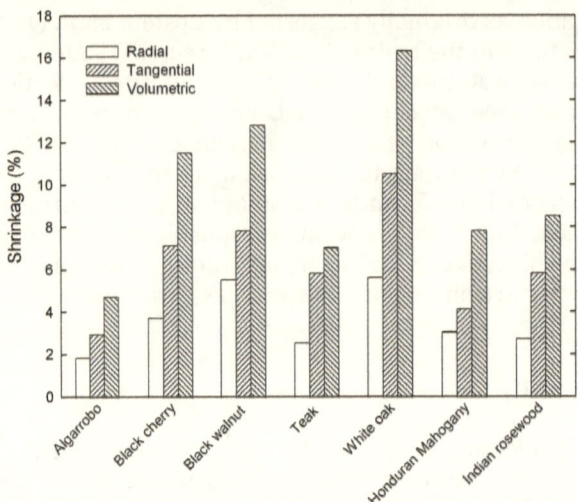

Figure 9.2. Comparison of shrinkage values of *Prosopis* lumber to other fine lumbers of the world. The corresponding Latin binomials are algarrobo (*Prosopis alba*), black cherry (*Prunus serotina*), walnut (*Junglans nigra*), teak (*Tectonia grandis*), white oak (*Quercus alba*), Honduran mahogany (*Swietenia macrophylla*), and Indian rosewood (*Dalbergia laifolia*). Values for *Prosopis alba* are from Turc and Cutter (1984) and the others from Chudnoff (1984).

Economically useful traits

Apart from environmental benefits such as shade, soil improvement and ornamental value, the two main economically useful traits of *Prosopis* are the pods, some of which are high in sugar and highly palatable to humans, and the lumber, which is excellent for flooring and high quality furniture.

Evidently, during evolution with great seasonal changes in moisture availability and thus water content of the conductive tissue, the wood of *Prosopis* developed a very low coefficient of movement with regard to moisture content. The values for *Prosopis alba* and *P. glandulosa* radial and tangential shrinkage of 1.8% and 2.9% (Turc and Cutter 1984) and of 2.2% and 2.6 % (Weldon 1986), respectively, are lower than all the woods listed in the compendium of tropical timbers (Chudnoff 1984), including teak (*Tectona grandis*), mahogany (*Swietenia macrophylla*), Indian rosewood (*Dalberghia latifolia*), and Brazilian rosewood (*Dalberghia nigra*; Fig. 9.2). Because low shrinkage values, and near equal radial and tangential shrinkages, are probably the best measure of wood stability, and because wood stability is one of the most important characteristics in furniture manufacture, *Prosopis* technically ranks with the world's best furniture species. When this stability is combined with the reddish-brown wood color and above average specific gravity (*ca.* 0.75) and hardness (770 kg cm^{-2} for *P. alba* and 1,010 kg cm^{-2} for *P. glandulosa* var. *glandulosa*), *Prosopis* lumber meets all the requisites to be included in the class of the world finest indoor furniture species. A simple computer search using mesquite and lumber as keywords will illustrate the value and variety of *Prosopis* wood products. (In this search it will be instructive to note that the economic analyses of *Prosopis* plantations for lumber (Felker and Guevara 2003; see discussion below) assumed a value of $800 per cubic meter which is equal to $2 per board ft). Unfortunately in Argentina, virtually none of the furniture is made from kiln dried wood or has finish and style that would be desirable by U.S. consumers. An example of the logs harvested for *Prosopis* furniture and of quality furniture by the company Fioramonte of Santiago del Estero, Argentina is illustrated in Fig. 9.3A, B.

Prosopis lumber could provide the basis for substantial value-added industries in arid lands that would indirectly contribute to increased food security, which is one of FAO's major objectives. To gauge the scale of this potential, it is important to note that North Carolina's wood furniture industry grosses about $4 billion annually. Furthermore, China has increased its wooden furniture exports to the U.S.A. from zero to almost $4 billion in the last 10 years (Buehlmann *et al.* 2002). The potential economic impact of new sustainable industries grossing even $100 million per year from countries in Sahelian Africa would be enormous.

Sweet *Prosopis* pods were an important component in the diets of indigenous people in North America (Felker 1979), South America (D'Antoni and Solbrig 1977), and the early people in India (Mann and Saxena 1980). A comparison of the protein and sugar content of North and South American species was made for plantation grown trees under various irrigation treatments in Riverside, California (Oduol *et al.* 1986). The mean sugar and protein concentration under wet, medium, and dry irrigation treatments is presented in Table 9.1 and illustrates great diversity in pod characteristics among *Prosopis* species. There was a negative correlation between pod sugar and pod protein concentration ($r = -0.63$, $p = 0.01$). This is attributable to the fact that the sugar content is located in the mesocarp while the protein is concentrated in the seeds. Thus a thick pod like *P. nigra* will have a high sugar content, while a pod with almost no mesocarp, such as *P. articulata* has very little sugar. Because the seed sizes are similar between species, the mass of protein per pod is approximately the same, but pods with more mesocarp have a lower percentage protein and *vice versa*. Given *P. alba* pod sugar concentrations of 37%, it is not surprising that flours based on milling the mesocarp portion of the pods have sucrose concentrations of 48 to 59% (Felker *et al.* 2003). While in the past *Prosopis* pods were a major form of sustenance for indigenous people and their livestock, in today's world economy it will be difficult to compete on a protein basis with soybean supplements or on an energy basis with molasses from sugarcane. Thus Felker *et al.* (2003) have proposed that *Prosopis* mesocarp flour will have its greatest potential from the spice type flavor and aroma it lends to baked food products.

Figure 9.3. Prosopis alba from Argentina. (A) Logs harvested in the Chaco Province to be used for furniture and flooring. (B) Furniture constructed by Fioramonte in Santiago del Estero.

Table 9.1. Concentrations of protein and sugar in the pods of *Prosopis* species grown in the University of California Riverside Agricultural Experiment Station (Oduol *et al.* 1986).

Species (Accession)	Entire pod sucrose concentration (%)	Entire pod protein concentration (%)
P. nigra 133	37.5 A	10.4 E
P. alba 137	37.3 A	11.0 E
P. alba 039	35.0 B	9.6 E
P. velutina 032	25.7 B	18.6 A
P. velutina 020	25.7 B	16.7 ABC
P. sp. 080	25.4 B	16.5 ABC
P. velutina 025	24.2 B	15.0 BCD
P. sp. 074	22.2 BC	13.0 D
P. glandulosa var. torreyana 001	20.1 BC	14.8 CD
P. glandulosa var. glandulosa 028	17.0 C	13.4 D
P. articulata 016	5.3 D	17.0 AB

P. sp. 080 and *P.* sp. 074 were collected from a region between *P. velutina* in southern Arizona and *P. glandulosa* var *glandulosa* in west Texas and were intermediate in morphological characters between these species. Means followed by the same letter are not significant at $P < 0.05$.

Unfortunately, a considerable percentage of *P. alba* trees in Argentina (Felker *et al.* 2001), of *P. pallida* trees in Peru (Alban *et al.* 2002), and virtually all of the *P. juliflora* trees in Haiti, Sahelian Africa, the Middle East, and the Indian subcontinent, have bitter pods that are not edible. Fortunately an intensive search in Yemen has identified one *P. pallida* tree that produced sweet pods (M. Al Nassiri, Director of Agricultural Research, Govt. of Yemen personal communication). This tree should be clonally multiplied and examined in other similar ecosystems. Preliminary evidence in some bitter *P. alba* suggests that saponins maybe responsible for the bitter flavor (G. Fabiani personal communication). From an evolutionary perspective it would be reasonable that plants would have evolved some anti-insect deterrent compound, such as might be provided by saponins, to avoid predation of high sugar content reproductive organs. Thus *Prosopis* utilization/genetic improvement programs must be cognizant of this limitation.

Potential opportunities and liabilities with *Prosopis* genetic improvement programs

When the extraordinary physiological properties described above such as high levels of photosynthesis and nitrogen fixation at temperatures of 45 °C, growth at seawater salinity and soil pH higher than 9, are combined with economically useful characters such as highly palatable pods with 35% sucrose and lumber of low-shrinkage suitable for high quality furniture, it would appear that all of the requisite genetic traits exist within *Prosopis* to meet the ideotype leguminous food crop tree proposed some 25 years ago (Felker and Bandurski 1979). Unfortunately all of the desirable characters do not yet exist within the same genotype.

Virtually all of the *Prosopis* that currently exists outside its native range was introduced by non-scientists and/or technical officers that evidently were impressed with the vigorous growth of *Prosopis* in its native habitat. These selections were not evaluated in replicated trials in the country of origin or destination prior to their release. After the initial release in the new location, the surviving seedlings were the ones that retained defense and survival components to avoid being eaten by wild and domestic stock or from being harvested (principally for fuelwood). Perhaps, this selection process inadvertently resulted in increased thorn size and lack of single erect stems that would be easy to harvest. As an example to support this hypothesis, *P. alba* almost never survives unprotected in wild in Argentina since its thorns are small and its foliage is highly palatable to goats and sheep, while in Sahelian Africa despite the need for forage, very few animals eat the foliage of *P. juliflora* as the leaves are unpalatable and the spines large.

There is a growing, and very legitimate, worldwide concern against the introduction of plants outside their native range which have the potential to become weedy in the new location. The indiscriminate worldwide introduction of *Prosopis* outside its native range without testing has legitimately contributed to this concern. However, it must be realized that the past exchanges were arranged by political functionaries or casual travelers and not by Ph. D. level geneticists under controlled conditions. Due to the pressing needs for plant species that produce food and/or provide raw materials for manufacturing in harsh arid lands, particularly of Africa, Latin America and south Asia, it would seem important to evaluate new *Prosopis* genetic materials. These materials must be evaluated under very controlled quarantine conditions in the country of destination and priority should be given to genetic materials that have improved characteristics in replicated trials in the country of origin. As a minimum safeguard, the following evaluation procedure is suggested: (a) evaluation of a limited number of elite clones or plants (perhaps 6 to 12 entries) that have resulted from previous field trials, to be compared with about 3 families or clones of the local strain; (b) use of about 6 single tree replicates per introduction; (c) with no more than about 90 trees total, this trial would occupy less than 0.5 ha; (d) location of the trial within the confines of a government or university controlled field site with limited access; (e) within the controlled field site, use of an animal proof fence around the trial to prevent the pods, containing the seed, from being taken off-site; (f) during the pod production season, weekly collection of all pods; and (g) quantitative measurements of form, growth, pod production pod quality.

After several years of quantitative data collection and a statistical comparison of the characteristics of the proposed new introductions to the existing germplasm, a national level review committee, with assistance from interested international organizations, would be in the position to make a quantitative risk/benefit analysis. At this time a decision could be made to destroy all or some of the introduced material (for instance using basal triclopyr/clopyralid applications), to release some or all of the material to other organizations for additional evaluation, or to continue evaluations with no release.

In spite of the risks involved with genetic improvement in locations where naturalized *Prosopis* has become a weed, or in locations such as Texas, where the existing native genetic material is often considered to be a weed, genetic improvement programs should be considered to: (a) incorporate specific highly desirable characters from known germplasm (such as erect form, lack of spines, highly palatable pods, resistance to extreme edaphic conditions) and (b) to provide the economic incentive with an agro-ecosystem of improved strains capable of generating the revenues necessary to control the weedy, non-useful ones.

A survey of the major genetic improvement needs for *Prosopis* around the world is presented in Table 9.2. The first region listed is the African Sahel and similar climates of eastern Africa. Here, the introduced *P. juliflora* is widely distributed and has been among the species most successful in reforestation for dune stabilization in Sudan (Bristow 1996) and Somalia (Zollner 1986), for fuelwood provision in Senegal (Diagne 1996), and for earlier reforestation efforts in the Sudan (El Houri 1986). In spite of the fact that it was used in the majority of CARE's (www.CARE.org) reforestation efforts in Niger, due to its long spines and aggressive spreading habit it has not been widely appreciated (Butterfield 1996). These negative perceptions have been increasing with the declaration that *Prosopis* is a weed in Sudan and major institutional complaints on its spread in Kenya. Thanks to the recent definitive taxonomic work of Harris *et al.* (2003) it has been shown that the naturalized species in the African Sahel is not *P. chilensis* (Molina) Stuntz as it was erroneously known in the Sudan, nor the highly valuable, non-weedy *P. pallida*, but *P. juliflora*. Field work by this author in Yemen has confirmed that *P. juliflora* is also the species in Yemen, which is thorny, aggressive, and has pods which are not palatable for human food use.

The very important recognition has been made that like the pods of *P. juliflora* from India, Pakistan, Haiti, etc., the *Prosopis* pods in Sahelian Africa are bitter and not suitable for the myriad of human use applications described for *P. pallida* in Peru (Grados and Cruz 1996) or for *P. alba* in Argentina (Burkart 1976). Given the extreme poverty in these harsh African countries, it is a tragedy that the pods that are produced so abundantly are, unlike their *P. pallida* and *P. alba* near relatives, not appropriate for human food use.

Thus it would seem appropriate that a major breeding/genetic initiative be made to replace the *P. juliflora* in Sahelian Africa, which is weedy and thorny, with non-palatable pods with types that are erect, have very small thorns, and have pods highly desirable for human food. As noted in the comparison trials in Haiti (Wojtusik *et al.* 1993), Cape Verde (Harris *et al.* 1996) and India (Harsh *et al.* 1996) genotypes in the *P. pallida* genetic pool seem to be able to offer these advantages. After an intensive search for *Prosopis* trees with sweet pods in Yemen, one tree out of 72 was found with sweet non-bitter pods (M. Al Nassiri personal communication) and would seem to be useful for asexual multiplication.

In a 1996 visit to South Africa, this author observed an incredible variety of *Prosopis* imported from North and South America. The short (height not exceeding 3.5 m), shrubby, thorny, multi-stemmed (often more than 3 stems at ground level), *P. velutina* accessions formed impenetrable stands (often more than 1 stem m^{-2}) in stream bed washes and was, by all standards, a very important and difficult problem. This particular phenotype had high pod production at early ages (as California pod production trials had demonstrated; Felker *et al.* 1984) that greatly contributed to the weedy spread. Cankers on the stems of the *P. velutina*, similar to that reported for this species by Lesney and Felker (1995) appeared to further stimulate branching and stunt the growth of this phenotype. On the other hand, tall (15 m), single stemmed *Prosopis alba* with trunk diameters exceeding 50 cm, known as tame mesquite were not reported to spread. The South Africans clearly have a serious problem. What can be learned from this experience is that if 3-to 4-year long replicated field trials had been conducted on the imported species in South Africa prior to release, the *P. velutina* accessions would never have been released, while the *P. alba* tame mesquite accessions might have been. The South Africans are legitimately very sensitive about any new introductions. However, perhaps the existing tame non-invasive *Prosopis alba* could be grafted onto the weedy ones in a type of biocontrol. Perhaps new single-stemmed, erect, non-spiny forms *Prosopis* when cultivated for lumber production would be so valuable (Felker and Guevara 2003) that the cost of eradicating the weedy ones could be absorbed by the new enterprise.

Prosopis cineraria, a native species to the deserts of India and Pakistan, has been revered in ancient writings in Sanskrit (Mann and Saxena 1980). This species is omnipresent in farmers' fields in the Thar desert where the soil fertility is increased below its canopy, where all of the leaves of the trees (up to 10 m tall) are annually harvested for livestock food, and where the pods are consumed for human food. *Prosopis juliflora* a faster growing species was introduced to what is currently India and Pakistan in the late 1800s and early 1900s and has been a mixed blessing as it provides critically needed fuelwood and livestock food on arid, saline, and sodic soils (Singh 1996; Varshney 1996). However, its long spines and its aggressive growth-form often bring complaints from farmers.

Table 9.2. Major genetic improvement needs for *Prosopis*.

Region	Species	Genetic improvement needs	Source of material
Senegal, Mauritania, Niger, Chad, Burkina Faso, Mali, Sudan, Ethiopia, Somalia, Kenya, Yemen, Saudi Arabia	*P. juliflora*	Pods that are edible by humans, greatly reduced spine size, erect form	*P. pallida* clones Ref: a
Southern Africa	*P. velutina,* *P. glandulosa*	Non invasive, non shrubby, low pod producing, greatly reduced spines	Possibly S. African "Tame mesquite," *P. alba*
South Asia	*P. juliflora*	Pods that are edible by humans greatly reduced spine size erect form	*P. pallida* clones Refs: a, b
	P cineraria	Increased pod production and quality	Native material Ref: c
U.S.A.	*P. glandulosa*	More erect form, smaller spines, pods with greater consumer acceptability	Native material Ref: d Cold hardy *P. alba* selections
Mexico	*P. glandulosa,* *P. laevigata,* and related	More erect form, faster growth rate, improved pod quality and production	Native material Ref: e
Haiti	*P. juliflora*	Pods edible by humans. Small spines and erect form	*P. pallida* Ref: f
Peru	*P. pallida*	Increased pod size, flavor and production more rapid growth resistance to leaf eating insects	Native material
Argentina	*P. alba,* *P. chilensis,* *P. flexuosa,* *P. nigra*	Straight form for lumber rapid growth to achieve high internal rate of returns for plantations to avoid current overharvest resistance to leaf chewing insects and diseases. increased pod size, flavor, and production	Native material Ref: g

Data are from: a) Alban *et al.* (2002); b) Harsh *et al.* (1996); c) Central Institute for Arid Horticulture, Kibaner (Pareek) and Central Arid Zone Research Institute, Jodhpur (Harsh); d) Felker and Ohm (2000); e) Frías Hernández personal communication; f) Wojtusik *et al.* (1993); g) Felker *et al.* (2001).

Both the Central Arid Zone Research Institute (CAZRI) in Jodhpur and the Central Institute for Arid Horticulture (CIAH), in Bikaner have been involved in genetic improvement of *P. cineraria* for some time (L. N. Harsh, CAZRI, personal communication; O. P. Pareek, CIAH, personal communication). At CIAH in Bikaner, O. P. Pareek has made numerous selections from the wild for pod characters and has had success in asexual propagation of those superior types.

For the introduced species Goel *et al.* (1997) have made selections and asexually propagated superior types of *P. juliflora* for growth rate and form. However, as occurs for this species in Sahelian Africa, Haiti, and Yemen, the pods of this species are not suitable for human food use due to a bitter, non-palatable flavor. Harsh *et al.* (1996) reported on a trial with more than 200 half-sib families of various South American species and found that the Peruvian *P. pallida* had the greatest overall ranking for erect form, lack of spines and growth rate (Fig. 9.4). These researchers (L. N. Harsh personal communication) also developed mini grafting techniques for 2-mm diameter *Prosopis* capable of converting genetically unimproved selections to improved selections by top working of the coppice regrowth. At the time of evaluation the trees had not produced pods, but as this species has highly palatable pods in Peru (Grados and Cruz 1996) it seems likely that some of them would have edible pods. It appears possible to develop selections or clones from the Peruvian germplasm for use in India that are erect, thornless, fast growing and have highly palatable pods. A very significant portion of the 1 billion people in India live in the semi arid and arid zones and have limited access to fuelwood (Fig. 9.5) and food for humans and livestock. Thus any technique, such as development of *P. pallida* germplasm, with potential to increase ease of fuelwood procurement and livestock and human food supplies needs to be vigorously pursued.

Despite the 20 million ha of *Prosopis* in the U.S.A. there has been little effort for genetic improvement. Form is not important for the growing Texas barbecue industry that harvests unmanaged native stands to produce chips and chunks for the nationwide retail market. The growing mesquite furniture and flooring industry often selects tall erect trees from river bottom sites but these elite trees are being rapidly depleted. With the goal to improve germplasm for trees for the growing mesquite furniture industry in Texas, a tall, straight tree contest was organized (Felker and Ohm 2000). The winning tree was straight and 5.3 m to the first branch. Seeds from these trees were used to establish a seed orchard at Texas A&M University-Kingsville.

A series of fast growing *P. alba* clones were selected for fast growth under heat drought conditions for use as biomass for renewable energy (Felker *et al.* 1983) as *Prosopis alba* was much more rapid growing than the native *P. glandulosa* or *P. velutina* and often thornless. One of these clones (B2V50) had a high productivity of 30 Mg ha^{-1} in the third year's growth in a non irrigated trial in Texas (Felker *et al.* 1989). Unfortunately

Figure 9.4. Erect three year old *Prosopis pallida* in progeny trials in Jodhpur, India, with Drs. Harsh and Tewari of CAZRI.

Figure 5. Woman with load of *Prosopis juliflora* firewood collected in Jodhpur India. Note shrubby, frequently harvested *Prosopis* in the background.

as the Texas winters are more severe than in California and Arizona the *P. alba* types were not adaptable to Texas. Ten year old progeny trials of *P. glandulosa* var. *glandulosa* found the mean annual basal diameter growth rate to vary from 1.45 to 1.81 cm yr^{-1} but these differences were not significant (Duff *et al.* 1994). In recent years, the U.S. private sector has promoted clones of patented and trademarked *Prosopis alba* and *P. glandulosa* for use in the ornamental nursery trade that are reputed to have superior cold tolerance and form for shade and foliage characteristics. The *Prosopis* Hybrid 'AZT' Thornless Hybrid Mesquite was developed by Arid Zone Trees (www.aridzonetrees.com), the *Prosopis* hybrid Phoenix was developed by Mountain State Wholesale (www.mswn.com) whose scion wood was from *P. alba* clone B2V50 mentioned above. This author patented *Prosopis alba* 'Laurie' (U.S. patent 9,072) and thornless *Prosopis glandulosa* 'Beth' (U.S. Patent 9,256).

A recent comparative economic analysis of *Prosopis* plantations in Argentina and the U.S.A. suggested the use of clones that would shorten the rotation age from 24 to 15 years and would increase the internal rate of return from 11.8 to 18.7% (Felker and Guevara 2003). Thus it would seem important to continue the work on clonal propagation of elite trees in the U.S.A.

Despite the enormous range, variety of species, and economic impact of *Prosopis* on the rural, poor areas of Mexico (Rodríguez-Franco and Maldonado-Aguirre 1996) there are no genetic improvement trials reported in the international literature for *Prosopis* in that country. In northern Mexico, extensive harvests continue for local firewood and for charcoal and mesquite floring production for export into the U.S.A. Near the city of Dolores Hidalgo, State of Guanajuato, this author has observed many small carpenter shops making mesquite furniture and trunks of relic *Prosopis* greater than 1 m in diameter in the midst of 7 to 30 cm diameter trees. Omnipresent mesquite doors and windows in older structures in this area attest to the prior presence of large *Prosopis* that no longer exist. Evidently, since the European arrival there has been an intensive harvest of *Prosopis* for firewood, charcoal, furniture doors, and for use in the mines. With no significant plantings, and evidently little investment into natural regeneration/stand management, the *Prosopis* has been overexploited for centuries. This author believes that range-wide germplasm collections for the most important species, replicated, half-sib field trials, followed by multi-purpose selection, cloning of elite individuals, and seed orchard establishment are essential to reverse this decline. Given the low land values and growing U.S. demand for mesquite solid wood products, with selected varieties/clones and plantation management, the internal rate of return should be sufficiently high to attract investors into commercial plantings (Felker and Guevara 2003).

In Haiti, the poorest country in the western hemisphere, *Prosopis* has been the major source of energy (Lee *et al.* 1992; Lea 1996). This is due to the charcoal energy base of the country and the fact that unlike many trees, *Prosopis* has almost no mortality after repeated harvest of the coppice growth. Unfortunately, the native *P. juliflora* has large spines, not so good form, and the pods are not really palatable for human consumption. Thus a range wide collection was made of the Haitian *Prosopis* resource and compared to a broad range of other species and families in a replicated trial (Lee *et al.* 1992; Wojtusik *et al.* 1993). Here it was found that Peruvian *Prosopis* (later found to be *P. pallida*; Harris *et al.* 2003) was erect, thornless, faster growing than the native *P. juliflora*, and probably produced sweet pods edible by humans. However the trials were not carried out long enough to evaluate this possibility. Clones were made of these elite trees (Wojtusik *et al.* 1993). The same best half-sib families in the Haitian trial were also the best in trials in Cape Verde (Harris *et al.* 1996) and in the Rajasthan desert in India (Harsh *et al.* 1996). Thus there is an immediate possibility for rapidly improving the multipurpose resource of this, the poorest country in the western hemisphere, by genetic improvement based on the Peruvian *Prosopis* genetic material. With the recently described multi purpose Peruvian clones that are easily rooted by cuttings (Alban *et al.* 2002), this should make for rapid progress.

In the northern coastal deserts of Peru, *Prosopis pallida* is highly revered among the local people for production of 35-40% sucrose pods. A significant cottage industry exists in this region for the preparation of a boiled down concentrate from the *Prosopis* pods, not unlike molasses in consistency, known as algarrobina (Grados and Cruz 1996; Bravo *et al.* 1998). The principal use of the algarrobina is in the preparation of an alcoholic beverage with pisco sour (a grape brandy), milk, and eggs. However, new products are under development based on the aroma and flavor characteristics of the 45% sucrose flour prepared from the pod mesocarp (Felker *et al.* 2003). While the technical properties of the wood of *P. pallida* are eminently suitable for fine furniture construction (as is done for *P. alba* and *P. glandulosa*) and while *P. pallida* has some of the best form and growth characteristics of all *Prosopis* species (Lee *et al.* 1992; Felker unpublished observations), this use is unknown in Peru.

A detailed analysis of form, growth rates, pod production and pod flavor on a 10 ha plantation from mixed *P. pallida* seed on the Universidad de Piura campus, Peru, lead to the cloning of 7 individual trees with superior performance (Alban *et al.* 2002). It is to be noted that in the final selection which was for pod flavor, 70% of the trees with good form, high pod production and fast growth were rejected due to bitter or very bitter pod flavor. Thus it is not surprising that the bitter *Prosopis* in tropical Africa (as described above) was introduced from a source with bitter, non-palatable pods. A seed orchard of the 7 elite trees was established. Casual observations in northern Peru reveal considerable variation in pod flavor, from mildly bitter to very sweet with out a bitter taste, and size, from about 20 to 35 cm in length. To locate trees with greatest potential for human food use applications in genetic improvement trials, a recent competition was sponsored by the Universidad de Piura and some trees were located that produced 40 cm long pods (L. Alban, G. Cruz, and N. Grados personal communication).

Hydroponic greenhouse trials in Argentina found that Peruvian *P. pallida* germplasm was the most promising for individual trees that could grow rapidly at full seawater (Velarde *et al.* 2003). It would be most useful to conduct new hydroponic screening trials for growth at high salinity using seeds from the 7 elite clones that were selected for form, growth and pod characters. It is fortunate that *P. pallida* is one of the easiest of the *Prosopis* to root from cuttings, and that when placed on a heated mist bench with hormones, over 90% of cuttings from greenhouse-grown stock plants have roots passing out of an 8 cm diameter pots in 3 weeks (Felker unpublished observations).

It is this author's firm conviction that the *P. pallida* germplasm pool in northern Peru will be crucial in the resolution of the weedy issue in Sahelian Africa, the Middle East, India and Pakistan that is centered around very thorny *P. juliflora* with pods that are not palatable for human food use. The multi-purpose selection methodology used by Alban *et al.* (2002) for wood and human pod uses will be fundamentally important in this work. The seven multi-purpose *P. pallida* clones selected by Alban *et al.* (2002) should be a good starting point in genetic improvement efforts that utilize robust, replicated trials comparing naturalized *P. juliflora* with *P. pallida*. Due to the strategic worldwide importance of this Peruvian *P. pallida* germplasm, intensive germplasm collection for economically important characters and subsequent field evaluation to select further improved clones is urgently needed.

Argentina with more than 20 *Prosopis* species is the world center of biodiversity for the genus, but not of the origin. The four arboreal species *P. alba*, *P. chilensis*, *P. flexulosa*, and *P. nigra* that occur from about 24 to 32° S latitude are the most important economically. In contrast to *P. pallida* that suffers significant damage from temperatures of only −3 °C, these species tolerate freezes of several hours duration with minimum temperatures of −10 °C with minimal damage. The major cash flow (as opposed to environmental benefits from soil improvement, etc.) is from furniture and flooring. Unfortunately the number of *Prosopis* plantations is minimal (probably less than 500 ha in all of Argentina; Government sources of Provinces of Chaco and Santiago del Estero). With more than 100,000 tons of logs recorded by the Provincial Government of the Chaco as harvested annually for furniture, the current situation is not sustainable. Historically, the high sugar content pods were important for indigenous peoples (Burkart 1976). As of 2003, pressed circular cakes from the ground *Prosopis* pods, known as Patay, were commonly sold in bus stations in interior provinces of Santiago del Estero and Tucuman. More refined ground products of the mesocarp as described by Felker *et al.* (2003), but without the often insect contaminated seeds, were in the process of being developed for export.

Due to the overwhelming rate of harvesting *vs.* planting, there is an urgent need to make fast growing selections that will have sufficiently high productivities to make plantations economically attractive. Felker and Guevara (2003) calculated that with good plantation management and improved seed, an internal rate of return of 11.8% would be achieved in a rotation age of 24 years. However, if the annual growth rate of 2.5 cm diameter yr^{-1} obtained by Felker *et al.* (1989) for clonal plantations in Texas could be achieved, the internal rate of return would increase to 22.8%. This would probably be sufficient to stimulate considerable forestry investment without government subsidies and begin to put *Prosopis* furniture and flooring industries in Argentina on a sustainable basis.

While Cony (1996) has calculated the heritabilities of various growth characteristics of *P. flexulosa* from half-sibling progeny trials, only Felker *et al.* (2001) have made multipurpose selections based on growth, pod production, and pod quality. A clonal seed orchard of these clones has been established at the Universidad Católica de Santiago del Estero and some of the trees have set pods the second year after planting (M. Ewens personal communication).

This author was continually perplexed at the slow growth rates of *P. alba* in Argentina *vs.* California and Texas, until Ewens of the Universidad Católica de Santiago del Estero more than doubled growth rates of unselected *P. alba* seedlings, to about 3 cm in diameter per year, with weekly applications of insecticides (Ewens personal communication). Evidently in its native habitat, a suite of chewing and sucking insects has co-evolved with *P. alba* that greatly impacts its growth. As weekly insecticide applications are impossible economically, as well as ecologically, resistance to these insects must be found in genetic improvement trials.

In Burkart's (1976) second monograph of *Prosopis*, he decried the overharvest of *Prosopis* and especially the harvest of taller, straighter trees that were leaving inferior genetically material behind to propagate by seeds. Argentina's economic crisis from 2000 to 2003 has exacerbated the over-harvest of *Prosopis* with poor cash flow landowners selling trees far below the cost of production.

This author's opinion is that Argentina needs both basic genetic studies on *Prosopis* and goal-oriented genetic improvement programs to rapidly achieve improved seeds, grafted seedlings or rooted cuttings that can make plantations sufficiently attractive to reverse the over harvest of native stands. It is important that the basic genetic studies include full-sib interspecific crosses, such as between *P. alba* and *P. nigra* so that the genetics of important characters such as the leaf insect resistance in *P. nigra* (but not *P. alba*), the stem boring resistance in *P. alba* (but not *P. nigra*), form and pod flavor traits can be mapped.

Genetics, new clones, and asexual reproduction

There is a critical need for some form of economically viable asexual propagation within *Prosopis* due to the high variability resulting from its self incompatible breeding mechanism. Due to this breeding mechanism, a "plus tree" in the forest will have at least as much genetic variation as an F1 seedling. Thus seeds from these superior "plus" trees will have at least as much variability as F2 seedlings. This variation manifests itself in the 2-to 3-fold range in the 95% confidence intervals for biomass production within half-sib families of *Prosopis* (Felker *et al.* 2001). Despite numerous attempts, there is no report of a tissue culture system for *Prosopis* with shoot subculture and subsequent multiplication from explants of field grown trees. Rooting of cuttings techniques have been reported for *Prosopis* (Leakey *et al.* 1984; Klass *et al.* 1990) but really are only viable for large scale plantations in *Prosopis pallida* and *P. juliflora* due to the much easier rooting in these species (Alban *et al.* 2002). Techniques have been developed to graft North and South American *Prosopis* species using moderately sized rootstock (Wojtusik and Felker 1993). Following the suggestion of L.N. Harsh in Jodhpur, India, we reexamined mini grafting of *Prosopis* and found that we could reliably graft 35-day old, 2-mm diameter seedlings of *P. alba* with about 75% success (Ewens and Felker 2003). At this time this is the only commercially viable method to asexually reproduce *P. alba* clones. Unfortunately, this eliminates the possibility of asexually propagating salt tolerant rootstock. However some research suggests that by continually regrafting the desired scions on young rootstock, higher rooting of cutting percentages can be obtained. Due to this constraint, separate clonal seed orchards have been established for highly salt tolerant and for multipurpose *P. alba* clones in Argentina (M. Ewens personal communication) and for multipurpose *P. pallida* clones in Piura, Peru. Hopefully due to the self incompatible nature of these clones, the resulting hybrid seed will be mixtures of both elite male and female parents.

Potential for combining economic development and management of the weediness

With the combination of nitrogen fixation, economically valuable products, heat-, drought-, salinity-tolerance, and genetic diversity, *Prosopis* offers an unparalled opportunity for arid lands. Unfortunately, these properties, in combination with non-holistic management, also have led to weediness. While some genetic combinations in *Prosopis* have led to very weedy strains, there is no reason why the ideotype of a deep rooted N fixing tree with edible pods that is resistant to extreme temperatures and edaphic conditions is not a useful goal for arid lands.

It is important to mention three important ecological factors that influence this weediness (a) the role of *Prosopis* in the N cycle of overgrazed ecosystems; (b) the dynamics of forest stand population/automortality/ stem diameter relationships in forest succession; and (c) the natural tendency for some species to be much less weedy than others.

From a steady state assessment of N fluxes in arid ecosystems, Felker (1998) and Geesing *et al.* (2000) noted that current livestock stocking rates were equal to the equilibrium point where the N inputs and outputs balanced. Further, this stocking rate was about 10-fold lower than could be supported on the basis of transpiration water use efficiency for C_3 or C_4 plants. The nearly 50% ecosystem loss of the N ingested by herbivores, due to volatilization from urine and feces, was suggested to be the major N loss to the system and the major steady state constraint to sustainable livestock production. Moreover this 50% elimination of the aboveground N in the ecosystem with each grazing cycle was probably a major contributor in the conversion of grasslands to the encroachment of N fixing *Prosopis* on arid ecosystems (Felker 1998). The management of multiple use, semi-arid ecosystems should recognize the drain on the ecosystem from volatilization of excreted N from grazing animals and incorporate tree legumes to balance N input/output ratios.

It is important to understand the relationships between tree spacing and stem diameter that occur as recently colonized dense stands of most tree species mature. Both hardwood and conifer stands may initially occur as closely spaced (< 1m) small trees (< 2m tall) on recently colonized sites. However as natural mortality or self thinning occurs, the stands thin-out and the trees achieve the large diameters and heights typical of commercially harvestable forests. When the stands mature and reach large diameters, very rarely do dense stands of small trees reoccur beneath their canopies. This type of self thinning control strategy has been proposed for *Prosopis* (Felker *et al.* 1990) to help in both achieving desirable diameters for lumber production and in preventing the re-encroachment of dense stands impenetrable to livestock or humans.

There is no doubt that some *Prosopis* species are much more weedy than others. For example *P. ruscifolia* in Argentina, *P. glandulosa* and *P. velutina* in some parts of southwestern U.S.A., and *P. juliflora* in some areas of the arid tropics have many undesirable characteristics and stands. As mentioned above, fungal pathogens that further stimulated multiple low branching in *P. velutina* in South Africa have made this species worse than this author has viewed in its native habitat. On the other hand *P. alba* which is in danger of being over harvested for lumber, has small thorns, leaves that are highly palatable to domestic stock and wildlife, and in its native Argentina is almost never found in open rangeland unprotected from livestock. Thus in addition to ecological considerations of N cycling and self thinning, genetics plays a role in the *Prosopis* weedy issue.

Typical net returns from arid lands are so low (*ca.* $5 ha^{-1} yr^{-1}; Felker and Guevara 2003) that it is difficult to finance expenditures necessary for out-of-control, weedy dense *Prosopis* stands. Felker and Guevara (2003) made a comparison of the rate of internal return provided by various *Prosopis* products (logs, lumber, pods), wildlife and grazing. These authors found that the only scenario that could provide an internal rate of return sufficiently high to attract commercial investors (*ca.* 10%) was that of growing trees for furniture quality *Prosopis* lumber that is currently valued about $850 m^{-3} ($2 board ft^{-1}). As the trees in this analysis were on 10 m by 10 m spacings, this scenario would be compatible with cattle, wildlife, and intercropping (in the first few years) and also prevent the encroachment of weedy stands. Thus it would seem important to include a lumber component in these ecosystems (1) to provide an attractive return on the investment, (2) to provide the financial means to convert weedy stands to a productive system, and (3) to provide an economically sustainable agricultural ecosystem on wide spacings compatible with multiple uses such as grazing and wildlife.

Conclusions

A comparison between the common Papilionoid cultivated legumes, such as soybeans, beans, cowpeas, and alfalfa and *Prosopis* from the Mimosoideae, reveals nearly an order of magnitude greater resistance in most of the fundamental physiological processes related to heat/drought and salinity in the latter genus. As noted above photosynthetic rates of 18 μmol m^{-2} s^{-1} at 45 °C, with leaf water potentials of -4.5 MPa, the ability to fix nitrogen at leaf air temperatures of 43 °C and water potentials of -3.8 MPa, the ability to grow in seawater 45 dS m^{-1} and at pH values of 10.4, all place *Prosopis* in a class apart from the Papilionoid legumes. The small size of its diploid genome (392 to 490 Mbp) and the ability to make wide interspecific crosses would facilitate the mapping of these traits and the possible utilization in more common annual legumes.

Perhaps no other species has the potential to create economic development in the most poverty stricken, and environmentally difficult areas of Sahelian Africa, the middle East and the deserts of India and Pakistan. These areas are characterized by daily summer maximum temperatures of approximately 42 °C, some with 6-month-long dry seasons, yearly potential evapotranspiration of about 2,000 mm and rainfalls less than 500 mm per year. In these areas *Prosopis* can grow and reproduce if there are favorable microsites, such as water courses, or permanent underground water that is within 4 m of the soil surface. As of this writing, *P. juliflora* has become naturalized to all of these areas where it is sometimes perceived as a weed. Thus there is no doubt as to its adaptability.

Unfortunately, unlike some of the North and South American *Prosopis* that have highly palatable sweet pods, through all this range the introduced naturalized species have pods that are not palatable for human use. Genetic improvement trials in Haiti, Cape Verde, and India found the same halfsibling families of Peruvian *Prosopis* to be more erect, faster growing, with much less thorniness than the naturalized *P. juliflora*. Recent trials in Peru found that less than 30% of the trees had pods that could be classified as sweet or very sweet, while the remaining 70% of the trees were classified as bitter or very bitter.

No other gene pool in the plant kingdom possesses the combination of:

(1) Genes for heat, drought, high pH, and salinity stress to permit routine active physiological functioning of critical processes such as photosynthesis and nitrogen fixation at temperatures higher than 40 °C typical of the African Sahel, India, etc.

(2) Many interbreeding species native to two continents capable of providing the reservoir of genes for disease and pest resistance.

(3) Economically useful traits such as highly edible, high sugar pods, and highly dimensionally stable lumber suitable for fine furniture.

The presence of unselected weedy species of this genus should not prevent scientists from taking a broad perspective of this genus to examine the potential of using these genes in conventional legumes and in deliberately creating, testing and clonally multiplying elite individuals for the world's harshest arid ecosystems.

Acknowledgements

The financial support of the Universidad Católica de Santiago del Estero, while the author was in Argentina, was gratefully appreciated.

Literature cited

Ahmad R, Ismail S, Moinuddin M, and Shaheen T. 1994. Screening of mesquite (*Prosopis* spp) for biomass production at barren sandy areas using highly saline water for irrigation. *Pakistan Journal of Botany* **26**: 265-282.

Alban L, Matorel M, Romero J, Grados N, Cruz G, and Felker P. 2002. Cloning of elite, multipurpose trees of the *Prosopis juliflora/pallida* complex in Piura,Peru. *Agroforestry Systems* **54**: 173-182.

Ansley RJ, Jacoby PW, Meadors CH, and Lawrence BK. 1992. Soil and leaf water relations of differentially moisture stressed honey mesquite (*Prosopis glandulosa* Torr.). *Journal of Arid Environments* 22: 147-159.

Ayers RS and Westcott DW. 1985. *Water quality for irrigation*. FAO Irrigation and drainage paper No. 29.FAO, Rome

Bailey AW. 1976. Nitrogen fixationin honey mesquite seedlings. *Journal of Range Management* **29**: 479-481.

Baker A, Sprent JI, and Wilson J. 1995. Effects of sodium chloride and mycorrhizal infection on the growth and nitrogen fixation of *Prosopis juliflora*. *Symbiosis* **19**: 39-51.

Basak MK and Goyal SK. 1975. Studies on tree legumes: nodulation pattern and characterization of the symbiont. *Annals Arid Zone* **14**: 367-370.

Bessega C, Ferreyra L, Julio N, Montoya S, Saidman B, and Vilardi JC. 2000. Mating system parameters in species of genus *Prosopis* (Leguminosae). *Hereditas* **132**: 19-27.

Bohnert HJ, Hua S, and Shen B. 1999. Molecular mechanisms of salinity tolerance. In: Shinozaki K and Yamaguchi-Shinozaki K (eds.) *Molecular responses to cold, drought heat and drought stress in higher plants*. R.G. Landes Company, Austin, TX. Pp: 29-60.

Bravo L, Grados N, and Saura-Calixto F. 1998. Characterizationof syrup and dietary fibre obtained from mesquite pods (*Prosopis pallida* L). *Journal ofAgricultural and Food Chemistry* **46**: 1727-1733.

Bristow S. 1996. The use of *Prosopis juliflora* for irrigated shelter belts in arid conditions in northern Sudan. In: Felker P and Moss J (eds.) *Prosopis: Semiarid Fuelwood and Forage Tree. Building Consensus for the Disenfranchised*. Center Semi-AridForest Resources Publ. Kingsville, TX. Available online at: http://www.udep.edu.pe/upadi/

Buehlmann U, Schuler A, and Nwagbara U. 2002. *Globalization and North Carolina's wood product industry.*
http://ahc.caf.wvu.edu/FactSheet/Furniture/forestweb.pdf

Burkart A. 1976. A monograph of the genus *Prosopis* (Leguminosae subfam. Mimosoideae). *Journal Arnold Arboretum* **57**: 217-249 and 450-525.

Butterfield R. 1996. *Prosopis* in Sahelian forestry projects: A case study from Niger. In: Felker P and Moss J (eds.) *Prosopis: Semiarid Fuelwood and Forage Tree. Building Consensus for the Disenfranchised*. Center Semi-Arid Forest Resources Publ. Kingsville, TX. Available online at: http://www.udep.edu.pe/upadi/

Craig GF, Bell DT, and Atkins CA. 1991. Response to salt and water logging stress of ten taxa of *Acacia* selected from naturally saline areas of Western Australia. *Australian Journal of Botany* **38:** 619-630.

Chudnoff M. 1984. *Tropical Timbers of the World*. Agricultural Handbook 607. U.S. Department of Agriculture, Forest Service, Washington, D.C.

Cony MA. 1996. Genetic variability in *Prosopis flexulosa* D.C., a native tree of the Monte phytogeographic province, Argentina. *Forest Ecology and Management* **87:** 41-49

Dahnke WC, Fanning C, Cattanach A, and Swenson LJ. 1992. *Fertilizing Alfalfa, Sweet Clover, Alsike Clover, Birdsfoot Trefoil, RedClover and Grass-Legume*. North Dakota State University Extension Service, SF-728 (Revised). http://www.ext.nodak.edu/extpubs/plantsci/soilfe rt/sf728w.htm

D'Antoni HL and Solbrig OT. 1977. Algarrobos in South American Cultures. Past and Present. In: Simpson B (ed.) *Mesquite: Its Biology in Two Desert Ecosystems*. Dowden Hutchinson and Ross Publ. Pp: 189-199.

De Soyza AG, Franco AC, Virginia RA, Reynolds JF, and Whiting WG. 1996. Effects of plant size on photosynthesis and water relations in the desert shrub *Prosopis glandulosa* (Fabaceae). *American Journal of Botany* **83:** 99-105.

Diagne O. 1996. Utilization and nitrogen fixation of *Prosopis juliflora* in Senegal. In: Felker P and Moss J (eds.) *Prosopis: Semiarid Fuelwood and Forage Tree. Building Consensus for the Disenfranchised*. Center Semi-Arid Forest Resources Publ. Kingsville, TX.Available online at: http://www.udep.edu.pe/upadi/

Doyle JJ and Luckow MA. 2003. The rest of the iceberg. Legume diversity and evolution in a phylogenetic context. *Plant Physiology* **131:** 900-910.

Drake M and Steckel JE. 1955. Solubilization of soil and rockphosphate as related to root cation exchange capacity. *Soil Science Society of America Proceedings* **19:** 449-450.

Duff AB, Meyer JM, Pollock C, and Felker P. 1994. Biomass production and diameter growth of nine half-sib families of mesquite (*Prosopis glandulosa* var. *glandulosa*) and a fast-growing *Prosopis alba* half-sib family grown in Texas. *Forest Ecology and Management* **67:** 257-266.

El-Houri AA. 1986. Some aspects of dry land afforestation in the Sudan with special reference to *Acacia tortilis* (Forsk), *A. senegal* (Willd) and *Prosopis chilensis* (Molina) Stuntz. *Forest Ecology and Management* **16:** 209-221.

El-Lakany MH and Luard EJ. 1982. Comparative salt tolerance of selected *Casuarina* species. *Australian Forest Research* **13:** 11-20.

Epstein E, Kingsbury RW, Norlyn JD, and Rush DW. 1979. Production of food crops and other biomass by seawater culture. In: Hollaender A, Aller JC, Epstein E, San Pietro A, and Zaborsky OR (eds.) *The Biosaline Concept: An Approach to the Utilization of Underexploited Resource*. Plenum Press, NY. Pp: 77-99.

Ewens M and Felker P. 2003. The potential of mini-grafting for large scale commercial production of *Prosopis alba* clones. *Journal of Arid Environments* **55:** 379-387.

Felker P. 1979. Mesquite-An all purpose leguminous arid land tree. In: Ritchie GA (ed.) *New Agricultural Crops*. American Association for the Advancement of Science Symposium. Pp: 89-132.

Felker P and Bandurski RS. 1979. Uses and potential uses of leguminous trees for minimal energy input agriculture. *Economic Botany* **33:** 172-184.

Felker P and Clark PR. 1980. Nitrogen fixation (acetylene reduction) and cross inoculation in 12 *Prosopis* (mesquite) species. *Plant and Soil* **57:** 177-186.

Felker P and Clark PR. 1982. Position of mesquite (*Prosopis* spp.) nodulation and nitrogen fixation (acetylene reduction) in 3 m long phraetophytically simulated soil columns. *Plant and Soil* **64**: 297-305.

Felker P, Cannell GH, Clark PR, Osborn JF, and Nash P. 1983. Biomass production of *Prosopis* species (mesquite), *Leucaena*, and other leguminous trees grown under heat/drought stress. *Forest Science* **29**: 592-606.

Felker P, Clark PR, Laag AE and Pratt PF. 1981. Salinity tolerance of the tree legumes mesquite (*Prosopis glandulosa* var *torreyana*, *P. velutina*, and *P. articulata*), algarrobo (*P. chilensis*), kiawe (*P. pallida*) and tamarugo (*P. tamarugo*) grown in sand culture on nitrogen free media. *Plant and Soil* **61**: 311-317.

Felker P, Clark PR, Osborn JF, and Cannell GH. 1984. *Prosopis* pod production–a comparison of North American, South American, and Hawaiian germplasm in young plantations. *Economic Botany* **38**: 36-51.

Felker P, Smith D, Wiesman C, and Bingham RL. 1989. Biomass production of *Prosopis alba* clones at 2 non-irrigated field sites in semiarid south Texas. *Forest Ecology and Management* **29**: 135-150.

Felker P, Meyer JM, and Gronski SJ. 1990. Application of self-thinning in mesquite (*Prosopis glandulosa* var *glandulosa*) to range management and lumber production. *Forest Ecology and Management* **31**: 225-232.

Felker P. 1998. The value of mesquite for the rural southwest. *Journal of Forestry* **96**: 16-20.

Felker P and Ohm R 2000. Results of tall, straight mesquite (*Prosopis glandulosa*) contest. *Western Journal of Applied Forestry* **15**: 37.

Felker P, López C, Soulier C, Ochoa J, Abdala R, and Ewens M. 2001. Genetic evaluation of *Prosopis alba* (algarrobo) in Argentina for cloning elite trees. *Agroforestry Systems* **53**: 65-76.

Felker P, Grados N, Cruz G, and Prokopiuk D. 2003. Economic assessment of production of flour from *Prosopis alba* and *P. pallida* pods for human food applications. *Journal of Arid Environments* **53**: 517-528.

Felker P and Guevara JC. 2003. Potential of commercial hardwood forestry plantations in arid lands-An economic analysis of *Prosopis* lumber production in Argentina and the United States under varying management and genetic improvement strategies. *Forest Ecology and Management* **186**: 271-286.

Figueiredo MVB, Vilar JJ, Burity HA, and de França FP. 1998. Alleviation of water stress effects in cowpea by *Bradyrhizobium* spp. inoculation. *Plant and Soil* **207**: 67-75.

Geesing D, Felker P, and Bingham RL. 2000. Influence of mesquite (*Prosopis glandulosa*) on soil nitrogen and carbon development: Implications for global C sequestration. *Journal of Arid Environments* **46**: 157-180.

Goel VL, Dogra PD, and Behl HM. 1997. Plus tree selection and their progeny evaluation in *Prosopis juliflora*. *Indian Forester* **123**: 196-205.

Grados N and Cruz G. 1996. New approaches to industrialization of algarrobo (*Prosopis pallida*) pods in Peru. In: Felker P and Moss J (eds.) *Prosopis: Semiarid Fuelwood and Forage Tree. Building Consensus for the Disenfranchised*. Center Semi-Arid Forest Resources Publ. Kingsville, TX. Available online at: http://www.udep.edu.pe/upadi/

Grewal SS and Abrol IP. 1986. Agroforestry on alkali soils: Effect of some management practices on initial growth, biomass accumulation and chemical composition of selected tree species. *Agroforestry Systems* **4**: 221-232.

Hansen JD and Dye AJ. 1980. Diurnal and seasonal patterns of photosynthesis of honey mesquite. *Photosynthetica* **14**: 1-7.

Harris PJC, Pasiecznik NM, Smith SJ, Billington JM, and Ramírez L. 2003. Differentiation of *Prosopis juliflora* (Sw.) DC and *P. pallida* (H B. Ex. Willd.) H.B.K. using foliar characters and ploidy. *Forest Ecology and Management* **180:** 153-164.

Harris PJC, Pasiecznik NM, Vera-Cruz MT, and Bradbury M. 1996. Prosopis genetic improvement trials in Cape Verde. In: Felker P and Moss J (eds.) *Prosopis: Semiarid Fuelwood and Forage Tree. Building Consensus for the Disenfranchised.* Center Semi-AridForest ResourcesPubl. Kingsville, TX. Available online at: http://www.udep.ed u.pe/upadi/

Harsh LN, Tewari JC, Sharma NK, and Felker P. 1996. Performance of *Prosopis* species in arid regions of India. In: Felker P and Moss J (eds.) *Prosopis: Semiarid Fuelwood and Forage Tree. Building Consensus for the Disenfranchised.* Center Semi-AridForest Resources Publ. Kingsville, TX. Available online at: http://www.udep.edu.pe/upadi/

Hasegawa PM, Bremen R, Zhu JK, and Bohnert HJ. 2000. Plant cellular and molecular responses to high salinity. *Annual Reviews Plant Physiology* **51:** 463-499.

Hua ST, Tsai VY, Lichens GM and Noma AT. 1982. Accumulation of amino acids in *Rhizobium* sp. strain WR1001 in response to sodium chloride salinity. *Applied and Environmental Microbiology* 44: 135-140.

Huang CY, Boyer JS and Vanderhoef LN. 1975. Acetylene reduction (nitrogen fixation) and metabolic activities of soybean having various leaf and nodule water-potentials. *Plant Physiology* **56:** 222-227.

Hunziker JH, Poggio L, Naranjo CA, and Palacios RA. 1975. Cytogenetics of some species and natural hybrids in *Prosopis* (Leguminosae). *Canadian Journal Genetics and Cytology* **17:** 253-262.

Hunziker JH, Saidman BO, Naranjo CA, Palacios RA, Poggio L, and Burghardt AD. 1986. Hybridization and genetic variation of Argentine species of *Prosopis*. *Forest Ecology and Management* **16:** 301-315.

Huss-Daniel K. 1978. Nitrogenase activity in intact plants of *Alnus incana*. *Physiologia Plantarum* 43: 372-376.

Jenny H. 1944. *Factors of soil formation*. McGraw-Hill Book Co., New York.

Johnson HB and Mayeux HS. 1990. *Prosopis glandulosa* and the nitrogen balance of rangelands: Extent and occurrence of nodulation. *Oecologia* **84:** 176-185.

Keys RN and Smith SE. 1994. Mating system parameters and population structure in pioneer populations of *Prosopis velutina* (Leguminosae). *American Journal of Botany* **81:** 1013-1020.

Khudairi AK. 1957. Root nodule bacteria of *Prosopis stephaniana*. *Science* **125:** 399.

Klass S, Bingham RL, Finkner-Templeman L, and Felker P. 1984. Optimizing the environment for rooting cuttings of highly productive clones of *Prosopis alba* (mesquite/algarrobo). *Journal of Horticultural Science* **60:** 275-284.

Lea JD. 1996. A review of literature on charcoal in Haiti. In: Felker P and Moss J (eds.) *Prosopis: Semiarid Fuelwood and Forage Tree. Building Consensus for the Disenfranchised.* Center Semi-AridForest Resources Publ. Kingsville, TX. Available online at: http://www.udep.ed u.pe/upadi/

Leakey RRB, Mesen JF, Tchoundjeu Z, Longman KA, DickJMcP, Newton A, Matin A, Grace J, Munro RC, and Muthoka PN. 1990. Low technology techniques for the vegetative propagation of tropical trees. *Commonwealth Forestry Review* **69:** 247-257.

Lee SG and Felker P. 1992. Influence of water/heat stress on flowering and fruiting of mesquite (*Prosopis glandulosa* var. *glandulosa*). *Journal of Arid Environments* **23:** 309-319.

Lee SG, Russell EJ, Bingham RL, and Felker P. 1992. Discovery of thornless, non-browsed, erect tropical *Prosopis* in 3-year-old Haitian progenytrials. *Forest Ecology and Management* **48:** 1-13.

Lesney MS and Felker P. 1995. Two field and greenhouse diseases of *Prosopis. Journal Arid Environments* **30:** 417-422.

Mann HS and Saxena SK. 1980. *Khejri*. Prosopis cineraria *in the Indian Desert. Its role in Agroforestry*. CAZRI monograph Vol 11, Central Arid Zone Research Institute, Jodhpur, India.

Martin MP. 1948. *Observations on the Nodulation of Leguminous Plants in the Southwest*. Plant Study Ser. 4, Regional Bulletin 107, Region 6, U.S. Department of Agriculture, Soil Conservation Service.

Meinzer OE. 1927. *Plants as indicators of groundwater*. U.S. Geological Survey. Water Supply Paper 577.

Mooney HA, Simpson BB, and Solbrig OT. 1977. Phenology, morphology, physiology. In: Simpson BB (ed.) *Mesquite-Its Biology in Two Desert Ecosystems*. Dowden Hutchinson Ross., Stroudsburg, PA. Pp: 26-43.

Mwanamwenge J, Loss SP, Siddique KHM, and Cocks PS. 1999. Effects of water stress during floral initiation, flowering and podding on the growth and yield of faba bean (*Vicia faba*). *European Journal of Agronomy* **11:** 1-11.

Ng BH. 1987. The effects of salinity on growth, nodulation and nitrogen fixation of *Casuarina equisetifolia*. *Plant and Soil* **103:** 123-125.

Nilsen ET, Rundel PW, and Sharifi MR. 1981. Summer water relations of the desert phreatophyte *Prosopis glandulosa* in the Sonoran Desert of Southern California. *Oecologia* **50:** 271-276.

Oduol PA, Felker P, McKinley CR, and Meier CE. 1986. Variation among selected *Prosopis* families for pod sugar and pod protein contents. *Forest Ecology and Management* **16:** 423-431

Olsen SR and Sommers LE. 1982. Phosphorus. In: Page AL, Miller RH, and Keeney DR (eds.) *Methods of Soil Analyses, Part 2. Chemical and Microbiological Properties*. Agronomy Monograph No.9. American Society of Agronomy, Crop Science Society of America, Soil Science Societyof America. Madison, Wisconsin. Pp: 403-430.

Pasiecznik NM, Felker P, Harris PJC, Harsh LN., Cruz G, Tewari JC, Cadoret K, and Maldonado LJ. 2001. *The Prosopis juliflora-Prosopis pallida complex. A monograph*. HDRA, Coventry, UK.

Philips WS. 1963. Depth of roots in soil. *Ecology* **44:** 424.

RagoneseAE. 1951. La vegetación de la República Argentina. II. Estudio fitosociológico de las salinas grandes. *Revista Investigaciones Agricolas* **V:** 1-233.

Ramírez L, de la Vega A, Razkin N, Luna V, and Harris PJC. 1999. Analyses of the relationships between species of the genus *Prosopis* revealed by the use of molecular markers. *Agronomie* **19:** 31-43.

Ramos MLG, Parsons R, Sprent JI, and James EK. 2003. Effect of water stress on nitrogen fixation and nodule structure of common bean. *Pequisa Agropecuria Brasileira*, Brasilia **38:** 339-347.

Rhodes D and Felker P. 1987. Mass screening *Prosopis* (mesquite) seedlings for growth at seawater salinity. *Forest Ecology and Management* **24:** 169-176.

Richards LA. 1954. *Saline and Sodic Soils*. USDA Salinity Handbook 54, Riverside, California.

Rodríguez-Franco C and Maldonado-Aguirre LJ. 1996. Overview of past, current and potential uses of mesquite in Mexico. In: Felker P and Moss J (eds.) *Prosopis: Semiarid Fuelwood and Forage Tree. Building Consensus for the Disenfranchised*. Center Semi-AridForest Resources Publ. Kingsville, TX. Available online at: http://www.udep.ed u.pe/upadi/

Rundel PW, Nilsen ET, Sharifi MR, Virginia RA, Jarrell WM, Kohl DH, and Shearer GB. 1982. Seasonal dynamics of nitrogen cycling for a *Prosopis* woodland in the Sonoran desert. *Plant and Soil* **67**: 343-353.

Saidman BO and Vilardi JC. 1987. Analyses of the genetic similarities among seven species of *Prosopis* (Leguminosae: Mimosoideae). *Theoretical Applied Genetics* **75**: 109-116.

Sharifi MR, Nilsen ET, Virginia R, Rundel PW, and Jarrell WM. 1983. Phenological patterns of current season shoots of *Prosopis glandulosa* var. *torreyana* in the Sonoran Desert of Southern California. *Flora* **173**: 265-277.

Shearer G, Kohl DH, Virginia RA, Bryant BA, Skeeters JL, Nilsen ET, Sharifi MR, and Rundel PW. 1983. Estimates of N-fixation from variation in the natural abundance of ^{15}N in Sonoran Desert Ecosystems. *Oecologia* **56**: 365-373.

Simpson B. 1977. Breeding systems of dominant perennial plants of 2 disjunct warm desert ecosystems. *Oecologia* **27**: 203-226.

Singh G. 1995. An agroforestry practice for the developmentof salt lands using *Prosopis juliflora* and *Leptochloa fusca*. *Agroforestry Systems* **29**: 61-75.

Singh G. 1996. The role of *Prosopis* in reclaiming high pH soils and in meeting firewood and forage needs of small farmers. In: Felker P and Moss J (eds.) *Prosopis: Semiarid Fuelwood and Forage Tree. Building Consensus for the Disenfranchised*. Center Semi-AridForest ResourcesPubl. Kingsville, TX. Available online at: http://www.udep.ed u.pe/upadi/

Singh G, Abrol IP, and Cheema SS. 1988. Agroforestry on alkali soil: Effect of planting methods and amendments on initial growth, biomass accumulation and chemical composition of mesquite (*Prosopis juliflora* (SW) DC) with inter-space planted with and without Karnal grass (*Diplachne fusca* Linn.P. Beauv.). *Agroforestry Systems* **7**: 135-160.

Singh G, Abrol IP, and Cheema SS. 1989a. Effects of gypsum application on mesquite (*Prosopis juliflora*) and soil properties in an abandoned sodic soil. *Forest Ecology and Management* **29**: 1-14.

Singh G, Abrol IP, and Cheema SS. 1989b. Effects of spacing and lopping on a mesquite (*Prosopis juliflora*)-Karnal grass (*Leptochloa fusca*) agroforestry system on an alkaline soil. *Experimental Agriculture* **25**: 401-408.

Thibodeau PS and Jaworski EG. 1975. Patterns of nitrogen fixation in soybean. *Planta* **127**: 133-147.

Turc CO and Cutter BE. 1984. Sorption and shrinkage studies of six Argentine woods. *Wood and Fiber Science* **16**: 575-582.

U.S. Department of Commerce. 1964. *Climatography of the United States No. 86-4. Supplement for 1951 through 1960. California*. U.S Government Printing Office, Washington, D.C.

Varshney A. 1996. Overview of the use of *Prosopis juliflora* for livestock feed, gum, honey and charcoal as well as combating drought and desertification: A regional case study from Gujarat, India. In: Felker P and Moss J (eds.) *Prosopis: Semiarid Fuelwood and Forage Tree. Building Consensus for the Disenfranchised*. Center Semi-Arid Forest Resources Publ. Kingsville, TX. Available online at: http://www.udep.ed u.pe/upadi/

Velarde M, Felker P, and Degano C. 2003. Evaluation of Argentine and Peruvian-*Prosopis* germplasm for growth at seawater salinities. *Journal ofArid Environments* **55**: 515-531.

Velarde M, Felker P, and Gardiner D. 2005. Influence of elemental sulfur, micronutrients, phosphorous, calcium, magnesium and potassium on growth of *Prosopis alba* on high pH soils in Argentina. *Journal of Arid Environments* **62**: 527-539.

Villagra-López GM, and Felker P. 1997. Influence of understory removal, thinning and P fertilization on N_2 fixation in a mature mesquite (*Prosopis glandulosa* var *glandulosa*) stand. *Journal of Arid Environments* **36:** 591-610.

Virginia RA and Jarrell WM. 1983. Soil properties in a mesquite-dominated Sonoran desert ecosystem. *Soil Science Society of America Journal* **47:** 138-144.

Virginia RA, Baird LM, La Favre JS, Jarrell WM, Bryan BA, and Shearer G. 1984. Nitrogen fixation efficiency, natural ^{15}N abundance and morhology of mesquite (*Prosopis glandulosa*) root nodules. *Plant and Soil* **79:** 273-284.

Weldon DE. 1986. Exceptional physical properties of Texas mesquite wood. *Forest Ecology Management* **16:** 149-153.

Wightman SJ and Felker P. 1990. Soil and foliar characterization for *Prosopis* clones on sites with contrasting productivity in semi-arid south Texas. *Journal of Arid Environments* **18:** 351-365.

Winicov I. 1998. New approaches to improving salt tolerance in crop plants. *Annals of Botany* **82:** 703-710.

Wojtusik T and Felker P. 1993. Inter-species graft incompatibility in *Prosopis*. *Forest Ecology and Management* **59:** 329-340.

Wojtusik T, Felker P, Russell EJ, and Benge MD. 1993. Cloning of erect, thornless, non-browsed nitrogen fixing trees of Haiti's principal fuelwood species (*Prosopis juliflora*). *Agroforestry Systems* **21:** 293-300.

Zollner D. 1986. Sand dune stabilization in Central Somalia. *Forest Ecology and Management* **16:** 223-232.

Chapter 10

BIOPHYSICAL EFFECTS OF ALTITUDE ON PLANT GAS EXCHANGE

William K. Smith and Daniel M. Johnson

Perspectives in Biophysical Plant Ecophysiology: A Tribute to Park S. Nobel, pp. 257-280
Edited by: E. De la Barrera and W.K. Smith
© 2009 by The Authors
Book Compilation © 2009 Universidad Nacional Autónoma de México

Introduction

On mountains around the world, individual plant species are commonly associated with relatively specific altitudinal limits that vary with latitude and longitude, depending primarily on the degree of continental *versus* oceanic influence (see Wardle 1974; Stevens and Fox 1991; Körner 1998 for recent reviews). Evidence from paleobotanical sources has also shown substantial upward and downward migrations of these altitude limits following major shifts in global climate regimes (*e.g.*, Kienast 1991; Lloyd and Graumlich 1997; Ferreyra *et al.* 1998; Luckman and Kavanagh 2000; Meshinev *et al.* 2000; Daniels and Veblen 2003).

Many studies of the altitudinal limits of plants have focused on forest trees, either the upper timberline where forest-like trees last appear, or the treeline, the altitude where stunted and deformed trees are last found beyond the timberline (*e.g.*, Smith *et al.* 2003). There are also predictions of future changes in the altitudinal limits of timberlines based on current scenarios of global warming and continued increases in atmospheric CO_2 (*e.g.*, Kienast 1991; Romme and Turner 1991).

Despite this long-term interest in the maximum altitude of occurrence, the specific, mechanistic causes of these limits are relatively unknown for any species, even timberlines and treelines (*e.g.*, see Körner 1998 and Smith *et al.* 2003, for recent perspectives). However, there has been a continuing focus on carbon gain and water loss, specifically the possibility that photosynthesis is more limited at higher elevation and transpirational water loss greater. It has also been proposed that it is not photosynthetic carbon gain that is limiting tree growth and survival at higher altitudes, but its metabolic processing (Körner 1998). Another perspective has been proposed that identifies the young, establishing seedling as the critical life stage for understanding the stability of timberline and treeline altitudes, with a strong dependence on ecological facilitation (Smith *et al.* 2003). Similarly, changes in specific abiotic factors with altitude that could limit plant growth at higher elevations (temperature, sunlight, water, and nutrients) are poorly understood. Although the most obvious limitation to growth and survival at higher elevations is seasonally cold temperatures, numerous other abiotic factors may also be limiting during summer when growth and reproduction are dependent on adequate carbon assimilation and processing (Tranquillini 1979; Smith and Knapp 1990; Maruta 1996). Thus, the interaction between summer and winter stress factors is now recognized as a potentially important determinant of timberline and treeline altitude (*e.g.*, Vostral *et al.* 2002; Smith *et al.* 2003).

The purpose here is to review current information concerning the abiotic, above-ground factors that may limit plant gas exchange as altitude increases, emphasizing photosynthetic CO_2 uptake and transpiration. In addition, simulation studies employing standard gas exchange and energy balance equations are combined with field measurements taken on leaf models located across an elevation gradient. Abiotic effects of altitude on photosynthesis and transpiration are evaluated for different air temperature lapse rates and seasonal temperatures characteristic of temperate and tropical mountains. Similar studies have been conducted previously, but with more limited objectives (Gale 1972; Smith and Geller 1979; Leuschner 2000).

Abiotics and altitude

Sunlight, temperature, water, and gas-phase nutrients (*e.g.*, CO_2 and O_2) are primary abiotic factors that most directly influence plant gas exchange capabilities, and can vary substantially with altitude, regional climate, and orographics (*e.g.*, maritime *versus* continental mountain ranges). In addition, many factors influencing leaf energy balance and temperature may also vary with elevation, including solar and longwave radiation, wind, and ambient humidity (Smith and Geller 1979; Piazena 1996; Saunders and Bailey 1997; Leuschner 2000; Marty *et al.* 2002). Probably the best known abiotic change with increasing elevation is the decline in air temperature in response to lower ambient pressure (Figs. 1 and 2). Ambient pressure decreases over 20% by 2 km and over 50% at 6 km (Fig. 10.2A), leading to a maximum, dry adiabatic lapse potential of 1.0 °C/100 m (Fig. 10.2B). Simulated dry (8.0 °C/km) versus wet (3.0 °C/km) lapse conditions resulted in more rapid decline in air temperature with altitude for both winter and summer temperatures. Also, dry lapse conditions in summer generated similarly cold air temperatures at higher elevations (> 4 km) that were very near values computed for wet lapse conditions during winter (Fig. 10.2B). Similar dry and wet lapse rates of 7.5 °C/km and 5.5 °C/km, respectively, have been used previously to evaluate transpiration potential for plants growing on mountains of temperate and tropical zones (Leuscher 2000).

Another well known change in abiotic factors with altitude is the decrease in partial pressure in gas-phase molecules such as CO_2 and O_2. Because ambient CO_2 concentration can have a strong, direct influence on photosynthesis via the leaf-to-air concentration gradient (driving force for diffusion), it has often been assumed to be a limiting factor for carbon gain and growth at high elevation (*e.g.*, Körner 1988). A lower ambient CO_2 concentration with altitude could result in a corresponding decrease in the leaf-to-air gradient, assuming a constant CO_2 concentration inside the leaf. For this reason, mountain ecosystems have been considered as natural field models for evaluating the effects of natural differences in atmospheric CO_2 concentrations. However, because molecular diffusion is more rapid at lower ambient pressure, a substantial compensatory effect on CO_2 uptake-potential occurs with greater elevation (*e.g.*, Gale 1972; Smith and Geller 1979; Körner 1999; Leuschner 2000). Although quantitative evaluations of these compensating effects on photosynthetic CO_2 uptake exist in the literature (Smith and Donahue 1991; Terashima *et al.* 1995), there is no comprehensive evaluation of the potentially important factors influencing plant gas exchange at higher altitudes. As well, correlations between altitude (lower CO_2 concentrations) and other indicators of photosynthetic potential such as stomatal size, or frequency, have been inconclusive (*e.g.*, Qiang *et al.* 2003).

A similar concern for the effects of lower O_2 concentrations with altitude has also been considered within the context of photosynthetic effects, based on the well-known influence of O_2 concentrations on photosynthesis (Terashima *et al.* 1995; Matsubara *et al.* 2002; Qiang *et al.* 2003). Other microclimatic factors such as changes in sunlight, ambient humidity, wind, and longwave radiation have been studied less comprehensively, and for only a few mountain systems. Only two studies have incorporated most of these abiotic factors to predict leaf-to-air water vapor differences and transpirational demands at high elevation (Smith and Geller 1979; Leuschner 1997). Thus, although numerous studies have considered changes in one, or a few, of the important abiotic factors influencing gas exchange physiology (*e.g.*, Friend and Woodward 1990; Barry 1992; Kitayama and Mueller-Dumbois 1994; Körner 1999), none have considered their concerted influence.

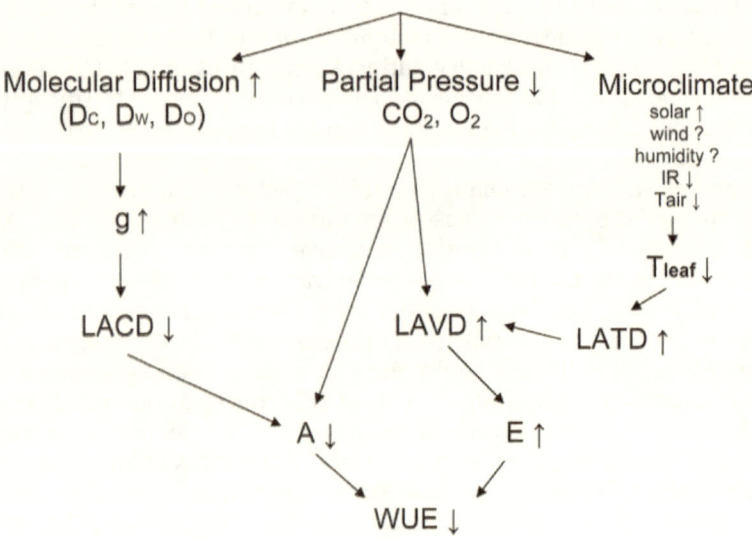

Figure 10.1. Schematic representation of abiotic factors influencing gas exchange (*A* and *E*) at high altitudes. Arrows denote increasing or decreasing values with greater altitude; acronyms are defined in text.

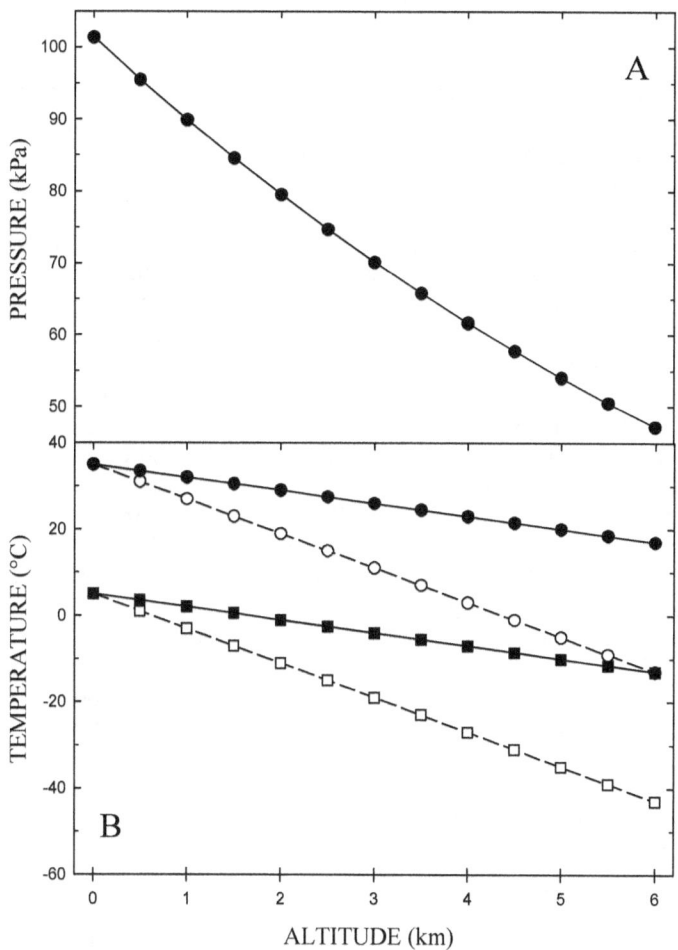

Figure 10.2. Change in ambient pressure (A) and air temperature (B) with altitude for different lapse rates and seasonal temperatures. Curves represent summer wet (●) (35 °C at sea level, −3 °C decrease per km), summer dry (○) (35 °C at sea level, −8 °C/km), winter wet (■) (5 °C at sea level, -3 °C/km), and winter dry (□) (5 °C at sea level, -8 °C/km) conditions.

Microclimate *versus* altitude

An important message concerning changes in abiotic factors with elevation is the realization that decreasing ambient pressure is the only colligative property associated with changes in abiotics. All others (*e.g.*, temperature, sunlight, wind, longwave radiation, water, and nutrient relations) can be strongly influenced by topography, microsite, and plant form at any altitude. Natural variability in these factors can substantially lower, or raise, the effective altitude of a microsite at any given altitude. In the northern hemisphere, south-facing, wind-sheltered microsites can effectively match conditions at an altitude thousands of meters lower, while similar north-facing microsites, sheltered from sun but not the cold night sky, could generate increases in effective altitude. Even smaller microsites around a fallen tree stem or an exposed boulder can result in effectively different altitudes based on differences in sunlight exposure and temperatures (Ball *et al.* 1997; Germino *et al.* 2002). Additionally, changes in leaf orientation can create different levels of sky and wind exposure, two primary factors influencing microclimate at any altitude (Germino and Smith 2002). Leaf and plant aggregation (close spacing) and height patterns can also influence microclimate due to the potentially strong boundary layer effects on temperature and ambient gas concentrations (Smith and Carter 1986). Thus, microclimate effects can significantly impact fundamental gas exchange processes at any altitude, with the exception of ambient pressure effects on molecular diffusion (Sage and Sage 2002). In contrast to the potential effects of microsite and plant form on effective altitude, individual plants cannot escape the ambient pressure of their respective altitudes (only negligible changes in ambient pressure due to weather fronts). Thus, lower ambient pressure and more rapid molecular diffusion are the only immutable abiotic factors associated with increasing altitude, one that is not dependent on microsite/microclimate effects.

Flux calculations

The equations employed in the following analysis follow Fick's and Ohm's Laws where Flux is proportional to the product of leaf conductance (g) and the difference in partial pressure between the air and diffusing surface (Nobel 1999). The analysis also incorporates the inverse relationship between g and the molecular diffusion coefficient (diffusivity) in air of the molecule in question (*e.g.*, CO_2 and water vapor). The equation for the change in diffusion coefficient with altitude is the same used in other studies showing and inverse proportionality between pressure and the diffusion rate, and an exponential relationship with temperature ($T^{1.8}$) (Gale 1972; Smith and Geller 1979; Leuschner 2000).

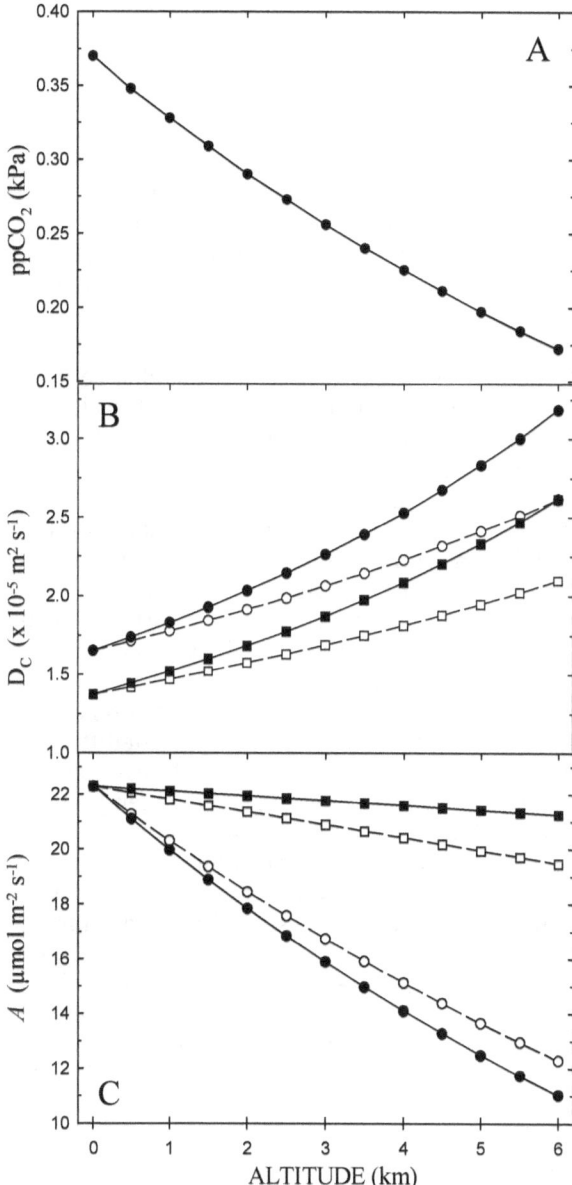

Figure 10.3. (A) Change in CO_2 partial pressure and (B) the diffusion coefficient for CO_2 in air (see Nobel 1991 for diffusion coefficient calculations) with altitude. Symbols in (B) are the same as in Figure 10.1. Predicted photosynthesis (C), based on a constant c_i/c_a ratio of 0.75 and either constant g (250 μmol m⁻² s⁻¹) or g proportional to D_C. Curves in (C) represent wet lapse rate (−3 °C/km) and constant g (●), dry lapse rate (−8 °C/km) and constant g (○), wet lapse rate and g proportional to D_C (■), and dry lapse rate and g proportional to D_C (□).

Photosynthetic CO_2 uptake (A)

As mentioned above, changes in plant CO_2 uptake is directly impacted by two contrasting properties of diffusion that change predictably in response to increasing altitude and lower ambient pressure (Fig. 10.1). Although the partial pressure of CO_2 declines with altitude, its diffusion rate (Diffusion Coefficient of CO_2 in air, D_C) increases (Fig. 10.3), leading to possible offsetting effects on the rate of photosynthetic CO_2 uptake through stomata. According to Fick's Law of Diffusion (Nobel 1999), both the leaf conductance term (proportional to $1/D_C$) and the leaf-to-air CO_2 difference (LACD) are influenced by pressure (altitude), although inversely. The obvious question concerns the relative effects of the change in D_C and LACD on photosynthetic CO_2 uptake (A) in response to higher altitude (Fig. 10.3). For the simulations presented here, an increase in altitude from sea level to 1 km results in a decrease in the partial pressure of CO_2 of approximately −12% (0.37 to 0.33 kPa) for an approximately equal increase in D_C of 11% (1.65 to 1.83 m^{-2} s^{-1} × 10^{-5}) under summer, wet lapse conditions, compared to approximately 11% under winter, dry lapse conditions (1.42 to 1.57 kPa) (Fig. 10.3). At an altitude of 4 km, corresponding changes were −38% (0.37 to 0.23 kPa) and 44% (1.65 to 2.38 m^{-2} s^{-1} × 10^{-5}) for simulated summer, wet lapse conditions and winter, dry lapse conditions, respectively. Much larger increases in both occur at 6 km. Moreover, smaller increases in D_W with elevation occur under dry lapse conditions regardless of simulated winter or summer conditions. Thus, wet lapse conditions and the smaller temperature changes associated with altitude resulted in more rapid increases in D_C due to the added effect of warmer temperature on D_C, regardless of the seasonal temperatures (cold winter or warm summer) simulated. Significant differences in D_C occurred at higher elevations under both the wet and dry lapse conditions in summer, and D_C under wet lapse conditions in winter equaled simulated values for dry lapse values in summer at an altitude of *ca.* 3,700 m (Fig. 10.2B).

Comparison of the change in leaf conductance due to altitudinal effects on D_C, *versus* the influence of declining ambient CO_2 levels on LACD, was accomplished by computing altitudinal effects on A with leaf conductance (g) either constant or as a function of D_C (Fig. 10.3C). Simulating a constant g (without changes due to lower ambient pressure) *versus* a greater g (due to an increasing D_C) resulted in relative small changes in photosynthetic CO_2 uptake under wet (<4%) or dry (<13%) lapse conditions (Fig. 10.3C). In contrast, simulated A values with a constant g (no effect of increasing D_C) resulted in an approximate 27% decline at 3 km and a 50% decline at 6 km. Thus, changes in LACD resulted in much smaller changes in photosyn-thesis with altitude than effects on g due to increasing D_C. Corroborative findings have also been reported whereby computed CO_2 uptake potential was undiminished with increasing altitude, due apparently to the compensatory effects of increasing D_C (Smith and Donahue 1991; Terashima *et al.* 1995; Sage *et al.* 1997; Matsubara *et al.* 2002).

Transpiration (*E*)

Air and plant temperatures can change substantially with altitude, although differences in microclimate and plant form characteristics can also overshadow altitudinal effects, as discussed above. For example, leaf temperatures of an alpine cushion plant have been measured well above (> 25 °C) and below (> 8 °C) air temperature due to a combination of plant form, boundary air layer effects, and microsite sky exposure (Warren-Wilson 1957). Microclimatic wind and air temperatures can vary substantially just above the ground surface (within *ca.* 10 cm) where still air and boundary layer effects generate steep temperature profiles (*e.g.*, Geiger 1957). Changes in air temperature can also influence ambient humidity levels around the plant, based on the well-known relationship between air temperature and evapotranspiration. However, without changes in the local sources for water vapor input (*e.g.*, plant biomass, standing water, wet soils), saturation vapor deficits of the air will follow the familiar exponential relationship between air dryness and temperature (influenced by lapse rate in air temperature and other microclimatic parameters). Moreover, it is important that the leaf-to-air vapor difference (LAVD) determines the driving force for transpiration, along with the absolute humidity of the surrounding air. Increases in leaf temperature due to greater incident sunlight at high elevations will cause an increase in internal leaf humidity (assumed to be near saturation) and, thus, LAVD, without any change in ambient humidity. Transpiration is also possible even under saturated air humidity because the leaf tends to warm substantially above air temperature, especially for low-stature plants under the influence of the soil/air boundary layer. However, the decline in air temperature due to the lapse rate would act to lower leaf temperatures as well. Thus, the influence of altitude on the leaf-to-air vapor difference is complex and dependent on numerous microclimatic factors affecting leaf energy balance and ambient humidity (Smith and Geller 1981).

Assuming constant sources of water vapor input to the ambient air at a given site, lower air temperatures will result in significantly dryer air at saturation as altitude increases. Only at saturated values of absolute humidity could lower temperatures act to reduce LAVD (assuming leaf temperature is above air temperature). As discussed previously, however, actual absolute air humidity (vapor pressure) may, or may not, vary with altitude-because of a strong dependence on water vapor sources. In contrast, if the leaf-to-air temperature difference (LATD) increases with altitude due to energy balance considerations, as reported in Smith and Geller (1981), then increases in LAVD with altitude will result, assuming similar vapor pressures. Because of the substantial rise in diffusion rates at higher altitudes (but with the offsetting effects of lower temperatures), these leaf warming effects on LAVD will compound the demands on transpiration at the same degree of stomatal opening. Few studies have provided measurements of LAVD across broad elevational gradients characteristic of high mountains (*e.g.*, Kitayama and Mueller-Dumbois 1994), and most of these data have come from measurements of pan evaporation, not LAVD.

Despite the above relationships, the fundamental influence of altitude on air and plant temperatures, or water vapor sources and, thus, absolute air humidity of a given site is relatively unknown (Fig. 10.1). Data for daily or hourly changes in ambient humidity at different microsites, or even standard weather box locations (2 m heights), are rare in the literature, and should be dependent on the amount of evaporating surface area *versus* altitude, *e.g.*, plant biomass, and soil and standing water (Blanken and Rouse 1994; LaFleur and Rouse 1995; Saunders and Bailey 1996; Konzelmann *et al.* 1997; Saunders *et al.* 1997). In particular, microclimate and plant form can cause important effects on ambient humidity, just as for surrounding air temperatures. These same factors may also impact photosynthesis indirectly via effects on the temperature sensitivity of photosynthesis, or the often strong response of stomata to ambient humidity.

D_W

Similar to altitudinal effects on *A*, the influence of altitude on plant transpiration can be compared by computing effects on D_W versus the leaf-to-air vapor difference (LAVD), the driving force for transpiration (Fig. 10.1). As opposed to CO_2 diffusion, not only is g_W a function of D_W (and thus pressure and temperature), but LAVD is directly affected by leaf temperature. For example, warm summer temperatures generated the greatest increases in D_W, regardless of simulated dry or wet lapse conditions (Fig. 10.4). Also, the greatest increase in D_W occurred under warm summer temperatures and a wet lapse rate, approaching 38% at 3 km elevation (2.6 *versus* 3.6 m^{-2} s^{-1} × 10^{-5}) and near 100% (2.6 *versus* 5.2 m^{-2} s^{-1} × 10^{-5}) at 6 km computed under summer/wet lapse conditions (Fig. 10.4A) compared to the much smaller difference between the corresponding D_C values (Fig. 10.4B). Also, just as for D_C, the percent change in D_W with altitude was substantially more dependent upon temperature lapse conditions (*i.e.*, dry versus wet) rather than typical differences in seasonal temperature regimes (Fig. 10.4B).

Air humidity

Does absolute air humidity (mass per unit volume, or partial pressure) increase or decrease with altitude, considering both dry and wet lapse conditions in air temperature? Data describing changes in absolute air humidity with altitude are rare in the literature, while those of LATD comparisons may not exist (to our knowledge) (Fig. 10.1). As already mentioned, changes in the absolute humidity of air with altitude involve a complex array of energy balance variables that can impact evaporating surfaces, including such landscape features as free-standing waterbodies, soil evaporation, and transpiration, as well as other sources such as cloud interception and dew deposition. In general, colder temperatures associated with higher altitudes (especially for dry lapse conditions), should decrease evaporation from all

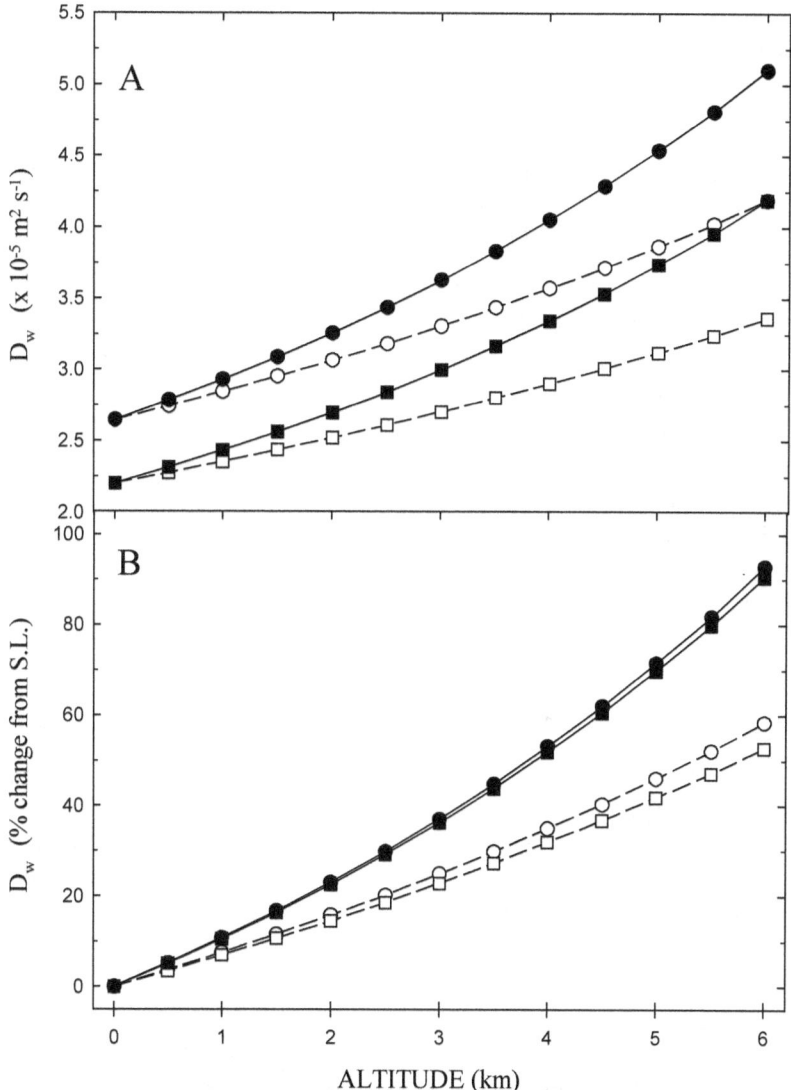

Figure 10.4. (A) Change in the diffusion coefficient for water vapor (D_W) in air with altitude and (B) the percent change in D_W from sea-level. Simulated conditions are as follows: summer wet (•) (35 °C at sea level, −3 °C/km), summer dry (○) (35 °C at sea level, −8 °C/km), winter wet (■) (5 °C at sea level, −3 °C/km), and winter dry (□) (5 °C at sea level, −8 °C/km) conditions.

sources due to a general reduction in the surface-to-air vapor gradient (see section below), although the increase in D_W should contribute to higher air humidities regardless of the amount of evaporating water sources. Once again, microclimate and sun exposure effects on warming of all evaporating surfaces (drivers of LATD and LAVD) may have a major, if not dominant, effect on ambient humidity levels, regardless of altitude. In addition, the contrasting structure between subalpine forest and alpine tundra may contribute substantially to microclimatic and plant form effects on absolute air humidity, LAVD, and, thus, transpiration. Regardless, changes in vapor pressure with altitude have been derived empirically for only a few mountain systems, *e.g.*, the temperate zone of Asia (Leuschner 2000). These results showed a substantial decline in vapor pressure (*ca.* 33%, 0.55 to 0.37 kPa) with increasing altitude between about 2 and 4 km elevations under wet lapse conditions. An even greater decline (*ca.* 48%, 0.21 to 0.11 kPa) occurred for dry lapse conditions, although values of vapor pressure were less than half wet lapse conditions. Thus, this decrease in ambient humidity with altitude would lead to a corresponding increase in LAVD, as well as D_W at higher altitude, and a predicted increase in transpiration caused by both factors. The following simulations of altitudinal effects employed the empirical equation provided in Kuz'min (1972) to compute air-humidity values according to altitude.

Leaf-model temperatures and LATD

At the same degree of stomatal opening, transpiration will typically increase if the leaf-to-air temperature difference (LATD), and thus LAVD, is increased under full sun exposure (Fig. 10.1). If greater levels of incident sunlight are characteristic of higher altitudes, as are lower air temperatures due to lapse rate, energy balance analysis also reveals that LATD will be greater, assuming all other energy exchange factors remain constant (Smith and Geller 1979; Gates 1980). However, greater wind movement and convective heat exchange, or decreases in longwave radiation from the sky and surroundings will result in leaf temperatures closer to air temperature. Thus, field measurements comparing LATD at different altitudes are needed to more clearly elucidate altitudinal effects on LATD and, thus, LAVD. Field data for leaf models (non-transpiring) are combined here with energy balance equations to provide a more fundamental understanding of the quantitative effects of altitude on leaf temperature, LAVD, and, thus, transpiration, that are independent of stomatal adjustments.

Dry leaf models were placed at different altitudes and accompanied by adjacent measurements of incident sunlight, wind, and longwave radiation to compare measured and computed LATD values without transpirational cooling. All models were painted with green paint of known reflectance (0.77) to solar irradiance (0.3 to 3 μm wavelengths) and longwave emissivity (0.96) (OpticalPaints Division, 3M Corporation, Stratford, CT, USA). Measurement intervals during mid-summer (clear skies) were selected for

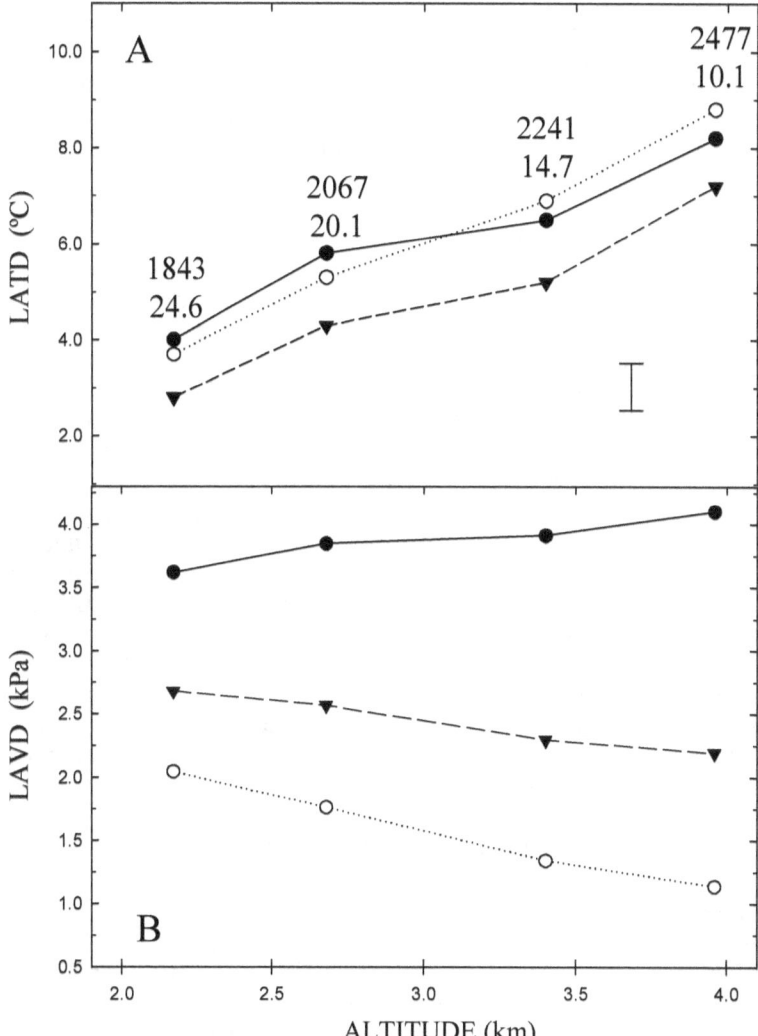

Figure 10.5. (A) Changes in measured leaf-to-air temperature differences (LATD) for leaf models in the field and (B) the computed leaf-to-air vapor pressure deficit (LAVD). Top numbers are mean PAR values during temperature measurements, while bottom numbers are mean model temperatures ($n = 2$). Closed symbols indicate measured LATD, open symbols are computed temperatures using energy balance equations , and closed triangles are computed model temperatures after including a transpiration term and a constant leaf conductance to water vapor ($g = 250$ mmol m^{-2} s^{-1}). (B) Changes in LAVD with altitude were calculated using measured LATD values for the leaf models and simulated lapse rates representing dry (○, −8.0 °C/km), intermediate (▲, −5.5 °C/km), and wet conditions (•, −3.0 °C/km). Vertical bar in (B) is the maximum 95% Confidence Interval computed from all measurements at all four altitudes (field sites).

comparison based on similar net longwave radiation exchange values (± 12%) and wind speed (± 6%). Also, model leaf temperatures were recomputed with a latent heat term representative of typical leaf transpiration. Accuracy in predicting the LATD term for the leaf models without latent heat transfer also provides a quantitative basis for adding the effects of transpiration on leaf temperature and, thus, LAVD.

In the south-central Rocky Mountains of southeastern Wyoming (USA), LATD for two identical leaf models (5 cm diameter, circular aluminum plates with 0.5 mm thickness; oriented horizontally) increased with altitude (four sites ranging from 2,104 m to 3,863 m) under natural conditions (Fig. 10.5A). Increased values of measured solar radiation and decreasing air temperatures (ca. 0.79 °C/100 m) with higher altitude (measure within 5 cm of each model) also reflected typical changes with altitude measured previously for this relatively dry, continental mountain range. Thus, measured values of LATD, under nearly identical energy balance conditions (no latent heat loss), corroborate that LATD increases with altitude due to greater insolation and lower air temperatures, under almost identical conditions of wind speed and convective heat exchange, net longwave radiation exchange, and no latent heat loss. In addition, standard energy balance analysis (Foster and Smith 1986; Jordan and Smith 1994) predicted mean LATD of each leaf model to within ± 0.6 °C of measured values (Fig. 10.5B). Adding the simulated transpiration term to the energy balance equation decreased LATD by a maximum of 1.4 °C at the highest elevation measured (3,955 m). In contrast, simulated LATD was less than measured and computed LATD (with simulated transpiration included) at all elevations under wet lapse conditions (Fig. 10.5A).

Leaf-to-air vapor difference (LAVD)

Combining data for LATD with a constant value of air humidity (Fig. 10.1) shows that computed LAVD increases only slightly (14%, 3.6 to 4.1 kPa) with altitude (2.2 to 3.9 km) under wet lapse conditions, and decreases with altitude under dry lapse conditions during summer (Fig. 10.5B). Because the measured model temperatures do not include a transpiration term, they were considered more indicative of high humidity conditions. Likewise, the model temperatures computed with a transpiration term were considered more reflective of dry lapse conditions. Much greater percent decreases occurred in LAVD under dry lapse conditions (>50%, 2.1 to 1.1 kPa) than the increases simulated under wet lapse conditions, although at much lower LAVD values (< 2.0 kPa) than wet lapse conditions (>3.5 kPa) (Fig. 10.5B). However, extrapolation to even higher elevations (6 km) resulted in computed increases in LAVD of near 50% (3.6 to 5.4 kPa).

Simulated transpiration (E)

Incorporating the measured temperatures of the leaf models (field data) resulted in significant increases in E with altitude (constant absolute humidity with altitude) due to altitudinal increases in LATD (Fig. 10.5A) and LAVD (Fig. 10.5B), and even greater increases due to increasing D_W and, thus, leaf conductance to water vapor (Fig. 10.6). For these measured and simulated conditions, the greatest effect on transpiration potential with increasing altitude resulted from increases in D_W (64% of the increase in E), while increases in LATD and LAVD resulted in an estimated increase in E of 36% over the same difference in elevation (Fig. 10.6).

Incorporating all of the interacting variables associated with higher elevation (discussed above), results in a substantial rise in simulated transpiration (E) with increased elevation (Fig. 10.6) when compared to that predicted for photosynthetic CO_2 uptake (Fig. 10.3C). Computed E increased 41% (17.7 to 24.9 mmol m^{-2} s^{-1}) from 2.2 to 3.9 km elevation, and nearly 55% when extrapolating to 6 km. However, the simulations in Figs. 3C and 6 assume identical, wind speeds, ambient SVD that declined as a function of the air temperature lapse rate, and measured net radiation values on clear days and nights (Smith and Geller 1979; Jordan and Smith 1994). Importantly, few studies have measured, or characterized theoretically, these abiotic variables for any mountain ecosystem. Moreover, the importance of microsite and plant form, and the accompanying microclimate generated, cannot be underestimated in their potential for altering the results in Figs. 3C and 6. For example, wind speed and sky exposure of leaf surfaces (as well as orientation and azimuth) for a given species is crucial to any evaluation of effects on SVD, LATD, LAVD, and E at any elevation. Thus, microsite and plant form effects could act to lower the effective, functional altitude of a given site at anytime of theyear or day, at least in terms of plant gas exchange.

There are existing data that show substantial increases in clear-sky solar radiation and decreases in longwave sky irradiance with greater altitude (Smith and Geller 1979; Bailey et al. 1989; Delacasiniere et al. 1993; Jordan and Smith 1994, 1995; Saunders and Bailey 1997; Leuschner 2000; Marty et al. 2002; Matzinger et al. 2003; Oliphant et al. 2003), as well as substantial effects of snow surface albedo on incident sunlight regimes. The lack of field data on sunlight incidence as a function of altitude is an example of the complexity of understanding altitudinal effects on LATD and transpiration. Increases in sunlight incidence with decreasing air temperatures has also been linked to the severe desiccating effects of winter conditions on evergreen, conifer trees (Hadley and Smith 1987; Boyce 1995; Lehner and Lutz 2003). Winter LATD and LAVD values occur that are even greater than summer, leading to high cuticular water loss and death in foliage exposed above the snowpack at high elevations (Anfodillo et al. 2002), plus the potential for xylem embolisms and excessive drought stress (Mayr et al. 2002, 2003). In addition, blowing snow (ice crystals) can also cause severe epicuticular wax erosion, exacerbating already high LATD, LAVD, and eva-

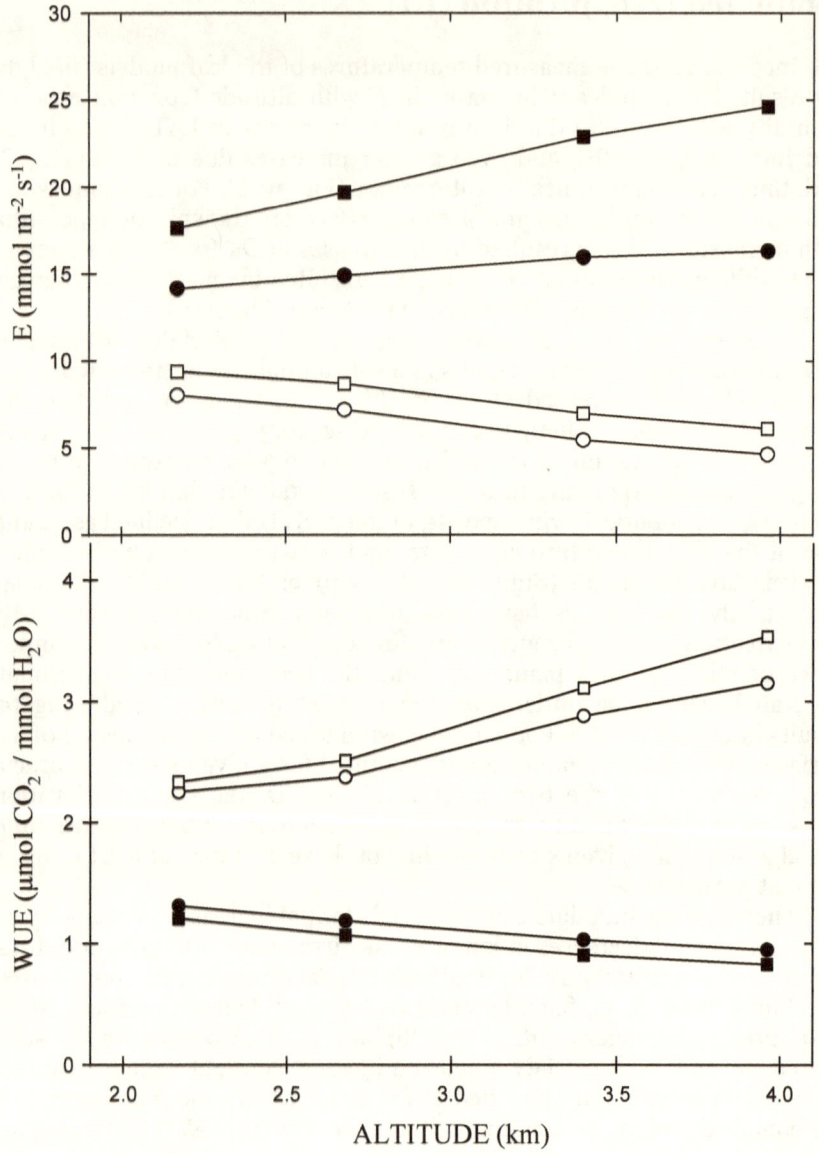

Figure 10.6. (A) Simulated transpiration (E) with changing altitude, computed as the product of LAVD and a constant g (400 mmol m^{-2} s^{-1}), or g as proportional to a changing D_W. Curves represent wet lapse rate (−3.0 °C/km) and constant g (•), dry lapse rate (−8.0 °C/km) and constant g (○), wet lapse rate and g proportional to D_W (■) and a dry lapse rate with g proportional to D_W (□). (B) Simulated changes in water use efficiency (WUE) with altitude for the same conditions simulated in (A).

porative demand in the upper canopy of subalpine forest trees (Hadley and Smith 1990).

As in the case for CO_2 diffusion at higher altitudes, a strong potential for increasing atmospheric evaporative demand is independent of any changes in soil water availability. In fact, observations of wilting in subalpine and alpine plants of the Southern Rocky Mountains (USA) have shown that severe wilting occurs typically on a daily and hourly basis for plants exposed to full sunlight and with roots in water saturated soil (as well as artificially watered), well above field capacity (Smith 1981). Once again, the colligative property of more rapid water vapor diffusion in response to lower ambient pressure will contribute to greater transpiration at higher altitudes, regardless of microsite effects. However, microclimate (generated by differences in microsite, plant form and aggregation) can also have a substantial influence on air and leaf temperatures, thus SVD, LAVD, and transpiration.

Water use efficiency (WUE)

Both photosynthesis and transpiration are coupled quantitatively within the computation of water use efficiency (WUE), the ratio of the amount of CO_2 uptake to water transpired, the consequence of a common stomatal pathway (Fig. 10.1). However, photosynthetic carbon gain can be either high or low at the same WUE for a species. Assuming a constant degree of stomatal opening and identical configuration on the leaf surface, greater elevation could result potentially in a substantial increase in transpiration, but little change in photosynthesis (Figs. 3C, 6) and, thus, declines in WUE with greater altitude. For example, WUE would decrease by over 50% between approximately 2.2 km and 4.0 km altitude when comparing the corresponding ratios of computed A and E values (Fig. 10.6B). Thus, assuming the representative conditions simulated in Figs. 3 and 6, including the second-order feedback between leaf temperature and transpiration, WUE would show large declines with elevation, indicating a much larger moisture requirement in support of the same level of photosynthetic carbon gain, due primarily to the increase in diffusion rates at higher altitudes (lower ambient pressure).

Such large declines in predicted WUE with altitude could ellicit stomatal closure or declines in the number of stomata per unit leaf area in order to curb high transpiration rates. Although Quiang et al. (2003) reported decreases in stomatal density at higher altitudes in larch, other possibilities such as leaf orientation and leaf arrangement (aggregation) on stems and among adjacent plants could also achieve similar reductions in E via leaf temperature and LATD, without sacrificing CO_2 uptake. It is these alternate solutions to lowering transpiration and increasing WUE that make evaluation of potential plant responses to increasing evaporative demand difficult, within the context of altitudinal studies, or elsewhere. Investigations are needed that employ a much more systematic measurement protocol whereby multiple biophysical factors are compared that influence transpiration, suchas LATD, in particular (Fig. 10.1).

Oxygen effects

Both photosynthesis and respiration can be influenced by lower O_2 concentrations at higher altitudes. In a similar manner for CO_2 uptake, the O_2 dependency of photosynthesis (under lower partial pressure of O_2) is compensated by the more rapid diffusion at lower ambient pressures and low internal conductance to CO_2 (Terashima *et al.* 1995; Kogami *et al.* 2001; Sakata and Yokoi 2002). A detailed analysis of O_2 impacts on photosynthetic potential at higher altitudes predicted that relatively small effects (< 5%) would occur due to a slower supply of O_2. However, most evaluations of the influence of altitude on plant gas exchange have assumed that the leaf conductance term (g) depends predominantly on stomatal limitations, and not the diffusion of CO_2 inside the leaf, *i.e.*, from cell walls (in contact with mesophyll air space) to the chloroplasts (Farquhar *et al.* 1980). Studies on the effects of altitude have employed this assumption to conclude that the internal leaf conductance must increase with altitude, disregarding the impacts of more rapid diffusion with altitude (Körner *et al.* 1988, 1991). However, this internal conductance to CO_2 is known to be strongly influenced by colder temperatures, and, thus, may be particularly limiting to photosynthetic carbon uptake at higher altitude, as suggested in numerous investigations (Eithier and Livingston 2004). Moreover, internal conductance to CO_2 can also be influenced strongly by drought, another-characteristic of high altitude documented here, and elsewhere (*e.g.*, Smith and Geller 1979; Terashima *et al.* 1995; Leuschner 2003). As emphasized previously, the diffusion of water vapor is a gas-phase phenomenon and not dependent on the liquid-phase, internal conductance. Thus, the potential impact of altitude is centered here on gas exchange properties, not diffusion in the liquid phase (*e.g.*, cell wall and cytosol) that are virtually unchanged by changes in ambient pressure (non-compressibility of liquids).

Because oxygen uptake supports respiratory metabolism, the decrease in oxygen concentration with altitude might also limit respiratory activities and growth. However, the same principles that apply to CO_2 uptake in the gas phase (described previously) should also reflect effects on O_2 exchange. Internal recycling of both CO_2 and O_2 (*e.g.*, mitochondria, chloroplasts, peroxisomes) will involve only liquid-phase diffusion and, thus, not be influenced by changes in ambient pressure. Albeit, the potential effects of such features as lower air temperatures and greater drought stress with greater altitude will still be expected to influence both respiratory and photosynthetic processes indirectly.

Conclusions and future prospects

As altitude increases, substantial increases in leaf conductance were predicted for both CO_2 and water vapor, regardless of the given degree of stomatal opening, due to increasing molecular diffusion rates at lower ambient pressure. In the case of photosynthetic CO_2 uptake, this increase in diffusivity counters the decrease in ambient CO_2 partial pressure and the leaf-to-air CO_2 difference, the driving force for diffusion. The same could be expected for the O_2 diffusion required to satisfy respiratory demands. However, these same changes for water vapor diffusion with altitude are complementary, increasing leaf transpiration at any given degree of stomatal opening and total stomatal pore area. Thus, much greater increases are predicted in transpiration compared to the slight declines predicted for photosynthetic CO_2 uptake. This is especially true for wet lapse conditions where air temperatures do not decrease as rapidly with altitude, generating even higher D_W and LAVD. However, these first-order comparisons include a number of estimated effects on leaf energy balance and temperature, including potentially important factors such as wind flow and boundary layer exchange properties that could be characteristic of particular altitudes. Most of these influencing factors are strongly associated with microtopographic and vegetation structure that may also be characteristic of a particular elevational zone (e.g., subalpine tree versus alpine tundra), and could exert strong influences on vertical and horizontal gradients in these abiotic factors. Regardless, the measures and simulated results presented here indicate that only small declines in CO_2 uptake potential are expected with increasing altitude, and, somewhat surprisingly, the greatest increases in transpiration are predicted under warmer, wet lapse conditions such as might be found in more tropical mountain ecosystems. Mechanistically, these increases in transpiration come from increases in molecular diffusion rates, with a smaller contribution due to increasing LAVD. In contrast, transpiration decreased under the dry lapse rates simulated. These results extend those reported in Smith and Geller (1979), Smith and Donahue (1991), and Tershima et al. (1995) by simulating different lapse conditions and seasonal temperatures, as well as indicate that greater internal conductance to CO_2 and accompanying enzymes may not be necessary at higher altitudes as proposed by some investigators (e.g., Körner 1989). Other studies correlating environmental factors with altitudinal limits of timberlines and treelines worldwide have also implicated water stress as a causative factor (Rochefort et al. 1994; Leuschner 1996; Maryta 1996; Rada et al. 1996; Sowell et al. 1996; Anfodillo et al. 1998; Moir et al. 1999; Boyce and Saunders 2000; Leuschner 2000; Ishida et al. 2001; Vostrol et al. 2002; Mayr et al. 2003; Wesche 2003). Similarly, paleobotanical studies have reported associations between declining timberline elevations and colder, dryer geological time periods (Innes 1991; Jobbagy and Jackson 1992; Liu et al. 2002; Daniels and Veblen 2003).

There are many abiotic and plant parameters that must be incorporated and integrated before a more comprehensive and fundamental synthesis of the influence of altitude on plant gas exchange will be possible. For example, measurement of standard gas exchange parameters, along with stomatal and leaf orientation characteristics, LATD, and microclimate for single species growing over a broad elevation span would provide a more comprehensive database for evaluating both the effects of altitude and corresponding plant response patterns. More specifically, species endemic to mountains with dry *versus* wet lapse rates could be compared for similarities in life form and phylogeny. Similarly, plant surveys that cross phylogenetic boundaries could also be compared, providing comparisons among many plant types and, thus, tests of possible evolutionary convergence. Measurements of factors influencing LATD, ambient humidity, and LAVD changes with altitude would be particularly important for understanding altitude effects on water stress. For example, leaf size and orientation, as well as whole plant form, effects on leaf temperatures are rare in the literature, although these parameters could have major impacts on LATD and LAVD. Use of evaporating leaf models with similar energy balance properties and under experimentally controlled conditions could provide valuable information when used in different mountain systems with varying lapse conditions.

Literature cited

Anfodillo T, Rento S, Carraro V, Furlanetto L, Urbinati C, and Carrer M. 1998. Tree water relations and climatic variations at the alpine timberline: seasonal changes of sap flux and xylem water potential in *Larix decidua* Miller, *Picea abies* (L.) Karst, and *Pinus cembra* L. *Annals of Forest Science* **55**: 159-172.

Anfodillo T, Pasqua D, Bisceglie D, and Urso T. 2002. Minimum cuticular conductance and cuticle features of *Picea abies* and *Pinus cembra* needles along an altitudinal gradient in the Dolomites (NE Italian Alps). *Tree Physiology* **22**: 479-487.

Bailey WG, Weick EJ, and Bowers JD. 1989. The radiation balance of alpine tundra, Plateau Mountian, Alberta, Canada. *Arctic and Alpine Research* **21**: 126-134.

Ball MC, Egerton JJR, Leuning R, Cunningham RB, and Dunne P. 1997. Microclimate above grass adversely affects spring growth of seedling snow gum (*Eucalyptus pauciflora*). *Plant, Cell and Environment* **20**: 155-166.

Barry R. 1992. *Mountain weather and climate, 2nd edn.* Routledge, London.

Boyce RL. 1995. Pattern of foliar injury to red spruce on Whiteface Mountain, New-York, during a high-injury winter. *Canadian Journal of Forest Research* **25**: 166-169.

Boyce RL and Saunders GP. 2000. Dependence of winter water relations of mature high elevation *Picea engelmannii* and *Abies lasiocarpa* on summer climate. *Tree Physiology* **20**: 1077-1086.

Carter GC and Smith WK. 1985. Influence of shoot structure on light interception and photosynthesis in conifers. *Plant Physiology* **79**: 1038-1042.

Carter GC and Smith WK. 1987. Stomatal conductance in three conifer species at different elevations during summer in Wyoming. *Canadian Journal of Forest Research* **18**: 242-246.

Carter GC and Smith WK. 1987. Microhabitat comparisons of transpiration and photosynthesis in three subalpine conifers. *Canadian Journal of Botany* **66**: 963-969.

Delacasiniere A, Grenier JC, Cabot T, and Twerneckfaga M. 1993. Altitude effect on the clearness index in the French Alps. *Solar Energy* **51**: 93-100.

Daniels L and Veblen T. 2003. Regional and local effects of disturbance and climate on altitudinal treelines in northern Patagonia. *Journal of Vegetation Science* **14**: 733-742.

Daniels L and Veblen T. 2003. Altitudinal treelines of the southern Andes near 40 degrees S-1. *Forest Chronicle* **79**: 234-241.

Etheir G and Livingston N. 2004. On the need to incorporate sensitivity to CO_2 transfer conductance into the Farquhar-von Caemmerer-Berry leaf photosynthesis model. *Plant, Cell and Environment* **27**: 137-153.

Farquhar G, Schulze E, and Kuppers M. 1980. Responses to humidity by stomata of *Nicotiana glacuca* L. and *Corylus avellana* L. are consistent with the optimization of carbon-dioxide uptake with respect to water-loss. *Australian Journal of Plant Physiology* **7**: 315-327.

Ferreyra M, Cingolani A, Ezcurra C, and Bran D. 1998. High-Andean vegetation and environmental gradients in northwestern Patagonia, Argentina. *Journal of Vegetation Science* **9**: 307-316.

Foster J and Smith WK. 1991. Stomatal conductance patterns and environment in high elevation phreatophytes of Wyoming. *Canadian Journal of Botany* **69**: 647-655.

Friend A and Woodward I. 1990. Evolutionary and ecophysical responses of mountain plants to the growing season environment. *Advances in Ecological Research* **20**: 59-124.

Gale J. 1972. Elevation and transpiration: some theoretical considerations with special reference to mediterranean-type climate. *Journal of Applied Ecology* **9**: 691-701.

Gates DM. 1980. *Biophysical Ecology*. Springer-Verlag, New York.

Geiger R. 1957. *The climate near the ground*. Harvard University Press, Cambridge.

Germino MA and Smith WK. 2002. Influence of microsite and plant form on photosynthetic responses to frost and high sunlight. *Plant Ecology* **162**: 65-94.

Germino MJ, Smith WK, and Resor AC. 2002. Conifer seedling distribution and survival in an alpine treeline ecotone. *Plant Ecology* **162**: 157-168.

Hadley JL and Smith WK. 1987. Influence of krummholz mat microclimate on needle physiology and survival. *Oecologia* **73**: 82-90.

Hadley JL and Smith WK. 1990. Influence of leaf surface wax and leaf area-to-water content ratio on cuticular transpiration in conifers. *Canadian Journal of Forest Research* **20**: 1306-1311.

Innes J. 1991. High altitude and high latitude tree growth in relation to past, present and future climate change. *Holocene* **1**: 168-173.

Ishida A, Nakano T, Sekikawa S, Maruta E, and Masuzawa T. 2001. Diurnal changes in needle gas exchange in alpine *Pinus pumila* during snow-melting and summer seasons. *Ecological Research* **16**: 107-116.

Jobbagy EG and Jackson RB. 2000. Global controls of forest line elevation in the northern and southern hemispheres. *Global Ecology and Biogeography* **9**: 253-268.

Jordan DN and Smith WK. 1994. Energy balance analysis of nighttime leaf temperatures and frost formation in a subalpine environment. *Agricultural and Forest Meteorology* **71**: 359-372.

Jordan DN and Smith WK. 1995. Microclimate factors influencing the frequency and duration of growth season frost in subalpine plants. *Agricultural and Forest Meteorology* **77**: 17-30.

Kienast F and Krauchi N. 1991. Simulated successional characteristics of managed and unmanaged low-elevation forests in central-Europe. *Forest Ecology and Management* **42**: 46-61.

Kienast F and Mohren G. 1991. Modeling forest succession in Europe-preface. *Forest Ecology and Management* **42**: 1-2.

Kitayama K and Mueller-Dumbois D. 1994. An altitudinal transect analysis of the windward vegetation on Haleakala, a Hawaiian mountain: (1) climate and soil. *Phytocoenologia* **24**: 111-134.

Konzelmann T, Calanca P, Muller G, Menzel L, and Lang H. 1997. Energy balance and evapotranspiration in a high mountain area during summer. *Journal of Applied Meteorology* **36:** 966-973.

Körner C. 1988. Does global increase of CO_2 alter stomatal density? *Flora* **181:** 253-257.

Körner C. 1989. The nutritional status of plants from high-altitudes. A worldwide comparison. *Oecologia* **81:** 379-391.

Körner C. 1998. A re-assessment of high elevation treeline positions and their explanation. *Oecologia* **115:** 445-459.

Körner C. 1999. *Alpine plant life: functional ecology of high mountain ecosystems.* Springer-Verlag, Berlin.

Körner C, Farquhar G, and Rokandic Z. 1988. A global survey of carbon isotope discrimination in plants from high-altitude. *Oecologia* **74:** 623-632.

Körner C, Farquhar G, and Wong S. 1991. Carbon isotope discrimination by plants follows latitudinal and altitudinal trends. *Oecologia* **88:** 30-40.

Körner C, Neumayer M, Palaez Menendez-Reidl S, and Smeets-Scheel A. 1989. Functional morphologyof mountain plants. *Flora* **182:** 353-383.

Kuz'min PP. 1972. *Melting of snowcover.* Published in Russian, 1961. Israel Program for ScientificTranslation, Jerusalem, Israel.

Lafleur PM, Rouse W, and Carlson D. 1992. Energy-balance differences and hydrologic impacts across the northern treeline. *International Journal of Climatology* **12:** 193-203.

Lehner G and Lutz C. 2003. Photosynthetic functions of cembran pines and dwarf pines during winter at timberline as regulated by different temperatures, snowcover and light. *Journal of Plant Physiology* **160:** 153-166.

Lafleur PM and Rouse W. 1995. Energy partitioning at treeline forest and tundra sites and its sensitivity to climate change. *Atmosphere-ocean* **33:** 121-133.

Leuschner C. 1996. Timberline and alpine vegetation on the tropical and warm-temperate oceanic islands of the world: Elevation, structure and floristics. *Vegetatio* **123:** 193-206.

Leuschner C. 1997. The concept of a potential natural vegetation (PNV): problems and suggested improvements. *Flora* **192:** 379-391.

Leuschner C. 2000. Are high elevations in tropical mountains arid environments for plants? *Ecology* **81:** 1425-1436.

Leuschner C and Schulte M. 1991. Mircroclimatological investigations in the tropical alpine scrub of Maui, Hawaii, USA: evidence for a drought-induced alpine timberline. *Pacific Science* **45:** 152-168.

Liu W, Fox J, and Xu Z. 2002. Nutrient fluxes in bulk precipitation, throughfall and stemflow in montane subtropical moist forest on Ailao Mountains in Yunnan, south-west China. *Journal of Tropical Ecology* **18:** 527-548.

Loyd A and Graumlich L. 1997. Holocene dynamics of treeline forests in the Sierra Nevada. *Ecology* **78:** 1199-1210.

Luckman B and Kavanagh T. 2000. Impact of climate fluctuations on mountain environments in the Canadian Rockies. *Ambio* **29:** 371-380.

Marty C, Philipona R, Fröhlich C, and Ohmura A. 2002. Altitude dependence of surface radiation fluxes and cloud forcing in the alps: results from the alpine surface radiation budget network. *Theoretical and Applied Climatology* **72:** 137-155.

Mayr S, Rothart B, and Damon B. 2003. Hydraulic efficiency and safety of leader shoots and twigs in Norway spruce growing at the alpine timberline. *Journal of Experimental Botany* **54:** 2563-2568.

Mayr S, Schwienbacher F, and Bauer H. 2003. Winter at the alpine timberline. Why does embolism occur in Norway spruce but not in stone pine? *Plant Physiology* **131:** 780-792.

Mayr S, Wolfschwenger M, and Bauer H. 2002. Winter-drought induced embolism in Norway spruce (*Picea abies*) at the alpine timberline. *Physiologia Plantarum* **115:** 74-80.

Meshinev T, Apostolova I, and Koleva E. 2000. Influence of warming on timberline rising: a case study on *Pinus peuce* Griseb. in Bulgaria. *Phytocoenologia* **30:** 431-438.

Moir W, Rochelle S, and Schoettle A. 1999. Microscale patterns of tree establishment near upper treeline, Snowy Range, Wyoming, USA. *Arctic, Antarctic, and Alpine Research* **31:** 379-388.

Murata E. 1996. Winter water relations of timberline larch (*Larix leptolepis* Gord.) on Mt. Fuji. *Trees* **11:** 119-126.

Nobel PS. 1999. *Physicochemical and Environmental Plant Physiology, 2nd edn.* Academic Press, San Diego.

Oliphant A, Spronken-Smith R, and Sturman A. 2003. Spatial variability of surface radiation fluxes in mountainous terrain. *Journal of Applied Meteorology* **42:** 113-128.

Qiang W, Wang X, Chen T, Feng H, An L, He Y, and Gang W. 2003. Variations of stomatal density and carbon isotope values of *Picea crassifolia* at different altitudes in the Qilian Mountains. *Trees-Structure and Function* **17:** 258-262.

Rada F, Azócar A, Briceño B, González J, and García-Núñez C. 1996. Carbon and water balance in *Polylepis sericea*, a tropical treeline species. *Trees-Structure and Function* **10:** 218-222.

Rochefort R, Little R, Woodward A, and Peterson D. 1994. Changes in sub-alpine tree distribution in western North America: a review of climatic and other causal factors. *Holocene* **4:** 89-100.

Romme W and Turner M. 1991. Implications of global climate changes for biogeographic patterns in the greater Yellowstone ecosystem. *Conservational Biology* **5:** 373-386.

Sage RF and Sage TL. 2002. Microsite characteristics of *Muhlenbergia richardsonis* (Trin.) Rydb., an alpine C_4 grass from the White Mountains, California. *Oecologia* **32:** 501-508.

Sage RF, Schappi B, and Körner C. 1997. Effect of atmospheric CO_2 enrichment on Rubisco content in herbaceous species from high and low altitude. *Acta Oecologica* **18:** 183-192.

Sakata T and Yokoi Y. 2002. Analysis of the O_2 dependency in leaf-level photosynthesis of two *Reynoutria japonica* populations growing at different altitudes. *Plant, Cell and Environment* **25:** 65-74.

Saunders IR and Bailey W. 1996. The physical climatology of alpine tundra, Scout Mountain, British Columbia, Canada. *Mountain Research and Development* **16:** 51-64.

Saunders IR and Bailey W. 1997. Longwave radiation modeling in mountainous environments. *Physical Geography* **18:** 36-50.

Saunders IR, Bailey W, and Bowers J. 1997. Evaporation regimes and evaporation modeling in an alpine tundra environment. *Journal of Hydrology* **195:** 99-113.

Smith WK and Donahue R. 1991. Simulated effect of altitude on photosynthetic CO_2 uptake potential in plants. *Plant, Cell and Environment* **14:** 133-136.

Smith WK and Geller G. 1979. Plant transpiration at high elevations: theory, field measurement, and comparisons with desert plants. *Oecologia* **41:** 109-122.

Smith WK and Geller G. 1980. Leaf and environmental parameters influencing transpiration at high elevation: Theory and field measurement. *Oecologia* **46:** 308-314.

Smith WK, Germino MJ, Hancock TE, and Johnson DM. 2003. Another perspective on altitudinal limits of alpine timberlines. *Tree Physiology* **23:** 1101-1112.

Smith WK and Knapp A. 1990. Ecophysiology of high elevation forests. In: Osmond CB and Pitelka L (eds.) *Plant Biology of the Great Basin and Range.* Ecological Studies Series, Springer-Verlag, London. Pp. 87-142.

Sowell JB, McNulty S, and Schilling B. 1996. The role of stem recharge in reducing the winter desiccation of *Picea engelmanii* (Pinaceae) needles at alpine timberline. *American Journal of Botany* **83:** 1351-1355.

Stevens G and Fox J. 1991. The causes of treeline. *Annual Review of Ecology and Systematics* **22:** 177-191.

Terashima I, Masuzawa T, Ohba H, and Yokoi Y. 1995. Is photosynthesis suppressed at higher elevation due to low CO_2 pressure? *Ecology* **76:** 2663-2668.

Tranquillini W. 1979. *Physiological Ecology of the Alpine Timberline.* Springer-Verlag, New York.

Vostral CB, Boyce RL, and Friedland AJ. 2002. Winter water relationsof New England conifers and factors influencing their upper elevational limits. I. Measurements. *Tree Physiology* **22:** 793-800.

Wardle P. 1974. Alpine timberlines. In: Ives J and Barry R (eds.) *Arctic and Alpine Environment.* Meuthuen Publishers, London. Pp: 371-402.

Warren-Wilson J. 1957. Observations on the temperatures of arctic plants and their environment. *Journal of Ecology* **45:** 499-531.

Wesche K. 2003. The importance of occasional droughts for afro-alpine landscape ecology. *Journal of Tropical Ecology* **19:** 197-207.

III. Ecosystem Processes and Climate Change

Chapter 11

THE USE OF DIGITAL CAMERAS IN PLANT ECOLOGY AND ECOPHYSIOLOGY

Eric A. Graham

Introduction
Comparing film to a digital sensor
Limitations of digital cameras
Moving from film to digital
Colorimetry and spectral analysis
Area measurements and counting
Internet image databases
New methods for digital cameras in ecology
Conclusions

Perspectives in Biophysical Plant Ecophysiology: A Tribute to Park S. Nobel, pp. 283-299
Edited by: E. De la Barrera and W.K. Smith
© 2009 by The Author
Book Compilation © 2009 Universidad Nacional Autónoma de México

Introduction

The use of visible-light digital cameras has increased dramatically in recent years, including both hand-held, high resolution cameras and Internet-linked, lower resolution video models. Most of this increase in use has been due to the reduction in cost, size, and the increase in the connectivity of digital cameras. This trend will most likely continue as cameras become even less expensive, smaller and more robust, and easier to interface with other technology. The rapid adoption of digital cameras to already existing photographic techniques, like the comparisons of digital images with historical photographs to detect changes in forest edges and the substitution of digital cameras for film cameras in hemispherical canopy photography, reflect the advantages of digital over film. Because plant canopy and leaf reflectance in visible wavelengths (400-700 nm) is predominantly influenced by chlorophyll and photosynthesis-related pigments, it is also possible to use digital cameras for quantitatively describing vegetation coverage and biomass, or nitrogen status and plant stresses. It is in this area where the use of digital cameras in plant sciences is expanding. Additionally, even simple computer-aided image analysis functions like shape recognition and the automatic counting of objects, as well as Internet-based sharing and storing of image data, will create new opportunities for using cameras for ecological and ecophysiological research. This brief review of literature will highlight some of the current work and possible future directions of such uses of digital, visible-light cameras in plant ecology and ecophysiology.

Comparing film to a digital sensor

There is a general trend of replacing film with images taken with digital sensors and stored on digital media. Two of the most understandable reasons for this are the elimination of both the cost and the delay associated with processing the exposed film used in standard cameras. Additionally, the variability of film stock and of film processing is not an issue when using a digital camera. A great advantage to using digital cameras over film cameras is that the images can be viewed immediately in the field, and retaken if necessary. Thus, the direct transfer of techniques using film cameras to those using digital cameras is attractive. Comparing the image quality of film versus digital is thus a current topic of interest, however many aspects of the systems used, such as the resolution of lenses and the speed of film, make the comparison of the two complicated. Nevertheless, the differences between film and digital cameras, like noise, spectral sensitivity, and resolution can affect the comparison of scientific data collected with the different instruments.

Film is composed of an emulsion of randomly distributed silver halide salts of various crystal (grain) sizes and thus can be considered as a continuous sensor, whereas digital sensors use uniform arrays of photodiodes that are of different sizes and spacing depending on the camera. For easier comparison, film grain size has been compared to the size of the array of red, green, and blue light sensors in the charge coupled device (CCD) or complimentary metal oxide semiconductor (CMOS) chip that constitutes the sensor on a digital camera (Baker 2005). Noise, or the signal-to-noise ration, in a digital signal may also be comparable to the grain size of film, because both cause artifacts in the resulting image. Digital cameras require interpolation of the red-, green-, and blue-filtered (RGB) photodiode outputs in order to reconstruct an image, which usually introduces color artifacts (Muresan *et al.* 2000; Orava *et al.* 2004) independent of the lens systems used (which can introduce additional artifacts; for example, Frazer *et al.* 2001). Another aspect of the digital chips used in single CCD cameras is the usual ratio of two green to one red and one blue sensors per light sensing unit (pixel) in an attempt to match the sensitivity of the chip to that of the human eye, which is more sensitive in the green region of the spectrum. This skews the output of digital cameras to that which is more pleasing to the human eye but does not preclude the partial reconstruction of spectra from the images (Marchant and Onyango 2002).

Dynamic range and resolution are also of concern when comparing film to digital imagery. Dynamic range, or the luminosity range over which film or digital sensors are sensitive, is variable depending on the film and sensors, however CCD sensors seem to be generally superior to film in this regard (Light 1996). The dynamic range of images taken using both digital and film (after digitizing) can also be extended by a technique of fusing multiple images taken at different exposures, resulting in high dynamic range images (Debevec and Malik 1997; Robertson *et al.* 1999) that can contain more information than single images in either format. The resolution of digital cameras compared to that of film, again depends on many aspects (Thomson 2004; Baker 2005), however consumer-grade cameras are now approaching or can match the resolution of 35 mm film cameras (Lenhardt and Kreuznach 2002), although there is still debate when comparing the two methods. The imaging industry standard method of measuring image resolution is by the modulation transfer function (MTF), which is based on measuring the contrast in images of test gratings of black and white pairs of lines (Baker 2005). Line pairs per millimeter resolution can be determined for any component of the optical system, from the lens to the film or digital sensor. An International Organization for Standardization document produced in 2000 specifically addresses standards for measuring resolution of digital cameras (ISO 12233:2000) and should result in more rigorous investigations into and standardizations of digital camera resolution (Williams and Burns 2001).

Limitations of digital cameras

All photographic systems have trade-offs and limitations for image acquisition. For instance, the depth of field is inversely related to the amount of light passing through the iris. Lens distortion, vignetting (where the luminance of the image drops off near the periphery), and chromatic aberrations (due to the different refractive properties of different wavelengths of light and which usually occur in the periphery of the image) affect photography regardless of whether it is film-based or digital. However there are aspects specific to digital photography that can influence their use as data collectors. For instance, on certain CCD sensors an array of microlenses is placed on top of the photodiode array in order to gather more light or reduce the non-sensing space between pixels. Because lenses tend to focus different wavelengths of light onto different focal planes, "purple fringing" can occur throughout the image frame in some digital cameras.

Another limitation to consumer-grade digital cameras is the inclusion of integrated infrared filters due to the sensitivity of the CCD or CMOS sensor units to the near infrared, limiting the responses of digital cameras to the visible spectrum. Thus, standard remote sensing methods such as Normalized Differential Vegetation Index (NDVI) used in aerial or satellite-based phenology and primary productivity studies that incorporate the reflection of infrared (790-890 nm) wavelengths cannot be used. Additionally, discreet-wavelength measurements such as are used in some canopy chlorophyll sensors (*e.g.*, SPAD which measures leaf transmittance at 650 and 940 nm; Soil Plant Analysis Development; Minolta Camera Co., Ltd, Japan), are difficult to make with standard digital cameras without the use of special filters.

The RGB-encoded image captured by a digital camera is also device-dependent, in that different digital cameras have different RGB responses because of different sensitivities and spectral responses of their image sensors (Fig. 11.1). This has serious consequences when trying to compare quantitatively images obtained by different cameras. However, methods for correcting color to correspond to the device-independent tristimulus values determined by the CIE standard colorimetric observer (CIE 1986) are numerous (*e.g.*, Hong *et al.* 2001). Additionally, the use of calibrated color charts within images (*e.g.* Munsell Digital ColorChecker SG, Gretag-Macbeth LLC, New York, USA) facilitate the color correction necessary for image comparisons between different cameras.

When an image is captured by a digital camera, pixel interpolation and color correction algorithms are often applied to the original signal, sometimes irreversibly changing the captured information. Color rendering operations used to transform scene colors to corresponding picture colors generally incorporate a dynamic range reduction because most output devices, such as computer monitors or color printers, cannot display the full dynamic range of a typical scene (Spaulding *et al.* 2003). In standard film photography when transferring the image to a print, for instance, such data is archived on the film negative, but for digital images, such data is usually lost. A RAW image file is becoming more commonly available as an optional output format from digital cameras. RAW files contain minimally processed data from the image sensors. Although there is no single RAW format, data is often available from pixels at 12 or 14 bits rather than the 8 bits of a processed RGB image and thus more information for data analysis is available. To retrieve a final, usable image from a RAW file, the data must be converted into an RGB image by demosaicing or "Digital Development."

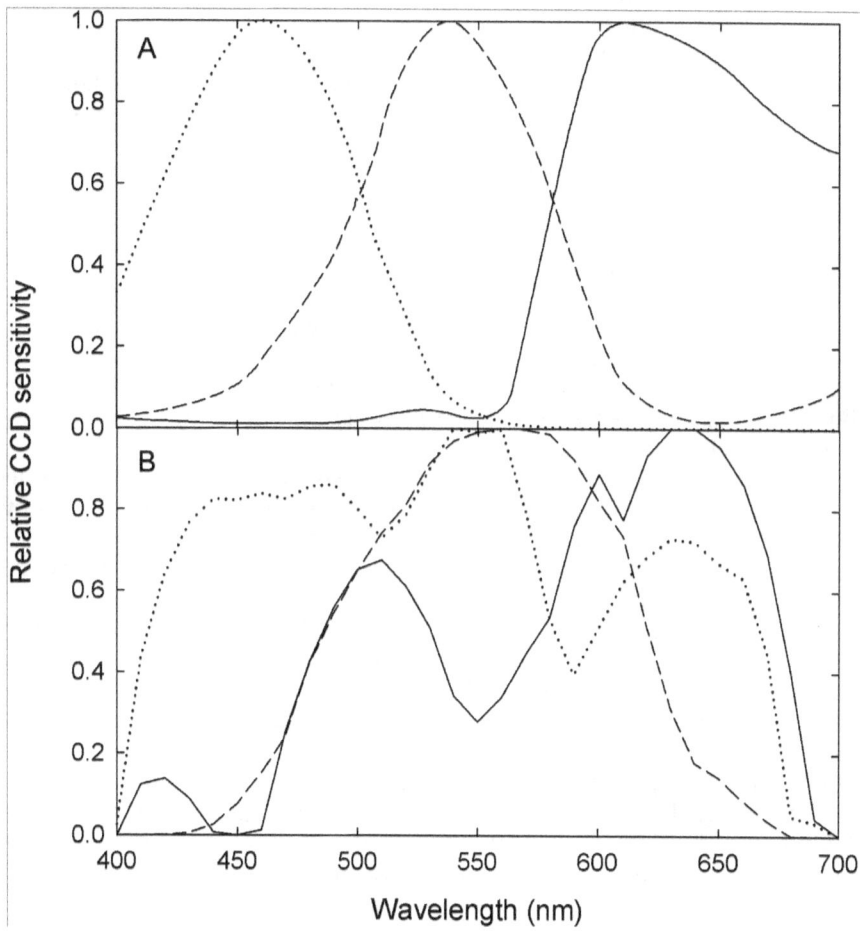

Figure 11.1. The relative wavelength responses of two digital cameras, (A) a high-end video camera (data redrawn from brochure) and (B) inexpensive cell-phone type camera (Graham unpublished data). Solid line indicates the red pixel response, dashed line indicates the green pixel response, and dotted line indicates the blue pixel response. Each red, green, and blue response was scaled separately.

Additionally, "lossy" compression algorithms used to reduce the amount of digital memory required to store images (*e.g.*, JPEG; ITU 1992) permanently reduces or distorts the information that may be extracted from stored digital images (Paola and Schowengerdt 1995). When the dynamic range of a camera is exceeded, as when parts of an image are over- or under-exposed, image information is also lost, and although methods exist for reconstructing saturated pixel values, they are limited (Zhang and Brainard 2004). Additionally, because there is only one image sensor per camera, damage to the sensor resulting in "hot" (always responding with saturated values) or "cold" (unresponsive) pixels will also affect every subsequent image captured by that camera.

Moving from film to digital

Regardless of the additional problems associated with digital imaging, the use of digital cameras is still an attractive alternative because of its many advantages. Two examples of techniques where the transition from film to digital has been successful are hemispherical canopy photography and historical repeat photography.

Hemispherical photography is an extensively used method for indirectly assessing canopy and understory characteristics such as canopy openness, leaf area, and transmittance of light. A hemispherical photograph is usually taken using a 180° field of view fisheye lens pointed up into the canopy near ground level and readily lends itself to the use of digital cameras (Leblanc *et al.* 2005). Several side-by-side comparisons of digital versus film have been made for hemispherical photography. Englund *et al.* (2000) noted that canopy openness and total site factor estimates were significantly higher in digital images compared with film photos (Fig. 11.2). The authors suggested that the differences between the two techniques might be a result of exposure of the film or differences in lens optical quality and field of view. Frazer *et al.* (2001) used a Nikon Coolpix 950 digital camera and noted substantial color blurring towards the periphery of the image that was attributed to chromatic aberration associated with the camera's lens, although this is something from which even film cameras can suffer from. Canopy openness measurements made with their system were 1.4 times greater than film estimates in most of their samples and cloud cover and sky brightness added an unpredictable effect on canopy openness. Hale and Edwards (2002) also compared film and digital cameras for hemispherical photography over a wide range of canopy openness conditions and had more favorable results. They found no systematic differences between the techniques over the range of canopy transmittance from 10 to 70%. Additionally, the digital camera was slightly more sensitive than the film camera to the proportion of diffuse radiation transmitted through the canopy. It was noted, however, that many errors can occur during any stage of the process of hemispherical photography and neither digital nor film, if used to take replicated photographs at one location, would produce exactly the same results.

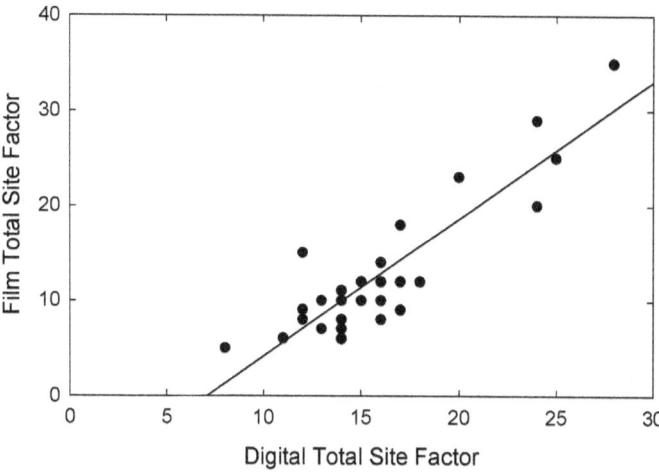

Figure 11.2. The relation between digital and film hemispherical photography for the Total Site Factor, a function of the relative contributions of direct and diffuse light. Data redrawn from Englund *et al.* (2000).

Other studies have had similar results, indicating that digital photography is a useful tool for measuring canopy closure (Guevara-Escobar *et al.* 2005), and canopy architectural parameters in boreal forests (Leblanc *et al.* 2005), although attention needs to be paid to effects of image size and the camera type for comparisons (Inoue *et al.* 2004). Techniques for standardizing the analysis, such as setting the thresholds based on sky luminance (Ishida 2004), will aid in the continued conversion from film to digital for hemispherical canopy photography.

Repeat and historical-comparison photography is used for the study of landscape changes. Because repeated photographs taken from the same position on the ground or from aircraft at different times can be compared qualitatively and quantitatively by measuring the differences between pairs, analysis of tree lines, vegetation coverage and composition, and changes in land use is possible. Although most studies employ film photography for comparison to historical film photographs (*e.g.*, Munroe 2003; Brown *et al.* 2006), images are usually first digitized and comparisons are made using computerized image processing (Okeke and Karnieli 2006), and computer-based geographical information systems (GIS) tools (Kadmon and Harari-Kremer 1999). This intermediate step of digitizing print images is now being replaced by images being captured directly with digital cameras to avoid the potential loss or distortion of information during the scanning of print photographs. For example, Booth *et al.* (2006) used an 11 megapixel, aircraft-mounted digital camera to create very large scale images having spatial resolutions as fine as 1 millimeter ground area per pixel and Nagler *et al.* (2005) demonstrated a low-cost method for creating aerial photo-mosaics for detecting land changes using an automated digital camera system.

Colorimetry and spectral analysis

Plant canopy and leaf reflectance in visible wavelengths (400-700 nm) is predominantly influenced by chlorophyll and photosynthesis-related pigments (Vogelmann 1993; Blackmer *et al.* 1994; Gitelson *et al.* 2003). It is not surprising then that visible-light digital cameras have been used for quantitatively describing vegetation coverage and biomass (Lukina *et al.* 1999; Yang *et al.* 2002), nitrogen status and plant stresses (Gamon and Surfus 1999; Wang *et al.* 2004), and, when filters are used, specific nutrient deficiencies (Graeff *et al.* 2001). Indeed, even underwater quantification of chlorophyll in algae has been determined using color analysis of images from a digital camera (Goddijn and White 2006). The spectral composition of natural light in forest environments also has a direct impact on many aspects of plant and animal ecology (Théry 2001). Thus, work using digital cameras as spectral sensors of natural light has many applications and algorithms for daylight spectral estimation using CCDs have been moderately successful (Chiao *et al.* 2000; Hernández-Andrés *et al.* 2004). For more controlled investigations, the spectral-calibration of the camera itself (Martínez-Verdú *et al.* 2002) can directly improve the quality of data collected from digital cameras. Additionally, the use of standard artificial light sources while collecting images in the field has also been suggested in order to compensate for the variation in the quality and quantitiy of natural light (Karcher and Richardson 2003), although care would have to be taken to avoid photomorphogenic and photoperiodic responses by plants to the artificial light.

To date, most of the plant studies that have involved the use of color analysis from images captured by digital cameras have concentrated on agricultural rather than ecological applications. Of the many studies that have examined plant canopy and leaf spectral reflectance, most have used hyperspectral or discreet-wavelength techniques. Of more direct application to visible light photography is the study by Gamon and Surfus (1999) that examined the red:green ratio of reflectance, where red referred to a broad band of wavelengths (600-699 nm) and green to a broad green band (500-599 nm). They compared this reflectance index with extracted pigment levels and found species-specific relationships that were also influenced by leaf structural properties. Indeed, comparisons of total or average RGB pixel intensities, the subtraction of images, shifts in the peak frequencies of the RGB channels, as well as color conversions from RGB to Hue, Saturation, and Luminance (HSL), Hue, Saturation, Brightness (HSB), and Lab color spaces (CIE 1986; Ohno 2000) are all simple techniques that can be used for color analysis of images. HSB and Lab color spaces are non-linear deformations of the RGB color space that can produce non-continuous data, although they have been used successfully in some color comparison studies (Graeff *et al.* 2001; Karcher and Richardson 2003).

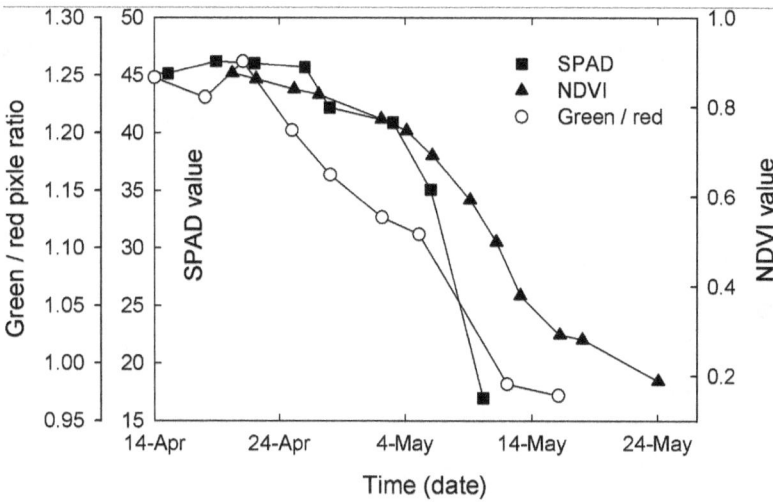

Figure 11.3. Senescence in a wheat canopy detected by the use of images over 1 m²
areas using a ratio of green to red pixel values compared to NDVI and spot-check
SPAD measurements. Data redrawn from Adamsen *et al.* (1999).

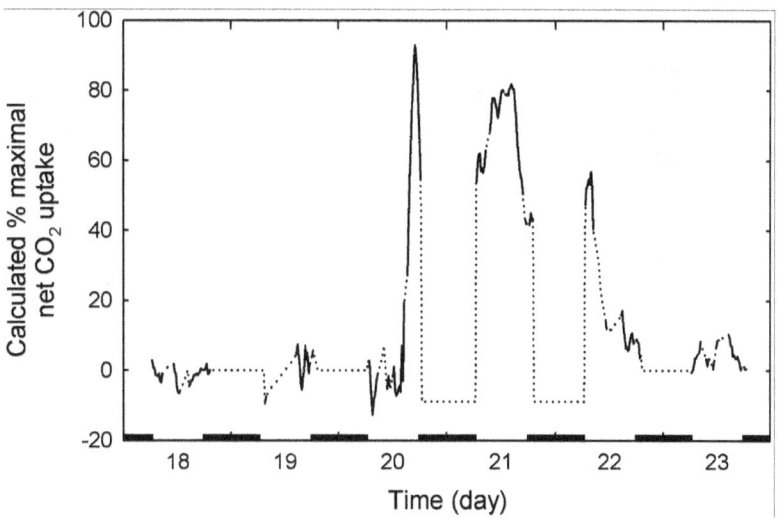

Figure 11.4. Net CO₂ uptake of a desiccation-tolerant moss calculated from a
change in its color due to a 1 mm rain event in the afternoon of 20 August 2003.

One such study involving the use of the saturation information contained in a digital image instead of colors or the ratios between colors is Souza-Echer and Pereira (2006) who were able to determine cloud cover state by use of an upwardly-facing digital camera.

One of the earliest users of digital cameras was Adamsen *et al.* (1999) who examined color changes due to senescence in a wheat canopy over 1 m^2 areas by using a ratio of green to red pixel values in digital images. They compared their camera results to NDVI and SPAD measurements and found that each of the indices was capable of quantifying canopy senescence on a near-daily time scale, although day-to-day variation was higher for the G/R index (Fig. 11.3).

The color signals captured by digital cameras have also been used in the measurement of yield and nutrient status of wheat (Hafsi *et al.* 2000; Jia *et al.* 2004a, b) and corn (Daughtry *et al.* 2000; Graeff *et al.* 2001) and for the measurement of the quality of turf grasses (Karcher and Richardson 2003).

Of the few ecological applications of color analysis involving digital cameras, Richardson *et al.* (2001) examined the spectral reflectance of pine needles along an elevational gradient, and Hamerlynck *et al.* (2000) compared total surface reflectance to the chlorophyll fluorescence characteristics of the desiccation tolerant moss, *Tortula ruralis*. Soil color has also been measured using a digital camera in order to characterize the soil's chemical and physical properties (Levin *et al.* 2005). Rather than using the standard Munsell color charts, an array of colored plastic chips was used within the images in order to standardize colors and make their comparisons quantifiable. In the MossCam Project, images of *Tortula princeps*, a desiccation-tolerant moss that changes in color depending on its state of hydration, are captured by a networked video camera every fifteen minutes in the San Jacinto Mountains of California (www.jamesreserve.edu). Correlation between the changes in the green:red ratio of images with gas exchange measurements made in the lab have allowed a prediction of instantaneous photosynthesis of this moss in the field (Graham *et al.* 2006; Fig. 11.4).

Two recent studies demonstrate the scales at which digital cameras can be effective for environmental studies. The first successfully correlated above-canopy CO_2 flux, measured with an eddy covariance tower, with images of the surrounding deciduous forest in the northeastern United States (Richardson *et al.* 2007). The second compared the HSL and RGB color spaces to better detect the occurrence of spring and subsequent total leaf area of a rhododendron species in understory microhabitats (Graham *et al.*, in press). Clearly, the color information that can be captured in digital images at any scale can effectively related to biological activity.

Area measurements and counting

At its most basic, area estimations from images captured with digital cameras are useful for determining the extent of species or canopy closure, as for repeat photography described above. Area identification and counting techniques also tend to overlap with those used in the colorimetry measurements mentioned in the previous section, however extra steps are required for these analyses. For example, Purcel (2000) compared captured images with under-canopy measurements of light for estimating total ground cover of soybean. Similarly, Boyd and Svejcar (2005) used the visual occlusion of a colored surface by willow seedlings to quantify growth and biomass accumulation. Although their results were based only on seedlings, they were performed with the intent of minimizing observer bias and allow archiving of images for different future analyses.

Counting is not a difficult task as long as objects can be successfully isolated in the image. Clover leaves were separated from grass leaves in images of a field by thresholding and edge detection methods in order to determine the percentage of clover cover and how it varies for management purposes (Bonesmo et al. 2004). Pitkänen (2001) tested several locally adaptive binarization (thresholding) methods for individual tree detection in three field plots of boreal forests in southern Finland using an airborne digital camera. Tree detection was 70-95% successful in stands with a density of less than 1,500 trees/ha. As an additional example, Adamsen and Coffelt (2005) expanded on the technique devised by Adamsen et al. (2000) used to count flowers of cultivated *Lesquerella fendleri* in order to count the flowers of both rape and crambe. They were able to determine not only the timing of the onset of flowering but also to detect differences between *Brassica* species and cultivars and to measure the duration of flowering because each species had a distinctive flowering pattern.

Internet image databases

The Internet allows the rapid exchange of digital data and thus has facilitated many distance-learning projects, including those involving the exchange of plant images taken with digital cameras. The Internet-based Distance Diagnostic and Identification System (DDIS) allows users to submit digital images captured in the field using hand-held digital cameras or those used in conjunction with a microscope for rapid diagnosis and identification of plants, diseases, insects, and animals (Xin et al. 2001; http://www.ddis.ifas.ufl.edu/). DDIS provides an avenue for specialists to share information based on digital images and maintains a digital image library that can be used in educational programs, assisted diagnosis, and data mining. Similarly, the Center for Internet Imaging and Database Systems (http://www.ciids.org) is involved in creating methods for people to share information via the Internet and supports Distance Diagnostics Through Digital Imaging programs where experts can diagnose agricultural problems via exchanged digital images (Brown et al., 1998). Regionally pertinent information is exchanged in accordance with the National Plant Diagnostic Network (http://www.npdn.org), a multi-state network which is set up to quickly detect introduced pests and pathogens. Work on the automatic classification of images, such as that by Saber and Tekalp (1998), who demonstrated algorithms for automatically classifying images based on object- or region-based color, shape, and texture features, have direct applicability to species recognition and the correct placement into and retrieval from Internet-based databases. On a mobile level, Nam *et al.* (2005) used a hand-held personal data assistant (PDA) equipped with a digital camera to allow users to sketch or photograph a leaf and then to send this information wirelessly to a server in order to identify the plant in question. Their work was based on a shape retrieval system using only a few points in the image and demonstrated the power that wireless systems can have in education when combined with digital imagery.

The ecological analysis of globally-available digital images is increasing with the large number of biologically-related, Internet-based cameras that are online. For example, at the date of this manuscript, almost 2000 Internet sites were returned when "forest webcam" was searched for using a popular internet search tool (Google Inc., Mountain View, CA). More sophisticated image processing techniques, however, will be required to analyze remote images because the quality of such images may not be controlled or other parameters known.

New methods for digital cameras in ecology

The methods employed by most of the ecological studies cited in the above sections tend to be based on relatively simple ratios of colors or on pixel counting. However, more creative and sophisticated algorithms and statistical techniques are available and the power of these methods has only recently been recognized by those in the field of ecological digital imaging. Many methods familiar to statisticians are widely used in areas outside of ecology, such as polynomial modeling (Hong *et al.* 2000), principal components analysis (Tzeng and Berns 2005), and improved spectral recovery algorithms (Sharma and Wang 2002). Currently, advanced computer models using digital image data as inputs, such as neural networks and fuzzy logic, are being employed in agricultural situations (Yang *et al.* 2003; Kavdir 2004; Meyer *et al.* 2004), often for the real-time application of herbicides. Such techniques could easily aid in ecological investigations.

On the more creative side of the use of digital cameras, Cox *et al.* (2003) used a digital camera to record time lapse images of a submergence-tolerant plant that responds to flooding with petiole elongation. By taking images as frequently as every ten minutes, they were able to quantitatively demonstrate the kinetics of both hyponastic growth and petiole elongation. Haraguchi *et al.* (2001) investigated methods for determining the areas of mung bean leaves determined from stereo images taken by two digital cameras separated in space. Although difficulties were encountered, if four points on a leaflet were clearly photographed in stereo, then the area could be calculated.

Conclusions

As technology advances and prices drop, digital cameras will replace film cameras in more and more ecological applications. The advantages to digital imagery, such as the quality control aspects of taking images in the field as well as the computer-assisted data analysis tools that are readily available, need to be weighed against the limitations of digital imaging, such as color inconsistencies and resolution limitations. With the option of having RAW file outputs from many newer digital cameras, even these limitations are becoming reduced. As ecologists increasingly embrace more sophisticated techniques, the digital camera will no doubt gain an even greater importance as a sensor in the tool kit for biological investigation.

Literature cited

Adamsen FJ and Coffelt TA. 2005. Planting date effects on flowering, seed yield, and oil content of rape and crambe cultivars. *Industrial Crops and Products* **21:** 293–307.

Adamsen FJ, Coffelt TA, Nelson JM, Barnes EM, and Rice RC. 2000. Method for using images from a color digital camera to estimate flower number. *Crop Science* **40:** 704–709.

Adamsen FJ, Pinter P.J., Barnes E.M., LaMorte R.L., Wall G.W., Leavitt S.W., and Kimball B.A. 1999. Measuring wheat senescence with a digital camera. *Crop Science* **39:** 719–724.

Baker LR. 2005. Digital cameras: spatial image quality matters. *The Imaging Science Journal* **53:** 38-45.

Blackmer T, Schepers JS, and Varvel GE. 1994 Light reflectance compared with other nitrogen stress measurements in corn leaves. *Agronomy Journal* **86:** 934–938.

Bonesmo H, Kaspersen K, and Bakken AK. 2004. Evaluating an image analysis system for mapping white clover pastures. *Acta Agriculturæ Scandinavica Section B, Soil and Plant Science* **54:** 76-82.

Booth DT, Cox SE, and Berryman RD. 2006. Precision measurements from very-large scale aerial digital imagery. *Environmental Monitoring and Assessment* **112:** 293–307.

Boyd CS and Svejcar TJ. 2005. A visual obstruction technique for photo monitoring of willow clumps. *Rangeland Ecology and Management* **58:** 434–438.

Brown EA, Hamilton RD, and Beckwith J. 1998. Distance diagnostics through digital imaging. University of Georgia, Gainesville, Ga. http://www.ces.uga.edu/Agriculture/plantpath/imaging/brochure.html.

Brown KA, Scatena FN, and Gurevitch J. 2006. Effects of an invasive tree on community structure and diversity in a tropical forest in Puerto Rico. *Forest Ecology and Management* **226:** 145–152.

Chiao C-C, Osorio D, Vorobyev M, and Cronin TW. 2000. Characterization of natural illuminants in forests and the use of digital video data to reconstruct illuminant spectra. *Journal of the Optical Society of America. Part A, Optics and Image Science* **17:** 1713–1721.

CIE 1986. *Colorimetry,* 2nd ed. Publication CIE No. 15.2. Central Bureau of the Commission Internationale de L'Eclairage, Vienna. 83 pp.

Cox MCH, Millenaar FF, de Jong van Berkel YEM, Peeters AJM, and Voesenek LACJ. 2003. Plant movement. Submergence-induced petiole elongation in *rumex palustris* depends on hyponastic growth. *Plant Physiology* **132:** 282–291.

Daughtry CST, Walthall CL, Kim MS, Brown de Colstoun E, and McMurtrey JE. 2000. Estimating corn leaf chlorophyll concentration from leaf and canopy reflectance. *Remote Sensing of the Environment* **74:** 229–239.

Debevec PE and Malik J. 1997. Recovering high dynamic range radiance maps from photographs. *SIGGRAPH 97*, 369--378.

Dymond JR, Trotter CM. 1997. Directional reflectance of vegetation measured by a calibrated digital camera. *Applied Optics* **36:** 4314–4319.

Englund SR, O'Brien JJ, and Clark DB. 2000. Evaluation of digital and film hemispherical photography and spherical densiometry for measuring forest light environments. *Canadian Journal of Forest Research* **30:** 1999–2005.

Frazer GW, Fournier RA, Trofymowc JA, and Hall RJ. 2001. A comparison of digital and film fisheye photography for analysis of forest canopy structure and gap light transmission. *Agricultural and Forest Meteorology* **109:** 249–263.

Gamon JA and Surfus JS. 1999. Assessing leaf pigment content and activity with a reflectometer. *New Phytologist* **143:** 105–117.

Gitelson AA, Gritz Y, and Merzlyak MN. 2003. Relationships between leaf chlorophyll content and spectral reflectance and algorithms for non-destructive chlorophyll assessment in higher plant leaves. *Journal of Plant Physiology* **160:** 271–282.

Goddijn LM and White M. 2006. Using a digital camera for water quality measurements in Galway Bay. *Estuarine, Coastal and Shelf Science* **66:** 429-436.

Graeff LS, Steffens D, and Schubert S. 2001. Use of reflectance measurements for the early detection of N, P, Mg, and Fe deficiencies in *Zea mays*. *Journal of Plant Nututrition and Soil Science* **164:** 445-450.

Graham EA, Hamilton MP, Mishler BD, Rundel PW, and Hansen MH. 2006. Use of a networked digital camera to estimate net CO_2 uptake of a desiccation tolerant moss. *International Journal of Plant Sciences* **167:** 751–758.

Graham EA, Yuen EM, Robertson GF, Kaiser WJ, Hamilton MP, Rundel PW. Budburst and leaf area expansion measured with a novel mobile camera system and simple color thresholding. *Environmental and Experimental Botany*, in press.

Guevara-Escobar A, Tellez J, and González-Sosa E. 2005. Use of digital photography for analysis of canopy closure. *Agroforestry Systems* **65:** 175–185.

Hafsi M, Mechmeche W, Bouamama L, Djekoune A, Zaharieva M, and Monneveux P. 2000. Flag leaf senescence, as evaluated by numerical image analysis, and its relationship with yield under drought in durum wheat. *Journal of Agronomy and Crop Science* **185:** 275–280.

Hale SE and Edwards C. 2002. Comparison of film and digital hemispherical photography across a wide range of canopy densities. *Agricultural and Forest Meteorology*, **112:** 51–56.

Hamerlynck EP, Tuba Z, Csintalan Z, Nagy Z, Henegry G, and Goodin D. 2000. Diurnal variation in photochemical dynamics and surface reflectance of the desiccation-tolerant moss, *Tortula ruralis*. *Plant Ecology* **151:** 55–63.

Haraguchi T, Hirota O, and Ahmed F. 2001. Development of a procedure to measure the structure of a plant community using stereo images taken by two digital cameras. *Journal of the Faculty of Agriculture Kyushu University* **45:** 405-413.

Hernández-Andrés J, Nieves JL, Valero EM, and Romero J. 2004. Spectral-daylight recovery by use of only a few sensors. *Journal of the Optical Society of America. Part A, Optics and Image Science* **21:** 13-23.

Hong GW, Luo MR, and Rhodes PA. 2001. A study of digital camera colorimetric characterization based on polynomial modeling. *Color Research and Application* **26:** 76–84.

Inoue A, Yamamotob K, Mizoue N, and Kawahara Y. 2004. Effects of image quality, size and camera type on forest light environment estimates using digital hemispherical photography. *Agricultural and Forest Meteorology* **126:** 89–97.

Ishida M. 2004. Automatic thresholding for digital hemispherical photography. *Canadian Journal of Forest Research* **34:** 2208–2216.

ITU 1992. Information technology – Digital compression and coding of continuous-tone still images – Requirements and guidelines. Recommendation T.81. International Telecommunication Union, Genève, Switzerland. 186 pp.

Jiaa L, Buerkertb A, Chena X, Roemheldc V, and Zhang F. 2004a. Low-altitude aerial photography for optimum N fertilization of winter wheat on the North China Plain. *Field Crops Research* **89:** 389–395.

Jiaa L, Chena X, Zhang F, Buerkertb A, and Roemheldc V. 2004b. Use of digital camera to assess nitrogen status of winter wheat in the Northern China Plain. *Journal of Plant Nutrition* **27:** 441–450.

Kadmon K and Harari-Kremer R. 1999. Studying long-term vegetation dynamics using digital processing of historical aerial photographs. Remote Sensing of the Environment **68:** 164–176.

Karcher DE and Richardson MD. 2003. Quantifying turfgrass color using digital image analysis. *Crop Science* **43:** 943–951.

Kavdır I. 2004. Discrimination of sunflower, weed and soil by artificial neural networks. *Computers and Electronics in Agriculture* **44:** 153–160.

Leblanc SG, Chen YM, Fernandes R, Deering DW, and Conley A. 2005. Methodology comparison for canopy structure parameters extraction from digital hemispherical photography in boreal forests. *Agricultural and Forest Meteorology* **129:** 187–207.

Lenhardt K, Kreuznach B. 2002. *Optics for Digital Photography*. Schneider Optics, Inc., Hauppauge, New York 11788, USA.

Levin N, Ben-Dor E, and Singer A. 2005. A digital camera as a tool to measure colour indices and related properties of sandy soils in semi-arid environments. *International Journal of Remote Sensing* **26:** 5475–5492.

Light DL. 1996. Film cameras or digital sensors? The challenge ahead for aerial imaging. *Photogrammetric Engineering and Remote Sensing* **62:** 285-291.

Lukina E, Stone M, and Raun W. 1999. Estimating vegetation coverage in wheat using digital images. *Journal of Plant Nutrition* **22:** 341–350.

Marchant JA and Onyango CM. 2002. Spectral invariance under daylight illumination changes. *Journal of the Optical Society of America. Part A, Optics and Image Science* **19:** 840–848.

Martínez-Verdú F, Pujol J, and Capilla P. 2002. Calculation of the color matching functions of digital cameras from their complete spectral sensitivities. *Journal of Imaging Science and Technology* **46:** 15–25.

Meyer GE, Hindman TW, Jones DD, and Mortensen DA. 2004. Digital camera operation and fuzzy logic classification of uniform plant, soil, and residue color images. *Applied Engineering in Agriculture* **20:** 519–529.

Munroe JS. 2003. Estimates of Little Ice Age climate inferred through historical rephotography, Northern Uinta Mountains, U.S.A. *Arctic, Antarctic, and Alpine Research* **35:** 489–498.

Muresan DD, Luke S, and Parks TW. 2000. Reconstruction of color images from CCD arrays. *Texas Instruments Digital Signal Processors Fest*, Houston TX., August.

Nagler P, Glenn EP, Hursh K, Curtis C, and Huete A. 2005. Vegetation mapping for change detection on an arid-zone river. Environmental Monitoring and Assessment **109:** 255–274.

Nam Y, Hwang E, and Kim D. 2005. CLOVER: A mobile content-based leaf image retrieval system. *Digital libraries: implementing strategies and sharing experiences, proceedings lecture notes in computer science* **3815:** 139-148.

Ohno Y. 2000. Paper for IS&T NIP16 Conference, Vancouver, Canada, Oct. 16–20, 2000. CIE Fundamentals for Color Measurements. Optical Technology Division, National Institute of Standards and Technology. 6 pp.

Okeke F and Karnieli A. 2006. Methods for fuzzy classification and accuracy assessment of historical aerial photographs for vegetation change analyses. Part I: Algorithm development. *International journal of remote sensing* **27:** 153-176.

Orava J, Jaaskelainen T, and Parkkinen J. 2004. Color errors of digital cameras. *COLOR Research and Application* **29:** 217-221.

Paola JD and Schowengerdt RA. 1995. The effect of lossy image compression on image classification. *Proceedings, 15th Annual International Geoscience and Remote Sensing Symposium*, Florence, Italy, July 10–14. pp 118–120.

Pitkänen J. 2001. Individual tree detection in digital aerial images by combining locally adaptive binarization and local maxima methods. *Canadian Journal of Forest Research* **31:** 832–844.

Purcell LC. 2000. Soybean canopy coverage and light interception measurements using digital imagery. *Crop Science* **40:** 834–837.

Richardson AD, Berlyn GP, Gregoire TG. 2001. Spectral reflectance of *Picea rubens* (Pinaceae) and *Abies balsamea* (Pinaceae) needles along an elevational gradient, Mt. Moosilauke, New Hampshire, USA. *American Journal of Botany* **88:** 667–676.

Richardson AD, Jenkins JP, Braswell BH, Hollinger DY, Ollinger SV, Smith M-L. 2007. Use of digital webcam images to track spring green-up in a deciduous broadleaf forest. *Oecologia* **152:** 323–334.

Robertson MA, Borman S, and Stevenson RL. 1999. Dynamic range improvement through multiple exposures. *International Conference on Image Processing*, Kobe Japan, Vol. 3. Pp. 159-163.

Saber E and Tekalp AM. 1998. Integration of color, edge, shape, and texture features for automatic region-based image annotation and retrieval. *Journal of Electronic Imaging* **7:** 684-700.

Sharma G and Wang S. 2002. Spectrum recovery from colorimetric data for color reproductions. Proceedings SPIE vol. 4663, Color imaging: device-independent color, color hardcopy, and applications VII, pp 20–25.

Souza-Echer MP and Pereira EB. 2006. A simple method for the assessment of the cloud cover state in high-latitude regions by a ground-based digital camera. *Journal of Atmospheric and Oceanic Technology* **23:** 437-447.

Spaulding KE, Woolfe GJ, and Joshi RL. 2003. Extending the color gamut and dynamic range of an sRGB image using a residual image. *COLOR Research and Application* **28:** 251-266.

Théry M. 2001. Forest light and its influence on habitat selection. *Plant Ecology* **153:** 251–261.

Thomson G.H. 2004. Analytical methods of assessing the image quality associated with digital and photographic imaging systems. *The Photogrammetric Record* **19:** 237-249.

Tzeng D-Y and Berns RS. 2005. A review of principal component analysis and its applications to color technology. *COLOR Research and Application* **30:** 84–98.

Vogelmann TC. 1993. Plant tissue optics. *Annual Review of Plant Physiology and Plant Molecular Biology* **44:** 231–251.

Wang ZJ, Wang JH, Liu LY, Huang WJ, Zhao CJ, and Wang CZ. 2004. Prediction of grain protein content in winter wheat (*Triticum aestivum* L.) using plant pigment ratio (PPR). *Field Crops Research* **90:** 311–321.

Williams D and Burns P. 2001. Diagnostics for digital capture using MTF. *PICS 2001: Image Processing, Image Quality, Image Capture Systems Conference,* Montreal, Quebec, Canada, 227-232.

Xin J, Beck HW, Halsey LA, Fletcher JH, Zazueta FS, and Momol T. 2001. Development of a distance diagnostic and identification system for plant, insect, and disease problems. *Applied Engineering in Agriculture* **17:** 561—565.

Yang C-C, Prasher SO, Landry J-A, and Kok R. 2002. A vegetation localization algorithm for precision farming. *Biosystems Engineering* **81:** 137-146.

Yang C-C, Prasher SO, Landrya J-A, and Ramaswamy HS. 2003. Development of a herbicide application map using artificial neural networks and fuzzy logic. *Agricultural Systems* **76:** 561–574.

Zhang Z and Brainard DH. 2004. Estimation of saturated pixel values in digital color imaging. *Journal of the Optical Society of America. Part A, Optics and Image Science* **21:** 2301–2310.

Chapter 12

REPRODUCTIVE ECOPHYSIOLOGY

Erick De la Barrera, Eulogio Pimienta-Barrios, and Jorge E. Schondube

Introduction

Phenology

Floral evocation

The peculiar water relations of flowers

Nectar

   Secretion

   Evolution

      PHYSIOLOGICAL CONSIDERATIONS

      SELECTION BY POLLINATORS

   The pollinators' perspective

      NECTAR DIGESTION BY BIRDS

      NECTAR DIGESTION BY BATS

Fruit development

Germination and establishment

Concluding remarks

Perspectives in Biophysical Plant Ecophysiology: A Tribute to Park S. Nobel, pp. 301-335
Edited by: E. De la Barrera and W.K. Smith
© 2009 by The Authors
Book Compilation © 2009 Universidad Nacional Autónoma de México

Introduction

Plant ecophysiology has advanced substantially over the last three decades allowing, for instance, the prediction of plant performance in the field and the development of ecophysiological techniques that can be used used as tools for establishing policies and making land management decisions in conservation, agricultural, and forestry contexts, as well as for dealing with important environmental issues such as global climate change (De la Barrera and Andrade 2005; Wikelski and Cooke 2006). However, a gap in our understanding of plant responses to environmental factors remains dealing with reproductive development, despite the fact that both vegetative and reproductive phenophases have shown to be sensitive to global warming (Bazzaz and Fajer 1992; Wang and Smith 2002). Reproduction is a costly process that consumes substantial amounts of water and carbon from plants in order to ensure the establishment of successive generations of reproducing individuals. This chapter considers plant ecophysiological responses during some stages of the reproductive cycle of angiosperms in order to present a brief survey of current knowledge of the processes involved (Fig. 12.1).

The first aspect that is considered is the importance of phenological studies. Plant development responds to predictable seasonal changes in light, air temperature, and water availability (Larcher 2003). However, anthropogenic alterations of climate might pose threats for plant reproduction. For instance, for perennial woody species that have a chilling requirement for bud break, milder winters will cause asynchronous reproductive development. In other cases, when plants respond to changes in daylength, the reduction of the period of low air temperatures could result in an early proliferation of insect pollinators that would reduce reproductive success.

Plants monitor the prevailing environment and their sexual reproduction is initiated when specific favorable conditions occur. In this respect, floral evocation, the series of physiological changes occurring in plant meristems that drive flowering, is externally regulated by light, air temperature, and water, but a physiological maturity of plants is also needed, *e.g.*, individuals must reach certain size before the onset of reproduction (García de Cortázar and Nobel 1992; Srivastava 2002; Oh and Lee 2007).

Although flowers evolved from leaves, recent research has revealed that their physiology, especially in terms of water and carbon relations during development, has some particularities. For instance, these developing reproductive organs demand substantial quantities of water and photosynthates from the plant in order to drive cell growth once fertilization has occurred. Another interesting feature is that such water is supplied by the phloem and not the xylem (De la Barrera and Nobel 2004a; Chapotin *et al.* 2003).

The accumulation of such photosynthates is a consequence of the phloem supply of water and high transpiration rates (De la Barrera and Nobel 2004a). The physiological mechanism for nectar secretion is described along with two possible mechanisms by which this energy reward for pollinators could have originated. In this respect, the evolutionary and ecological success of angiosperms stems from their interaction with animals. Therefore, the role of animals as gene dispersors is also considered. However, beyond the description of pollination syndromes, the discussion focuses on the energetic and digestive constraints of nectarivorous animals. A special consideration is made for the digestive limitations of certain passerines, which apparently exert selective pressures on nectar composition.

Following an effective pollination, fruit development can ensue. Because fruit development is hormonally regulated, the roles of various phytohormones are considered in the present chapter. The production of mature viable seeds marks the end of the reproductive development for plants. However, because seedlings usually exhibit high mortality rates, the actual causes and some physiological strategies for mortality reduction are also discussed.

As a result of human activity, global climate change poses a great threat to plant biodiversity (De la Barrera and Andrade 2005). Because plant reproduction depends on timely interactions with the environment and with certain animals that may also be affected by climate change, such considerations are discussed whenever appropriate (Cotton 2008).

Phenology

Plant phenological data includes the tracking of the periods of vegetative (leaf growth, leaf fall, shoot growth) and reproductive (flowering and fruiting) phenophases, including structural and physiological plant traits, related with the environmental changes that occur periodically throughout the year, and recently with global climate changes (Pimienta *et al.* 2002; Cotton 2008; Marcati *et al.* 2008). The importance of understanding phenology, which can be considered as the primigenial environmental biology, has been recognized throughout human history. The migrations of early hunter-gatherer human groups responded to the phenology of edible plants (Diamond 1997, 2002). A progressive accumulation of knowledge about phenology could have led to the eventual rise of sedentary communities, as suggested by archaeological evidence of edible plant parts predating the origin of agriculture (Diamond 2002). More recently, phenological observations have been recorded for at least one thousand years in China and Japan and for about three hundred years in England, while most countries maintain records of the phenology of cultivated species along with the prevailing weather, especially when dealing with the occurrence of agricultural disasters, and the seasonality of agricultural pests (Billings 1985; Larcher 2003; García Acosta *et al.* 2004). For instance, in Denmark a database of crop phenology, climate, and pests permits a centralized agricultural management system that has practically rid this country of the use of chemical supplies for fertilization and pest control (Rasmussen *et al.* 2006).

The seasonality of weather is the main driver of phenological changes. Such environmental predictability has enabled plant adaptation to extreme temperatures ranging from below the freezing point, as is the case for

Opuntia humifusa in Canada and the northern United States, to those above 50 °C, as is the case for *Agave deserti* and *Ferocactus acanthodes* from the Sonoran Desert (Nobel 1988). Phenological patterns can also respond to the amount and regularity of rainfall as is illustrated by various drought-coping strategies found in nature (see for example Chapters 4, 7, and 8 in this volume). Additionally, plant phenology can respond to light, day length in particular, or to a combination of environmental factors. For instance, *Opuntia ficus-indica*, a cactus with an ample distribution produces new cladodes and flower buds during the spring, following the period of low winter temperatures (Fig. 12.2A, C). During the summer the availability of water drives fruit growth and maturation and cladode expansion. Contrasting is the case for *Stenocereus queretaroensis*, a cactus from west-central Mexico for which flowering occurs during the late winter and fruits are produced during the dry and warm spring (Fig. 12.2B, C). Reproductive development for this species is supported by water and sugars that were stored in its succulent stems during the summer when rain is available (Pimienta-Barrios and Nobel 1995). Moreover, both species differ in their germination strategies. In particular, the hard testa for seeds of *O. ficus-indica*, which can be released during the late-spring and throughout the summer, prevents germination until some scarification occurs one or more years later (Pimienta Barrios 1990; Pimienta *et al.* 2004). In contrast, seeds of *S. queretaroensis* are released towards the end of the dry spring, readily germinating during the summer of the same year at the time when the rainy season begins (De la Barrera and Nobel 2003). In fact, the temperature optimum for germination in this species coincides with the prevalent summer temperatures in its native habitat. Moreover, while adequate soil water potentials may be available during the winter, lower temperatures inhibit their germination.

Flowering for temperate trees that require a period of chilling is triggered by mild spring temperatures (Larcher 2003). In this respect, the time of ripening for *Prunus avium* from the eastern Alps can be delayed up to two months for individuals growing at an altitude of 1600 m compared to those at sea level (Larcher 2003). Similar can be the temperature-driven delay of flowering for *Salix smithiana* and vegetative growth for *Picea abies* between plants from northern Spain or Portugal and those from higher latitude from northern Norway (Larcher 2003). In addition, while global warming is not caused by urban activity (Parker 2004), local warming in cities can cause early flowering for trees as it has been documented in cities of China and Europe (Roetzer *et al.* 2000; Larcher 2003; Lu *et al.* 2006).

Light, specifically changes in day length or photoperiod, can also be an important driver of phenological changes even at low latitudes (Borchert 1983). Such has been found to be the case for various species from Guanacaste, Costa Rica, where photoperiod merely changes one hour throughout the year, from 11.5 h in the winter to 12.5 h in the summer (Rivera and Borchert 2001). However, triggering of reproductive development for evergreen and stem-succulent species is more frequent between December and February (Rivera and Borchert 2001).

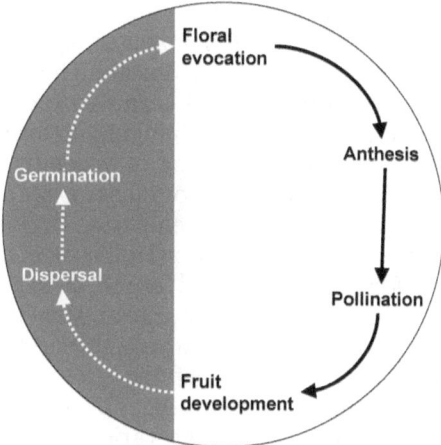

Figure 12.1. Plant lifecycle highlighting various stages of reproductive development. The shaded area indicates the stages of vegetative development that link the reproductive cycles of successive generations.

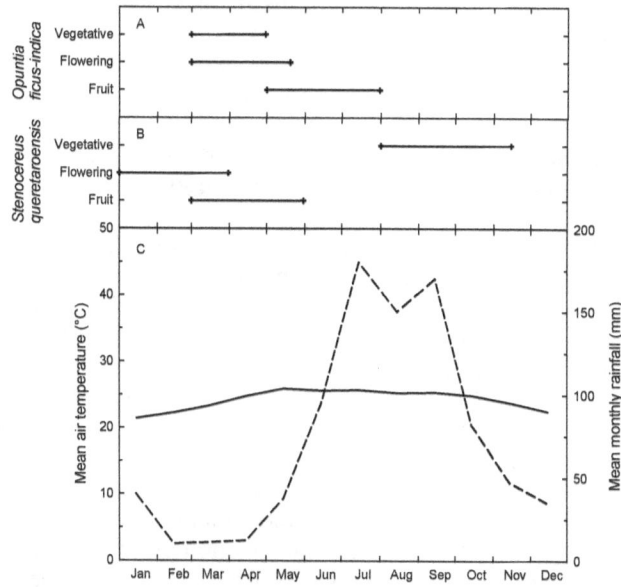

Figure 12.2. Phenology for the cacti (A) *Opuntia ficus-indica* and (B) *Stenocereus queretaroensis* from semi-arid regions of west-central Mexico. Plant developmental stages are contrasted with (C) a typical climate from the neotropical dry forests where these species are native. Data are from Pimienta Barrios *et al.* (2000) and Nobel and De la Barrera (2004).

The relation between phenology and climate is not as clear in tropical regions as it is in higher latitudes. For instance, species of the temperate genera *Fagus* and *Quercus* can grow in tropical highland regions where the winter minimum temperature ranges between −10 and 12 °C, and tree ecotypes change from deciduous, to leaf exchanging, to evergreen (Borchert *et al*. 2005). These differences in the timing of vegetative and reproductive phases for tropical dry forest species are often associated with soil moisture and rainfal patterns. For a majority of temperate deciduous trees, the reproductive phases occur in the wet season (Holbrook *et al*. 1995). However, for species as *Spondias purpurea*, bud break of vegetative shoots and flower development occur during the dry season, showing a simliar pattern to other species of tropical dry regions, whose flowering and fruiting coincide with the dry season and whose vegetative growth is not initiated until flowering ceases (Janzen 1967; Pimienta-Barrios and Ramírez-Hernández 2003). These observations suggest that various environmental factors, including temperature, can act as environmental cues (Frankie *et al*. 1974). For tree species studied in Mexico, Borchert *et al*. (2005) found that the extreme seasonality of water availability, rather than temperature, is the main factor driving phenology (contrast, for instance, the variability of precipitation *versus* the stability of mean air temperature from Fig. 12.2C). Although some evidence exists underscoring the importance of pollinators, as opposed to the physical environment, as drivers of plant phenology (Lobo *et al*. 2003), in many cases water availability, which can be highly unpredictable in many tropical forests, may be the overriding environmental factor that drives phenology in seasonal tropical forests, as has been documented for tropical dry forests from Costa Rica to Sonora, Mexico (Borchert 1994; Borchert *et al*. 2004), and during drier El Niño Southern Oscillation events in Panama (Wright *et al*. 1999).

The close interaction between plant phenology and climate has enabled an accurate estimation of past climate based on phenological records in temperate regions (Borchert 1998; Chuine *et al*. 2004). However, as discussed above, climate is not such an accurate predictor of phenology in tropical regions, mainly because of the difficulty of predicting water availability, which depends on air temperature, precipitation, and soil properties (Young and Nobel 1986; Mendoza *et al*. 2002; Larcher 2003; Nobel 2005). General circulation models predict a greater impact of changes in water regimes than in temperatures for tropical regions, so a better understanding of the drivers of tropical plant phenology is in need (Garduño 1997; Magaña *et al*. 1997; Borchert 1998; Adem *et al*. 2000; Menzel 2002). A possible approach for bridging the phenology of individual trees with studies and simulations at large geographic scales could be the use of remote sensing tools. Technological improvements have enabled the estimation of primary productivity by oceanic phytoplankton and by terrestrial plants and have allowed the determination of various physiological parameters such as water stress in leaves (Hunt Jr. *et al*. 1987; Álvarez-Añorve *et al*. 2008). The selection of very narrow wavelength bands is now possible with current remote sensing technology a feature that is already being used to detect flowering (Frey 2007; Smith *et al*. 2008).

Floral evocation

The evolution of flowers, some 100 million years ago was determinant in the success of angiosperms, allowing their rapid diversification (Cronquist 1977; Theissen and Meizer 2007). In turn, flower initiation and differentiation involves defined developmental programs that control the growth of reproductive organs, following changes in plant meristems (Fosket 1994; Meyerowitz 1994; Taiz and Zeiger 2002). In this respect, two distinct developmental programs have been described: 1) the conversion of a vegetative meristem into a reproductive meristem that will lead to the production of flower buds; or 2) the conversion of the vegetative meristem into one that produces an inflorescence which will, in turn, give origin to flowers (Fosket 1994). In many cases, especially for long lived species, plants need to reach a certain age or size to a stage termed maturity, before reproduction can occur. For instance, cladodes of *Opuntia ficus-indica* only flower after their dry mass to surface area ratio reaches certain threshold (García de Cortázar and Nobel 1992). In other cases, reproductive development is triggered by air temperature and light acting as environmental cues. The response of plant reproduction to light is mediated by phytochrome in close relation with a plant's circadian clock (Taiz and Zeiger 2002; Oh and Lee 2007). Until recently, evidence of the existence of a flowering hormone, florigen, was only indirect and mostly speculative (Chailakhyan and Krikorian 1975; Fosket 1994; Taiz and Zeiger 2002). However, current research on *Arabidopsis* has partially unraveled the molecular mechanisms of flowering (Oh and Lee 2007). In this respect, four independent genetic pathways have been identified that lead to reproductive development: 1) cold induced flowering, 2) light induced flowering, 3) a so called autonomous flowering that results from endogenous signals, and 4) gibberellic acid mediated flowering (Oh and Lee 2007). A major breakthrough in this field of research was the discovery that flowering, driven by either pathway, is actually controlled by a mere three genes (Oh and Lee 2007; Turk *et al.* 2008). Further studies of one of them, Floral Locus T (FT), have revealed that this protein is highly mobile and that it is found systemically prior to flowering, suggesting that this is an important component of the elusive florigen (Corbesier *et al.* 2007; Turk *et al.* 2008). Although recent evidence strongly supports the molecular control, *i.e.*, endogenous, of flowering, as illustrated by the very early flowering of *Arabidopsis* mutants that over-express FT, these developmental changes do not occur unless adequate environmental signals are available (Corbesier and Coupland 2006).

Many spring-flowering species, especially those from high latitudes, require a period of low air temperatures, before the differentiation of floral buds during warm air temperatures and longer days can occur, a process known as vernalization (Byrne and Bacon 1992; Larcher 2002; Sung and Amasino 2004; Oh and Lee 2007). The recurring example of *Opuntia ficus-indica* clearly illustrates this fact. As discussed above, both floral bud differentiation and the initiation of new cladodes occurs simultaneously during the dry spring (Fig. 12.2A). Interestingly, like for the olive (*Olea europea*), floral differentiation for various species of *Opuntia* occurs at the end of the winter during a relatively short period of time (50 to 60 days) in the same year that the fruit develop (Pimienta-Barrios 1990). In contrast, for temperate fruit crops such as peach, apple, and almond, floral initiation and differentiation start during the summer, one year before fruit development (Ryugo 1986). The differentiation of floral buds for *O. ficus-indica* is favored by a lower air temperature than the one favoring the production of new vegetative organs for mature detached cladodes kept under controlled environmental conditions (Nobel and Castañeda 1998). Similarly to the case for temperate fruit trees, a substantial decrease in flower formation is observed for *O. ficus-indica* when mild winters are experienced (Badeck *et al.* 2004; Avitia García and Castillo González 2007). While mild winters can lead to floral differentiation for peach trees, high rates of fruit abortion are often observed (Zegbe Domínguez 2005). In some cases, the effect of mild winter temperatures can be ameliorated by nitrogen fertilization (Nobel 1983; Pimienta Barrios *et al.* 1990). The horticultural literature and practice have been aware of the need of cold accumulation for several years and have, in fact, developed methods for measuring such cold accumulation in terms of a Chill Units index (CU; Byrne and Bacon 1992). Briefly, a chill unit is summed hourly with value of 1 when air temperatures are below 7 to 8 °C. The CU value drops to zero when air temperature drops below 1° or when they exceed 14 °C. In turn, higher temperatures will yield negative CU index values.

Vernalization has an adaptive importance because, by being able to sense a prolonged period of cold, plants can avoid flowering after brief periods of low air temperature that can occur during the fall. Otherwise if flowers/fruits were present during the winter, they would be susceptible to frost- or cold-damage because developing fruits are unable to cold-acclimate and tend to tolerate milder temperatures than underlying stems (Nobel and De la Barrera 2003). At the genetic level, vernalization involves epigenetic changes in the meristem that are different from the effects of cold hardening (Sung and Amasino 2004). In particular, the accumulation of cold leads to the repression of the Floral Locus C (FLC), a gene that is a repressor of flowering (Sung and Amasino 2004; Oh and Lee 2007). Thus, vernalization is a necessary condition for reproductive development for certain species, but an additional environmental cue might be necessary to actually trigger flowering.

Leaf-bound phytochrome allows plants to sense day length enabling them to "decide" when to flower (Taiz and Zeiger 2002; Oh and Lee 2007). For instance, coffee and strawberry only flower during short days, while oat and ryegrass require long days to trigger reproduction (Fosket 1994). Flowering for bean, cucumber, and tomato, in turn, is considered to be a day neutral process because day length does not influence reproduction. Because day length is very predictable between years, flowering by light-sensitive species is also highly predictable. Examples of this are the blossoming of cherries in Japan, of tulips in Holland, and the synchronous flowering of various deciduous trees from tropical dry forests in Costa Rica and Mexico (Borchert and Rivera 2001; Rivera and Borchert 2001; Lobo *et al.* 2003; Larcher 2005; Oh and Lee 2007). Also, flower production for autumn cultivated plants of *Cucurbita pepo* ranged from 5 to 30 flowers per plant in a dose-response experiment of nutrient availability (Orozco-Martínez and De la Barrera unpublished observations). However, flower production peaked at 60 days after sowing regardless of treatment, when the photoperiod amounted to 11 h for this herbaceous crop.

Global climate change poses potential threats for the reproductive development of various plant species. For instance, for plants that require vernalization, an increase in winter temperatures might result in the reduction or inhibition of a plant's reproductive effort. In other cases, when reproductive development is triggered by warmer, spring-like, temperatures a hastening of floral bud initiation might occur, as has been observed for some urban trees from China and northern Europe (Larcher 2003; Lu *et al.* 2006). Global warming might not be an obvious threat for the reproduction of species that respond to photoperiod rather than to air temperature. However, the story could become complicated when the role of pollinators is considered (Bazzaz and Fajer 1992). Indeed, owing to global warming, mild spring temperatures can occur earlier in the year triggering the proliferation of insect pollinators before flowers are available (Bradshaw and Holzapfel 2007). In this case a two-fold disruption of the plant-animal interaction may occur. First, the lack of food for pollinators might result in a reduction in their population size, with an expected reduction in the pollination effort the following season. And, second, the lack or shortage of pollinators during flowering, as a result of the air temperature/photoperiod decoupling, would result in a reduced fruit set and consequently in a reduced recruitment for affected species.

The peculiar water relations of flowers

Floral structures evolved from leaves, but physiological attributes greatly differ between these organs. Such differences stem from the fact that flowers are not self-sufficient photosynthetic organs that substantially contribute to dry mass accumulation. On the contrary, flowers and fruits are strong sinks demanding substantial quantities of water and photosynthates from the plant.

The xylem is the main conductor of water in plants (Fig. 12.3). Because this vascular tissue is constituted by dead water-conducting cells and lignified fibers, *i.e.*, no energy input can be performed by the plant to drive the movement of water, flow must occur spontaneously from regions of higher water potential to regions of lower water potential (Fig. 12.3). Indeed, long-distance water movement in the xylem can be explained by the cohesion-tension theory, by which a water potential gradient is created by transpiration at the leaf surface (Fig. 12.3; Taiz and Zeiger 2002; Nobel 2005).

Water is needed to maintain turgor in plant cells which is the driver of cell growth and, in some cases, of flower opening and other floral movements (van Doorn and van Meeteren 2003; Nobel 2005; Azad *et al.* 2007). Also, flowering often involves the rapid expansion of reproductive and accessory structures, so that the mobilization of substantial amounts of water is required to drive cell growth. For example, the reproductive structures and tepals of the columnar cactus *Stenocereus stellatus* undergo a substantial growth during the single night that flowers are open (Casas *et al.* 1999). Also, the petals of the yellow alder (*Turnera ulmifolia*) triple their length at 24 h before anthesis (Ball 1933), and floral structures fully expand within 5 days from anthesis for some agaves (Molina-Freaner and Eguiarte 2003). However, a most dramatic example of the massive consumption of water is the flowering of *Agave deserti*. The inflorescences of this semelparous monocot from the Sonoran Desert can exceed a height of 3.5 m and grow approximately 10 cm per day (Nobel 1977a). Such growth consumes 18 kg of water that are supplied by the leaves. While 4 kg of such remobilized water are incorporated into the inflorescence structures to maintain cell turgor, nearly 80% of the water is lost via transpiration. Similarly, the massive flowers of *Opuntia ficus-indica* transpire 3 g of water per day, or 15% of their mass, at the time of anthesis (De la Barrera and Nobel 2004a, b).

Because substantial amounts of water are needed to drive the expansion of cells of developing organs and water only flows from regions of higher water potential to regions of lower water potential, a water potential gradient needs to be created to drive water towards developing reproductive organs. One way that such a gradient can be attained is by the accumulation of hygroscopic mucilage, which has been related to the maintenance of cell turgor during periods of water shortage, including for developing flowers (Goldstein and Nobel 1991; Chapotin *et al.* 2003). In addition, transpiration by flowers can be aided by the relatively thin cuticle of petals (Nobel 1988; Bughart and Riederer 2003) or by the presence of non-functional stomates, as is the case for *Ferocactus acanthodes*. Petal stomates for this barrel cactus always remain partially open (Nobel 1977b). While only 44 g of water are required during the three month-long fruit development, 60% of such water is transpired during the three days that flowers remain open. However, in dry environments where soil water can be limiting, morphological changes can occur that reduce water loss, as is the case for the skypilot, *Palemonium viscosum* (Galen *et al.* 1999). Individuals of this species occurring at dryer sites tend to present smaller corollas than counterparts from moister sites, so that halving of the surface area of the corolla leads to resulting in a 30% reduction of transpiration, although a reduction in the plants' photosynthetic capacity is also observed.

As stated above, flow in the xylem occurs spontaneously from regions of higher water potential to regions with a lower water potential (Nobel 2005; Fig. 12.3). However, for many species, the water potential of reproductive organs is higher than that for underlying vegetative organs (Fig. 12.3; Table 12.1). In addition, for some species, such as *Vitis vinifera*, an anatomical restriction also occurs. For this plant, xylem function is reduced and eventually ceases as fruit development progresses (Bondada *et al.* 2005). If the water potential for reproductive organs is higher (less negative) than for underlying vegetative organs, the supply of water is not

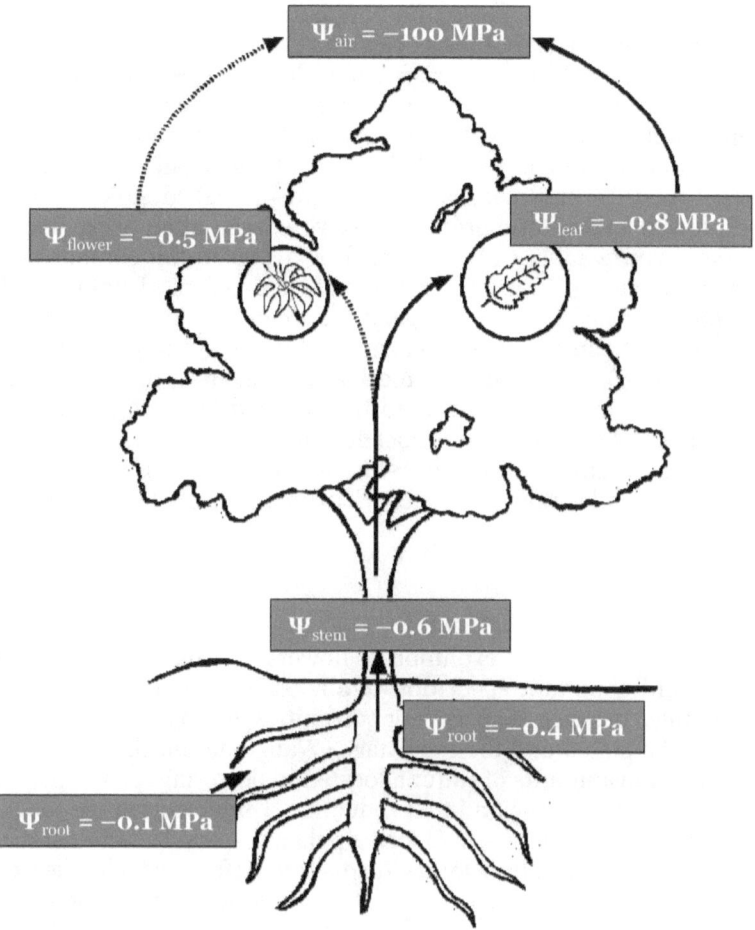

Figure 12.3. The xylem transports the water taken up by the roots upward through the stem to the leaves, where it is transpired (solid lines). The driving force for such a flow is a gradient in the water potential of the xylem solution. Considering the example portrayed, water potential can be −0.1 MPa for wet soil, −0.6 MPa in a stem, and about −100 MPa for air at 50% relative humidity. Such differences permit a spontaneous flow of water from higher to lower water potentials from the soil through the xylem to the surrounding air. However, the water potential of flowers is greater (less negative) than for adjacent stems, so water flow into the reproductive organs goes against the energy gradient for water. In this case, the phloem, which has live cells, is the supplier of water, along with solutes, to developing reproductive organs. Figure was redrawn from De la Barrera and Nobel (2004a).

thermodynamically feasible through the xylem (Fig. 12.3; Nobel 2005). Thus, the phloem must be the sole supplier of water to developing reproductive organs (Goldstein and Nobel 1991; Nobel and De la Barrera 2000; De la Barrera and Nobel 2004a).

Phloem, the second vascular tissue of plants, has living cells that use energy to drive the flow of a solution of photosynthates as explained by the Münch pressure-flow hypothesis (Taiz and Zeiger 2002; De la Barrera and Nobel 2004a; Nobel 2005). Briefly, the phloem uses active transport to load photosynthates through its cell membranes causing a large enough concentration difference with the apoplast that drives water into the phloem's lumen. In turn, the added water volume locally increases the hydrostatic pressure producing a flow towards so called sink regions. Certainly, phloem hydrostatic pressures are very high, often exceeding 2 MPa (a pressure comparable to that exerted by a 200 m-tall column of water; Fisher and Cash-Clark 2000). In some cases droplets of phloem solution can be observed on developing organs of species such as *Hylocereus undatus* (Nerd and Neuman 2004) and certain species of the genus *Eucalyptus* (Pate *et al.* 1998), illustrating the mechanical weakness of developing plant tissues that are unable to withstand the elevated hydrostatic pressures of the phloem (De la Barrera and Nobel 2004a). In this respect, in addition to reproductive organs, the phloem is the main supplier of water for young roots of maize and young stems of *O. ficus-indica* (Nobel *et al.* 1994; Prichard *et al.* 2000).

Nectar

A consequence of the evolution of flowers that has probably triggered the rapid and abundant speciation of angiosperms is their close relation with animals, because dispersal opportunities for plants are largely restricted to the pollen and the seed stage (Wang and Smith 2002). In addition to the large amounts of water allocated to flowering, plants invest considerable amounts of carbon in reproduction. Underscoring the magnitude of such resource investment in favoring pollen dispersal is, for instance, the increased floral respiration for *Datura stramonium*, which releases CO_2 eddies that attract their moth pollinator, *Manduca sexta* (Guerenstein *et al.* 2004). Similarly, an increased metabolic activity by philodendron results in an elevated plant temperature and the release of volatile compounds that lure pollinators (Nagy *et al.* 1973). More frequenly studied is the production of nectar, which can consume a high proportion of photosynthates. For instance, nectar production by *Medicago sativa* and *Asclepias syriaca* respectively amount to 20 and 35% of the fixed carbon of an entire growing season (Southwick 1984).

Table 12.1. Tissue water potential for photosynthetic and reproductive organs of various plants.

Species (Common name or Family)	Organs	Water potential		Source
		Photosynthetic organ	Reproductive organ	
Capparis spinosa (Caper)	— / Flower	—	-1.00	Rhizopolou et al. 2006
Citrus reticulata (Tangerine)	Leaf / Fruit	-0.79	-0.34	Huang et al. 2000
Gossypium hirsutum (Cotton)	Leaf / Petals	-0.96	-0.35	Trolinder et al. 1993
Luehea speciosa (Tilliaceae)	Leaf / Flower	-1.5	-0.5	Chapotin et al. 2003
Lycopersicon esculentum (Tomato)	Leaf / Fruit	-0.86	-0.42	Guichard et al. 2005
Malus domestica (Apple)	— / Fruit	—	-1.7	Yamada et al. 2004
Nopalea cochenillifera (Cactus pear)	Stem / Fruit	-0.6	0.44	Nobel & De la Barrera 2000
Opuntia ficus-indica (Cactus pear)	Stem / Fruit	-0.63	-0.39	"
O. megacantha (Cactus pear)	Stem / Fruit	-0.56	-0.40	"
O. robusta (Cactus pear)	Stem / Fruit	-0.62	-0.42	"
O. streptacantha (Cactus pear)	Stem / Fruit	-0.61	-0.42	"
O. undulata (Cactus pear)	Stem / Fruit	-0.57	-0.34	"
Phaseolus vulgaris (Bean)	Leaf / Floral bud	-0.65	-0.45	Tsukaguchi et al. 2003
Pyrus serotina (Asian pear)	Leaf / Fruit	-2.21	-1.33	Behboudian et al. 1994
Tabebuia roseae (Bignoniaceae)	Leaf / Flower	-1.5	-0.7	Chapotin et al. 2003

Secretion

Nectar, which originates from the phloem solution, is secreted through the nectaries, superficial glands generally occurring near the inside base of flowers (Elias et al. 1975; Nepi et al. 1996; Razem and Davis 1999). Some species have extrafloral nectaries, which could reduce pollen predation (Elias et al. 1975; Cuautle and Rico-Gray 2003; Rogers et al. 2003). Such glands are usually vascularized only by phloem (Elias et al. 1975; Razem and Davis 1999). The nectaries of pea (Pisum sativum) have a specialized nectar-secreting parenchyma that occurs between the phloem and an epidermis with modified stomata that lack subsidiary cells and become permanently closed as the nectary matures (Razem and Davis 1999). At anthesis, the starch granules in the parenchyma are broken down and nectar is secreted. Unlike the case for the phloem solution, which is mainly constituted by sucrose (ca. 500 mM; De la Barrera and Nobel 2004a), nectars can present various concentrations of the hexoses fructose and glucose, constituents of sucrose (De la Barrera and Nobel 2004a; Lotz and Schondube 2006). The amount and activity of the enzyme acid invertase determine the relative nectar concentration of sucrose versus its hexose components (Nicolson 2002). Plants with high activity of this enzyme have hexose rich nectars, whereas plants with low levels of the enzyme have sucrose rich ones (Yelle et al. 1991; Chetelat et al. 1993; Nicolson 2002). In this respect, for fruits the activity of the invertase is regulated by one gene that follows simple Mendelian inheritance rules, although no information is available regarding the nectar (Muller-Rober et al. 1992; Chetelat et al. 1993; Nicolson 2002; De la Barrera and Nobel 2004a; Lotz and Schondube 2006).

Evolution

PHYSIOLOGICAL CONSIDERATIONS—Even though it is amply recognized that the prevalence of nectar in extant angiosperms was been driven by ecological interactions (Darwin 1862; Pacini and Nicolson 2007), it most likely originated as a physiological by-product (Lorch 1978). In this respect, considering the particularities of floral water relations, two physiologically plausible mechanisms for the origin of nectar have recently been proposed (De la Barrera and Nobel 2004a).

First, the "leaky phloem" hypothesis takes in consideration that water input to flowers occurs via the phloem and not the xylem. In this respect, the vascular tissues of developing reproductive organs would be mechanically weak and would not be able to withstand the elevated hydrostatic pressures that occur in the phloem. As a result, the phloem solution could leak as nectar through pathways of low resistance, such as nectaries. In support of this hypothesis are the aforementioned observations of phloem leakage in various developing organs.

Second, the "sugar excretion" hypothesis also considers that water input to flowers occurs through the phloem. In this case, because solutes are supplied in addition to water, and because a considerable volume of water is transpired by the developing reproductive organs, a solute accumulation might ensue that could potentially interfere with floral physiological functions. In this case, excess solutes would have to be compartmentalized, as occurs for starch and mucilage, or excreted as nectar.

In either case, a strong selection by pollinators of nectar producing flowers might have fixed such a trait, enabling the various extant specialized plant-pollinator associations.

SELECTION BY POLLINATORS—To determine the importance of pollinators on the evolution of nectar composition, it is necessary to figure out the ancestral state of nectar composition and the identity of the original pollinators on a clade by clade basis (Lotz and Schondube 2006). In this respect, beetles were early animal pollinators, as has been determined for extinct plant species and as is the current case for basal angiosperms (Thien *et al.* 2000; Endress 2001). Bees are pollinators for many species, including the predominantly nocturnal flowers of *Hylocereus undatus*, a hemiepiphytic cactus whose flowers are open during only one night and for a few hours of the following day when bee pollination occurs (Ortiz 1999; Valiente-Banuet *et al.* 2007). In addition, very specialized plant-pollinator interactions have evolved, leading to mutualistic associations in which both the animal and the plant may undergo special adaptations, as do some wasps whose entire lifecycle occurs inside developing fruits of certain *Ficus* species (Molbo *et al.* 2003), the long bill of hummingbirds or the proboscis of butterflies and moths that feed from tubular flowers (Temeles and Kress 2003; Guerenstein *et al.* 2004), or the floral physiognomy of the insect mimic *Gillesia graminea* (Alliaceae; Rudall *et al.* 2002). In other cases the benefits of an association with nectarivorous animals, such as ants, are indirect. While ants are not polinators, they visit the flowers of some cacti in the genera *Coryphanta, Ferocactus, Opuntia* and *Stenocereus*, and can be associated with species of the legume *Acacia* that have extrafloral nectaries. The protection from herbivores that ants can convey result in increased fruit set and reduced fruit abortion (Pimienta-Barrios and del Castillo 2002; Wagner and Kay 2002).

While moths are considered specialists in terms of plant morphology, usually feeding from tubular flowers, the nectar composition for plants pollinated by these sphingids ranges from only hexoses to solely sucrose (Fig. 12.4A). Nevertheless, although nectar composition for plants that are pollinated by bees also ranges amply, about 60% of 72 of such species have sucrose concentrations below 50% (Fig. 12.4A). The case for plants pollinated by birds is very interesting (Fig. 12.4B). Specifically, for nearly 90% of 140 hummingbird-pollinated plants, nectar contains higher proportions of sucrose than of hexoses. In contrast, only 2 of 75 species of passerine-pollinated flowers produce nectar with more than 50% sucrose. This pattern of nectar-sugar composition seems to be closely related to the sugar preferences by these two groups of birds (Martínez del Rio *et al.* 1992; Lotz and Schondube 2006).

The sugar composition of nectar was thought to be a conservative trait in contrast with the rapid changes of floral morphology observed in response to pollinators (vanWyk 1993; Baker *et al.* 1998; Perret *et al.* 2001). However, hybridization experiments with tomatoes (*Lycopersicon esculentum*, rich in hexoses, and the sucrose accumulating *L. chmielewskii* and *L. hisutum* that have a low invertase activity) reveal that segregation for sucrose accumulation is consistent with the action of a single recessive gene, suggesting that the sugar composition of fruit is an evolutionarily labile trait (Chetelat *et al.* 1993). Although explicit experiments regarding the genetics of nectar composition are lacking, recent phylogenetic studies sugest that sugar composition of nectar is also labile. For instance, the pantropical legume genus *Erythrina* has more than 100 species that are exclusivelly pollinated by either passerines or hummingbirds (Galetto *et al.* 2000; Etcheverry and Trucco Alemán 2005). Contrasting pollinator type with nectar composition over a phylogeny for this genus has revealed that every shift of pollinator from passerine to hummingbird was accompanied with a chage of nectar composition from hexose-rich to sucrose-rich (Bruneau 1997). In addition, for 23 species of plants from the Canary Islands representing seven lineages, insect pollinated plants produce sucrose-rich nectar, while passerine pollinated plants produce hexose-rich nectar (Dupont *et al.* 2004). For these plants, the ancestral nectars contained sucrose, so it appears that the evolution of hexose-rich nectars was a response to visits by opportunistic nectarivorous birds (Dupont *et al.* 2004). The genus *Erythrina* and the Canarian plant lineages offer an unparalleled opportunity to investigate the biochemical bases that accompany pollinator shifts in plants.

Two hypotheses have recently been proposed regarding evolutionary changes of nectar composition as responses to selective pressure exerted by different pollinator type (Lotz and Schondube 2006). First, that non-specialized passerine birds act as a selective pressure for plants to produce hexose rich nectar. In this respect, the monophyletic asucrotic, *i.e.*, lacking sucrase, Sturnid-Muscicapid group of the Passeriformes is extremely speciose and its members could have played an important role in selecting against sucrose rich nectar and fruit pulp. Also, because hexoses are readily assimilated while sucrose is not, birds with intermediate abilities to digest sucrose should present higher energy intakes while feeding on glucose-fructose mixtures than when feeding on a sucrose rich diet. Such is the case for the nectarivorous Cinnamon-bellied Flowerpiercer (*Diglossa baritula*) that ingests 10% less energy when feeding on sucrose than when feeding on a glucose-fructose diet (Schondube and Martínez del Rio 2003). The second hypothesis considers that most nectarivorous passerines would show lower food intake rates when feeding on sucrose than when feeding on hexoses. However, sunbirds are an exception to this prediction as they have sucrase activity levels similar to those of hummingbirds (McWhorter and Schondube unpublished observations). If the second hypothesis is correct, birds with intermediate sucrase activity levels should act as selective agents for hexose rich nectar.

The pollinators' perspective

NECTAR DIGESTION BY BIRDS—While hummingbirds consistently display behavioral preferences for sucrose over hexoses (Hainsworth and Wolf 1976; Stiles 1976; Martínez del Rio 1990a, Martínez del Rio *et al.* 1992), several fruit-eating passerines display marked preferences for hexoses over sucrose, or even complete aversion to sucrose (Martínez del Rio *et al.* 1988; Martínez del Rio and Stevens 1989; Martínez del Rio *et al.* 1989, Martínez del Rio 1990a; Brugger and Nelms 1991; Martínez del Rio *et al.* 1992; Brugger *et al.* 1993). These bird preferences attractively fit an early notion of a distinct dichotomy in the composition of floral nectar between hummingbird- and passerine-pollinated plants (Fig. 12.4B; Baker and Baker 1983; Martínez del Rio *et al.* 1992; Schondube and Martínez del Rio 2003). However, it turns out that there is no such passerine-hummingbird dichotomy. Indeed, the aforementioned Sturnid-Musciapid clade is the only one within the Passeriformes lacking the enzyme sucrase and thus cannot assimilate sucrose at all (Lotz and Schondube 2006). In this respect, high frequencies of sucrose-dominant nectar in South African passerine-pollinated flowers have been observed (Barnes *et al.* 1995) and several South African passerines either show no preference for hexoses over sucrose or actually prefer sucrose (Lotz and Nicolson 1996; Franke *et al.* 1998; Jackson *et al.* 1998), while other species show a reversal from hexose preference at low concentrations to sucrose preference at higher concentrations (Schondube and Martínez del Rio 2003; Fleming *et al.* 2007). Outside of the Passeriformes, not only the hummingbirds, but also two nectar-feeding parrots assimilate sucrose with 90 to 100% efficiency (Karasov and Cork 1996; Downs 1997).

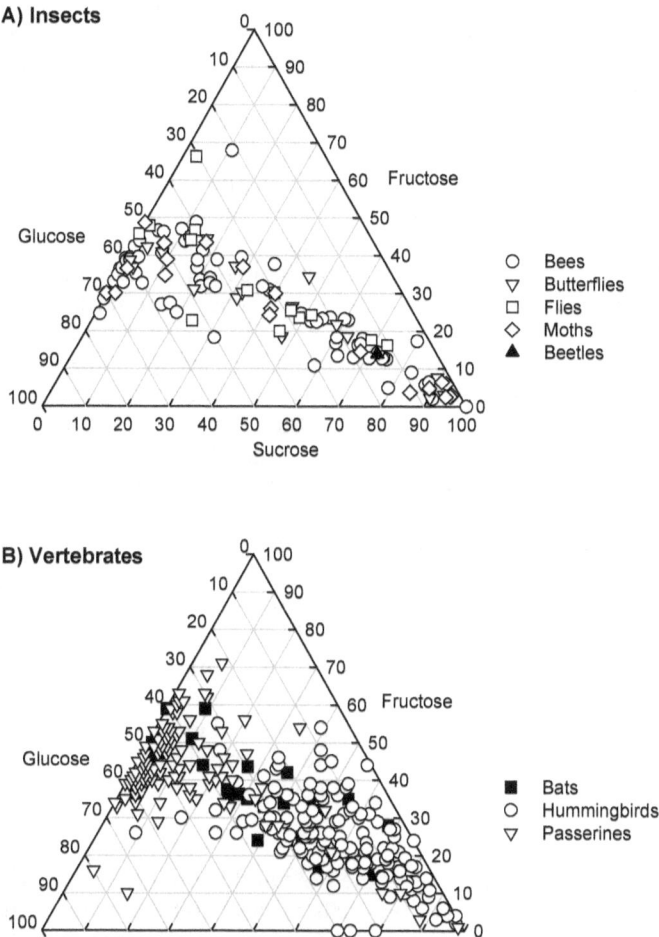

Figure 12.4. Sugar composition of nectar for (A) 238 plant species pollinated by insects and (B) 369 plant species pollinated by vertebrates. When the composition for an individual species could not be determined, an average composition of nectar was plotted for a particular type of pollinator. Such was the case for beetles (A; solid triangle). For these primitive pollinators, the graph can be interpreted by drawing a line parallel to the left axis intercepting the Sucrose axis at 72.1%. The Glucose concentration can be found to be 13.95% by drawing a line parallel to the right axis. The average Fructose content (13.95%), which can also be calculated as 100−[Glucose+Fructose], can be found by drawing a line parallel to the bottom axis. Data were compiled from Baker *et al.* (1998), Perret *et al.* (2001), Galetto and Bernardello (2003), Chalcoff *et al.* (2006), and Wolf (2006).

So, are the observed behavioral preferences and aversions of birds for different sugar compositions caused by physiological limitations or by more superficial taste preferences? Aversion, at least, appears to be caused by an inability to assimilate particular sugars, as is the case for the fruit-eating European starling (*Sturnus vulgaris*) that completely lacks the enzyme sucrase (Martínez del Rio and Stevens 1989). These birds suffer from osmotic diarrhea when fed sucrose solutions, owing to an accumulation of high concentrations of unabsorbed sugar in the intestine, and develop an intense behavioral aversion to this sugar. Hummingbirds, on the other hand, digest and absorb sucrose with almost 100% efficiency (Martínez del Rio 1990a, 1990b). Less clear from a physiological standpoint is why hummingbirds and other nectar-feeding birds display a preference for sucrose over the sugars that it is hydrolyzed into, glucose and fructose. Lotz and Schondube (2006) hypothesized that in most cases specialized nectar feeding birds do not act as a selective pressure for sucrose rich nectars. Supporting this hypothesis is the fact that the sucrase activity levels of hummingbirds and sunbirds, which are ten times higher than those of other birds, seem to be the result of a physiological adaptation to their sucrose rich diets (Schondube and Martínez del Rio 2004; McWhorter and Schondube unpublished observations). Thus, it has been proposed that the presence of sucrose rich nectars in bird pollinated plants could be the result of plant physiological processes and/or selective pressures exerted by insects before the plants were pollinated by birds (Lotz and Schondube 2006). In this case, species of nectar-eating birds with a high capacity to digest sucrose, could have released the selective pressure of non-specialized nectarivorous birds for hexoses, allowing some clades of plants to produce sucrose rich nectars. Plants that secrete hexose-rich nectars avoid the cost of synthesizing invertase and may be selected for if the pollinators do not prefer hexoses.

NECTAR DIGESTION BY BATS— If the Lotz and Schondube (2006) hypothesis that specialized nectar-eating birds with a high capacity to digest sucrose could have released the selective pressure of non-specialized nectarivorous birds for hexoses is true, nectar from plants pollinated by other groups of specialized pollinators should present sucrose-rich nectars, or a wide array of sugar nectar compositions. Specialized pollinators, with the capacity to digest sucrose, can use nectars with all possible sugar compositions with the same digestive efficiency (Schondube and Martínez del Rio 2003; Fleming *et al.* 2004). As a result, in the absence of generalist pollinators, the plants visited by these animals could have nectars with any possible mixtures of sucrose, glucose or fructose. So this hypothesis can be tested by analyzing the sugar composition of plants visited by other groups of specialized pollinators, like phyllostomid bats, which should not present a clear pattern of sugar composition.

The family Phyllostomidae, endemic to the New World, includes two clades of specialized nectar-feeding bats (Wetterer *et al.* 2000). These bats have high levels of sucrase activity (Hernandez and Martínez del Rio 1992; Schondube *et al.* 2001) and do not show preferences for nectars made out of sucrose or hexoses (Rodríguez-Peña *et al.* 2007). In addition, they assimilate both sucrose- or hexose-rich nectars with the same efficiency (Ayala-Berdon *et al.* 2008). Also, by being nocturnal, these bats do not compete with other pollinators for the nectar of the flowers they visit. Analyses of the floral nectar of plants pollinated by phyllostomid bats, independently of their phylogenetic origin, show a wide diversity in their sugar composition (Fig. 12.4B; Baker *at al.* 1998, Rodriguez-Peña *et al.* 2007). This suggests that bats are not acting as a selective pressure for nectar composition, and that for plants visited by these animals, nectar composition seems to be solely determined by plant physiology.

The behavioral preferences of pollinators for the different sugars present in floral nectar show highly diverse patterns. Some of this behavioral variation can be explained by known physiological constraints of animals, although much of the variation remains mysterious. It is clear that plant physiology, by determining the sucrose content of floral nectar, plays a central role in controlling the behavior of pollinators, and affects which organisms can use the energy present in nectar. However, the role that pollinators play in shaping the sugar composition and concentration of floral nectar is just beginning to be understood.

Fruit development

An effective pollination can lead to the development of seeds and, consequently, of fruits. Given that many plant species also rely on animal vectors for seed dispersal, fruits have evolved that provide sugars and, in some cases, water. Further, the germination for some species requires acid scarification that mimics passage through a digestive tract, illustrating a close coevolution of plants and animals (Baskin and Baskin 1998; Wang and Smith 2002). However, because developing fruits are strong sinks of photosynthates, tradeoffs between the number and size of fruits per plant and between the number and size of seeds per fruit have been amply documented (Stephenson 1981). Indeed, this fact has been translated into the horticultural practice of fruit or flower thinning in order to yield larger fruits (*e.g.*, Avitia García and Castillo Martínez 2007; Chapter 6).

Competition for maternal resources between seeds within a fruit and between fruits within a branch is mediated by phytohormones (Srivastava 2002; Taiz and Zeiger 2002). In this respect, the levels of all such plant growth regulators fluctuates during fruit development as it has been described in detail for tomato (Gillaspy *et al.* 1993; Fig. 12.5). For instance, cytokinins, which are known to promote cell division, peak at the time of fertilization and their levels decrease as fruit growth progresses (Fig. 12.5; Gillaspy *et al.* 1993; Taiz and Zeiger 2002). In turn, the levels of the cell growth mediator auxin peak during the time of fruit expansion for tomato (Fig. 12.5). Fertilization can induce a rapid cell division for various fruits (Avitia García and Castillo González 2007). However, the actual fruit expansion depends on the sink strength of fruits, competition with neighboring sib fruits, and the availability of resources (Lee 1988). For instance, fertilized fruits of *Opuntia ficus-indica*, whose seeds—the source of the sink signal—are aborted, yield fruits that accumulate 30% less dry mass than fruits from a control group (Nobel and De la Barrera 2004b). Moreover, the effect of resource limitation can be illustrated by fertilized fruits growing on detached cladodes, *i.e.*, they are water-limited, which accumulate 60% less dry mass than fruits growing on intact plants (Nobel and De la Barrera 2004b).

As mentioned above, gibberellins regulate one of the four known genetic pathways for flowering (Oh and Lee 2007). In this respect, levels of gibberellins have been detected to increase at the time of fertilization or shortly after for *Opuntia ficus-indica* (Inglese *et al.* 1998), grape (Agüero *et al.* 2000), and tomato (Gillaspy *et al.* 1993; Fig. 12.5), and play an important role in fertilization and seed formation for *Arabidopsis*, pea and grape (Swain *et al.* 1997; Agüero *et al.* 2000; Singh *et al.* 2002). However, the application of exogenous gibberellins causes seed abortion for developing fruits of grape, pea, and *Opuntia ficus-indica* (García-Martínez and Hedden 1997; Agüero *et al.* 2000; De la Barrera and Nobel 2004b). Indeed, seed-produced gibberellic acid seems to act as an inhibitor of further fruit development in trees, which is believed to mediate competition for maternal resources in plants (Haig and Westoby 1988). For tomato, the levels of gibberellins have a second peak at the time of seed formation and decrease at the time of fruit expansion, concomitant with a second peak of auxin levels (Fig. 12.5). Another well documented role of gibberellic acid is the induction of seed germination (Srivastava 2002; Taiz and Zeiger 2002). In this case, viviparous germination of seeds in developing fruits may be inhibited during the second peak of auxins which have an antagonistic effect to that of gibberellins in germination (Fig. 12.5; Baskin and Baskin 1998; Taiz and Zeiger 2002; De la Barrera and Nobel 2004b).

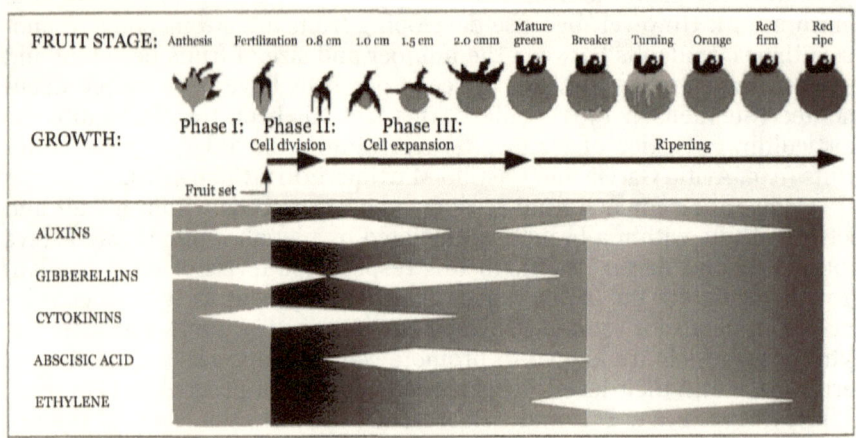

Figure 12.5. Changes in phytohormone levels throughout fruit development for tomato. Copyright by the American Society of Plant Biologists, modified from Gillaspy *et al.* (1993) with permission.

Abscisic acid can be detected in developing fruits of tomato from the onset of fruit expansion to the time of color break (Fig. 12.5). This hormone is a regulator of stomatal opening (Taiz and Zeiger 2002). As discussed above, plant reproduction is a costly process in terms of water. For the case of developing fruits, water is required to drive the cell expansion leading to plant organ growth. In many cases, young fruits are able to conduct CO_2 uptake until the time of color break with an inevitable transpirational water loss. For instance, fruits of *Opuntia ficus-indica* have a daily transpiration of about 0.6 g of water (half of their daily supply) at 20 days after anthesis, when stomatal conductance reaches 15% of that at the time of fertilization (De la Barrera and Nobel 2004b). Here, the presence of abscisic acid could reduce transpirational water loss in favor of fruit growth.

Ethylene is the last phytohormone involved in fruit development (Fig. 12.5). The discovery of the regulatory function of this gas was serendipitous, stemming from the observation of off-season leafless urban trees that grew in the vicinity of the ethylene lamps that were used for city illumination (Srivastava 2002; Taiz and Zeiger 2002). Research about the transduction cascade of ethylene mediated ripening has revealed reciprocal regulation of developmental processes between the gas and auxin (Srivastava 2002; Taiz and Zeiger 2002; Trainotti *et al.* 2007). Currently the ethylene mediated ripening of so called climacteric fruits is used to manage shelf-life, as is the case for the year-round available banana.

Once a fruit has reached its final size, ripening involves changes in fruit color, a de-polymerization of storage carbohydrates, and the degradation of cell walls that entices its ingestion by animal dispersers (Gillaspy *et al.* 1993; Agüero *et al.* 2000; Taiz and Zeiger 2002; De la Barrera and Nobel 2004b).

Germination and establishment

While germination and establishment are not *per se* stages in the reproductive cycle of a plant, successful germination and eventual establishment contribute to the fitness of an individual. They can be considered as the link between the reproductive cycles of successive plant generations (Fig. 12.1). Because the environmental biology of seed germination has been extensively reviewed (Baskin and Baskin 1998; Rojas-Aréchiga and Vázquez-Yanes 2000), especially for cacti, the focus of this section is the survival and performance of very young seedlings. In this respect, the seedling stage constitutes the bottleneck for the recruitment of most long-lived plants and mortality is especially high for very young individuals.

A straightforward way to assess the magnitude of maternal investment in seedling survival is by measuring the amount of resources that are allocated to seeds as seed mass. This parameter ranges over ten orders of magnitude from a few micrograms for certain orchids to more than 20 kg for the double coconut (Moles *et al.* 2005). Ecological studies suggest that parental investment in seed mass can decrease seedling mortality. In this respect, a so called "reserve effect" has been proposed as a mechanism explaining maternal investment in seedling development with implications in survival (Westoby *et al.* 1996). In particular, smaller seeds have to invest a larger proportion of their cotyledon reserves in the construction of vegetative organs than do their larger counterparts. This can be observed as greater relative growth rates for seedlings originated from smaller seeds than their counterparts from larger seeds (Paz *et al.* 2005). For instance, a meta-analysis of about thirty different species (Moles and Westoby 2004a) revealed that small seeds (0.02 mg in mass) had an expected 40% mortality one week after emergence and 95% after six weeks of emergence, while the mortality for larger seeds (3, 500 mg) was substantially lower, 0.8% after one week of emergence and 4.7% after six weeks. A similar trend is observed for seeds from eight species of the tropical rainforest genus *Psychotria* for which seedling survival and size for one-year-old seedlings are generally greater for individuals originated from larger seeds than those from smaller seeds (Paz and Martínez-Ramos 2003).

Seedling mortality can be very high and have multiple causes. Nevertheless, the most prevailing causes for such mortality are, in decreasing order, drought, herbivory, and pathogen attack (Moles and Westoby 2004b). In either case the reserve effect can explain, at least in part, an increased survival for seedlings from larger seeds. For the case of drought, it has been observed that tolerance increases for seedlings of *Stenocereus benecki* from heavier seeds (Ayala-Cordero *et al.* 2006). Solutes from cotyledons could be mobilized into seedling cells to osmotically control turgor. In other cases, the effect of increased cotyledon reserves can be morphological, as has been observed for seedlings of *Stenocereus beneckei* and *S. queretaroensis* whose volume to surface area ratio is positively correlated with cotyledon reserves. By altering their morphology, seedlings are able to increase the storage of water while decreasing the surface area available for water loss (Ayala-Cordero *et al.* 2004; Gallardo-Vásquez and De la Barrera 2007).

Regarding herbivory, the reserve effect hypothesis predicts that because larger seeds allocate a smaller proportion of their reserves to the production of vegetative organs, they are able to resprout following an attack by herbivores, as has been observed for various species (Foster 1986; Bonfil 1998; Green and Juniper 2004). However, seeds with larger cotyledons are more prone to predation than their smaller counterparts (Baskin and Baskin 1998; Alexander *et al.* 2001; Moles and Westoby 200b).

Plant responses to patogens are complex and costly. For instance, the systemic acquired resistance (SAR) consists of a plant-wide accumulation of chitinases and other hydrolitic enzymes in a process that is apparently induced by salicylic acid (Taiz and Zeiger 2002). Another resource demanding feature of SAR is the involvement of the enzyme phenylalanine ammonia-lyase (PAL) that mediates the synthesis of phenolic compounds that aid in protection against pathogen and herbivore attack (Camm and Towers 1973; Fosket 1994). (Actually, PAL also plays a major role in plant carbohydrate metabolism, including the synthesis of cell walls and certain responses to drought, so its involvement should also be considered when studying seedling drought tolerance; Camm and Towers 1973; Hura *et al.* 2007; Hura *et al.* 2008). Thus, it makes sense that seedlings that originated from larger seeds will be better suited for powering the carbohydrate demanding SAR metabolic machinery.

Cotyledons, however, are not mere reservoirs for carbohydrates and mineral nutrients that maintain seedling physiological functions until reaching autotrophy (but see Kitajima 2002; Lamont and Groom 2002; Kennedy *et al.* 2004; Moles *et al.* 2005). Recent studies, mainly about herbivory, have shown that in addition to an impact on seedling survival, cotyledon integrity during early development for seedlings has repercussions that only become evident at the reproductive stage (Kitajima 2003; Green and Juniper 2004; Hanley and May 2006; Hanley and Fegan 2007). In this respect, the influence of cotyledons on seedling performance can be indirectly observed by studies of very young seedlings. Further, for some species, tolerance to drought and high irradiation, as well as resistance to herbivory, decrease as the reliance of seedlings on cotyledon reserves decreases resulting in an enhanced survival of younger individuals (del-Val and Crawley 2005; Gallardo-Vásquez and De la Barrera 2007).

While germination can occur under relatively ample environmental conditions, the eventual emergence and survival of seedlings can, in some cases, require specific and relatively stable microenvironments (Harper *et al.* 1961; Harper *et al.* 1965; Baskin and Baskin 1998; De la Barrera and Nobel 2003). For instance, seedlings are very infrequently observed in the field for many species from arid or semi-arid environments, as is the case for the North American succulents *Agave deserti, Carnegia gigantea, Ferocactus acanthodes, Stenocereus queretaroensis* (Turner *et al.* 1966; Jordan and Nobel 1979; Nobel 1988; Gallardo-Vásquez and De la Barrera 2007). For these species a close association with nurse plants or other nurse objects can increase survival, through the amelioration of the physical microenvironment to which seedlings are exposed and, in many cases, provide shelter against predation (Schupp 1988; Franco and Nobel 1988, 1989; Nobel 1989; Valiente-Banuet and Ezcurra 1991; Nobel *et al.* 1992; Leirana Alcocer and Parra-Tabla 1999; Ibáñez and Schupp 2001; Munguía-Rosas and Sosa 2007; Peters *et al.* 2008). In this case, a tradeoff between a temperature and herbivory "safe" environment and a vigorous photosynthesis (although many species that frequently use nurse plants/objects inherently have very slow growth rates) is caused by low light availability under the canopy of nurse plants and might favor the survival of larger seeds, as is the case for seedlings of rainforest species growing in the understory (Kitajima 2002; Paz and Martínez-Ramos 2003; Myers and Kitajima 2007). Although the environmental and herbivory-avoiding advantages of interactions with nurse plants have been amply documented to be beneficial, recent phylogenetic studies of nurse-protegee interactions reveal that not all nurse plants are created equal (Valiente-Banuet *et al.* 2006; Valiente-Banuet and Verdú 2008). While any seed can germinate under any nurse species, as seedlings grow, a shift from facilitation to competition (where the younger interactor inevitably dies) can occur when the interacting species are taxonomically related. In this case, further studies of mechanisms of plant-plant interactions are in need.

Concluding remarks

The scientific career of Park Nobel has been most influential in the development of plant ecophysiology. Although reproductive aspects of plants were not the main focus of his research program, his contributions to the field are substantial. Whenever available examples from work conducted in his laboratory bodies were used throughout this chapter preferentially over other examples from the literature.

Plant reproduction is a costly process for which resources are invested in order to increase the chances of the recruitment of a new generation of reproductive individuals. This process is greatly influenced by enviromental cues complexly interacting with endogenous signals, such as phytohormones. A thorough revision of all processes and interactions involved would require at least an entire book and was not the focus of this chapter. However, the main aspects of plant-environment interactions during reproductive development were considered. Of particular importance are light and air temperature that control phenology, although water limitations can also play an important and sometimes disruptive role.

An issue of major concern for natural scientists in general is the impending global climate change. Expected climate alterations in prevailing air temperature and rainfall regimes will most likely affect plant reproduction and present a two-fold challenge for plant ecophysiologists. First, studies of plant reproductive ecophysiology will allow the assessment of the vulnerability of plants to various scenarios of climate change. Such assessments could even enable the implementation of management practices leading to the conservation of biodiversity. The second challenge is more complex. Food supply for humans is largely dependant of successful plant reproduction, as is the case for cereals and fruits. In addition to climate alterations, a dramatic increase of the human population, particularly in developing countries, is expected, so the contributions of ecophysiology in aiding sufficient food production are of paramount importance.

Acknowledgements

We thank rewarding discussions with Drs. Alejandro Casas, Rodrigo Méndez, Mauricio Quesada, and Enrico Yepez, that helped in shaping the resulting manuscript, and assistance with the preparation of figures by Rodrigo Orozco. Financial support was provided by DGAPA-UNAM, grant PAPIIT IN221407.

Literature cited

Adem J, Mendoza VM, Ruiz A, Villanueva EE, and Garduño R. 2000. Recent numerical experiments on three-months extended and seasonal weather prediction with a thermodynamic model. *Atmósfera* **13:** 53-83.

Agüero C, Vigliocco A, Abdala G, Tizio R. 2000. Effect of gibberellic acid and uniconazol on embryo abortion in the stenospermocarpic grape cultivars Emperatriz and Perlon. *Plant Growth Regulation* **30:** 9-16.

Alexander HM, Cummings CL, Kahn L, and Snow AA. 2001. Seed size variation and predation of seeds produced by wild and crop-wild sunflowers. *American Journal of Botany* **88**: 623-627.

Álvarez-Añorve M, Quesada M, and De la Barrera E. 2008. Remote sensing for plant functional groups detection: Physiology, ecology, and spectroscopy in tropical systems. In: Kalacska M and Sánchez-Azofeifa GA (eds.) *Hyperspectral Remote Sensing of Tropical and Sub-Tropical Forests*. CRC Press.

Anderson EF. *The Cactus Family*. Timber Press.

Arizmendi MC, Domínguez CA, and Dirzo R. 1996. The role of an avian nectar robber and of hummingbird pollinators in the reproduction of two plant species. *Functional Ecology* **10**: 119-127.

Avitia García E and Castillo González AM. 2007. *Desarrollo Floral en Frutales*. Universidad Autónoma Chapingo, Chapingo.

Azad AK, Sawa Y, Ishikawa T, and Shibata H. 2007. Temperature-dependent stomatal movement in tulip petals controls water transpiration during flower opening and closing. *Annals of Applied Biology* **150**: 81-87.

Ayala-Berdón J, Schondube JE, Stoner KE, Rodríguez-Peña N, and Martínez del Rio C. 2008. The intake responses of three species of leaf-nosed Neotropical bats. *Journal of Comparative Physiology B* **178**: 477-485.

Ayala-Cordero G, Terrazas T, López-Mata L, and Trejo C. 2004. Variación en el tamaño y peso de la semilla y su relación con la germinación en una población. *Interciencia* **29**: 692-697.

Ayala-Cordero G, Terrazas T, López-Mata L, and Trejo C. 2006. Morpho-anatomical changes and photosynthetic metabolism of *Stenocereus beneckei* seedlings under soil water deficit. *Journal of Experimental Botany* **57**: 3165-3174.

Badeck FW, Bondeay A, Böttcher K, Doktor D, Lucht W, Schaber J, and Sitch S. 2004. Responses of spring phenology to climate change. *New Phytologist* **162**: 295-309.

Baker HG and Baker I. 1982. Chemical constituents of nectar in relation to pollination mechanisms and phylogeny. In: Niteki MH (ed.) *Biochemical Aspects of Evolutionary Biology*. University of Chicago Press, Chicago. Pp. 131-171.

Baker HG and Baker I. 1983. Floral nectar constituents in relation to pollinator type. In: Jones CE and Little RJ (eds.) *Handbook of Experimental Pollination Biology*. Van Nostrand Reinhold, New York. Pp. 117-141.

Baker HG, Baker I, and Hodges SA. 1998. Sugar composition of nectars and fruits consumed by birds and bats in the tropics and subtropics. *Biotropica* **30**: 559-586.

Ball NG. 1933. A physiological investigation of the ephemeral flowers of *Turnera ulmifolia* L. var. *elegans* Urb. *New Phytologist* **32**: 13-36.

Barnes K, Nicolson SW, and van Wyk BE. 1995. Nectar sugar composition in *Erica*. *Biochemical Systematic Ecology* **23**: 419-423.

Baskin CC and Baskin JM. 1998. *Seeds: Ecology, Biogeography, and Evolution of Dormancy and Germination*. Academic Press, San Diego.

Bazzaz FA and Fajer ED. 1992. Plant life in a CO_2 rich environment. *Scientific American* **266**: 68-74.

Behboudian MH, Lawes GS, and Griffiths KM. 1994. The influence of water deficit on water relations, photosynthesis and fruit growth in Asian pear (*Pyrus serotina* Rehd.). *Scientia Horticulturae* **60**: 89-99.

Billings WD. 1985. The historical development of physiological plant ecology. In: Chabot BF and Mooney HA (eds.) *Physiological Ecology of North American Plant Communities*. Chapman and Hall, New York. Pp. 1-15.

Bondada BR, Matthews MA, Shackel KA. 2005. Functional xylem in the post-veraison grape berry. *Journal of Experimental Botany* **56:** 2949-2957.

Bonfil C. 1998. The effects of seed size, cotyledon reserves, and herbivory on seed-ling survival and growth in *Quercus rugosa* and *Q. laurina* (Fagaceae). *American Journal of Botany* **85:** 79-87.

Borchert R. 1983. Phenology and control of flowering in tropical trees. *Biotropica* **15:** 81-89.

Borchert R. 1994. Soil and stem water storage determine phenology and distribu-tion of tropical dry forest trees. *Ecology* **75:** 1437-1449.

Borchert R. 1998. Responses of tropical trees to rainfall seasonality and its long-term changes. *Climatic Change* **39:** 381-393.

Borchert R and Rivera G. 2001. Photoperiodic control of seasonal development and dormancy in tropical stem-succulent trees. *Tree Physiology* **21:** 213-221.

Borchert R, Meyer SA, Felger RS, and Porter-Bolland L. 2004. Environmental con-trol of flowering periodicity in Costa Rican and Mexican tropical dry forests. *Global Ecology and Biogeography* **13:** 409-425.

Borchert R, Robertson K, Schwartz MD, and Williams-Linera G. 2005. Phenology of temperate trees in tropical climates. *International Journal of Biometeorol-ogy* **50:** 57-65.

Bradshaw WE and Holzappfel CM. 2007. Genetic response to rapid climate change: it's seasonal timing that matters. *Molecular Ecology* **17:** 157-166.

Brugger KE and Nelms CO. 1991. Sucrose avoidance by American Robins (*Turdus migratorius*): Implications for control of bird damage in fruit crops. *Crop Pro-tocols 10:* 455-460.

Brugger KE, Nol P, and Phillips CI. 1993. Sucrose repellency to European starlings: Will high-sucrose cultivars deter bird damage to fruit? *Ecological Applications* **3:** 256-261.

Bruneau A. 1997. Evolution and homology of bird pollination syndromes in *Erythrina* (Leguminosae). *American Journal of Botany* **84:** 54-71.

Bughart M and Riederer M. 2003. Ecophysiological relevance of cuticular transpi-ration of deciduous and evergreen plants in relation to stomatal closure and leaf water potential. *Journal of Experimental Botany* **54:** 1941-1949.

Byrne DH and Bacon TA. 1992. Chilling accumulation: its importance and estima-tion. *The Texas Horticulturist* **18:** 5, 8-9.

Camm EL and Towers GHN. 1973. Phenylalanine Ammonia Lyase. *Phytochemistry* **12:** 961-973.

Casas A, Valiente-Banuet A, Rojas-Martínez A, and Dávila P. 1999. Reproductive biology and the process of domestication of the columnar cactus *Stenocereus stellatus* in Central Mexico. *American Journal of Botany* **86:** 534-542.

Chailakhyan MK and Krikorian AD. 1975. Forty years of research on the hormonal basis of plant development: Some personal reflections. *The Botanical Review* **41:** 1-29.

Chalcoff VR, Aizen MA, and Galetto L. 2006. Nectar concentration and composition of 26 species from the temperate forest of South America. *Annals of Botany* **97:** 413-421.

Chapotin SM, Holbrook NM, Morse SR, and Gutiérrez MV. 2003. Water relations of tropical dry forest flowers: pathways for water entry and the role of extracel-lular polysaccharides. *Plant, Cell and Environment* **26:** 623-630.

Chetelat RT, Klann E, DeVera JW, Yelle S, and Bennett AB. 1993. Inheritance and genetic mapping of fruit sucrose accumulation in *Lycopersicon chmielewskii*. *Plant Journal* **4:** 643-650.

Chuine I, Yiou P, Viovy N, Seguin B, Daux V, and Ladurie ElR. 2004. Grape ripen-ing as past climate indicator. *Nature* **432:** 289-290.

Corbesier L and Coupland G. 2006. The quest for florigen: a review of recent progress. *Journal of Experimental Botany* **57:** 3395-3403.

Corbesier L, Vincent C, Jang S, Fornara F, Fan Q, Searle I, Giakountis A, Farrona S, Gissot L, Turnbull C, and Coupland G. 2007. FT protein movement contributes to long-distance signaling in floral induction in *Arabidopsis*. *Science* **316:** 1030-1033.

Cotton PA. 2008. Avian migration phenology and global climate change. *Proceedings of the National Academy of Science* **100:** 12219-12222.

Cronquist A. 1977. *Introducción a la Botánica*. Compañía Editorial Continental, México.

Cuautle M and Rico-Gray V. 2003. The effect of wasps and ants on the reproductive success of the extrafloral nectaried *Turnera ulmifolia* (Turneraceae). *Functional Ecology* **17:** 417-423.

Darwin CR. 1862. *On the Various Contrivances by which British and Foreign Orchids are Fertilised by Insects, and on the Good Effects of Intercrossing*. John Murray. Digital version produced by van Wyhe J (2003), available online at http://darwin-online.org.uk/.

De la Barrera E and Andrade JL. 2005. Challenges to plant megadiversity: How environmental physiology can help. *New Phytologist* **5-8.**

De la Barrera E and Nobel PS. 2003. Physiological ecology of seed germination for the columnar cactus *Stenocereus queretaroensis*. *Journal of Arid Environments* **53:** 297-306.

De la Barrera E and Nobel PS. 2004a. Nectar: properties, floral aspects, and speculations on origin. *Trends in Plant Science* **9:** 65-69.

De la Barrera E and Nobel PS. 2004b. Carbon and water relations for developing fruits of *Opuntia ficus-indica* (L.) Miller, including effects of drought and gibberellic acid. *Journal of Experimental Botany* **55:** 719-729.

del-Val E and Crawley MJ. 2005. Are grazing increaser species better tolerators than decreasers? An experimental assessment of defoliation tolerance in eight British grassland species. *Journal of Ecology* **93:** 1005-1016.

Diamond J. 1997. *Guns, Germs, and Steel: the Fates of Human Societies*. Norton.

Diamond J. 2002. Evolution, consequences and future of plant and animal domestication. *Nature* **418:** 700-707.

Downs CT. 1997. Sugar digestion efficiencies of Gurney's sugarbirds, malachite sunbirds, and black sunbirds. *Physiological Zoology* **70:** 93-99.

Dupont YL, Hansen DM, Rasmussen JT, and Olesen JM. 2004. Evolutionary changes in nectar sugar composition associated with switches between bird and insect pollination: The Canarian bird- flower element revisited. *Functional Ecology* **18:** 670-676.

Elias TS, Rozich WR, and Newcombe L. 1975. The foliar and floral nectaries of *Turnera ulmifolia* L. *American Journal of Botany* **62:** 570-576.

Endress PK. 2001. The flowers in extant basal angiosperms and inferences on ancestral flowers. *International Journal of Plant Sciences* **162:** 1111-1140.

Etcheverry AV and Trucco Alemán CE. 2006. Reproductive biology of *Erythrina falcata* (Fabaceae: Papilionoidea). *Biotropica* **37:** 54-63.

Fischer BD and Cash-Clark CE. 2000. Gradients in water potential and turgor pressure along the translocation pathway during grain filling in normally watered and water-stressed wheat plants. *Plant Physiology* **123:** 139-147.

Fleming PA, Hartman-Bakken B, Lotz CN, and Nicolson SW. 2004. Concentration and temperature effects on sugar intake and preferences in a sunbird and a hummingbird. *Funcional Ecology* **18:** 223-232.

Fleming PA, Hofmeyr SD, and Nicolson SW. 2007. Role of insects in the pollination of *Acacia nigrescens* (Fabaceae). *South African Journal of Botany* **73:** 49-55.

Fosket DE. 1994. *Plant Growth and Development: A Molecular Approach*. Academic press.

Foster SA. 1986. On the adaptive value of large seeds for tropical moist forest trees: a review and synthesis. *Botanical Review* 52: 260-299.

Franco AC and Nobel PS. 1988. Interactions between seedlings of *Agave deserti* and the nurse plant *Hilaria rigida*. *Ecology* 69: 1731-1740.

Franco AC and Nobel PS. 1989. Effect of nurse plants on the microhabitat and growth of cacti. *Journal of Ecology* 77: 870-886.

Franke E, Hackson S, and Nicolson S. 1998. Nectar sugar preferences and absorption in a generalist African frugivore, the Cape White-eye *Zosterops pallidus*. *Ibis* 140: 501-506.

Frankie GW, Barker HG, and Opler PA. 1974. Comparative phenological studies of trees in tropical wet and dry forests in the lowlands of Costa Rica. *Journal of Ecology* 62: 881-919.

Frey FM. 2007. Phenotypic integration and the potential for independent color evolution in a polymorphic spring ephemeral. *American Journal of Botany* 94: 437-444.

Galen C, Sherry RA, and Carroll AB. 1999. Are flowers physiological sinks or faucets? Costs and correlates of water use by flowers of *Polemonium viscosum*. *Oecologia* 118: 461-470.

Galetto L and Bernardello G. 2003. Nectar sugar composition in angiosperms from Chaco and Patagonia (Argentina): an animals visitors's matter? *Plant Systematics and Evolution* 238: 69-86.

Galetto L, Bernardello G, Isele IC, Vesprini J, Speroni G, and Berduc A. 2000. Reproductive biology of *Erythrina crista-galli* (Fabaceae). *Annals of the Missouri Botanical Garden* 87: 127-145.

Gallardo-Vásquez JC and De la Barrera E. 2007. Environmental and ontogenetic influences on growth, photosynthesis, and survival for young pitayo (*Stenocereus queretaroensis*) seedlings. *Journal of the Professional Association for Cactus Development* 9: 118-135.

García Acosta V, Pérez Zevallos JM, Molina del Villar A. 2004. *Desastres agrícolas en México. Catálogo Histórico. I. Época Prehispánica y Colonia 958-1822*. Fondo de Cultura Económica.

García de Cortázar V and Nobel PS. 1992. Biomass and fruit production for the prickly pear cactus, *Opuntia ficus-indica*. *Journal of the Americal Society of Horticultural Science* 117: 558-562.

García-Martínez JL and Hedden P. 1997. Gibberellins and fruit development. In: Tomás-Barberán FA and Robins RJ (eds.) *Phytochemistry of Fruit and Vegetables*. Clarendon Press, Oxford. Pp. 263-285.

Garduño R. 1997. Past and future climates simulated with the Adem thermodynamic model. *Quaternary International* 43/44: 19-24.

Gillaspy G, Ben-David H, and Gruissem W. 1993. Fruits: a developmental perspective. *Plant Cell* 5: 1439-1451.

Goldstein G and Nobel PS. 1991. Changes in osmotic pressure and mucilage during low-temperature acclimation of *Opuntia ficus-indica*. *Plant Physiology* 97: 954-961.

Green PT and Juniper PA. 2004. Seed mass, seedling herbivory and the reserve effect in tropical rainforest seedlings. *Functional Ecology* 18: 539-547.

Guerenstein P, Yepez EA, van Haren J, Williams DG, Hildebrand J. 2004. Floral CO_2 emissions may signal food abundance to nectar-feeding moths. *Naturwissenschaften* 91: 329-333.

Guichard S, Gary C, Leonardi C, and Bertin N. 2005. Analysis of growth and water relations of tomato fruits in relation to air vapor pressure deficit and plant fruit load. *Journal of Plant Growth Regulation* **24:** 201-213.

Haig D and Westoby M. 1988. Inclusive fitness, seed resources, and maternal care. In: Lovett Doust J and Lovett Doust L (eds.) *Plant Reproductive Ecology: Patterns and Strategies.* Oxford University Press, New York. 60-79.

Hainsworth FR and Wolf LL. 1976. Nectar characteristics and food selection by hummingbirds. *Oecologia* **25:** 101-113.

Hanley ME and Fegan EL. 2007. Timing of cotyledon damage affects growth and flowering in mature plants. *Plant, Cell and Environment* **30:** 812-819.

Hanley ME and May OC. 200. Cotyledon damage at the seedling stage affects growth and flowering potential in mature plants. *New Phytologist* **169:** 243-250.

Harper JL, Clatworthy JN, McNaughton IH, and Sagar GR. 1961. The evolution and ecology of closely related species living in the same area. *Evolution* **15:** 209-227.

Harper JL, Williams JT, and Sagar GR. 1965. The behaviour of seeds in soil. I. The heterogeneity of soil surfaces and its role in determining the establishment of plants from seed. *The Journal of Ecology* **53:** 273-286.

Hernández A and Martínez del Rio C. 1992. Intestinal disaccharidases in five species of Phillostomid bats with contrasting feeding habits. *Comparative Biochemistry and Physiology* **103B:** 105-111.

Huang XM, Huang HB, and Gao FF. 2000. The growth potential generated in citrus fruit under water stress and its relevant mechanisms. *Scientia Horticulturae* **83:** 227-240.

Hunt Jr. ER, Rock BN, and Nobel PS. 1987. Measurement of leaf relative water content by infrared reflectance. *Remote Sensing of Environment* **22:** 429-435.

Hura T, Grzesiak S, Hura K, Thiemt E, Tokarz K, and Wedzony M. 2007. Physiological and biochemical tools useful in drought-tolerance detection in genotypes of winter triticale: Accumulation of ferulic acid correlates with drought tolerance. *Annals of Botany* **100:** 767-775.

Hura T, Hura K, and Grzesiak S. 2008. Contents of total phenolics and ferulic acid, and PAL activity during water potential changes in leaves of maize single-cross hybrids of different drought tolerance. *Journal of Agronomy and Crop Science* **194:** 104-112.

Ibáñez I and Schupp EW. 2001. Positive and negative interactions between environmental conditions affecting *Cercocarpus ledifolius* seedling survival. *Oecologia* **129:** 543-550.

Inglese P, Chessa I, La Mantia T, and Nieddu G. 1998. Evolution of endogenous gibberellins at different stages of flowering in relation to return bloom of cactus pear (*Opuntia ficus-indica* L. Miller). *Scientia Horticulturae* **73:** 45-51.

Jackson S, Nicolson SW, and Lotz CN. 1998. Sugar preferences and "side bias" in Cape Sugarbirds and Lesser Double-collared Sunbirds. *Auk* **115:** 156-165.

Janzen HD. 1967. Synchronizatin of sexual reproduction of trees withing the dry season in Central America. *Evolution* **21:** 620-637.

Jordan PW and Nobel PS. 1979. Infrequent establishment of seedlings of *Agave deserti* (Agavaceae) in the Northwestern Sonoran Desert. *American Journal of Botany* **66:** 1079-1084.

Karasov WH and Cork SJ. 1996. Glucose absorption by a nectarivorous bird: The passive pathway is paramount. *American Journal of Physiology* **267:** G16-G26.

Kennedy PG, Hausmann, Wenk EH, and Dawson TE. 2004. The importance of seed reserves for seedling performance: an integrated approach using morphological, physiological, and stable isotope techniques. *Oecologia* **141:** 547-554.

Kitajima K. 2002. Do shade-tolerant tropical tree seedlings depend longer on seed reserves? Functional growth analysis of three Bignoniaceae species. *Functional Ecology* **16:** 433-444.

Kitajima K. 2003. Impact of cotyledon and leaf removal on seedling survival in three tree species with contrasting cotyledon functions. *Biotropica* **35:** 429-434.

Lamont BB and Groom PK. 2002. Green cotyledons of two *Hakea* species control seedling mass and morphology by supplying mineral nutrients rather than organic compounds. *New Phytologist* **153:** 101-110.

Larcher W. 2003. *Physiological Plant Ecology: Ecophysiology and Stress Physiology of Functional Groups* (4th edn). Springer-Verlag, New York.

Lee TD. 1988. Patterns of fruit and seed production. In: Lovett-Doust J and Lovet Doust L (eds.) *Plant Reproductive Ecology: Patterns and Strategies*. Oxford University Press. Pp. 179-202.

Leirana-Alcocer J and Parra-Tabla V. 1999. Factors affecting the distribution, abundance and seedling survival of *Mammillaria gaumeri*, an endemic cactus of coastal Yucatán, Mexico. *Journal of Arid Environments* **41:** 421-428.

Lobo JA, Quesada M, Stoner KE, Fuchs EJ, Herrerias-Diego Y, Rojas-Sandoval J, and Saborio-Rodríguez G. 2003. Factors affecting phenological patterns of Bombacaceous trees in seasonal forests in Costa Rica and Mexico. *American Journal of Botany* **90:** 1054-1063.

Lorch J. 1978. The discovery of nectar and nectaries and its relation to views on flowers and insects. *Isis* **69:** 514-533.

Lotz CN and Nicolson SW. 1996. Sugar preferences of a nectarivorous passerine bird, the lesser double-collared sunbird (*Nectarinia chalybea*). *Functional Ecology* **10:** 360-365.

Lotz CN and Schondube JE. 2006. Sugar preferences in nectar- and fruit-eating birds: Behavioral patterns and physiological causes. *Biotropica* **38:** 3-15.

Lu P, Quiang Y, Liu J, Lee X. 2006. Advance of tree-flowering dates in response to urban climate change. *Agricultural and Forest Meteorology* **138:** 120-131.

Magaña V, Conde C, Sánchez O, and Gay C. 1997. Assessment of current and future regional climate scenarios for Mexico. *Climate Research* **9:** 107-114.

Marcati CR, Dias Milanez CE, and Rodrigues Machado S. 2008. Seasonal development of secondary xylem and phloem in *Schizolobium parahyba* (Vell.) Blake (Leguminosae: Caesalpinioideae). *Trees-Structure and Function* **22:** 3-12.

Martínez del Rio C. 1990a. Sugar preferences in hummingbirds: The influence of subtle chemical differences on food choice. *Condor* **92:** 1022-1030.

Marínez del Rio C. 1990b. Dietary, phylogenetic, and ecological correlates of intestinal sucrase and maltase activity in birds. *Physiological Zoology* **63:** 987-1011.

Martínez del Rio C and Stevens BR. 1989. Physiological constraint on feeding behavior: Intestinal membrane disaccharidases of the starling. *Science* **243:** 794-796.

Martínez del Rio C, Daneke D, and Andreadis PT. 1988. Physiological correlatoes of preference and aversion for sugars in three species of birds. *Physiological Zoology* **61:** 222-229.

Martínez del Rio C, Karasov WH, and Levey DJ. 1989. Physiological basis and ecological correlates of sugar preferences in a North American frugivore, the Cedar Waxwing. *Auk* **106:** 64-71.

Martínez del Rio C, Baker HG, and Baker I. 1992. Ecological and evolutionary implications of digestive processes: Bird preferences and the sugar constituents of floral nectar and fruit pulp. *Experientia* **48:** 544-550.

Mendoza VM, Villanueva EE, and Adem J. 2002. Simulation of the annual thermal and hydrological cycle in Mexico. *Geofísica Internacional* **41:** 163-178.

Menzel A. 2002. Phenology: its importance to the global change community. *Climatic Change* **54:** 379-385.

Meyerowitz EM. 1994. The genetics of flower development. *Scientific American* **271:** 56-65.

Molbo D, Machado CA, Sevenster JG, Keller L, Herre EA. 2003. Cryptic species of fig-pollinating wasps: Implications for the evolution of the fig-wasp mutualism, sex allocation, and presicion of adaptation. *Proceedings of the National Academy of Sciences.* **100:** 5867-5872.

Moles AT and Westoby M. 2004a. Seedling survival and seed size: a synthesis of the literature. *Journal of Ecology* **92:** 372-383.

Moles AT and Westoby M. 2004b. What do seedlings die from and what are the implications for evolution of seed size. *Oikos* **106:** 193-199.

Moles AT, Ackerly D, Webb CO, Tweddle JC, Dickie JB, Pitman AJ, and Westoby M. 2005. Factors that shape seed mass evolution. *Proceedings of the National Academy of Sciences* **102:** 10540-10544.

Molina-Freaner F and Eguiarte LE. 2003. The pollination biology of two paniculate *Agaves* (Agavaceae) from northwestern Mexico: contrasting roles of bats as pollinators. *American Journal of Botany* **90:** 1016-1024.

Muller-Rober B, Sonnewald W, and Willmitzer L. 1992. Inhibition of the ADP-glucose pyrophosphorylase in transgenic potatoes leads to sugar storing tubers and influences tuber formation and expression of tuber storage protein genes. *EMBO Journal* **11:** 1229-1238.

Munguia-Rosas MA and Sosa VJ. 2007. Nurse plant *vs.* nurse objects: effects of woody plants and rocky cavities on the recruitment of the *Pilosocereus leucocephalus* columnar cactus. *Annals of Botany* **101:** 175-185.

Myers JA and Kitajima K. 2007. Carbohydrate storage enhances seedling shade and stress tolerance in a neotropical forest. *Journal of Ecology* **95:** 383-395.

Nagy KA, Odell DK, and Seymour RS. 1973. Temperature regulation by the inflorescence of philodendron. *Science* **178:** 1195-1197.

Nepi M, Pacini E, Willemse MTM. 1996. Nectary biology of *Cucurbita pepo*: ecophysiological aspects. *Acta Botanica Neerlandica* **45:** 41-51.

Nerd A and Neuman PM. 2004. Phloem water transport maintains stem growth in a drought-stressed crop cactus (*Hylocereus undatus*). *Journal of the American Society for Horticultural Science* **129:** 486-490.

Nicolson SW. 2002. Pollination by passerine birds: why are the nectars so dilute? *Comparative Biochemistry and Physiology. Part B.* **131:** 645-652.

Nobel PS. 1977a. Water relations of flowering of *Agave deserti.Oecologia* **27:** 117-133.

Nobel PS. 1977b. Water relations and photosynthesis of a barrel cactus, *Ferocactus acanthodes*, in the Colorado desert. *Oecologia* **27:** 117-133.

Nobel PS. 1983. Nutrient levels in cacti in relation to nocturnal acid accumulation and growth. *American Journal of Botany* **70:** 1244-1253.

Nobel PS. 1988. *Environmental Biology of Agaves and Cacti.* Cambridge University Press.

Nobel PS. 1989. Temperature, water availability, and nutrient levels at various soil depths: Consequences for shallow-rooted desert succulents, including nurse plant effects. *American Journal of Botany* **76:** 1486-1492.

Nobel PS. 2005. *Physicochemical and Environmental Plant Physiology.* Academic Press/Elsevier.

Nobel PS and Castañeda M. 1998. Seasonal, light, and temperature influences on organ initiation for unrooted cladodes of the prickly pear cactus *Opuntia ficus-indica. Journal of the American Society for Horticultural Science* **123:** 47-51.

Nobel PS and De la Barrera E. 2000. Carbon and water balances for young fruits of platyopuntias. *Physiologia Plantarum* **109**: 160-166.

Nobel PS and De la Barrera E. 2003. Tolerances and acclimation to low and high temperatures for cladodes, fruits, and roots of a widely cultivated cactus, *Opuntia ficus-indica*. *New Phytologist* **157**: 297-306.

Nobel PS and De la Barrera E. 2004. CO_2 uptake by the cultivated hemiepiphytic cactus, *Hylocereus undatus*. *Annals of Applied Biology* **144**: 1-8.

Nobel PS, Miller PM, and Graham EA. 1992. Influence of rocks on soil temperature, soil water potential, and rooting patterns for desert succulents. *Oecologia* **92**: 90-96.

Nobel PS, Andrade JL, Wang N, and North GB. 1994. Water potentials for developing cladodes and fruits of a succulent plant, including xylem-versus-phloem implications for water movement. *Journal of Experimental Botany* **45**: 1801-1807.

Oh M and Lee I. 2007. Historical perspective breakthroughs in flowering field. *Journal of Plant Biology* **50**: 249-256.

Ortiz YD. 1999. *Pitahaya: a new crop for Mexico*. Noriega-Limusa.

Pacini E and Nicolson SW. 2007. Introduction. In: Nicolson SW, Nepi M, and Pacini E (eds.) *Nectaries and Nectar*. Springer. Pp. 1-18.

Parker DE. 2004. Large-scale warming is not urban. *Nature* **432**: 290.

Pate J, Shedley E, Arthur D, and Adams M. 1998. Spatial and temporal variations in phloem sap compostion of plantation-grown *Eucalyptus globulus*. *Oecologia* **117**: 312-322.

Paz H and Martínez-Ramos M. 2003. Seed mass and seedling performance within eight species of *Psychotria* (Rubiaceae). *Ecology* **84**: 439-450.

Paz H, Mazer SJ, and Martínez-Ramos M. 2005. Comparative ecology of seed mass in *Psychotria* (Rubiaceae): within- and between-species effects of seed mass on early performance. *Functional Ecology* **19**: 707-718.

Perret M, Chautems A, Spichiger R, Peixoto M, and Savolainen V. 2001. Nectar sugar composition in relation to pollination syndromes in Sinningieae (Gesneriaceae). *Annals of Botany* **87**: 267-273.

Peters E, Martorell C, and Ezcurra E. 2008. Nurse rocks are more important than nurse plants in determining the distribution and establishment of globose cacti (*Mammillaria*) in the Tehuacán Valley, Mexico. *Journal of Arid Environments* **72**: 593-601.

Pimienta Barrios E. 1990. *El Nopal Tunero*. Universidad de Guadalajara, Guadalajara.

Pimienta-Barrios E and del Castillo R. 2002. Reproductive biology. In: Nobel PS (ed.) *Cacti: Biology and Uses*. University of California Press. Pp. 75-90.

Pimienta Barrios E, Muñoz Urias A, García D. 1990. Efecto de la fertilización química y orgánica en la diferenciación de cladodios reproductivos y flores en nopal (*Opuntia ficus-indica*) tunero. *Tiempos de Ciencia* **20**: 56-62.

Pimienta-Barrios E and Nobel PS. 1995. Reproductive characteristics of pitayo (*Stenocereus queretaroensis* (Weber) Buxbaum) and their relationships with soluble sugars and irrigation. *Journal of the American Society for Horticultural Science* **120**: 1082-1086.

Pimienta-Barrios Eu, Pimienta-Barrios En, Salas-Galván ME, Sañudo-Hernández J, and Nobel PS. 2002. Growth and reproductive characteristics of the columnar cactus *Stenocereus queretaroensis* and their relationships with environmental factors and colonization by arbuscular mycorrhizae. *Tree Physiology* **22**: 667-674.

Pimienta-Barrios Eu, Pimienta-Barrios En, and Nobel PS. 2004. Ecophysiology of the pitayo de Querétaro (*Stenocereus queretaroensis*). *Journal of Arid Environments* **59**: 1-17.

Pimienta-Barrios E and Ramírez-Hernández BC. 2003. Phenology, growth, and physiological responses to light for ciruela mexicana (*Spondias purpurea* L.). *Economic Botany* **57**: 481-490.

Pimienta-Barrios Eu, Zañudo-Hernández J, Yepez EA, Pimienta-Barrios En, and Nobel PS. 2000. Seasonal variation of net CO_2 uptake for cactus pear (*Opuntia ficus-indica*) and pitayo (*Stenocereus queretaroensis*) in a semiarid environment. *Journal of Arid Environments* **44**: 73-83.

Prichard J, Winch S, and Gould N. 2000. Phloem water relations and root growth. *Australian Journal of Plant Physiology* **27**: 539-548.

Rasmussen IA, Askergaard M, and Olesen JE. 2006. The Danish organic crop rotation experiment for cereal production 1997-2004. In: Raupp J, Pekrun C, Oltmanns M, and Köpke U (eds.) *Long-term Field Experiments in Organic Farming*. International Society of Organic Agriculture Research Scientific Series. Pp. 117-134.

Razem FA and Davis AR. 1999. Anatomical and ultrastructural changes of the floral nectary of *Pisum sativum* L. during flower development. *Protoplasma* **206**: 57-72.

Rhizopoulou S, Ioannidi E, Alexandredes N, and Argiropoulos A. 2006. A study of functional and structural traits of the nocturnal flowers of *Capparis spinosa* L. *Journal of Arid Environments* **66**: 635-647.

Rivera G and Borchert R. 2001. Induction of flowering in tropical trees by a 30-min reduction in photoperiod: evidence from field observations and herbarium specimens. *Tree Physiology* **21**: 201-212.

Roetzer T, Wittenzeller M, Haeckel H, and Nekovar J. 2000. Phenology in central Europe: differences and trends of spring phenophases in urban and rural areas. *International Journal of Biometeorology* **44**: 60-66.

Rogers WE Seimann E, and Lankau RA. 2003. Damage induced production of extrafloral nectaries in native and invasive seedlings of Chinese tallow tree (*Sapium sebiferum*). *American Midland Naturalist* **149**: 413-417.

Rodríguez-Peña N, Stoner KE, Schondube JE, and Martínez del Rio C. 2007. Effects of sugar composition and concentration on food selection by saussure's long-nosed bat (*Leptonycteris curasoae*) and the long-tongued bat (*Glossophaga soricina*). *Journal of Mammalogy* **88**: 1466-1474.

Rojas-Aréchiga M and Vázquez-Yanes C. 2000. Cactus seed germination: A review. *Journal of Arid Environments* **44**: 85-104.

Rudall PJ, Bateman RM, Fay MF, Eastman A. 2002. Floral anatomy and systematics of Alliaceae with particular reference to *Gilliesia*, a presumed insect mimic with strongly zygomorphic flowers. *American Journal of Botany* **89**: 1867-1883.

Ryugo K. 1986. Promotion and inhibition of glower initiation and fruit set by plant manipulation and hormones, a review. *Acta Horticulturae* **179**: 301-307.

Schondube JE and Martínez del Rio C. 2003. Concentration-dependent sugar preferences in nectar-feeding birds. Mechanisms and consequences. *Functional Ecology* **17**: 445-453.

Schondube JE and Martínez del Rio C. 2004. Sugar and protein digestion in flowerpiercers and hummingbirds. A comparative test of adaptive convergence. *Journal of Comparative Physiology B* **174**: 263-273.

Schondube JE, Herrera-M LG, and Martínez del Rio C. 2001. Diet and the evolution of digestion and renal function in phyllostomid bats. *Zoology* **104**: 59-73.

Schupp EW. 1988. Factors affecting post-dispersal seed survival in a tropical forest. *Oecologia* **76**: 525-530.

Singh DP, Jermakow AM, and Swain SM. 2002. Gibberellins are required for seed development and pollen tube growth in *Arabidopsis*. *The Plant Cell* **14**: 3133-3147.

Smith SDW, Ané C, and Baum DA. 2008. The role of pollinator shifts in the floral diversification of *Iochroma* (Solanaceae). *Evolution* **62**: 793-806.

Southwick EE. 1984. Photosynthate allocation to floral nectar: a neglected energy investment. *Ecology* **65**: 1775-1779.

Srivastava LM. 2002. *Plant Growth and Development: Hormones and Environment*. Academic Press.

Stephenson AG. 1981. Flower and fruit abortion: proximate causes and ultimate functions. *Annual Review of Ecology and Systematics* **12**: 253-279.

Stiles FG. 1976. Taste preferences, color preferences, and flower choice in hummingbirds. *Condor* **78**: 10-26.

Sung S and Amasino RM. 2004. Vernalization and epigenetics: how plants remember winter. *Current Opinion in Plant Biology* **7**: 4-10.

Swain SM, Reid JB, and Kamiya Y. 1997. Gibberellins are required for embryo growth and seed development in pea. *The Plant Journal* **112**: 1329-1338.

Taiz L and Zeiger E. 2002. *Plant Physiology. 3rd. edn.* Sinauer.

Temeles EJ and Kress WJ. 2003. Adaptation in a plant- hummingbird association. *Science* **300**: 630-633.

Theisen G and Melzer R. 2007. Molecular mechanisms underlying origin and diversification of the angiosperm flower. *Annals of Botany* 100: 603-619.

Thien LB, Azuma H, and Kawano S. 2000. New perspectives on the pollination biology of basal angiosperms. *International Journal of Plant Sciences* **161**: S225-S235.

Trainotti L, Tadiello A, Casadoro G. 2007. The involvement of auxin in the ripening of climacteric fruits comes of age: the hormone plays a role of its own and has an intense interplay with ethylene in ripening peaches. *Journal of Experimental Botany* **58**: 3299-3308.

Trolinder NL, McMichael BL, and Upchurch DR. 1993. Water relations of cotton flower petals and fruit. *Plant, Cell and Environment* **16**: 755-760.

Tsukaguchi T, Kawamitsu Y, Takeda H, Suzuki K, and Egawa Y. 2003. Water status of flower buds and leaves as affected by high temperature in heat-tolerant and heat-sensitive cultivars of snap bean (*Phaseolus vulgaris* L.). *Plant Production Science* **6**: 24-27.

Turk F, Fornara F, and Coupland G. 2008. Regulation and identity of Florigen: FLOWERING LOCUS T moves center stage. *Annual Review of Plant Biology* **59**: 573-598.

Turner RM, Alcorn SM, Olin G, and Booth JA. 1966. The influence of shade, soil, an water on Saguaro seedling establishment. *Botanical Gazette* **127**: 95-102.

Valiente-Banuet A and Ezcurra E. 1991. Shade as a cause of the association between the cactus *Neobuxbaumia tetetzo* and the nurse *Mimosa luisana*. *Journal of Ecology* **79**: 961-971.

Valiente-Banuet A and Verdú M. 2008. Temporal shifts from facilitation to competition occur between closely related taxa. *Journal of Ecology* **96**: 489-494.

Valiente-Banuet A, Santos Gally R, Arizmendi MC, Casas A. 2007. Pollination biology of the hemiepiphytic cactus *Hylocereus undatus* in the Tehuacán Valley, Mexico. *Journal of Arid Environments* **68**: 1-8.

Valiente-Banuet A, Vital Rumebe A, Verdú M, and Callaway R. 2006. Modern Quaternary plant lineages promote diversity through facilitation of ancient Tertiary lineages. *Proceedings of the National Academy of Sciences* **103**: 16812-16817.

van Doorn WG and van Meeteren U. 2003. Flowering opening and closure: a review. *Journal of Experimental Botany* **54**: 1801-1812.

van Wyk BE. 1993. Nectar sugar composition in southern African Papilionoideae (Fabaceae). *Biochemical Systematical Ecology* **21**: 271-277.

Wagner D and Kay A. 2002. Do extrafloral nectaries distract ants from visiting flowers? An experimental test of an overlooked hypothesis. *Evolutionary Ecology Research* **4**: 293-305.

Wang BC and Smith TB. 2002. Closing the seed dispersal loop. *Trends in Ecology and Evolution* **17**: 379-385.

Westoby M, Leishman M, and Lord J. 1996. Comparative ecology of seed size and dispersal. *Philosophical Transactions of the Royal Society, London B* **351**: 1309-1318.

Wetterer AL, Rockman MV, and Simmons NB. 2000. Phylogeny of phyllostomid bats (Mammalia: Chiroptera): data from diverse morphological systems, sex chromosomes, and restriction sites. *Bulletin of the American Museum of Natural History* **248**: 1-200.

Wikelski M and Cooke SJ. 2006. Conservation physiology. *Trends in Ecology and Evolution* **21**: 38-46.

Wolf D. 2006. Nectar sugar composition and volumes of 47 species of Gentianales from a Southern Ecuadorian montane forest. *Annals of Botany* **97**: 767-777.

Wright SJ, Carrasco C, Calderón O, and Paton S. 1999. The El Niño southern oscillation, variable fruit production, and famine in a tropical forest. *Ecology* **80**: 1632-1647.

Yamada H, Takechi K, Hoshi A, and Amano S. 2004. Comparison of water relations in watercored and non-watercored apples induced by fruit temperature treatment. *Scientia Horticulturae* **99**: 309-318.

Yelle S, Chetelat RT, Dorais M, DeVerna JW, Bennet AB. 1991. Sink metabolism in tomato fruit. IV. Genetic and biochemical analysis of sucrose accumulation. *Plant Physiology* **95**: 1026-1035.

Young DR and Nobel PS. 1986. Predictions of soil-water potentials in the northwestern Sonoran desert. *Journal of Ecology* **74**: 143-154.

Zegbe Domínguez JA. 2005. Seasonal changes of nutriments in leaves and fruit drop of 'Criollo' peach in Zacatecas, Mexico. *Revista Fitotecnia Mexicana* **28**: 71-75.

Chapter 13

PRECIPITATION PULSES AND ECOSYSTEM CARBON AND WATER EXCHANGE IN ARID AND SEMI-ARID ENVIRONMENTS

Enrico A. Yepez and David G. Williams

Perspectives in Biophysical Plant Ecophysiology: A Tribute to Park S. Nobel, pp. 337-361
Edited by: E. De la Barrera and W.K. Smith
© 2009 by The Authors
Book Compilation © 2009 Universidad Nacional Autónoma de México

Introduction

Water availability is the principal factor controlling population, community, and ecosystem dynamics in arid and semi-arid environments (Schwinning *et al.* 2004). Precipitation in these water-limited environments comes in discreet pulses varying in size, frequency and intensity (Loik *et al.* 2004), thereby producing unique hydrological patterns that constrain the location and residence time of soil water available for plant uptake (Noy-Meir 1973). Due to high rates of evapotranspiration, the residence time of soil water during the growing season is short. Yet, these moisture pulses can trigger a suite of short-lived, but very active biological processes that, integrated through time, can shape many aspects of plant physiology, ecosystem function, and biogeochemistry (Smith and Nobel 1986; Austin *et al.* 2004; Huxman *et al.* 2004c). The ecological sensitivity to pulsed inputs of moisture varies tremendously among different ecosystem components and across different levels of biological organization (Schwinning and Sala 2004). Further complicating these patterns is that the short-term dynamics of soil water availability in the landscape varies along gradients of vegetation cover (Breshears *et al.* 1998; Midwood *et al.* 1998) and soil texture and topography (McAuliffe 2003). At a local scale, rainfall interception and alteration of microclimate by plant canopies (Carlyle-Moses 2004), biotic modification of soil hydraulic characteristics (Johnson-Maynard *et al.* 2002) and root water uptake and redistribution (Hultine *et al.* 2004) enhance spatial and temporal variability of soil water.

Variation of soil moisture availability in space and time promotes coexistence of plant species that differ in their modes of water use and uptake (Burgess 1995). Plant photosynthetic carbon assimilation (A) leads to an inevitable loss of water via transpiration (T) through stomatal pores (Nobel 1999). Thus, attributes that increase plant water-use efficiency (WUE; A/T) also may impose limitations on carbon gain. Consequently, plants continuously adjust WUE in response to changing evaporative demand by the atmosphere and availability of resources in the soil (Nobel 1999). The way in which plants optimize this photosynthesis to transpiration tradeoff illustrates key mechanisms of adaptation and survival in semi-arid environments (Schwinning and Ehleringer 2001). Water-use efficiency is also a useful concept when applied to the productivity potential of an entire ecosystem (Ponton *et al.* 2006). But at the ecosystem scale, temporal dynamics of net carbon gain and water exchange are considerably more complex because soil organisms and other biophysical processes also contribute to ecosystem gas exchange (Huxman *et al.* 2004c).

At large spatial scales, the synthetic parameter rain-use efficiency (RUE = annual net primary production/mean annual precipitation; Le Houerou 1984) is widely used as an integrator of processes and constraints on net biological productivity (Lauenroth and Sala 1992; Prince *et al.* 1998; Paruelo *et al.* 1999; Huxman *et al.* 2004b). RUE declines across different major biomes as mean annual precipitation rises, but is fairly constant and high across most arid and semi-arid ecosystems (Lieth 1975; Huxman *et al.*

2004b). Such patterns imply the existence of a universal constraint imposed by water scarcity at low precipitation averages and more complex, multiple limiting factors (*e.g.*, light, nutrient availability) when water is abundant (Huxman *et al.* 2004b). Variation in RUE among different sites within water-limited arid and semi-arid environments is related to contrasting plant functional types (Lauenroth and Sala 1992), lags in physiological recovery following drought (Ogle and Reynolds 2004), interactions with other organisms nutrient cycles that respond to intra- and interannual variability of precipitation (Austin *et al.* 2004) and the level of ecosystem degradation (Ehleringer 2001).

In this chapter we provide an overview of ecophysiological processes that interact at the ecosystem scale to shape patterns of carbon and water exchange in semi-arid environments. The emphasis is on the short-term response of whole ecosystems to pulsed inputs of precipitation and the need to understand the dynamics of gas exchange fluxes associated with different ecosystem components. We then present some examples showing how stable isotope ratio measurements are used to trace component sources of ecosystem fluxes to provide insights into processes controlling ecosystem water-carbon interactions.

Vegetation responses to pulsed precipitation

Plants influence the ecosystem carbon and water cycles through physiological regulation of photosynthesis, respiration and transpiration and by their effects in modulating macroclimate. Furthermore, patterns of soil water uptake and use by plants in semi-arid ecosystems has influence in life form coexistence and therefore in ecosystem carbon and water relations. Plant effects on ecosystem carbon-water interactions are greatest soon after precipitation pulses when resource availability and metabolic activity is high. However, plants in arid and semi-arid regions are not always in an optimal state to take advantage of water and associated resources when they first become available (Schwinning and Sala 2004). Following rainfall inputs, plants can continue to function in a down-regulated state or invest internal resources to enhance root and leaf activity to exploit the new resources (Schwinning *et al.* 2003). Species-specific responses to threshold sizes of precipitation input determine variation in physiological dynamics within plant communities. Below these thresholds, plants respond passively (*e.g.*, opening stomata without energy investments) or do not respond at all to increased water availability, and above the threshold plants respond actively to acquire more resources (*e.g.*, root construction, enzyme production; Schwinning *et al.* 2002; Williams and Snyder 2003). Following optimization theory, plants that do not respond to pulses of precipitation incur no metabolic cost, but benefits for the plant are also marginal. However, energy investments necessary to fully exploit a pulse are risky since the return on the investment depends on soil moisture duration, which in semi-arid environments is highly variable and largely unpredictable (Schwinning *et al.* 2003; Williams and Snyder 2003; Loik *et al.* 2004).

Different plant lifeforms of semi-arid environments have evolved specialized spatial and temporal patterns of water uptake. Shallow rooted grasses and herbaceous dicots are capable of using frequent and short-duration water pulses, while extensively rooted woody plants generally use deep and more stable sources of water (Sala *et al.* 1989; Ehleringer 1991). This niche separation is found in several regions of the world (Scholes and Archer 1997), but may not apply to early developmental stages of woody plants (Weltzin and McPherson 1997). In general, the response of trees and shrubs to pulsed inputs of precipitation is more variable but overall less pronounced than in herbaceous species, since deeper soil horizons maintain more stable sources of water (Sala and Lauenroth 1982; Loik *et al.* 2004). Additionally, the way plants partition and use water resources vary according to the size of precipitation events (Schwinning *et al.* 2003). Small rainfall events (<5 mm) may favor shallow-rooted herbaceous species (Sala and Lauenroth 1982) and deeper-rooted plants likely are favored only by large events that wet deeper layers in the soil. Golluscio *et al.* (1998) observed that shrubs were responsive to large precipitation events only when precipitation was preceded by drought conditions, thus highlighting the relevance of moisture conditions preceding rainfall.

The extent of the response in trees or shrubs to precipitation pulses also varies according to particular ecophysiological characteristics. For example, rooting depths and functionality (Nobel 1994), size/age hydraulic constraints inherent to different stages of development (*e.g.*, Magnani *et al.* 2000; Eamus 2003) and canopy characteristics (*e.g.*, roughness; Tuzet *et al.* 1997) all affect the speed of recovery following rain events. A threshold-like response in size/age may trigger a transition from a precipitation pulse-driven ecosystem to one partially decupled from rain (Williams and Ehleringer 2000). In fact, in mature stages, trees may have the ability to avoid drought by accessing deeper sources of water (*e.g.*, phreatophytes) and thus decoupling tree function from inputs of rain (Scott *et al.* 2003).

In general, the ecophysiological responses of plants to infrequent inputs of water are variable and to a large extent controlled by antecedent conditions and lag effects (Ogle and Reynolds 2004). In their long-term analysis Lauenroth and Sala (1992) observed that annual NPP of the shortgrass steppe in North America had a coefficient of variation of 44%, and recovery of NPP took up to 4 years following drought. Yan *et al.* (2000) observed that patterns of water status and leaf-level gas exchange in shrubs from the Chihuahuan desert were a function of the length of the period preceding rainfall rather than to the pulse size when precipitation occurred. Similarly, shrubs and grasses from the Colorado Plateau did not respond equally to spring and summer pulse events due to conditions prior to the pulse and the physiological controls on gas exchange when the stress was released (Schwinning *et al.* 2002). Although not fully understood, it appears that the physiological and morphological condition of dominant vegetation in arid and semi-arid environments underlie thresholds and lags in ecosystem recovery following precipitation events (Schwinning and Sala 2004).

Ecosystem response to pulsed precipitation

General circulation models predict that precipitation intensity may increase during this century due to intensification of the hydrological cycle under warmer climates (Karl and Trenberth 2003). Such changes, in combination with regional and local climatic features and topographies, may result in modified precipitation regimes characterized by fewer precipitation days, longer periods between precipitation events and higher temporal variation. These changes will directly impact the structure and function of terrestrial ecosystems. In a mesic grassland Knapp et al. (2002) observed that altered precipitation frequencies (but not amount totals) affected key ecosystem processes to a degree comparable to that observed in experiments with elevated CO_2. Accordingly, arid and semi-arid ecosystems are likely to be more sensitive to changes in precipitation than to other global changes (Weltzin et al. 2003). The present and future trends of precipitation variability underscore the urgent need to understand ecosystem and component responses to moisture variability at an appropriate scale of integration. In order to understand how arid and semi-arid ecosystems respond to inputs of precipitation within the context of ecosystem science and global change studies, knowledge is needed on how plants and other organisms respond to such environmental controls as an integrated system.

Annual rain-use efficiency (RUE; Le Houerou 1984) is an important index relating carbon and water cycles at the ecosystem scale. However, this index lacks resolution on the contributing processes that occur over the course of individual wetting and drying cycles. Dynamics of carbon and water exchange at the ecosystem level following pulsed inputs of rain is an integration of several soil and plant metabolic reactions and biophysical constraints whose rates and magnitudes of response are shaped by multiple factors interacting at different scales (Huxman et al. 2004c). Recent measurements of ecosystem-scale gas exchange following pulsed precipitation show that species-specific plant and soil processes are independently influenced by pulse frequency/magnitude and inter-pulse duration but, as a whole, produced unique patterns of ecosystem gas exchange (Huxman et al. 2004a; Scott et al. 2006a). For example, carbon exchange following a pulse has two phases. Soon after the pulse, CO_2 is lost to the atmosphere because the net flux is dominated by a high CO_2 efflux from the soil associated with CO_2 mass displacement from soil pores and rapid metabolic up regulation of soil microbial populations (Xu et al. 2004; Jarvis et al. 2007). While, at some point thereafter, plants may recover their photosynthetic capacity and the net carbon flux of the ecosystem reflects a net carbon gain. Therefore, ecosystem responses to precipitation pulses are better described by the instantaneous patterns of net ecosystem carbon exchange (NEE) and evapotranspiration (ET; Hastings et al. 2005; Scott et al. 2006a). NEE is the balance of gross photosynthesis and plant and microbial respiration fluxes, expressed as:

$$NEE = P_{gross} - R_{plant} - R_{microbial} \qquad \text{(Eq. 13.1)}$$

where P_{gross} is the gross carbon gain of photosynthesis in the light, R_{plant} is the total respiratory CO_2 loss from the plant (about half of P_{gross}) and $R_{microbial}$ is the carbon loss through respiration by the soil microbiota. Evapotranspiration is the sum of plant transpiration (T) and soil evaporation (E) and may equal or exceed annual precipitation in arid and semi-arid environments (e.g., when groundwater is accessed). Furthermore, similar to the functional adaptations at the organismal level (Smith and Nobel 1986; Schwinning and Ehleringer 2001), the tradeoff of water for carbon at the ecosystem level (an ecosystem water-use efficiency index, EWUE = NEE/ET) may provide the basis for linking ecological and hydrological mechanisms to the productivity potential of a site. Although these parameters are integrative, they provide resolution on seasonal and short-term biological processes controlled by resource supply and climatic constraints (Chapin et al. 2002). However, further understanding of the controls on individual component fluxes at the appropriate spatial and temporal scales is needed to advance theory on ecosystem function. For example, assuming that run on and run off are minor components of the ecosystem water budget (likely in semi-arid basins; Wilcox et al. 2003), the short-term rain-use efficiency of a particular ecosystem can be interpreted as the product of three interrelated variables: the proportion of precipitation that is transpired by the plant community, the precipitation fraction that is lost as evaporated water from the soil (without stimulating photosynthetic activity but rapidly drying the surface) and the water-use efficiency of the vegetation (e.g., productivity per unit of water transpired; Noy-Meir 1985).

Transpiration usually explains a significant proportion of the ET flux in semi-arid environments (Table 13.1) and is strongly related to P_{gross} because of the common pathway of water vapor and CO_2 exchange through stomatal pores (Nobel 1999). Thus, the T/ET ratio reflects ecosystem level responses to precipitation in a synthetic way. Estimates in the literature of T/ET for semi-arid ecosystems vary widely, in part due to a lack of standardization of measurement approaches (Wilson et al. 2001) and the difficulties of directly measuring T/ET under field conditions (Table 13.1). Reynolds et al. (2000) argued that some reconciliation among disparate estimates of T/ET might be reached by determining the T/ET trend over long-term periods relying on models of ecosystem water balance (Paruelo and Sala 1995; Kemp et al. 1997). Such an approach provides relevant information for understanding the year-to-year variability controlling ecosystem water balance, but still lacks resolution on the dynamics of single infrequent precipitation pulses. Nevertheless, observations of T/ET illustrate key mechanistic constraints that are likely to control NEE, since these parameters are interconnected by temporal and spatial dynamics of soil moisture availability.

Following a precipitation event (or a cluster of significant events), arid and semi-arid ecosystems experience a very dynamic period of soil moisture conditions and canopy resistances (Fig. 13.1a; Schlesinger et al. 1987; Scott et al. 2006a). In the absence of new events, this wetting and drying period lasts just a few days allowing only a short-duration window for bio-

Table 13.1. Estimations of water loss indicated as the ratio of transpiration to evapotranspiration (T/ET) in various arid and semi-arid ecosystems.

Location	Vegetation type	T/ET % (mean)	Method	Ref.
Sonoran desert	Mixed	27 to 41	Modeling	2
Sonoran desert	Mixed	5 to 25	Soil water budget (SWB)	3
Sonoran desert	Shrubland	7	Lisymeter	4
Sonoran desert	Shrubland	80	Stable isotopes	5
Chihuahuan desert	Shrubland	72	Hygrometers in bare and vegetated	6
Chihuahuan desert	Shrubland	6-60 (34)	Long-term modeling	1
Chihuahuan desert	Grassland	1-54 (34)	Long-term modeling	1
Chihuahuan desert	Shrubland	58 to70	Eddy covariance and sap flow	7
Chihuahuan-Sonoran transition	Mesquite woodland	65 to100 (90)	Eddy covariance and stable isotopes	8
Mojave desert	Mixed shrub	35	Integrated leaf transp. and soil evap.	9
Mojave desert	Mixed shrub	15 to 37	Water balance	10
Death valley	Mixed shrub	53	Separated leaf transp. and soil evap.	11
Great basin	*Atriplex*	54	Integrated leaf transp. and SWB	12
Great basin	Sagebrush	55 to 77	Modeling	13
Central Plains	Shortgrass steppe	68 to 78	Stable isotopes and SWB	14
Central Plains	Shortgrass steppe	40 to 75 (51)	Modeling	15
Patagonian steppe	Mixed	38	Modeling, Remote sensing	16
Californian chaparral	Chaparral	8 to 59	Modeling	17
Mediterranean grassland	Grassland	31 to 60	Modeling	18
Morocco	Olive orchard	72 to 86	Eddy covariance and stable isotopes	19
Sahel	Fallow woodland	20	Stable isotopes	20

Numbers in parenthesis are average values. Table was expanded from 1) Reynolds *et al.* 2000, with data from: 2) Young and Nobel 1986; 3) Evans *et al.* 1981; 4) Sammings and Gay 1979; 5) Liu *et al.* 1995; 6) Schlessinger *et al.* 1987; 7) Scott *et al.* 2006; 8) Yepez *et al.* 2007; 9) Smith *et al.* 1995; 10) Lane *et al.* 1984; 11) Stark 1970; 12) Caldwell *et al.* 1977; 13) Campbell and Harris 1977; 14) Ferreti *et al.* 2003; 15) Lauenroth and Bradford 2006; 16) Pauelo *et al.* 2000; 17) De Jong and Hayhoe 1984; 18) Floret *et al.* 1982; 19) Williams *et al.* 2004; and 20) Brunel *et al.* 1997.

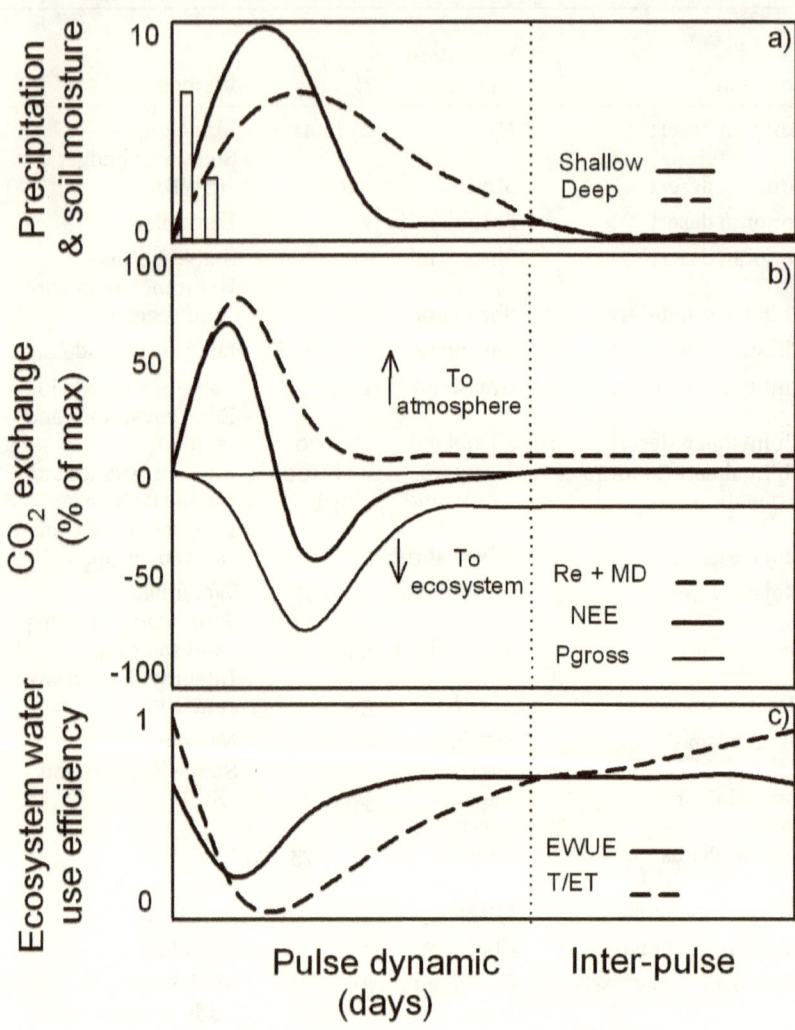

Figure 13.1. Hypothetical short-term response of CO₂ exchange and ecosystem water-use efficiency over a wetting event and subsequent drying cycle. a) Precipitation (vertical bars) and soil moisture content in shallow and deep layers. b) Net ecosystem carbon exchange (NEE), gross photosynthesis (P_{gross}) and ecosystem CO₂ efflux, autotrophic and heterotrophic respiration (R_e) and CO₂ mass displacement (MD) from the soil pores. c) Ecosystem water-use efficiency (NEE/ET) and the transpiration to evapotranspiration ratio (T/ET).

logical activity to occur (Noy-Meir 1973; Austin *et al.* 2004). Consequently, if water circulates in the transpiration stream (*e.g.*, higher T/ET), precipitation pulses may contribute to gross photosynthesis and a larger fraction of available precipitation would be used for production. The extent of the response depends on the pre-pulse conditions, microclimate, the organisms present, the phenological phase and the canopy and root structure of the dominant vegetation. Thus, a more mechanistic explanation for the dynamic interaction between biogeochemical and ecophysiological controls on ET and NEE in semi-arid ecosystems can be provided by measuring flux of ecosystem components during individual wetting and drying cycles. For example, transpiration during photosynthetic carbon gain and the physical process of soil evaporation and soil CO_2 efflux produce unique patterns of EWUE in pulse-driven ecosystems (Fig. 13.1c). Soon after the pulse, EWUE is low because of high soil evaporation rate (*e.g.*, low T/ET), low photosynthetic carbon gain and a substantial soil CO_2 efflux (Fig. 13.1b). In the initial stage of the pulse cycle water is lost without participation in photoautotrophic processes. Later in the pulse dynamic the soil surface dries, limiting microbial activity and reducing soil evaporation as photosynthesis compensates. At this stage, if the magnitude of the pulse and speed of photosynthetic recovery is high, NEE will reflect a net carbon gain for the ecosystem (Huxman *et al.* 2004c). During the inter-pulse period, T/ET is high (although rates are low) and EWUE may be more a function of physiological limitations imposed by drought (*e.g.*, limited photosynthetic capacity, low stomatal conductance, or leaf shedding). Such a trend was recently demonstrated in stands dominated by two different grass species in semi-arid grassland after an experimental irrigation event (Huxman *et al.* 2004a), where ET and T/ET followed species-specific patterns through time but correlated with physiological activity and LAI. Water was lost primarily via soil evaporation in the first day or two following the irrigation, but T/ET, carbon uptake and leaf growth increased thereafter. The differences between the responses of the two species reflected differing patterns of ecophysiological up-regulation to the pulse. To illustrate these patterns with experimental data, Fig. 13.2 shows the response of *Eragrostis lehmanniana* stands to a 39 mm irrigation event early in the growing season of 2003. In this case, during the first week following the moisture pulse the principal component of ET was soil evaporation with T/ET varying from 0.2 to 0.40 as the pulse dynamic progressed (Fig. 13.2d). Notably, as the transpiration flux increased, NEE also reflected a net CO_2 gain of the ecosystem which suggested a rapid physiological up-regulation in response to the pulse.

A strong response to the moisture pulse was also reflected in the pattern of EWUE which remained low during periods of poor photosynthetic carbon gain (day-1) and high CO_2 losses (day 1), but was at maximum later in the pulse dynamic when respiratory and evaporation fluxes decreased and NEE peaked for carbon gain to the ecosystem (Fig. 13.2b, e). As in the Huxman et al. (2004c) study, stands that had higher T/ET ratios throughout the pulse showed a higher EWUE and gained more carbon resulting from pulse (not shown).

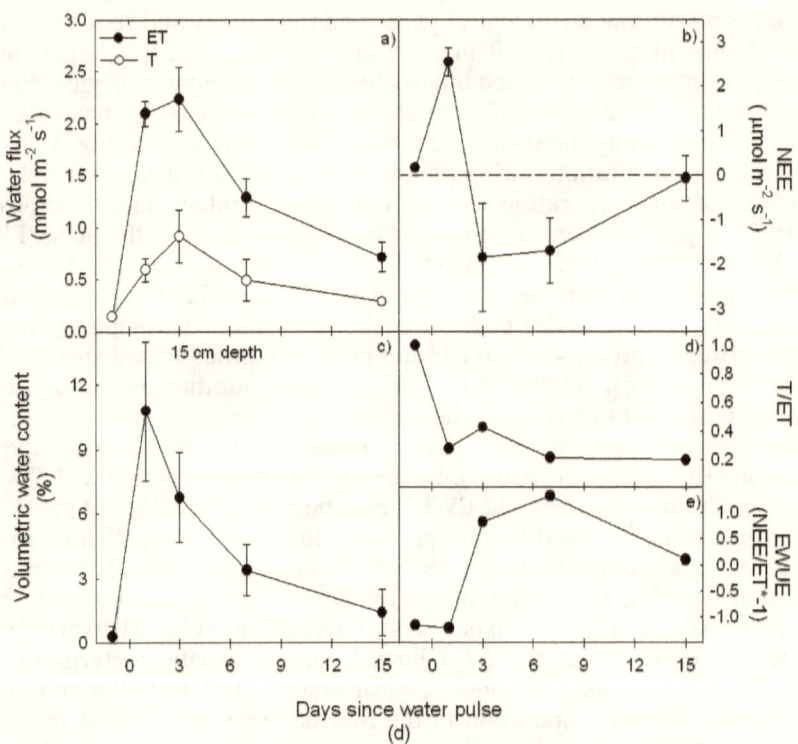

Figure 13.2. Stand-level measurements of water (a; *ET* and *T*) and CO_2 exchange (b; NEE) in experimental stand of the C4 grass *Eragrostis lehmanniana* following an irrigation of 39 mm. Volumetric water content in the soil (c) was measured with reflectrometers. Fluxes of CO_2 and water were measured with an infrared-gas analyzer arranged in a close loop with the headspace of 4000 l chamber placed and sealed above the grass stand. The *T/ET* ratio (d) was calculated based on the isotopic composition of water vapor withdrawn from the chamber at different time intervals to produce Keeling plots (see Box 13.1, Fig. 13.4 and Yepez *et al.* 2005 for details). Transpiration rates were calculated by multiplying *T/ET* by *ET*. The ecosystem water-use efficiency was calculated as NEE/*ET* and multiplied by −1. Error bars are 1 standard error of the mean (*n*=3).

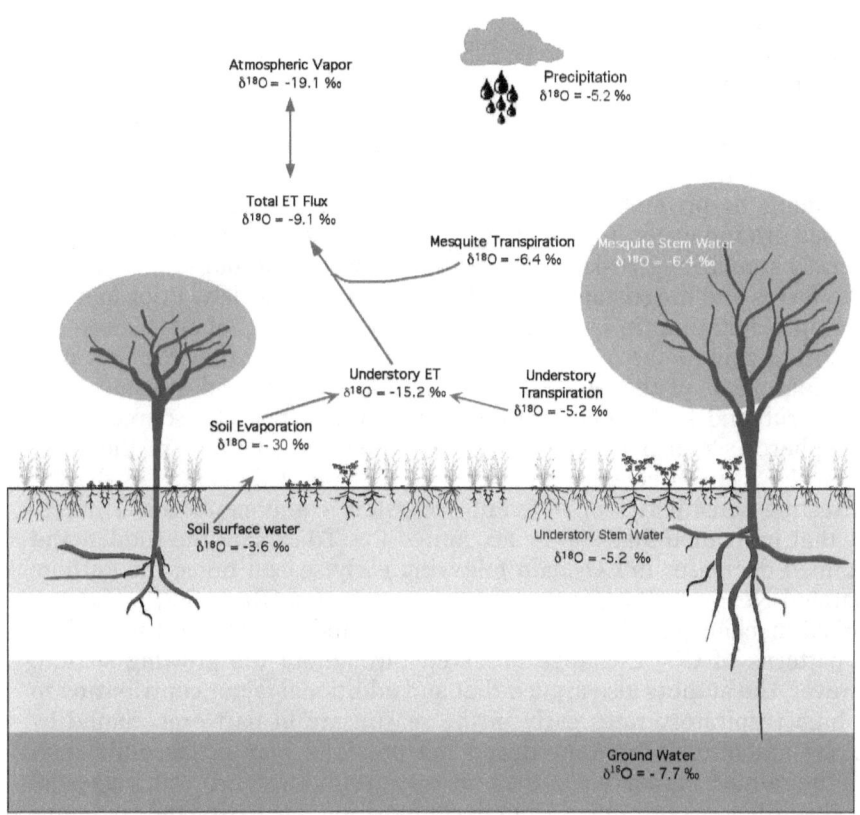

Figure 13.3. Isotopic composition of the principal pools and fluxes of the ecosystem water cycle of a semi-arid mesquite woodland late in the summer growing season of 2001. Notice the large isotopic ratio differences between the tree and understory transpiration fluxes in relation to soil evaporation as a result of equilibrium and kinetic fractionations during the process of evaporation (Box 13.1). Such differences form the basis for an isotopic mass balance partitioning to calculate *T/ET* as described in the text (Eq. 13.2). Data source: Yepez *et al.* 2003.

Variation from the trends depicted in Fig. 13.1 may be found in more complex communities composed of multiple life forms (*e.g.*, savannas or shrublands). Woody plants alter the structure and function of the ecosystem through spatial resource concentration (Reynolds 1999), modification of the microclimate (Breshears *et al.* 1998) and access to more extensive water sources (Scott *et al.* 2003). These effects establish a mosaic of environmental and physiological stress conditions prior to a precipitation pulse that produce unique dynamics of net ecosystem fluxes (Eamus 2003).

Across a shrub encroachment gradient within a semi-arid riparian area, Scott *et al.* (2006b) illustrated important differences in *ET* and NEE responding to precipitation as the encroaching vegetation increasingly accessed ground water. In that study, the trends depicted in Fig. 13.1 where generally similar in grassland, scrubland and woodland but the magnitude of the fluxes and the sustainability of the rates varied widely. Prior to monsoonal precipitation (in late spring and early summer), all sites showed a net carbon gain as the canopies greened out, but soon after the first precipitation events of the summer NEE approached zero and in some cases (*e.g.*, shrubland and woodland) the system was a net CO_2 source to the atmosphere (see also Yepez *et al.* 2007). Remarkably, later in the rainy season, NEE in all ecosystems suggested a strong CO_2 sink. Clearly, the contrasting sensitivities of different ecosystem components made necessary that individual fluxes to be accounted for. To explain the sudden and sustained decreases in CO_2 gain following early season rains, the authors separated NEE into P_{gross} and ecosystem respiration ($R_{plant} + R_{microbial}$) and invoked mechanisms similar to the ones described for Fig. 13.1 to explain the patterns of CO_2 exchange observed throughout the growing season. However, the authors also argued that and additional factor contributing to the high respiratory rates early in the season are in part exacerbated by copious amounts of litter produced the previous year in disequilibrium with the rainfall inputs due to the access to ground water. It was suggested that this litter was prevented from decomposition due to dry and cold temperate during the non-growing season and therefore was readily available for decomposing when temperature was optimal and moisture arrived with the first rains (McLain and Martens 2006). From this and other similar studies in arid and semi-arid ecosystems (Huxman *et al.* 2004; Xu *et al.* 2004; Hastings *et al.* 2005; Scott *et al.* 2006a, b; Jarvis *et al.* 2007; Yepez *et al.* 2007), it appears that the strong and sudden CO_2 losses to the atmosphere following early season precipitation is a defining characteristic of seasonally dry ecosystems (see also Saleska *et al.* 2003).

Box 13.1. Isotopic composition of water vapor sources.

Evapotranspiration vapor, δ_{ET}

Determination of the isotopic composition of the evapotranspiration flux (δ_{ET}) is possible using "Keeling plots" of water vapor, mass balance mixing relationships similar to those used in studies of CO_2 research (Keeling 1961; Pataki *et al.* 2003), where:

$$\delta_{ebl} = C_{pbl}(\delta_{pbl} - \delta_{ET}) \cdot (1/C_{ebl}) + \delta_{ET} \qquad \text{(Eq. B1)}$$

In this case, δ_{ebl} is the isotopic composition of water vapor from the ecosystem boundary layer, C_{pbl}, and δ_{pbl} are the vapor concentration and the isotopic composition of the background atmosphere (planetary boundary layer), C_{ebl} is the vapor concentration within the mixed ecosystem boundary layer, and δ_{ET} is isotopic composition of the evapotranspiration vapor flux. This expression is linear with a slope of $C_{pbl}(\delta_{pbl}-\delta_{ET})$ and a y-intercept of δ_{ET} (Yakir and Sternberg 2000).

Vapor from soil evaporation, δ_E

During the process of evaporation the isotopic composition of water is modified. By diffusing through boundary layers, fractionation processes act against the heavy isotopes resulting in an evaporation water flux depleted in heavy isotopes relative to the water at the evaporating surface in the soil (Gat 1996).

The intensity of such depletion is a function of the isotopic composition of the vapor in the atmosphere, relative humidity and equilibrium and kinetic fractionations associated with a phase change and diffusion (Craig and Gordon 1965). Assuming similar temperatures in the soil surface and the surrounding atmosphere and that no further isotopic fractionation occurs during fully turbulent transport away from the surface, a simple expression of the overall fractionation is provided in Moreira *et al.* (1997):

$$R_E = \left(\frac{1}{\alpha_k}\right) \frac{(R_s / \alpha^*) - R_a h}{1 - h} \qquad \text{(Eq. B2)}$$

where R_E is isotopic composition of water evaporated from the soil and thus, equals δ_E (^{18}O) by $(R_E/0.0020052-1)*1000$ and δ_E (^2H) by $(R_E/0.00015576-1)*1000$, R_s is the molar ratio from the liquid water at the evaporating surface, and R_a is the ratio of the atmospheric vapor. δ^* is the temperature dependent equilibrium fractionation factor (^{18}O $\delta^* = (1.137(10^6/T_{soil}^2) - 0.4156 (10^3/T_{soil}) - 2.0667)$ / 1000 +1 and ^2H $\delta^* = (24.844(10^6/T_{soil}^2) - 76.248 (10^3/T_{soil}) + 52.612)$ / 1000 +1; with T in Kelvin units; Majoube 1971), δ_k is the Kinetic fractionation factor for molecular diffusion in air, 1.0285 and 1.025 for oxygen and hydrogen respectively (Merlivat 1978) or 1.0189 (~19 ‰) for oxygen and 1.017 (~17 ‰) for hydrogen in a turbulent boundary layer (Flanagan *et al.* 1991; Wang and Yakir 2000; see also Cappa *et al.* 2003) and h is the relative humidity of the air.

continues on following page

Box 13.1 (continued)

Isotopic composition of the transpiration vapor sources, δ_T

During transpiration an isotopic steady state (ISS) can be attained, in which the vapor leaving the leaf has the same isotopic composition of water moving into the leaves from the xylem (Flanagan *et al.* 1991; but see Hardwood *et al.* 1999). This condition may occur despite the common enrichment of heavy isotopes in the leaf water as a product of kinetic and equilibrium fractionations (see above). Because there is no isotopic fractionation during water uptake by roots and transport to sites of evaporation in leaves (Ehleringer and Dawson 1992; Brunel *et al.* 1995; but see Ellsworth and Williams 2007), under ISS, the net effect on the isotopic composition of water used by the plant and then transpired is null and therefore the isotopic composition of the stem water could a reliable surrogate for δ_T (Yakir and Sternberg 2000). However, field and laboratory experiments investigating factors controlling the isotopic composition of leaf water suggest that, during typical diurnal regimes of atmospheric humidity, leaf water is not always at ISS and that transpiration at ISS occurs only after ambient conditions are relatively stable (Flanagan *et al.* 1991; Harwood *et al.* 1998; Farquhar and Cernusak 2005). A careful consideration of potential deviations from ISS is necessary when using the isotopic mass balance depicted in Eq. (2) because failure to account for deviations from ISS can translate in significant errors in the final estimates of T/ET under certain circumstances (Yepez *et al.* 2005, 2007; Lai *et al.* 2006).

Ecosystem flux partitioning using stable isotopes

We highlighted the importance of understanding the short-term and seasonal dynamics of net ecosystem fluxes and their components at congruent scales of integration. We now present examples illustrating the application of stable isotopes to estimate the relative contribution of component ecosystem fluxes.

The stable isotopes of water are useful for tracing sources of *ET* because water vapor from plant transpiration and soil evaporation each have unique isotopic signatures (Fig. 13.3; Wang and Yakir 2000). Thus, by knowing the isotopic signature of each *ET* source we can separate the relative contributions of *T* and *E* at the ecosystem scale. This is done using a mixing equation:

$$ T/ET = \frac{\delta_{ET} - \delta_E}{\delta_T - \delta_E} \qquad \text{(Eq. 13.2)} $$

where δ_{ET} is the isotopic composition of evapotranspiration, δ_E is the isotopic composition of soil evaporation and δ_T is the isotopic composition of transpiration. δ_{ET} is estimated using the 'Keeling plot' approach (Box 13.1; Yakir and Sternberg 2000).

Figures 13.4 and 13.5 illustrate the application of the Keeling plot approach for partitioning *ET* in three semi-arid ecosystems of contrasting physiognomies. In the first case (Fig. 13.4) we collected vapor samples for

Keeling plots from a large (4000 liter) gas exchange chamber placed over experimental plots dominated by a perennial *Eragrostis lehmanniana* following an experimental irrigation with δ^2H-labeled water (see above). Results from the Keeling plot analysis several days after the watering pulse revealed that early in the pulse dynamic soil evaporation was high (~70 %) and that it represented the largest fraction of ET throughout the duration of the pulse (Fig. 13.2d; Yepez *et al.* 2005). In contrast, water vapor to fit Keeling plots was collected from five heights (0.1 to 9 m) above the ground surface in an olive orchard six days after a large (100 mm) irrigation event (Williams *et al.* 2004). An estimate of δ_{ET} of −8.5 ‰ was determined by regressing the δ^{18}O composition of water vapor on the reciprocal of the vapor concentration collected from each height (Fig. 13.5a; Table 13.2). *T/ET* was 0.73 in this case, indicating that transpiration from olive trees dominated evapotranspiration even when the soil surface was moist (Williams *et al.* 2004). Similarly, six days after a large natural rainfall event in a fairly dense mesquite (*Prosopis velutina*) woodland in Arizona, the *T/ET* ratio was quite high (0.85). But in this case, the authors were able to analyze the vapor profile from underneath the mesquite canopy separately from the above canopy profile. Upper and lower profiles yielded unique estimates of δ_{ET} (Fig. 13.5b, c). These differences were used to partition *ET* from the whole ecosystem separately from that of the understory layer. Together with overstory and understory eddy covariance measurements of *ET* (Scott *et al.* 2003), the *ET* flux was partitioned into soil evaporation, understory transpiration and tree-canopy transpiration (Table 13.3; Yepez *et al.* 2003). In a later study, Yepez *et al.* (2007) used the same approach to assess the seasonal variably of *T/ET* at the mesquite woodland and combined the results with measurements of carbon exchange to elucidate the mechanisms controlling the intraseasonal rain-use efficiency. From that study it was concluded that T/ET was as low as 0.65 during wet periods immediately after rain events but was generally above 0.80 trough most of the growing season (with a seasonal mean of 0.90; Table 13.1). In relation to the amount of rain received during that 6-month growing season (253 mm) the isotopic ET partitioning further suggested that the soil E flux amounted 78 mm (31% of the precipitation received) while the mesquite trees and the C_4 grassy understory represented 38 and 31% of the precipitation input respectively. The ability to estimate *T/ET* and combine such estimates with scale-congruent measurements of *ET* and NEE allowed to calculate stand-level *T* an after relating this flux with an estimate of P_{gross} an ecosystem-level vegetation water-use efficiency (T/P_{gross}). Notably, early in the season, this index reflected the WUE of mesquite trees as the trees leafed out, but was higher during the peak growing season when the C_4 component was fully developed.

Table 13.2. Summary of parameters for the isotopic partitioning of *ET* on three semi-arid ecosystems (mean ± SEM), δ_T is the isotopic composition of the transpiration flux assuming isotopic steady state (*e.g.*, based on xylem water), δ_E is the modeled isotopic composition of evaporation vapor based on Eq. B2 and the statistical parameters of "Keeling plots," y-intercepts (bold) correspond to δ_{ET} (Box 13.1).

		Parameter
δ_T		
Olive orchard	$\delta^{18}O$	-3.1 ± 0.2
Mesquite Woodland	$\delta^{18}O$	
δ_T Mesquite Trees		-6.4 ± 0.5
δ_T Bulk Understory		-5.3 ± 2
δ_T Bulk Vegetation		-6.3 ± 1
Desert grassland	δ^2H	121 ± 2
δ_E		
Olive orchard	$\delta^{18}O$	-23.3 ± 0.6
Mesquite Woodland	$\delta^{18}O$	-30 ± 0.3
Desert grassland	δ^2H	78 ± 2

δ_{ET}		Regression
Olive orchard		
Total *ET*	$\delta^{18}O$	y=(-29.78 ± 3.47) − **8.49** ±0.97
Mesquite Woodland		
Total *ET*	$\delta^{18}O$	y=(-97.64 ± 9.41) − **9.16** ± 1.06
Understory	$\delta^{18}O$	y=(-40.76 ± 13.67) − **15.20** ± 1.43
Desert grassland		
Total *ET*	δ^2H	y=(-1018.77 ± 93.98) + **93.37** ± 7.61

Table 13.3. Results from the evapotranspiration flux partitioning of three semi-arid ecosystems based on isotopic mass balance expression (Eq. 13.2) and assuming that transpiration occurred at isotopic steady state. $T/ET \pm$ SE according to the variability of the sources and the regression coefficients of the Keeling plots (Phillips and Gregg 2001). D_{max} is the daytime maximum vapor pressure deficit during the collection days.

Ecosystem	Isotope	T/ET	Days after mois-ture input	D_{max} (kPa)	Location
Olive orchard	$\delta^{18}O$	0.73 ± 0.05	6	2.8	Marrakech, Morocco
Mesquite woodland	$\delta^{18}O$		6	4.1	San Pedro River, Arizona
Trees		0.71 ± 0.05			
Understory		0.17 ± 0.06			
Total		0.85 ± 0.05			
Desert grassland	$\delta^{2}H$	0.36 ± 0.17	3	8.6	Santa Rita Exp. Range, Arizona

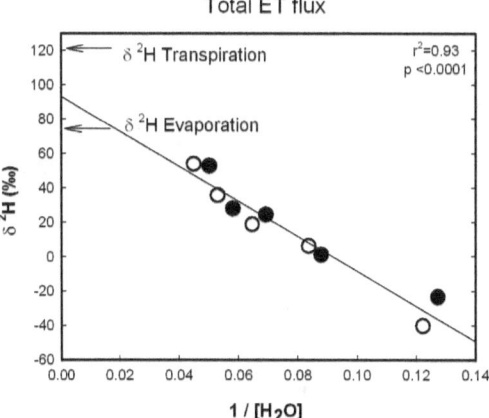

Desert Grassland

Total ET flux

Figure 13.4. Daytime Keeling plot from a desert grassland. Data were produced by placing a large chamber over a stand of *Eragrostis lehmanniana* 3 days after a $\delta^{2}H$ labeled experimental irrigation at the Santa Rita Experimental Range in southeastern Arizona. Closed symbols represent vapor collection periods in the morning and open symbols represent collections in the afternoon. Regression coefficients are shown in Table 13.2 (See Yepez *et al.* 2005 for details).

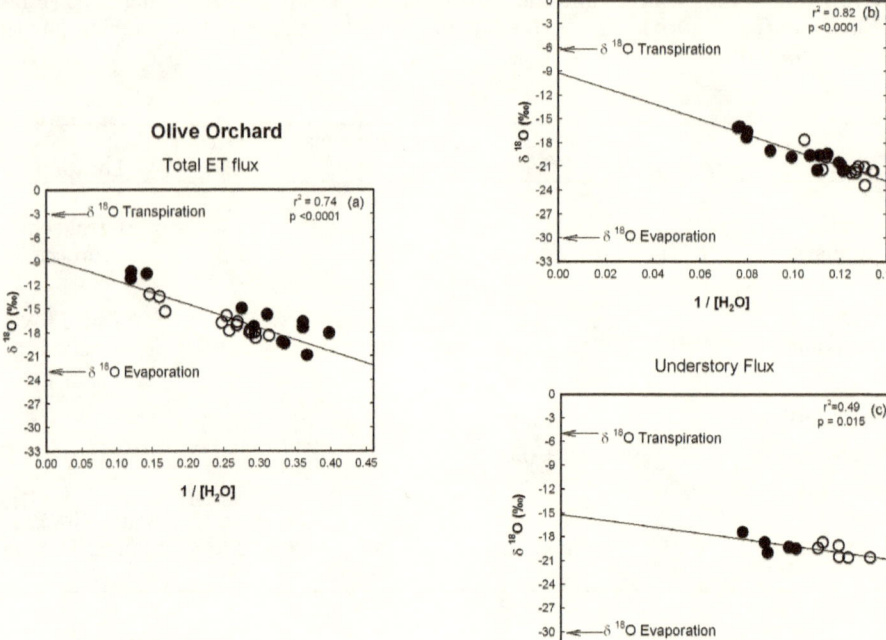

Figure 13.5. Daytime Keeling plots of water vapor collected within the turbulent boundary layer in two semi-arid woodlands six days after moisture input. Plots represent: a) total *ET* flux of an olive orchard monoculture in Marrakech Morocco, on day 313, 2002, b) total *ET* flux of a mesquite woodland in southeastern Arizona, on day 265, 2001, and c) *ET* flux of the mesquite woodland understory dominated by a C4-bunch grass and annual herbs. Closed symbols represent vapor collection periods in the morning and open symbols represent collections in the afternoon. Regression coefficients are shown in Table 13.2.

Relying on Keeling plot analysis and mixing models CO_2 flux partitioning is also possible (Pataki *et al.* 2003). In such case, the approach has been mainly used to partition the nighttime respiratory flux (Werner *et al.* 2007). For example, at the same mesquite woodland described above, Williams *et al.* (2006) demonstrated that partitioning nighttime respiration is possible if the isotopic composition of the carbon from the contributing sources is know. The y-intercept of the regression between the $\delta^{13}C$ of CO_2 collected in a height profile *vs.* the corresponding concentration (*i.e.*, the result from a Keeling plot; Box 13.1) indicated that the isotopic composition of ecosystem respiration was −28.3 ‰ prior to the monsoon rains and shifted to −21.7 during the peak monsoon. During this growing season (2002), the $\delta^{13}C$ signal from mesquite respiration varied between −26 and −28‰ and varied between −24 and −21 ‰ for the litter and soil organic matter. Thus, as suggested by Scott *et al.* 2006b (see above), the shift to a more positive $\delta^{13}CO_2$ values of ecosystem respiration indicated by the Keeling plots following monsoon rains, point to a larger contribution to the nighttime respiratory flux by heterotrophic respiration.

These results reiterate the need to understand the relative contributions of component fluxes to gain advanced knowledge about the functioning of ecosystems as a function of variable rainfall events and conform to the idea of interpreting the rain-use efficiency of an ecosystem as the product of the transpired proportion of precipitation, the precipitation fraction that is evaporated back to the atmosphere and the water-use efficiency of the vegetation (Noy-Meir 1985).

Concluding remarks

Anticipated changes in regional and global precipitation will likely alter key processes affecting productivity of arid and semi-arid ecosystems. Rain-use efficiency (RUE) is an integrated index of carbon-water interactions useful for understanding annual variation of ecosystem function. However, RUE does not capture potentially important short-term responses to precipitation. Detailed understanding of the plant and microbial responses to wetting and drying cycles common of arid and semi-arid environments is necessary to explain the dynamic of carbon and water exchange at the ecosystem level.

In order to describe the mechanisms underlying ecosystem responses to climate and pulsed resource availability in semi-arid environments we need greater temporal resolution in measurements of water and carbon exchange. Similarly, detailed information regarding the disparate variation of contributing fluxes (*i.e.*, respiration *vs.* photosynthesis and transpiration *vs.* evaporation) is needed.

Recent developments with stable isotope techniques provide the means to partition *ET* and respiratory fluxes at the ecosystem scale. Application of these techniques will promote greater understanding of the ecophysiological constraints on major biogeochemical cycles and their likely responses to global change.

Acknowledgements

Research presented here was partially funded by SAHRA (Sustainability of semi-Arid Hydrology and Riparian Areas) under the STC Program of the National Science Foundation, Agreement No. EAR-9876800 and the Upper San Pedro Partnership. EAY gratefully acknowledges the University of Wyoming for its hospitality while writing this chapter and CONACYT-Mexico for the graduate scholarship no 150496.

Literature cited

Austin AT, Yahdjian L, Stark JM, Belnap J, Porporato A, Norton U, Ravetta D, and Sean Schaeffer. 2004. Water pulses and biogeochemical cycles in arid and semi-arid ecosystems. *Oecologia* **141:** 221-235.

Breshears DD, Nyhan JW, Heil CE, and Wilcox BP. 1998. Effects of woody plants on microclimate in a sem-arid woodland: soil temperature and evaporation in canopy and intercanopy patches. *International Journal of Plant Sciences* **159:** 1010-1017.

Brunel JP, Walker GR, Dighton JC, and Monteny B. 1997. Use of stable isotopes of water to determine the origin of water used by the vegetation and to partition evapotranspiration. A case study from HAPEX-Sahel. *Journal of Hydrology* **188-189:** 466-481.

Brunel JP, Walker GR, and Kennett-Smith A. 1995. Field validation of isotopic procedures for determining source water used by plants in a semi-arid environment. *Journal of Hydrology* **167:** 351-368.

Burguess TL. 1995. Desert grassland, mixed shrub savanna, shrubs steppe or semi-desert shurb? The dilemma of coexisting growth froms. In: McClaran MP and Van Devender TR (eds.) *The Desert Grassland*. Arizona University Press, Tucson. Pp. 31-67.

Caldwell MM, White RS, Moore RT, and Camp LB. 1977. Carbon balance, productivity; water use of cold-winter desert shrub communities dominated by C3 and C4 species. *Oecologia* **29:** 275-300.

Campbell GS and Harris GA. 1977. Water relations and water use patterns for Artermisa tridentata Nutt. in wet and dry years. *Ecology* **58:** 652-659.

Cappa CD, Hendricks MB, DePalo DJ, and Cohen RC. 2003. Isotopic fractionation of water during evaporation. *Journal of Geophysical Research* **108:** 525.

Carlyle-Moses DE. 2004. Throughfall, stemflow, and canopy interception loss fluxes in a semi-arid Sierra Madre Oriental matorral community. *Journal of Arid Environments* **58:** 181-202.

Chapin III FS, Matson PA, and Mooney HA. 2002. *Principles in Terrestrial Ecosystem Ecology*. Springer-Verlag, New York.

Craig H and Gordon LI. 1965. Deuterium and oxygen-18 variations in the ocean and the marine atmosphere. In: Tongiori E. (ed.) *Proceedings of the Conference on Stable Isotopes in Oceanographic Studies and Paleotemperatures*. Laboratory of Geology and Nuclear Science, Pisa. Pp. 9-130.

De Jong R and Hayhoe HN. 1984. Diffusion-based soil water simulation for native grassland. *Agricultural Water Management* **9:** 47-60.

Eamus D. 2003. How does ecosystem water balance affect net primary productivity of woody ecosystems. *Functional Plant Biology* **30:** 187-205.

Ehleringer JR. 2001. Productivity of deserts. In: Roy J, Saugier B, and Mooney HA (eds.) *Terrestrial Global Productivity*. Academic Press, New York. Pp. 345-362.

Ehleringer JR and Dawson TE. 1992. Water uptake by plants: perspectives from stable isotope composition. *Plant, Cell and Environment* **15**: 1073-1082.

Ehleringer JR, Phillips SL, Schuster WSF, and Sandquist DR. 1991. Differential utilization of summer rains by desert plants. *Oecologia* **88**: 430-434.

Ellsworth PZ and Williams DG. 2007. Hydrogen isotope fractionation during water uptake by woody xerophytes. *Plant and Soil* **291**: 93–107.

Evans DD, Sammis TW, and Cable DR. 1981. Actual evapotranspiration under desert conditions. In: Evans D and Thames JL (eds.) *Water in Desert Ecosystems*. Dowen Hutchinson and Ross, Inc., Stroudsburg.

Farquhar GD and Cernusak LA. 2005. On the isotopic composition of the leaf water in the non-steady state. *Functional Plant Biology* **32**: 293-303.

Ferretti DF, Pendall E, Morgan JA, Nelson JA, LeCain D, and Moiser AR. 2003. Partitioning evapotranspiration fluxes from a Colorado grassland using stable isotopes: Seasonal variations and ecosystem implications of elevated atmospheric CO_2. *Plant and Soil* **254**: 291-303.

Flanagan LB, Comstock JP, and Ehleringer JR. 1991. Comparison of modelled and observed environmental influences on the stable oxygen and hydrogen isotope composition of leaf water in *Phaseolus vulgaris* L. *Plant Physiology* **96**: 588-596.

Floret C, Pontainer R, and Rambal S. 1982. Measurement and modeling of primary production and water use in a south Tunisia steppe. *Journal of Arid Environments* **5**: 77-90.

Gat JR. 1996. Oxygen and hydrogen isotopes in the hydrological cycle. *Annual Review of Earth and Planetary Sciences* **24**: 255-62.

Golluscio RA, Sala OE, and Lauenreoth WK. 1998. Differential use of large summer rainfall events by shrubs and grasses: a manipulative experiment in the Patagonian steppe. *Oecologia* **115**: 17-25.

Harwood KG, Gillon JS, Roberts A, and Griffiths H. 1999. Determinants of isotopic coupling of CO_2 and water vapour within a *Quercus petraea* forest canopy. *Oecologia* **119**: 109-119.

Helliker BR, Roden JS, Cook C, and Ehleringer JR. 2002. A rapid and precise method for sampling and determining the oxygen isotope ratio of atmospheric water vapor. *Rapid Communications in Mass Spectrometry* **16**: 929-932.

Hultine KR, Scott RL, Cable WL, and Williams DG. 2004. Hydraulic redistribution by a dominant, warm desert phreatophyte: seasonal patterns and response to precipitation pulses. *Functional Ecology* **18**: 530-538.

Huxman TE, Cable JM, Ignace DD, Eilts JA, English NB, Weltzin J, and Williams DG. 2004a. Response of net ecosystem gas exchange to a simulated precipitation pulse in a semi-arid grassland: the role of native versus non-native grasses and soil texture. *Oecologia* **141**: 295-305.

Huxman TE, Smith MD, Fay PA, Knapp AK, Shaw MR, Loik ME, Smith SD, Tissue DT, Zak JC, Weltzin JF, Pockman WT, Sala OE, Haddad B, Harte J, Koch GW, Schwinning S, Small E, and Williams DG. 2004b. Productivity and precipitation: ecosystem vary in sensitivity but converge to an overall maximun rain use-efficiency. *Nature* **429**: 651-654.

Huxman TE, Snyder KA, Tissue D, Leffler AJ, Ogle K, Pockman WT, Sandquist DR, Potts D, and Schwinning S. 2004c. Precipitation pulses and carbon fluxes in semi-arid and arid ecosystems. *Oecologia* **141**: 254-268.

Jarvis P, Rey A, Petsikos C, Wintage L, Rayment M, Pereira J, Banza J, Davis J, Miglietta F, Borgetti M, Manca G, and Valentini R. 2007. Drying and wetting of Mediterranean soils stimulates decomposition and carbon dioxide emission: the "Birch effect". *Tree Physiology* **27**: 929-940.

Johnson-Maynard JL, Graham RC, Wu L, and Shouse PJ. 2002. Modification of soil structural and hydraulic properties after 50 years of imposed chaparral and pine vegetation. *Geoderma* **110**: 227-240.

Karl TR and Trenberth KE. 2003. Modern Global Climate Change. *Science* **302**: 1719-1723.

Keeling CD. 1961. The concentration and isotopic abundances of carbon dioxide and marine air. *Geochimica and Cosmochimica Acta* **24**: 277-298.

Kemp PR, Reynolds JF, Pachepsky Y, and Chen J. 1997. A comparative study of soil water dynamics in a desert ecosystem. *Water Resources Research* **33**: 73-90.

Knapp AK, Fay PA, Blair JM, Collins SL, Smith MD, Carlisle JD, Harper CW, Danner BT, Lett MS, and McCarron JK. 2002. Rainfall variability, carbon cycling, and plant species diversity in a mesic grassland. *Science* **298**: 2202-2205 .

Lai CT, Ehleringer JR, Bond BJ, and Paw UKT. 2006. Contributions of evaporation, isotopic non-steady state transpiration, and atmospheric mixing on the $\delta^{18}O$ of water vapor in Pacific Northwest coniferous forest. *Plant, Cell and Environment* **29**: 77-94.

Lane LJ, Rommey EM, and Hakonson TE. 1984. Water balance calculations and net production of perennial vegetation in the northern Mojave Desert. *Journal of Range Management* **37**: 12-18.

Lauenroth WK and Bradford JB. 2006. Ecohydrology and the partitioning AET between transpiration and evaporation in a semi-arid steppe. *Ecosystems* **9**: 756-767.

Lauenroth WK and Sala OE. 1992. Long-term production of North American short grass steppe. *Ecological Monographs* **2**: 397-403.

Le Houerou HN. 1984. Rain use-efficiency: a unifying concept in arid-land ecology. *Journal of Arid Environments* **7**: 213-247.

Lieth H. 1975. Modeling the primary productivity of the world. In: Lieth H and Wittaker RH (eds.) *Primary Productivity of the Biosphere*. Springer, New York. Pp. 237-263.

Lin G and Sternberg L da SL. 1993. Hydrogen isotopic fractionation by plant roots during water uptake in coastal wetland plants. In: Ehleringer JR, Hall AE, and Farquhar GD (eds.) *Stable Isotopes and Plant Carbon/Water Relations*. Academic Press, New York. Pp. 497-510.

Liu BL, Hoines S, Campbell AR, and Sharma P. 1995. Water movement in desert soil traced by hydrogen and oxygen isotopes, chloride and chlorine 36, southern Arizona. *Journal of Hydrology* **168**: 91-110.

Loik ME, Breshears DD, Lauenroth WK, and Belnap J. 2004. A multi-scale perspective of waterpulse in dryland ecosystems: climatology and ecohydrology of the western USA. *Oecologia* **141**: 269-181.

Magnani F, Mencuccini M, and Grace J. 2000. Age-related decline in stand productivity: the role of structural acclimatation under hydraulic constraints. *Plant, Cell and Environment* **23**: 251-263.

Majoube M. 1971. Fractionnement en oxygene-18 et en deuterium entre l'eau et sa vapaeur. *Journal de Chimie Physique* **68**: 1423-1436.

Mcauliffe JR. 2003. The interface between precipitation and vegetation: The importance of soils in arid and semi-arid environments. In: Weltzing JF and McPherson GR (eds.) *Changing precipitation regimes and terrestrial ecosystems*. University of Arizona Press. Tucson. Pp. 9-27.

McLain JET and Martens DA. 2006. Moisture control on trace gas fluxes in semi-arid riparian soils. *Soil Science Society of America Journal* **70:** 367–77.

Merlivat L. 1978. Molecular diffusivities of H_2 ^{18}O in gases. *Journal of Chemical Physics* **69:** 2864-2871.

Midwood AJ, Boutton TW, Archer SR, and Watts SE. 1998. Water use by woody plants on contrasting soils in a savanna parkland: assessments with δ^2H and $\delta^{18}O$. *Plant and Soil* **205:** 13-24.

Moreira MZ, Sternberg L da SL, Martinelli LA, Victoria RL, Barbosa EM, Bonates LCM, and Nepstad DC. 1997. Contribution of transpiration to forest ambient vapour based on isotopic measurements. *Global Change Biology* **3:** 439-450.

Nobel PS. 1994. Root-Soil Responses to water pulses in dry environments. In: Caldwell MM, Pearcy RW, and Caldwell T (eds) *Exploitation of Environmental Heterogeneity by Plants.* Academic Press, New York. Pp. 285-304.

Nobel PS. 1999. *Physicochemical and Environmental Plant Physiology.* Academic Press, San Diego.

Noy-Meir I. 1973. Desert ecosystems: Environment and producers. *Annual Review of Ecology and Systematics* **4:** 23-51.

Noy-Meir I. 1985. Desert ecosystems structure and function. In: Evenari M, Noy-Meir E, and Goodall DW (eds.) *Ecosystems of the World, 12A, Hot Deserts and Arid Shrublands.* Elsevier, Amsterdam. Pp. 93-104.

Ogle K and Reynolds JF. 2004. Plant responses to precipitation in desert ecosystems: integrating functional types, pulses, thresholds and delays. *Oecologia* **141:** 182-294

Paruelo JM, Lauenroth WK, Burke IC, and Sala OE. 1999. Grassland precipitation-use efficiency varies across a resource gradient. *Ecosystems* **2:** 64-68.

Paruelo JM and Sala O. 1995. Water losses in the Patagonian steppe: A modeling approach. *Ecology* **76:** 510-520.

Paruelo JM, Sala OE, and Beltran AB. 2000. Long-term dynamics of water and carbon in semi-arid ecosystems: a gradient analysis in the Patagonian steppe. *Plant Ecology* **150:** 133-143.

Pataki DE, Ehleringer JR, Flanagan LB, Yakir D, Bowling DR, Still CJ, Buchmann N, Kaplan JO, and Berry JA. 2003. The application and interpretation of keeling plots in terrestrial carbon cycle research. *Global Biogeochemical Cycles* **17:** 1022.

Phillips DL and Gregg JW. 2001. Uncertainty in source partitioning using stable isotopes. *Oecologia* **127:** 171-179.

Ponton S, Flanagan LB, Alstad KP, Johnson BG, Morgenstern K, Kljun N, Black TA, and Barr AG. 2006. Comparison of ecosystem water-use efficiency among Douglas-fir forest, aspen forest and grassland using eddy covariance and carbon isotope techniques. *Global Change Biology* **12:** 1-17.

Prince SD, Brown de Colstoun E, and Kravitz LL. 1998. Evidence from rain-use efficiency does not indicate extensive Sahelian desertification. *Global Change Biology* **4:** 359-374.

Reynolds JF, Kemp PR, and Tenhunen JD. 2000. Effects of long-term variability on evapotranspiration and soil water distribution in the Chihuahuan Desert: A modeling analysis. *Plant Ecology* **150:** 145-159.

Reynolds JF, Ross VA, Kemp PR, de Soyza AG, and Tremmel DC. 1999. Impact of drought on desert shrubs: effects of seasonality and degree of resource island development. *Ecological Monographs* **63:** 69-106.

Sala OE, Golluscio RA, Lauenroth WK, and Soriano A. 1989. Resource partitioning between shrubs and grasses in the Patagonian steppe. *Oecologia* **81:** 501-505.

Sala OE and Laurenroth WK. 1982. Small rain events: an ecological role in semi-arid regions. *Oecologia* **53:** 301-304.

Sammings TW and Gay LY. 1979. Evapotranspiration from an arid zone plant community. *Journal of Arid Environments* **2:** 313-321.

Schlersinger WH, Fonteyn PJ, and Marion GM. 1987. Soil moisture and plant transpiration in the Chihuahuan Desert New Mexico. *Journal of Arid Environments* **12:** 119-126.

Scholes RJ and Archer SR. 1997. Tree-grass interactions in savannas. *Annual Review of Ecology and Systematics* **28:** 517-544.

Schwinning S, Davis K, Richarson L, and Ehleringer JR. 2002. Deuterium enriched irrigation indicates different forms of rain use in shrubs/grass species of the Colorado plateau. *Oecologia* **130:** 345-355.

Schwinning S and Ehleringer JR. 2001. Water use trade-off and optimal adaptations to pulse-driven arid ecosystems. *Journal of Ecology* **89:** 464-480.

Schwinning S and Sala O. 2004. Hierarchy of responses to resource pulses in arid and semi-arid ecosystems. *Oecologia* **141** : 211-220.

Schwinning S, Sala OE, Loik ME, and Ehleringer JR. 2004. Thresholds, memory, and seasonality: understanding pulse dynamics in arid/semi-arid ecosystems. *Oecologia* **141:** 191-193.

Schwinning S, Starr BI, and Ehleringer JR. 2003. Domiant cold desert plants do not partition warm season precipitation by event size. *Oecologia* **136:** 252-260.

Scott RL, Human TE, Cable WL, and Emmerich WE. 2006a. Partitioning of evapotranspiration and its relation to carbon dioxide exchange in a Chihuahuan Desert shrubland. *Hydrological Processes* **20:** 3227-3243.

Scott RL, Huxman TE, Williams DG, and Goodrich DC. 2006b. Ecohydrological impacts of woody plant encroachment: seasonal patterns of water and carbon dioxide exchange within a semi-arid riparian environment. *Global Change Biology* **12:** 311-324

Scott RL, Watts C, Garatuza J, Edwards E, Goodrich DC, Williams DG, and Shuttleworth WJ. 2003. The understory and overstory partitioning of energy and water fluxes in a semi-arid woodland ecosystem. *Agricultural and Forest Meteorology* **114:** 127-139.

Smith SD, Herr CA, Leary KL, and Piorkowski JM. 1995. Soil-plant water relations in a Mojave Desert mixed shrubs community: a comparison of three geomorphic surfaces. *Journal of Arid Environments* **29:** 339-351.

Smith SD and Nobel PS. 1986. Deserts. In: Baker NR and Long SP (eds.) *Photosynthesis in Contrasting Environments*. Elsevier, Amsterdam. Pp. 13-62.

Stark N. 1970. Water balance in some warm desert plants in a wet year. *Journal of Hydrology* **10:** 113-126.

Tuzet A, Castell JF, Perrier A, and Zurfluh O. 1997. Flux heterogeneity and evapotranspiration partitioning in a sparse canopy: The fallow savanna. *Journal of Hydrology* **188-189:** 482-493.

Wang XF and Yakir D. 2000. Using stable isotopes of water in evaporation studies. *Hydrological Processes* **14:** 1407-1421.

Werner C, Unger S, Pereira J, Ghashghaie J, and Maguas C. 2007. Temporal dynamics in d13C of ecosystem respiration in response to environmental changes. In: Dawson TE and Siegwolf TW (eds) *Stable Isotopes as Indicators of Ecological Change*. Academic Press/Elsevier, San Diego. 399-405.

Williams DG, Cable W, Hultine K, Hoedjes JCB, Yepez EA, Simonneaux V, Er-Raki S, Boulet G, de Bruin HAR, Chehbouni A, Hartogensis OK, and Timouk F. 2004. Components of evapotranspiration in an olive orchard determined by eddy covariance, sap flow and stable isotope techniques. *Agricultural and Forest Meteorology* **125:** 241-258.

Williams DG and Ehleringer JR. 2000. Intra- and interspecific variation for summer precipitation use in pinyon-juniper woodlands. *Ecological Monographs* **70:** 517-537.

Williams DG, Scott RL, Huxman TE, Goodrich DC, and Lin G. 2006. Sensitivity of riparian ecosystems in arid and semi-arid environments to moisture pulses. *Hydrological Processes* **20:** 3191-3205

Williams DG and Snyder KA. 2003. Variation in the response of woody plants to heterogeneity in soil water in arid and semi-arid environments. In: Weltzin JF and McPherson GR (eds.) *Changing Precipitation Regimes and Terrestrial Ecosystems.* University of Arizona Press, Tucson. Pp. 28-46.

Weltzin JF, Loik ME, Schwinning S, Williams DG, Fay PA, Haddad BM, Harte J, Huxman TE, Knapp AK, Lin G, Pockman WT, Shaw R, Small E, Smith M, Smith SD, Tissue DT, and Zak JC. 2003. Assessing the response of terrestrial ecosystem to potential changes in precipitation. *Bioscience* **53:** 941-952.

Weltzin JF and McPherson GR. 1997. Spatial and temporal soil moisture resource partitioning by trees and savannas in a temperate savanna, Arizona, USA. *Oecologia* **112:** 156-164.

Wilcox BP, Seyfried MS, and Breshears DD. 2003. The water balance on rangelands. In: Stewart BA and Howell TA (eds.) *Encyclopedia of Water Science.* Marcel Dekker, New York. Pp. 791-794.

Wilson KB, Hanson PJ, Mulholland PJ, Baldocchi DD, and Wullschleger SD. 2001. A comparison of methods for determining forest evapotranspiration and its components: Sap-flow, soil water budget, eddy covariance and catchment water balance. *Agricultural and Forest Meteorology* **106:** 153-168.

Xu L, Baldocchi DD, and Tang J. 2004. How soil moisture, rain pulses, and growth alter the response of ecosystem respiration to temperature. *Global Biogeochemical Cycles* **18:** GB4002.

Yakir D and Sternberg L da S. 2000. The use of stable isotopes to study ecosystem gas exchange. *Oecologia* **123:** 297-311.

Yan S, Wan C, Sosebee RE, Wester DB, Fish EB, and Zartman RE. 2000. Responses of photosynthesis and water relations to rainfall in the desert shrub creosote bush (*Larrea tridentata*) as influenced by municipal biosolids. *Journal of Arid Environments* **46:** 397-412.

Yepez EA, Huxman TE, Ignace D, English NB, Weltzin JF, Castellanos AE, and Williams DG. 2005. Dynamics of transpiration and evaporation following a moisture pulse in semi-arid grassland: a chamber-based isotope method for partitioning flux components. *Agricultural and Forest Meteorology* **132:** 359-376.

Yepez EA, Williams DG, Scott RL, and Lin G. 2003. Partitioning overstory and understory evapotranspiration in a semi-arid woodland ecosystem from the isotopic composition of water vapor. *Agricultural and Forest Meteorology* **119:** 53-68.

Yepez EA, Scott RL, Cable WL, and Williams DG. 2007. Intraseasonal Variation in Water and. Carbon Dioxide Flux Components in a Semi-arid Riparian Woodland. *Ecosystems* **10:** 1100-1115.

Young DR and Nobel PS. 1986. Predictions of soil-water potentials in the northwestern Sonoran desert. *Journal of Ecology* **74:** 143-154.

Chapter 14

ECOPHYSIOLOGICAL RESPONSES OF DESERT PLANTS TO ELEVATED CO₂: ENVIRONMENTAL DETERMINANTS AND CASE STUDIES

*Stanley D. Smith, David T. Tissue,
Travis E. Huxman, and Michael E. Loik*

Introduction
Carbon assimilation
    *Acclimation of photosynthesis and respiration
    to growth in elevated CO₂*
        ACCLIMATION OF PHOTOSYNTHESIS
        ACCLIMATION OF RESPIRATION
    Effects of elevated CO₂ on stomatal conductance
    *Environmental stress, resource availability,
    and elevated [CO₂]*
    *Functional types: the role of growth potential
    in controlling the [CO₂] response*
Case studies
    CAM succulents
    Desert shrubs
        EVERGREEN SHRUBS
        DECIDUOUS SHRUBS
    Herbaceous annuals
Summary and conclusions

Perspectives in Biophysical Plant Ecophysiology: A Tribute to Park S. Nobel, pp. 363-390
Edited by: E. De la Barrera and W.K. Smith
© 2009 by The Authors
Book Compilation © 2009 Universidad Nacional Autónoma de México

Introduction

Anthropogenic activities have significantly affected the composition of Earth's atmosphere through increasing carbon dioxide and other trace gas concentrations (Vitousek *et al.* 1997). Except for human land use, no global change factor has been more rapid and substantial than the increase in atmospheric CO_2 partial pressure ($[CO_2]$). From the beginning of the industrial age until today, $[CO_2]$ has risen from approximately 28 to 38 Pa, a 30% rise in the last 150 years. This rise is continuing, with a doubling from pre-industrial $[CO_2]$ projected by 2050, and a doubling of current-day $[CO_2]$ by the end of this century. Increased $[CO_2]$ is one of the primary factors forcing greater global atmospheric temperatures (Karl and Trenberth 2003), and is expected to further alter Earth's climate systems in the coming decades (Schneider 1992).

In addition to its effect on climate, increased $[CO_2]$ has direct effects on two important physiological processes in plants. First, it increases photosynthetic rates by increasing the diffusional gradient of CO_2 from the atmosphere to the site of carboxylation in photosynthetic cells. Second, it decreases stomatal conductance, presumably due to the effect of reduced CO_2 "demand" in the leaf. Due to the combination of increased photosynthesis and decreased water loss via decreased stomatal conductance, plants generally exhibit significantly increased leaf-level water-use efficiency (WUE, the ratio of carbon gain per unit water lost). Furthermore, photosynthetic CO_2-assimilation is well correlated with leaf nitrogen concentration at ambient CO_2, a relationship that scales across diverse plant life forms and biome types (Reich *et al.* 2003). Elevated CO_2 can alter this relationship by increasing photosynthetic CO_2-assimilation per unit leaf nitrogen investment (NUE, the ratio of carbon gain per unit nitrogen investment). Therefore, elevated CO_2 has the potential to increase the efficiency with which key resources—water water and nitrogen—are utilized by plants.

Increased WUE and NUE at elevated CO_2 are anticipated to be important in resource-limited environments, where water and/or nitrogen consistently limit plant growth. This is particularly true for water-limited environments, in that elevated CO_2, through significant increases in plant WUE, should result in greater plant growth, improved water balance at a number of scales (leaf-to-landscape), and potentially extended duration of growth into dry seasons (Smith *et al.* 1997). The predictions, and results, with N-availability are more equivocal, with no clear differential response to elevated CO_2 in high- or low-nutrient habitats (Nowak *et al.* 2004a). Nevertheless, several conceptual models have predicted that resource-limited ecosystems will respond more strongly to elevated CO_2 (percent increase relative to ambient CO_2) than will non-resource-limited ecosystems (Melillo *et al.* 1993). For example, two ecosystem types that have been shown to respond strongly to elevated CO_2 include deserts (Smith *et al.* 2000) and nutrient-poor grasslands (Niklaus *et al.* 2001), where biological activities are significantly constrained by pronounced water- and nutrient-deficits.

Some of the early work on the responses of desert plants to elevated CO$_2$ was conducted by Park Nobel and his students and post-docs (*e.g.*, Nobel and Hartsock 1986). Clearly, Park saw that this was an important topic before elevated CO$_2$ research was "in vogue". It is therefore appropriate, as a chapter in this *Festschrift*, that we summarize advances that have been made in elevated CO$_2$ research over the past several decades. Although none of the authors in this chapter conducted elevated CO$_2$ research while in Park's lab, we subsequently conducted a significant amount of elevated CO$_2$ research in our respective careers, often forming collaborations that we developed in Park's lab. In each of our careers, and in our specific approaches to elevated CO$_2$ research, we applied unique skills that we developed in his lab to the experimental problem at hand. As a result of this "bias", we will concentrate this review on plants from resource-limited environments, particularly deserts, and we will also explicitly address the interactions between elevated CO$_2$ and environmental stress, a relatively neglected area of elevated CO$_2$ research. In the first part, we will concentrate on key process studies with respect to various environmental stresses, and then in the second part we will address some Case Studies in which Park Nobel and his former students and post-docs have made important contributions to our understanding of how elevated CO$_2$ may impact plant performance in resource-limited environments.

Carbon assimilation

Acclimation of photosynthesis and respiration to growth in elevated CO$_2$

ACCLIMATION OF PHOTOSYNTHESIS—Most plants exposed to elevated CO$_2$ exhibit short-term increases in photosynthetic rates largely due to the stimulation of ribulose-1,5-bisphosphate carboxylase/oxygenase (Rubisco) carboxylation. Rubisco is CO$_2$ substrate-limited so an increase in [CO$_2$] results in greater substrate availability, and also suppresses oxygenation (and thus photorespiration), resulting in increased net photosynthesis (A_{net}) (Long 1991). However, over the long term (*i.e.*, days to years) photosynthetic responses to elevated CO$_2$ may be affected by biochemical, morphological and physiological feedbacks that balance carbon assimilation (*i.e.*, source) with growth and reproductive demands (*i.e.*, sink) (Tissue and Oechel 1987; Tissue *et al.* 1993). There may also be climatic feedbacks (*e.g.*, more frequent droughts or extreme temperature episodes) that limit any enhancement of photosynthetic productivity due to elevated CO$_2$ (Naumburg *et al.* 2004). A reduction in photosynthetic capacity after long-term exposure to elevated CO$_2$, termed down-regulation or acclimation, readjusts source-sink relationships, but rarely eliminates the whole-plant positive response of photosynthesis to elevated CO$_2$ (Sage 1994; Körner 2003). This is due to the benefits that even a short stimulation of photosynthetic rate can have on leaf area production, which despite acclimation responses provide greater whole-plant carbon balance at elevated CO_2. Therefore, continued photosynthetic enhancement due to elevated CO_2 is expected in plants from most biomes, including arid and semiarid systems.

Long-term reductions in photosynthetic capacity are often attributed to an imbalance in source-sink relationships. Indeed, the maintenance of active sinks (*i.e.*, regions of active growth or metabolic activity) is necessary to sustain the stimulation of photosynthesis in elevated CO_2. An imbalance in the source-sink relationship may result in the accumulation of leaf carbohydrates, which ultimately trigger feedback mechanisms that reduce leaf photosynthetic capacity. One feedback mechanism includes reductions in gene transcription and production of Rubisco protein, thereby reducing Calvin cycle activity and further sugar production (Moore *et al.* 1999). Both the maximum carboxylation rate of Rubisco (V_{cmax}) and maximum rate of electron transport (J_{max}) are reduced in plants exhibiting photosynthetic down-regulation (Wullschleger 1993; Sage 1994; Medlyn *et al.* 1999; Tissue *et al.* 1999). Depending upon growth form and environmental conditions, this long-term down-regulation of photosynthesis may occur within 6-9 days after initial exposure to elevated CO_2. Regulation of photosynthesis on a shorter time scale may be accomplished by reducing Rubisco activity through decarbamylation of the enzyme (Sage *et al.* 1988).

ACCLIMATION OF RESPIRATION—The effects of elevated CO_2 on leaf respiration rates are less well understood than the effects on photosynthesis, particularly the mechanisms that regulate respiratory responses to changes in [CO_2]. In some studies, a short-term effect of elevated CO_2 on dark respiration has been observed, in which respiration rates are immediately and significantly reduced (Amthor *et al.* 1992; Drake *et al.* 1999; Baker *et al.* 2000; Hamilton *et al.* 2001). However, this immediate suppression of dark respiration has not been observed in other studies (Tjoelker *et al.* 1999; Amthor 2000; Amthor *et al.* 2001; Jahnke 2001; Tissue *et al.* 2002). To date, there is no mechanism to explain a direct suppression of dark respiration by elevated CO_2, although it has been hypothesized to be due to the inhibition of respiratory enzymes such as succinate dehydrogenase and cytochrome c oxidase (Azcon-Bieto *et al.* 1994; González-Meler *et al.* 1996; González-Meler and Siedow 1999); however, this has not been demonstrated in whole plants.

Long-term effects on respiration may occur after extended growth at elevated CO_2 and may be mediated through elevated CO_2 effects on growth rate, nonstructural carbohydrate concentration, and tissue composition. Contrasting results, in which elevated CO_2 was shown to increase (Thomas and Griffin 1994; Wang *et al.* 2001), decrease (Wullschleger *et al.* 1992; Bunce and Ziska 1996), or not affect (Lewis *et al.* 1999; Tissue *et al.* 2002) leaf respiration rates have confounded our efforts to predict long-term plant responses to CO_2 enrichment. The complicating practice of calculating respiration on a leaf area or leaf mass basis further constrains our understanding of the direction of long-term leaf respiratory response to elevated CO_2 (Poorter *et al.* 1992). A meta-analysis of leaf respiration responses to elevated CO_2 (Wang and Curtis 2002) indicated that leaf respiration on a mass basis was significantly reduced (18%), but there was no significant effect when respiration rates were calculated on an area basis.

Several explanations for the differential long-term respiratory response in leaves have been suggested. For example, growth in elevated CO$_2$ often reduces leaf nitrogen and protein concentrations, and increases leaf carbohydrate concentrations, all of which affect oxidative respiration in leaves due to maintenance activities (Amthor 1991). Ontogenetic and canopy position confound these responses to [CO$_2$], resulting in a complex pattern of leaf biochemistry in natural plant canopies (Tissue *et al.* 2001). Some studies indicate that increased number of mitochondria per unit cell area for plants grown in elevated CO$_2$ may affect respiratory responses to elevated CO$_2$ (Robertson *et al.* 1995; Griffin *et al.* 2001), but not always (Tissue *et al.* 2002).

Effects of elevated CO$_2$ on stomatal conductance

A fundamental prediction about the effect of elevated CO$_2$ is that reductions will occur in stomatal conductance (g_s) because of the greater supply of CO$_2$ for photosynthetic assimilation. In other words, stomata will not have to open as much for the same amount of CO$_2$ uptake. It has been suggested that the resultant water savings at the leaf and whole-plant levels will mean less water uptake from soils, and therefore an increase in soil water content (Jackson *et al.* 1994; Field *et al.* 1995). Much research has tested this prediction for plants from various ecosystem types, but ecosystem-level changes in water balance remain poorly understood (Shaw *et al.* 2005). A full review is beyond the scope of this chapter – here we focus on the effects of elevated CO$_2$ on stomatal responses for desert plants for which soil water availability occurs in highly distinct pulses (Loik *et al.* 2004a). The ephemeral nature of soil water availability in these systems could confound the predictions of reduced g_s and increased soil water availability. For a more comprehensive review for other vegetation types, see Drake *et al.* (1999), Ghannoum *et al.* (2000), Morgan *et al.* (2004b), and Nowak *et al.* (2004a).

Research in both glasshouse and field experimental manipulations (*e.g.*, the Nevada Desert FACE Facility; Jordan *et al.* 1999) suggests that ambient soil water availability exerts strong control over the response of g_s to atmospheric CO$_2$ enrichment. Using both FACE and glasshouse CO$_2$ enrichment methods, Huxman *et al.* (1998c) showed that stomatal conductance in *Larrea tridentata* was most responsive in wet years. In a wet El Niño year, g_s was reduced at elevated CO$_2$ during most of the spring/early summer growing season, leading to significantly higher water-use efficiency and midday water potentials in mid-summer (Fig. 14.1). However, these water savings apparently did not result in an extension of photosynthetic activity into the dry season or during hotter, higher-*VPD* times of the day (Hamerlynck *et al.* 2000a). These results highlight the potential importance or confounding effects of external and internal factors (*i.e.*, soil water content and plant water potential, respectively) in mediating the response of stomatal conductance to elevated CO$_2$.

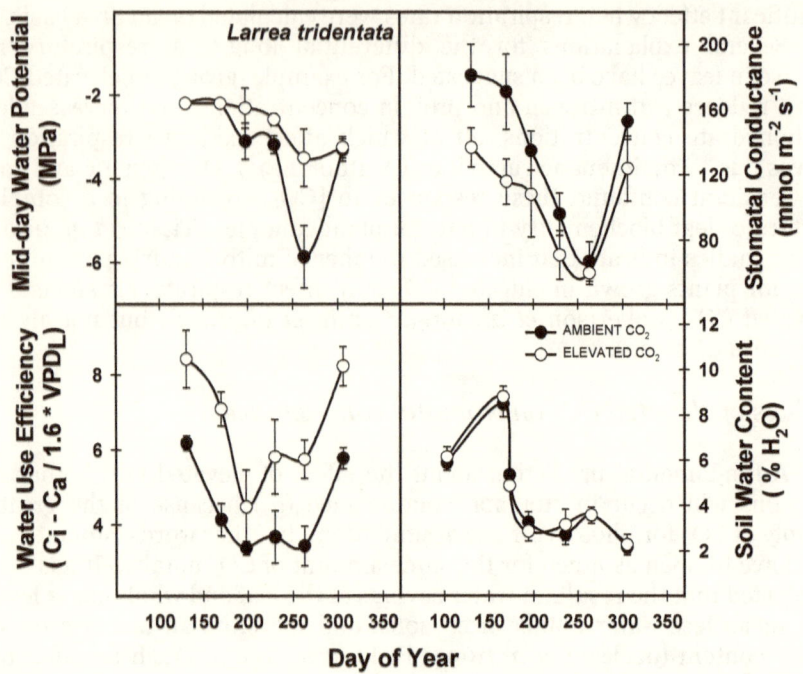

Figure 14.1. Plant water potential (upper left), stomatal conductance (upper right), water-use efficiency (lower left), and beneath-shrub soil water content (lower right) for the Mojave Desert shrub *Larrea tridentata* growing at ambient (closed circles) and elevated CO_2 (open circles) at the Nevada Desert FACE Facility during a wet El Niño year (1998). Water potential was determined at mid-day, stomatal conductance at mid-morning (peak daily value), water-use efficiency is calculated from environmental data and estimates of internal CO_2 concentration, and soil water content (0-20 cm depth) was determined by TDR under the canopy of *Larrea* shrubs. Unpublished data from T.E. Huxman and S.D. Smith.

Although *Larrea* is a useful model for shrub responses to elevated CO_2 —especially in the Mojave Desert which is dominated by the shrub growth form—other shrub species can exhibit contrasting responses of g_s to elevated CO_2. For example, Naumburg *et al.* (2003) compared g_s for the drought-deciduous shrub *Ambrosia dumosa*, the winter-deciduous shrub *Krameria erecta,* and the evergreen *Larrea* during a wet El Niño and subsequent dry years at the NDFF. Daily integrated photosynthesis was significantly enhanced by elevated CO_2 for all three species, but only *Larrea* exhibited a decrease in g_s. This response to elevated CO_2 only occurred prior to the onset of the summer dry season and following late-summer rainfall, when plant water potentials were relatively high. Given consistent increases in photosynthesis, this resulted in consistently higher water-use efficiency at elevated CO_2 (Fig. 14.1).

Responses of g_s to elevated CO$_2$ may also depend on the type of photo-synthetic pathway utilized. For example, Nowak *et al.* (2001) compared g_s and temperature responses of the C$_3$ bunchgrass *Achnatherum hymenoides*, the C$_4$ bunchgrass *Pleuraphis rigida*, and the C$_3$ shrub *Larrea* to elevated CO$_2$. For each species, conductance tended to be lower at elevated CO$_2$ in the spring growing season of the wet El Niño year of 1998, higher when soils dried later in 1998, and there were few treatment effects in the extremely dry year of 1999, when rates were all very low. Responses were similar for measurements made in early morning or at midday. When averaged over all observations, reductions in g_s were greater for the C$_4$ grass (35% for *Pleuraphis*) than for the C$_3$ grass (20% for *Achnatherum*) and C$_3$ shrub (17% for *Larrea*), with greater apparent differences in the dry *versus* the wet year. Such a differential response would not be expected based on the photosynthetic physiology of C$_3$ and C$_4$ plants (*i.e.*, C$_3$ plants would be expected to have greater declines in g_s at elevated CO$_2$).

Do the impacts of elevated CO$_2$ on g_s extend from the leaf to the whole plant and plot scales in arid systems? Hamerlynck *et al.* (2002) examined the responses of the drought-deciduous *Lycium andersonii*, for which phenological shifts occur in growth form, where short-shoots maximize early-season leaf area and long-shoots determine the annual growth increment. The two shoot types exhibited similar rates of net CO$_2$ assimilation, but g_s was reduced by 27% at elevated CO$_2$. This was accompanied by the production of larger leaves on short-shoots, and more leaves on long-shoots. There were no differences in midday water potential for shrubs at ambient *versus* elevated CO$_2$, yet there was an overall reduction in whole-plant water use (Hamerlynck *et al.* 2002). In related work, Pataki *et al.* (2000) compared whole-plant water use for *Ephedra nevadensis* (a leafless shrub) and *Larrea* exposed to elevated CO$_2$ in a glasshouse and under FACE in the field. For *Ephedra*, daily sap flux and mid-morning g_s for whole branches were lower on elevated CO$_2$ plots. These were associated with a 33% reduction in the ratio of surface area available for transpiration relative to sapwood area (LSR). LSR was reduced 60% for *Larrea* grown in elevated CO$_2$ in the glasshouse, but there was no difference when grown in elevated CO$_2$ at the FACE site in the field. Moreover, g_s was reduced under elevated CO$_2$ for *Larrea* in the field even though soil moisture was high (Pataki *et al.* 2000).

In general, reduced g_s did not appear to increase soil moisture in relatively arid land regions with low leaf area. For example, Nowak *et al.* (2004b) found no consistent differences in soil moisture content (20, 50 and 185 cm depth intervals) between elevated and ambient CO$_2$ plots. Indeed, soil water content in the surface 50 cm was greater in the elevated CO$_2$ plots on only one sampling date, and was lower on six other sampling dates. Greater aboveground production (and leaf area for transpiration) apparently resulted in greater soil moisture extraction on the elevated CO$_2$ plots, which countered the potential water savings from leaf-level reductions in g_s and transpiration at elevated CO$_2$. Alternatively, the lack of a CO$_2$ response in soil moisture may simply be due to the fact that about two-thirds of the ET budget in Mojave Desert ecosystems is lost as surface evaporation rather than transpiration (Smith *et al.* 1995), and so reductions in leaf transpiration at elevated CO_2 may simply not scale to the landscape level (perennial plant cover is less than 20% at the NDFF; Jordan *et al.* 1999).

Overall, we conclude that arid ecosystems may not behave according to the generalized prediction that 20 to 50% reductions in g_s will lead to soil water conservation in a future higher-CO_2 atmosphere (Field *et al.* 1995). Not surprisingly, the responses of leaf g_s, plant water potential, ET, and soil water content to elevated CO_2 vary across species, seasons, and ecosystem types (Morgan *et al.* 2004b). Subsequently, we emphasize the importance of species, growth form, resource limitation, plant age, development stage, photosynthetic pathway, and timing and amount of moisture inputs when interpreting the responses of g_s to elevated CO_2 (Morgan *et al.* 2004b; Nowak *et al.* 2004a).

Environmental stress, resource availability, and elevated [CO₂]

The changes in leaf water relations and carbon status resulting from growth at elevated CO_2 have significant implications for the maintenance of plant function during periods of low resource availability or environmental stress. Because elevated CO_2 both increases net photosynthetic rates and reduces leaf water loss, greater carbohydrate loads and improved leaf water relations are often seen at elevated CO_2 (Bowes 1993), which can enhance plant carbon and water balances through periods of temperature, nutrient or biotic stress (Osmond *et al.* 1987).

Cell expansion in leaves is extremely sensitive to water status (Boyer 1970), where even mild water deficit can cause leaf turgor to fall below critical yield thresholds required to maintain growth or produce fully expanded leaves (Acevedo *et al.* 1971; Bradford and Hsiao 1982). Even minor enhancements in leaf turgor during water deficit, such as those seen at high CO_2, should promote greater plant performance during water limitation (Hsiao and Jackson 1999). The ability to continue growth during periods of low water availability further provides the means to better cope with stressful conditions, by allowing for continued extraction of resources from the environment. For example, in addition to improved leaf water relations at high CO_2, a greater allocation of biomass to roots allows for a greater soil volume to be explored by an individual plant (Moore and Field 2006; Norby and Iverson 2006); however, there is considerable variation across systems in root responses to elevated CO_2 (Jastrow *et al.* 2005; Hui and Jackson 2006; Luo *et al.* 2006). Because water deficit can interact with nutrient or temperature stress to affect plant performance (Kozlowski 2000; Lynch and St. Clair 2004; Bohnert *et al.* 2006), this has implications for plant function in a number of settings. In this regard, elevated CO_2 has been observed to mitigate chilling-induced water stress in some species and maintain greater turgor due to greater carbohydrate availability (Boese *et al.* 1997).

Deserts are stressful environments where changes in plant function resulting from growth at elevated CO$_2$ may have important population, community and ecosystem effects (Smith *et al.* 1999). Species distributions in the North American deserts are controlled by interactions between high and low temperature, along with seasonal periods of water availability, all of which affect the potential for seedling establishment (Smith and Nowak 1990). Short-duration extreme high and low temperature events are important sources of mortality or reduced productivity for many annual and perennial species from desert regions (Jordan and Nobel 1979; Pockman and Sperry 1997). Taub *et al.* (2000) illustrated, using 16 species of monocots and dicots, that many plants maintain greater Photosystem II efficiencies (*Fv/Fm*) at extreme high temperatures following growth at elevated CO$_2$. Critical temperatures for tissue mortality appear to shift up by as much as 3 °C for some species. This greater efficiency at high temperatures and high CO$_2$ in part comes from both greater maximum fluorescence values and reductions in dark-adapted fluorescence, which indicates that the CO$_2$ effect is in part a function of maintaining membrane integrity through high-temperature events (*e.g.*, acclimation of membrane lipid composition).

The ability to survive freezing events explains the geographic distribution of many species. From desert regions, the distributions of *Carnegiea gigantea, Yucca brevifolia, Opuntia fragilis*, and *Larrea tridentata* can be explained in large part by the intensity of freezing events (Steenbergh and Lowe 1976; Smith *et al.* 1983; Loik and Nobel 1993; Pockman and Sperry 1997). Plant growth at elevated CO$_2$ may influence the responses of many species to extreme low temperatures in that it may influence both the process of acclimation and the response to freezing itself (Loik *et al.* 2000). Elevated CO$_2$ has been suggested to interfere with the acclimation process, and in *Picea mariana* and *Eucalyptus pauciflora* growth at elevated CO$_2$ increases the susceptibility of plant tissues to freezing events (Margolis and Vezina 1990; Barker *et al.* 2005). But in a number of semi-succulent and alpine species, low temperature tolerance or performance at low temperatures appears to be enhanced by elevated CO$_2$ (Boese *et al.* 1997; Loik *et al.* 2000, 2004b). Changes in either direction have important implications for species interactions, distributions, and ecosystem function.

Functional types: the role of growth potential in controlling the [CO$_2$] response

Plants of different growth habit, or form, should potentially respond differently to elevated [CO$_2$]. Based on both physiological theory and experimental evidence, growth stimulation at elevated CO$_2$ should depend on both the kinetics of carboxylation (*i.e.*, photosynthetic pathway and leaf N-concentration) and the capacity of the plant to maintain a steep concentration gradient of assimilates between source leaves and various growth-induced sinks (Diaz 1995). As an example of the latter, Körner *et al.* (1995) observed symplastic phloem loaders, which are less efficient than apoplastic loaders, to more quickly reach carbohydrate saturation in the phloem-feeding tissues when grown at elevated CO$_2$. Therefore, from this perspective of the entire leaf-to-sink organ pathway, we review the response of different functional groups to growth at elevated CO_2, concentrating on (1) different photosynthetic pathways, (2) inherently fast- *versus* slow-growing plant species, and (3) woody *versus* herbaceous growth forms, with specific reference to woody plant invasions in arid and semiarid regions.

It has long been held that C_3 plants are much more responsive to elevated CO_2, in both photosynthesis and growth, than are C_4 plants, which is primarily a consequence of the lack of CO_2-saturation at the carboxylation site of Rubisco in C_3 plants. This has been shown in many controlled-environment experiments, as well as longer-term field experiments in a salt marsh (Arp *et al.* 1993), tallgrass prairie (Owensby *et al.* 1999), and in shortgrass steppe (Morgan *et al.* 2004a). However, Wand *et al.* (1999) conducted a meta-analysis of C_3-C_4 growth responses of non-domesticated grasses to a doubling of [CO_2]. Their results were not consistent with expectations based strictly on photosynthetic theory, with both C_3 and C_4 plants showing higher carbon assimilation rates (33% and 25%, respectively) and increased total biomass (44% and 33%, respectively) when grown in elevated CO_2. They concluded that the stimulation of carbon assimilation by C_3 species at elevated CO_2 is often reduced by stress, particularly nutrient stress. Indeed, BassiriRad *et al.* (1997) observed a C_4 grass from the Chihuahuan Desert to have dramatically enhanced nitrate and phosphate uptake rates at elevated CO_2 compared to two C_3 shrubs, even though it had an overall lower growth response than the C_3 species due to obvious photosynthetic constraints. Drought may also be important, as CO_2 enrichment may alleviate drought effects in C_4 species to a greater extent than in C_3 species (Ward *et al.* 1999). The mechanistic underpinnings for these responses are not entirely clear, but they do suggest that we cannot linearly extrapolate responses based on photosynthetic pathway theory to plants in natural environments, particularly arid environments where drought and nutrient stresses commonly occur.

A variety of studies have examined plant species that vary in growth potential, either from a strictly physiological perspective (*i.e.*, plants with faster growth rates and therefore stronger sink strength should be more responsive to elevated CO_2) or from a community perspective (*i.e.*, early successional plants should be more responsive because they grow faster). From the physiological perspective, a number of studies have shown fast-growing species to be more responsive to elevated CO_2 than slow-growing species (Poorter 1998; Poorter and Navas 2003). These observations have generally been supported by comparative studies in grassland or old field systems, which tend to show legumes and forbs to be more responsive to elevated CO_2 than are grasses (Lüscher *et al.* 1998; Reich *et al.* 2001). In a comparative analysis of eight deciduous and evergreen woody species, Cornelissen *et al.* (1999) found specific leaf area at ambient CO_2 to explain 88% of the variation in growth response to elevated CO_2 (*i.e.*, deciduous species with higher SLA and thus lower maintenance requirements responded with greater growth when exposed to higher CO_2). Similarly, within 10 species in the genus *Acacia*, the absolute increase in growth at elevated CO_2 was greater in the faster-growing species (Atkin *et al.* 1999).

However, when potentially extrapolating these results to drier ecosystems, the caveat of stress again becomes an issue. In his study of fast- *versus* slow-growing plants, Poorter (1998) also showed that plants presented with abundant nutrients were much more responsive to elevated CO$_2$ than are nutrient-stressed plants, an observation that has been repeated in many studies. Indeed, Arp *et al.* (1998) observed that while faster-growing species do respond more vigorously to elevated CO$_2$ when grown in optimum growing conditions, inherently slow-growing species are potentially more successful when grown under nutrient stress conditions.

The question of how the above functional groups will respond in arid and semiarid ecosystems is interesting mainly because the differential advantage of fast growth may be largely nullified when those plants are confronted with environmental stress. In an analysis of how elevated CO$_2$ may affect semiarid rangelands, Polley (1997) proposed the greatest response will occur in C$_3$/C$_4$ mixed grasslands, and at the transition between grasslands and woodlands. Specifically, studies have suggested that historic increases in [CO$_2$] (*ca.* 30% increase since the beginning of the 19th century) have potentially stimulated the encroachment of C$_3$ woody species into historically C$_4$-dominated grasslands (Johnson *et al.* 1993; Polley *et al.* 2002). However, Archer *et al.* (1995) have questioned that conclusion, based on physiological, community-level, and paleobotanical data. Nevertheless, Bond and Midgley (2000) have proposed that elevated CO$_2$ will continue to stimulate woody plant invasion into semiarid grasslands and savannas by stimulating tree sapling growth and allowing them to more quickly escape the grass layer 'topkill zone'. This may allow woody canopies to escape the grass layer and thus survive periodic fires that tend to effectively maintain grassland structure. Clearly, this is an important ecological question that needs additional experimental work, particularly examining grassland/shrubland interfaces under future global change scenarios.

In conclusion, a majority of work on plant functional group responses has indicated that C$_3$ plants are more responsive to elevated CO$_2$ than are C$_4$ species (and presumably also CAM species, but see CAM case study below), and fast-growing species with greater sink strength and higher SLA's also respond more strongly than do slow-growing species. However, these relationships appear to become less pronounced, or to disappear altogether, when plants are confronted with environmental stress. Therefore, it is not surprising that in a review of multiple FACE sites, no statistically significant trends emerged when examining either photosynthetic or growth responses of various functional groups to elevated CO$_2$ under varying climatic conditions (Nowak *et al.* 2004a). Therefore, we conclude that current functional group classifications are useful for examining the mechanistic underpinnings of plant responses to elevated CO$_2$, but they have limited utility when examining community- or ecosystem-level responses to global change (Reich *et al.* 2001).

Case studies

CAM succulents

Initially, CAM plants, such as desert cacti, were assumed to be minimally responsive to elevated CO_2 due to the CO_2 concentrating mechanism of PEP carboxylase (PEPCase), the primary carboxylating enzyme for these plants. However, some well-watered CAM plants utilize Rubisco during day-time CO_2 fixation (Osmond 1978), thereby suggesting that they might be responsive to elevated CO_2.

Park Nobel's lab spent nearly a decade studying the responses of CAM plants to elevated CO_2. A majority of their experiments were conducted outdoors with plants rooted in the ground (*i.e.*, not in pots) grown at ambient and twice-ambient $[CO_2]$ (*ca.* 36-37 and 72-73 Pa, respectively) in open-top chambers (*see* Drennan and Nobel (2000) and Chapter 3 in this volume). Although they utilized a wide variety of species in their studies, the primary taxa that they examined were *Opuntia ficus-indica*, a highly productive stem succulent that is widely cultivated in Mexico and the subtropics, and leaf-succulent *Agave* spp. (*A. deserti* from the Sonoran Desert and *A. salminiana*, a cultivated plant from central Mexico; Nobel 1988).

In all of their studies with CAM plants, elevated CO_2 was found to substantially increase CO_2 assimilation and growth, similar to that commonly observed in C_3 plants and contrary to early expectations. *Opuntia ficus-indica* and *Agave* spp. increased daily net CO_2 uptake by an average of 50% and water-use efficiency (gas-exchange based) by 85%, while biomass increased by 34% (Table 14.1). However, as an apparent photosynthetic acclimation response, there were decreases in chlorophyll content (an average of 16%), PEPCase concentration (35%), Rubisco concentration (18%), and the K_M for PEPCase (30%). These declines were accompanied by a 52% increase in the ratio of activated-to-total Rubisco and a small (6%) increase in the K_M of Rubisco (Table 14.1). Therefore, increased $[CO_2]$ apparently facilitated efficient utilization of both PEPCase and Rubisco, which decreased investment of nitrogen in carboxylating enzymes while still maintaining increased photosynthetic CO_2 assimilation. These CAM plants also increased the proportion of daily CO_2 uptake that occurred during the daylight period (Nobel and Hartsock 1986; Cui *et al.* 1993; Graham and Nobel 1996), which may have contributed to the strong positive photosynthetic response to elevated CO_2 that is typical of C_3 plants.

Photosynthetic down-regulation is thus observed in *Opuntia ficus-indica* during long-term exposure to elevated CO_2 (Cui *et al.* 1993; Nobel *et al.* 1994). Wang and Nobel (1996) observed increased phloem transport of sucrose out of basal cladodes of *O. ficus-indica* (*i.e.*, indicating strong sink strength in daughter cladodes), and increased soluble starch synthase and sucrose synthase activity in daughter cladodes. Based on these data, reductions in carboxylating enzymes in chlorenchyma tissues may have allowed more nitrogen to be available for carbohydrate processing in sink tissues, which in turn alleviated P$_i$ limitation of photosynthesis or the suppression of genes governing photosynthesis due to end product inhibition (*e.g.*, Moore *et al.* 1999).

Desert shrubs

EVERGREEN SHRUBS—Evergreen shrubs retain physiologically active leaf area throughout the year, and therefore must adjust to environmental factors during times that are both conducive and non-conducive to photosynthesis and growth. The evergreen growth habit is viewed as characteristic of resource-limited environments, adapting to low resource availability through retention of a leaf canopy and efficient resource use. Non-succulent evergreens in desert environments are best represented by xerophytic shrubs, particularly the warm desert dominant *Larrea tridentata* (creosotebush), which is very efficient in utilizing limited resources and extremely tolerant of environmental stress, particularly a lack of water (Smith *et al.* 1997). Because elevated CO$_2$ tends to increase photosynthetic rates while simultaneously reducing g_s, and therefore substantially increasing instantaneous water-use efficiency, it has been assumed that plants which are regularly exposed to water stress may indeed be highly responsive to elevated CO$_2$ (Strain and Bazzaz 1983). However, the inherent trade-off between growth and stress tolerance in xerophytes may limit their ability to exploit additional CO$_2$ even when other resources are not readily available.

Table 14.1. Physiological and growth responses for *Agave* spp. (*deserti* or *salminiana*) and *Opuntia ficus-indica* to elevated [CO$_2$]. Data are presented as an E/A ratio (elevated CO$_2$/ambient CO$_2$) for each parameter; in each case elevated CO$_2$ was a doubling of ambient CO$_2$ concentration, to 70 Pa. Data are from Drennan and Nobel (2000), except for the enzyme activity data, which are from Israel and Nobel (1994).

Parameter	*Agave*	*Opuntia*
Biomass	1.30	1.37
Integrated 24h CO$_2$ uptake (A_{24})	1.49	1.46
Daytime CO$_2$ uptake (A_{light})	1.41	1.87
Water-use efficiency (gas exchange based)	2.10	1.60
Chlorophyll content (area based)	0.80	0.80
PEPCase content (area based)	0.67	0.62
Rubisco content (area based)	0.87	0.77
K_M for PEPCase	0.85	0.56
K_M for Rubisco	1.04	1.09

The physiological and growth responses of *Larrea* at elevated CO_2 have been studied in both field (Nevada Desert FACE Facility; Smith *et al.* 2000) and glasshouse plants. Elevated CO_2 increases A_{net} (Huxman *et al.* 1998c; Hamerlynck *et al.* 2000b; Naumburg *et al.* 2003), and tends to result in a decline in g_s (Nowak *et al.* 2001; Naumburg *et al.* 2003) (Table 14.2). However, elevated CO_2 has a differential photosynthetic response in *Larrea* plants during times of high soil moisture and during water stress, with a clear pattern of photosynthetic down-regulation (lower A_{max} and V_{cmax}) during the wet season but not during the dry season (Huxman *et al.* 1998c; Table 14.2). Furthermore, wet-dry annual cycles in the field have resulted in a doubling of new shoot production in *Larrea* during a wet El Niño year, but then little enhancement of shoot growth in subsequent dry years (Smith *et al.* 2000; Housman *et al.* 2006), again suggesting a possible remobilization of limited nitrogen to actively growing sink tissues. Although g_s has been observed to consistently decrease at elevated CO_2 (Table 14.2) despite 6 and 11% increases in stomatal density and aperture, respectively (Reid *et al.* 2003), this apparently does not result in decreased plant transpiration (Pataki *et al.* 2000). This is probably because reduced leaf area per unit stem length (Table 14.2) due to greater internode development results in a substantial decrease in LSR in *Larrea* at elevated CO_2 (Pataki *et al.* 2000), while root hydraulic conductivity remains unchanged (Huxman *et al.* 1999a).

The view that greater water-use efficiency should result in greater stress tolerance has been only partially borne out in our experiments. In an experiment with glasshouse-grown *Larrea*, Hamerlynck *et al.* (2000a) found that elevated CO_2 enhanced photosynthesis to a greater degree in droughted plants than in well-watered plants. When exposed to extreme heat (53 °C), plants grown at elevated CO_2 showed complete recovery of photosynthetic processes when the heat stress was removed, whereas ambient CO_2-grown plants did not exhibit full recovery, suggesting that elevated CO_2 may improve heat tolerance in *Larrea*. However, in field-grown plants, Naumburg *et al.* (2004) found nighttime freezing and summer high temperatures to affect photosynthetic processes similarly at both ambient and elevated CO_2. In each case, the degree of photosynthetic enhancement under elevated CO_2 was directly proportional to the response of g_s to CO_2 and temperature. Nevertheless, *Larrea* seedlings germinated at the Nevada Desert FACE Facility showed enhanced survivorship, but not photosynthesis or growth, at elevated CO_2 during the first post-germination drought cycle (Housman *et al.* 2003). This is clearly an area that needs additional research emphasis, given the contrasting results from these studies.

Table 14.2. Photosynthetic and growth responses (E/A ratios) to elevated [CO_2] for the evergreen shrub *Larrea tridentata* during wet and dry years (from Huxman *et al.* 1998c; Hamerlynck *et al.* 2000b), the deciduous shrub *Lycium andersonii* for short- and long-shoots in a wet year (Hamerlynck *et al.* 2002), and the deciduous shrub *Encelia farinosa* (Zhang and Nobel 1996). Elevated [CO_2] are 55 Pa for *Larrea* and *Lycium* at the Nevada Desert FACE Facility, and 75 Pa for potted plants of *Encelia* in a growth chamber. Parameters are expressed as the E/A ratio (elevated/ambient) of daily average net CO_2 assimilation (A_{net}), stomatal conductance (g_s), CO_2 assimilation at light and CO_2 saturation (A_{max}), maximum carboxylation rate of Rubisco (V_{cmax}), light-saturated photosynthetic electron transport rate (J_{max}), sucrose concentration during the day in mature leaves ([Sucrose]), starch concentration during the day in mature leaves ([Starch]), leaf area, and plant dry weight; leaf area and plant dry weight parameters are on a per-shoot basis for *Larrea* and *Lycium* on a per-plant basis for *Encelia*.

Parameter	Larrea		Lycium		Encelia
	Wet	Dry	Short	Long	
A_{net}	1.30	1.33	1.09	0.89	
g_s	0.74	0.89	0.75	0.70	
A_{max}	0.87	1.21			0.79
V_{cmax}	0.82	0.90	0.76	0.98	
J_{max}			0.76	0.82	
[Sucrose]					1.51
[Starch]					1.90
Leaf area	0.81	1.02	1.36	1.28	1.27
Plant dry weight	[*ca.* 2.0]	[*ca.* 1.0]			

* Shoot-based productivity values (from Smith *et al.* 2000).

DECIDUOUS SHRUBS—Many deciduous shrubs, particularly those that lose their leaves in response to a predictable dry season (as opposed to winter deciduousness) have larger, more mesophytic leaves than desert evergreen shrubs (Smith *et al.* 1997). Two warm desert shrubs have been studied with regard to their responsiveness to elevated CO_2 – *Encelia farinosa* (brittlebush) from the Sonoran Desert, and *Lycium andersonii* (wolfberry) from the Mojave Desert. *Lycium* did not exhibit a significant effect of elevated CO_2 on A_{net} at the Nevada Desert FACE Facility, but did exhibit declines in g_s similar to those observed in the evergreen *Larrea* (Table 14.2). This contrasts with another drought-deciduous shrub from the same site, the microphyllous *Ambrosia dumosa*, which did show significant increases in A_{net} but no decline in g_s at elevated CO_2 (Naumburg *et al.* 2003). *Lycium* also exhibited strong down-regulation of photosynthesis, particularly in short shoots, which supply energy to long shoots, where active growth occurs. Down-regulation was particularly pronounced in short shoots early in the growing season when photosynthesis was high but long shoots were not yet strong carbon sinks (Hamerlynck *et al.* 2002). The lack of a stimulation of photosynthesis in long shoots further suggests end-product inhibition of photosynthesis in these structures, despite increased leaf area at elevated CO_2 (Table 14.2). Indeed, *Encelia* also exhibited a down-regulation of photosynthesis (lower A_{max} despite an increase in A_{net}), which was accompanied by large increases in the concentration of sucrose and especially starch in mature leaves (Table 14.2; Zhang and Nobel 1996). A lack of stimulation of root growth at elevated CO_2 in *Encelia* (Drennan and Nobel 1996) further indicates no pronounced increase in sink strength belowground in this species at elevated CO_2.

These combined results from desert shrubs suggest that we cannot make life form generalizations with regard to their responses to elevated CO_2 (see Section II.D above). We predicted that elevated CO_2 would increase stress tolerance in the evergreen *Larrea*, and increase growth and photosynthesis in the more mesophytic deciduous shrubs when water and nitrogen were readily available in the soil. However, we did not observe a sustained increase in A_{net} in *Lycium* and photosynthetic down-regulation in *Larrea* was only observed in the wet season. Thus, plant responses to elevated CO_2 may be species-specific and determined by annual patterns in growth dynamics and resource availability that have not been accounted for in our relatively short-term experiments. And, as noted above, climatic patterns may buffer responses to elevated CO_2, and episodic extremes may override or interact with physiological effects.

Herbaceous annuals

Annual plants are a major source of species diversity in deserts (Shreve and Wiggins 1964) and significantly affect year-to-year production and water use (Turner and Randall 1989). The episodic nature of surface-soil water availability has shaped the evolution of their life history characteristics, population dynamics and community composition (Venable *et al.* 1993; Pake and Venable 1995). In the deserts of North America, many important invasive plant species with significant ecosystem effects have annual life history strategies (D'Antonio and Vitousek 1992; Brooks 1999a, b). How global change may influence the diverse flora of annual plants in these arid regions, or may alter the impact of exotic invasive species is of great concern for land managers.

Desert annuals tend to have relatively high leaf-level photosynthetic rates, along with greater investment of nitrogen in their leaf tissues as compared to congeners from more mesic environments (Mooney *et al.* 1981). The rapid growth rates that derive from these traits suggest that this functional type may be highly responsive to changes in $[CO_2]$ as compared to the balance of the desert flora, mainly consisting of much slower growing woody perennials (Poorter and Navas 2000).

An herbaceous forb and an annual invasive grass exposed to elevated CO_2 in FACE conditions in the Mojave Desert showed contrasting leaf-level photosynthetic responses to elevated CO_2 (Huxman and Smith 2001). Prior to and following flowering, the invasive grass *Bromus madritensis* ssp. *rubens* increased both V_{cmax} and J_{max} at elevated CO_2, but showed no consistent change in g_s over the course of a growing season. By contrast, the forb *Eriogonum inflatum* illustrated significant photosynthetic down-regulation at elevated CO_2, especially late in the life cycle. Also, this species showed consistent reductions in g_s. Therefore, both species maintained greater leaf-level photosynthetic rates at elevated CO$_2$ compared to ambient CO$_2$, but as a result of different mechanisms.

Increased photosynthetic performance resulted in changes in whole plant function. During an extremely wet El Niño growing season in the Mojave Desert, individual aboveground biomass at a 50% increase in [CO$_2$] was more than two-fold that at ambient [CO$_2$]. The responsiveness of different plant species appears to be predictable based on differences in their intrinsic growth potential and the effects of rising CO$_2$ on tissue construction costs – plants with a greater ability to reduce construction costs had a greater biomass stimulation at elevated CO$_2$ (Nagel et al. 2004). Interestingly, increases in annual plant biomass production appear to be a greater function of the direct effects of CO$_2$ on leaf-level processes, rather than indirect effects, such as increasing the period of high soil water availability into later parts of the year.

Annual plants in the Mojave and Sonoran Deserts exhibit rapid growth and seed production during fairly short growing seasons, and as a result soil seed banks accumulate that buffer environmental variation and promote long-term persistence on the landscape (Venable and Lawlor 1980; Cabin et al. 1998, 2000; Cabin and Marshall 2000). Annual plants are often described as "stress escapers" due to this aspect of their life history strategy – being present only in the form of seeds during periods of extended stress (Smith et al. 1997). This 'escape in time' strategy is the key life history characteristic promoting both persistence and species coexistence in a variable environment (Venable and Lawlor 1980). Rising [CO$_2$] influences a number of factors that are important in seed production and persistence. The quantity and the quality of seeds produced are both important factors in seed bank persistence over time. Desert plants produce seed numbers in relation to plant size (Bell et al. 1979; Ehleringer 1985), and nutrient status affects the quality and quantity of seeds produced in many annuals (Venable 1992).

The large effects of elevated CO$_2$ on biomass production described above predict large reproductive responses from annuals. Indeed, total seed production from the entire annual plant community was nearly doubled when [CO$_2$] was increased 50% above ambient levels (Smith et al. 2000). However, the reproductive response to elevated CO$_2$ was species-specific, and related to biomass responsiveness, changes in plant tissue quality, and impacts on the timing of reproduction (Huxman et al. 1999b; Nagel et al. 2004). The native species *Lepidium lasiocarpum* and *Vulpia octoflora* had increased seed production per plant at elevated CO$_2$, especially in 'fertile-island' microsites associated with the canopies of perennial plants (Smith et al. 2000). Similarly, the invasive annual *Bromus madritensis* showed increases in total seed production per plant at elevated CO$_2$, but had decreases in individual seed mass. These changes in seed mass are consistent across the major microsites on the landscape, with the exception of within the canopy of the C$_4$ grass *Pleuraphis rigida* (Fig. 14.2). Interestingly, this microsite is the only location where the physical benefits of a 'fertile island' are present but the perennial plant canopy is inactive during the growing season of the annual plants.

Decreases in seed mass at elevated CO_2 have been shown to result in up to a 20% reduction in growth rates of seedlings emerging from these seeds (Huxman *et al.* 1998a). It is not clear how increased plant performance at elevated CO_2 and changes in seed quality affect long-term patterns of population dynamics in these species, but considering the role of seeds as the primary adaptation allowing for persistence through drought (Smith *et al.* 1997), and considering how seed production is an important factor that allows for species coexistence in aridlands (Venable *et al.* 1993), the implications are significant.

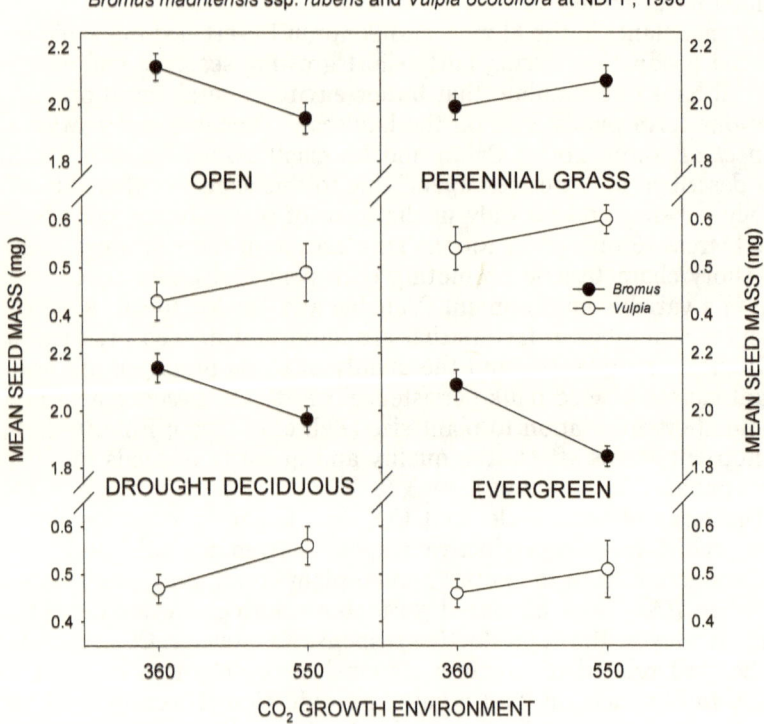

Figure 14.2. Seed mass of two Mojave Desert annual grasses, the invasive *Bromus madritensis* ssp. *rubens* (closed circles) and the native *Vulpia octoflora* (open circles), at ambient CO_2 (360; abscissa) and elevated CO_2 (550) in a wet El Niño year (1998) at the Nevada Desert FACE Facility. Seed mass data are for plants from open interspace microsites (upper left), within a perennial bunchgrass (upper right), beneath a drought-deciduous shrub (lower right), and beneath the evergreen shrub *Larrea* (lower right).

Table 14.3. Response of net photosynthesis (A_{net}), leaf nitrogen concentration (area based), and aboveground net primary production (ANPP) to growth at elevated CO_2 (depicted as E/A ratios) for forest, grassland, and desert ecosystems. Data are means for different species at different free-air CO_2 enrichment (FACE) sites, with the dominant growth form measured given in parentheses below. Data are from Nowak *et al.* (2004a).

Biome Type (growth form)	A_{net}	Leaf N	ANPP
Forests (trees)	1.46	1.06	1.19
Grasslands (herbaceous)	1.16	0.94	1.14
Desert (shrubs)	1.15	0.89	1.38

Summary and conclusions

In this review, we examined the hypothesis that elevated CO_2 will increase the efficiency with which key resources, such as water and nitrogen, are utilized by plants in arid ecosystems. Utilization patterns of these resources are key to understanding aridland ecosystem function in a future high CO_2 world. In a comparative analysis of plant response to elevated CO_2 across free-air CO_2 enrichment (FACE) sites in the U.S., Nowak *et al.* (2004a) found that desert shrubs exhibited approximately the same photosynthetic stimulation as herbaceous species from grasslands, both of which were lower than forest trees (Table 143). This appears to be at least partially correlated with greater down-regulation in leaf nitrogen concentration at elevated CO_2 in deserts and grasslands. However, of potentially greater interest is the observation that the response of annual aboveground production to elevated CO_2 is highest in deserts (Table 14.3), suggesting that plant growth is driven by important carbon and nitrogen allocation processes, as well as photosynthesis. Indeed, almost all of the productivity increases observed in the desert FACE site occurred in wet years, when soil resources were more abundant (S.D. Smith unpublished data) and there was minimal sink control of new growth. In contrast, growth in forests and grasslands was less variable on an annual basis, presumably leading to more predictable sink controls in these communities.

A second broad objective of this review was to examine different functional types of desert plants, and to ascertain whether there were broadly predictable responses of different growth forms to elevated CO_2. We found daily photosynthesis to be enhanced by elevated CO_2 in both evergreen and deciduous shrubs despite significant photosynthetic down-regulation in both the evergreen *Larrea* and the deciduous *Encelia*. We also found that reduced g_s in the dominant shrub *Larrea* and in some annuals at elevated CO_2 did not result in enhanced soil moisture, as has been observed in semiarid ecosystems. We also found a consistent stimulation of photosynthesis and growth in several desert CAM plants (*Agave* and *Opuntia*) at levels commonly observed in C_3 plants – this was attributed to a more efficient utilization of both PEPCase and Rubisco, and more day-time CO_2 uptake at elevated CO_2. We conclude that the response of desert plants to elevated CO_2 is probably species-specific and determined by phenological patterns and source-sink dynamics in concert with the timing of resource availability, all of which are highly variable on an inter-annual basis in these episodic environments. While providing important insights into these patterns, our relatively short-term experiments have yet to account for long-term patterns that may ultimately drive community and ecosystem change, particularly if continuing increases in [CO_2] are accompanied by pronounced climate change.

Literature cited

Acevedo E, Hsiao TC, and Henderson DW. 1971. Immediate and subsequent growth responses of maize leaves to changes in water status. *Plant Physiology* **48:** 631-636.

Amthor JS. 1991. Respiration in a future, higher-CO_2 world. *Plant, Cell and Environment* **14:** 13-20.

Amthor JS. 2000. Direct effects of elevated CO_2 on nocturnal *in situ* leaf respiration in nine temperate deciduous tree species is small. *Tree Physiology* **20:** 139-144.

Amthor JS, Koch GW, and Bloom AJ. 1992. CO_2 inhibits respiration in leaves of *Rumex crispus* L. *Plant Physiology* **98:** 757-760.

Amthor JS, Koch GW, Willms JR, and Layzell DB. 2001. Leaf CO_2 uptake in the dark is independent of coincident CO_2 partial pressure. *Journal of Experimental Botany* **52:** 2235-2238.

Archer S, Schimel DS, and Holland EA. 1995. Mechanisms of shrubland expansion: land use, climate, or CO_2? *Climatic Change* **29:** 91-99.

Arp WJ, Drake BG, Pockman WT, Curtis PS, and Whigham DF. 1993. Interactions between C_3 and C_4 salt marsh species during four years of exposure to elevated atmospheric CO_2. *Vegetatio* **104/105:** 133-143.

Arp WJ, van Mierlo JEM, Berendse F, and Snijders W. 1998. Interactions between elevated CO2 concentration, nitrogen and water: effects on growth and water use of six perennial plant species. *Plant, Cell and Environment* **21:** 1-11.

Atkin OK, Schortemeyer M, McFarlane N, and Evans JR. 1999. The response of fast- and slow-growing *Acacia* species to elevated atmospheric CO_2: an analysis of the underlying components of relative growth rate. *Oecologia* **120:** 544-554.

Azcon-Bieto J, Gonzalez-Meler MA, Doherty W, and Drake BG. 1994. Acclimation of respiratory O_2 uptake in green tissues of field grown native species after long-term exposure to elevated atmospheric CO_2. *Plant Physiology* **106:** 1163-1168.

Baker JT, Allen Jr. LH, Boote KJ, and Pickering NB. 2000. Direct effects of atmospheric carbon dioxide concentration on whole canopy dark respiration of rice. *Global Change Biology* **6:** 275-286.

Barker DH, Loveys BR, Egerton JJG, Gorton H, Williams WE, and Ball MC. 2005. CO_2 enrichment predisposes foliage of a eucalypt to freezing injury and reduces spring growth. *Plant, Cell and Environment* **28:** 1506-1515.

BassiriRad H, Reynolds JF, Virginia RA, and Brunelle MH. 1997. Growth and root NO^-_3 and PO^{3-}_4 uptake capacity of three desert species in response to atmospheric CO_2 enrichment. *Australian Journal of Plant Physiology* **24:** 353-358.

Bell KL, Hiatt HD, and Niles WE. 1979. Seasonal changes in biomass allocation in eight winter annuals of the Mohave Desert. *Journal of Ecology* **67:** 781-787.

Boese SR, Wolfe DW, and Melkonian JJ. 1997. Elevated CO$_2$ mitigates chilling-induced water stress and photosynthetic reduction during chilling. *Plant, Cell and Environment* **20:** 625-632.

Bohnert HJ, Gong QQ, Li PH, and Ma SS. 2006. Unraveling abiotic stress tolerance mechanisms – getting genomics going. *Current Opinions in Plant Biology* **9:**180-188.

Bond WJ and Midgley GF. 2000. A proposed CO2-controlled mechanism of woody plant invasion in grasslands and savannas. *Global Change Biology* **6:** 865-869.

Bowes G. 1993. Facing the inevitable: Plants and increasing atmospheric CO$_2$. *Annual Review of Plant Physiology and Plant Molecular Biology* **44:** 309-332.

Boyer JS. 1970. Leaf enlargement and metabolic rates in corn, sorghum and sunflower at various water potentials. *Plant Physiology* **46:** 233-235.

Bradford KJ, and Hsiao TC. 1982. Physiological responses to moderate water stress. In: Lange OL, Nobel PS, Osmond CB, and Ziegler H (eds.) *Physiological Plant Ecology II: Water Relations and Carbon Assimilation.* Encyclopedia of Plant Physiology, New Series, Springer-Verlag, Berlin. Pp. 263-324.

Brooks ML. 1999a. Alien annual grasses and fire in the Mojave Desert. *Madroño* **46:** 13-19.

Brooks ML. 1999b. Habitat invasibility and dominance by alien annual plants in the western Mojave Desert. *Biological Invasions* **1:** 325-337.

Bunce JA and Ziska LH. 1996. Responses of respiration to increases in carbon dioxide concentration and temperature in three soybean cultivars. *Annals of Botany* **77:** 507-514.

Cabin RJ and Marshall DL. 2000. The demographic role of soil seed banks. I. Spatial and temporal comparisons of below- and above-ground populations of the desert mustard *Lesquerella fendleri. Journal of Ecology* **88:** 283-292.

Cabin RJ, Marshall DL, and Mitchell RJ. 2000. The demographic role of soil seed banks. II. Investigations of the fate of experimental seeds of the desert mustard *Lesquerella fendleri. Journal of Ecology* **88:** 293-302.

Cabin RJ, Mitchell RJ, and Marshall DL. 1998. Do surface plant and soil seed bank populations differ genetically? A multipopulation study of the desert mustard *Lesquerella fendleri* (Brassicaceae). *American Journal of Botany* **85:** 1098-1109.

Cornelissen JHC, Carnelli AL, and Callaghan TV. 1999. Generalities in the growth, allocation and leaf quality responses to elevated CO$_2$ in eight woody species. *New Phytologist* **141:** 401-409.

Cui M, Miller PM, and Nobel PS. 1993. CO$_2$ exchange and growth of the Crassulacean acid metabolism plant *Opuntia ficus-indica* under elevated CO$_2$ in open-top chambers. *Plant Physiology* **103:** 519-524.

D'Antonio CM and Vitousek PM. 1992. Biological invasions by exotic grasses, the grass/fire cycle, and global change. *Annual Review of Ecology and Systematics* **23:** 63-87.

Diaz S. 1995. Elevated CO2 responsiveness, interactions at the community level and plant functional types. *Journal of Biogeography* **22:** 289-295.

Drake BG, Azcon-Bieto J, and Berry J. 1999. Does elevated atmospheric CO$_2$ concentration inhibit mitochondrial respiration in green plants? *Plant, Cell and Environment* **22:** 649-657.

Drennan PM and Nobel P.S. 1996. Temperature influences on root growth for *Encelia farinosa* (Asteraceae), *Pleuraphis rigida* (Poaceae), and *Agave deserti* (Agavaceae) under current and doubled CO$_2$ concentrations. *American Journal of Botany* **83:** 133-139.

Drennan PM and Nobel PS. 2000. Responses of CAM species to increasing atmospheric CO$_2$ concentrations. *Plant, Cell and Environment* **23:** 767-781.

Ehleringer JR. 1985. Annuals and perennials of warm deserts. In: Chabot BF, and Mooney HA (eds.) *Physiological Ecology of North American Plant Communities*. Chapman and Hall, New York. Pp. 162-180.

Field CB, Jackson RB, and Mooney HA. 1995. Stomatal responses to increased CO_2 – implications from the plant to the global scale. *Plant, Cell and Environment* **18:** 1214-1225.

Ghannoum O, von Caemmerer S, Ziska LH, and Conroy JP. 2000. The growth response of C_4 plants to rising atmospheric CO_2 partial pressure: a reassessment. *Plant, Cell and Environment* **23:** 931-942.

González-Meler MA and Siedow JN. 1999. Direct inhibition of mitochondrial respiratory enzymes by elevated CO_2: does it matter at the tissue or whole-plant level? *Tree Physiology* **19:** 253-259.

González-Meler MA, Ribas-Carbo M, Siedow JN, and Drake BG. 1996. Direct inhibition of plant mitochondrial respiration by elevated CO_2. *Plant Physiology* **112:** 1349-1355.

Graham EA and Nobel PS. 1996. Long-term effects of a doubled atmospheric CO_2 concentration on the CAM species *Agave deserti*. *Journal of Experimental Botany* **47:** 61-69.

Griffin KL, Anderson OR, Gastrich MD, Lewis JD, Lin G, Schuster W, Seemann JR, Tissue DT, Turnbull MT, and Whitehead D. 2001. Plant growth in elevated CO_2 alters mitochondrial number and chloroplast fine structure. *Proceedings of the National Academy of Science (USA)* **98:** 2473-2478.

Hamerlynck EP, Huxman TE, Charlet TN, and Smith SD. 2002. Effects of elevated CO_2 (FACE) on the functional ecology of the drought-deciduous Mojave Desert shrub, *Lycium andersonii*. *Environmental and Experimental Botany* **48:** 93-106.

Hamerlynck EP, Huxman TE, Loik ME, and Smith SD. 2000a. Effects of extreme high temperature, drought and elevated CO_2 on photosynthesis of the Mojave Desert evergreen shrub, *Larrea tridentata*. *Plant Ecology* **148:** 183-193.

Hamerlynck EP, Huxman TE, Nowak RS, Redar S, Loik ME, Jordan DN, Zitzer SF, Coleman JS, Seemann JR, and Smith SD. 2000b. Photosynthetic responses of *Larrea tridentata* to a step-increase in atmospheric CO_2 at the Nevada Desert FACE Facility. *Journal of Arid Environments* **44:** 425-436.

Hamilton JG, Thomas RB, and DeLucia EH. 2001. Direct and indirect effects of elevated CO_2 on leaf respiration in a forest ecosystem. *Plant, Cell and Environment* **24:** 975-982.

Housman DC, Naumburg E, Huxman TE, Charlet TN, Nowak RS, and Smith SD. 2006. Increases in desert shrub productivity under elevated CO_2 vary with water availability. *Ecosystems* **9:** 374-385.

Housman DC, Zitzer SF, Huxman TE, and Smith SD. 2003. Functional ecology of shrub seedlings after a natural recruitment event at the Nevada Desert FACE Facility. *Global Change Biology* **9:** 718-728.

Hsiao TC and Jackson RB. 1999. Interactive effects of water stress and elevated CO_2. In: Luo Y and Mooney HA (eds.) *Carbon Dioxide and Environmental Stress*. Academic Press, San Diego. Pp. 3-29.

Hui DF and Jackson RB. 2006. Geographical and interannual variability in biomass partitioning in grassland ecosystems: a synthesis of field data. *New Phytologist* **169:** 85-93.

Huxman KA, Smith SD, and Neuman DS. 1999a. Root hydraulic conductivity of *Larrea tridentata* and *Helianthus annuus* under elevated CO_2. *Plant, Cell and Environment* **22:** 325-330.

Huxman TE and Smith SD. 2001. Photosynthesis in an invasive grass and native forb at elevated CO₂ during an El Niño year in the Mojave Desert. *Oecologia* **128:** 193-201.

Huxman TE, Hamerlynck EP, Jordan DN, Salsman KA, and Smith SD. 1998a. The effects of parental CO₂ environment on seed quality and subsequent seedling performance in *Bromus rubens. Oecologia* **114:** 202-208.

Huxman TE, Hamerlynck EP, Loik ME, and Smith SD. 1998b. Gas exchange and chlorophyll fluorescence responses of three south-western *Yucca* species to elevated CO₂ and high temperature. *Plant, Cell and Environment* **21:** 1275-1283.

Huxman TE, Hamerlynck EP, Moore BD, Smith SD, Jordan DN, Zitzer SF, Nowak RS, Coleman JS, and Seemann JR. 1998c. Photosynthetic down-regulation in *Larrea tridentata* exposed to elevated atmospheric CO₂: interaction with drought under glasshouse and field (FACE) exposure. *Plant, Cell and Environment* **21:** 1153-1161.

Huxman TE, Hamerlynck EP, and Smith SD. 1999b. Reproductive allocation and seed production in *Bromus madritensis* ssp. *rubens* at elevated CO₂. *Functional Ecology* **13:** 769-777.

Israel AA and Nobel PS. 1994. Activities of carboxylating enzymes in the CAM species *Opuntia ficus-indica* grown under current and elevated CO₂ concentrations. *Photosynthesis Research* **40:** 223-229.

Jackson RB, Sala OE, Field CB, and Mooney HA. 1994. CO₂ alters water-use, carbon gain, and yield for the dominant species in a natural grassland. *Oecologia* **98:** 257-262.

Jahnke S. 2001. Atmospheric CO₂ concentration does not directly affect leaf respiration in bean or poplar. *Plant, Cell and Environment* **24:** 1139-1151.

Jastrow JD, Miller RM, Matamala R, Norby RJ, Boutton TW, Rice CW, and Owensby CE. 2005. Elevated atmospheric carbon dioxide increases soil carbon. *Global Change Biology* 11:2057-2064.

Johnson HB, Polley HW, and Mayeux HS. 1993. Increasing CO₂ and plant-plant interactions: effects on natural vegetation. *Vegetatio* **104/105:** 157-170.

Jordan DN, Zitzer SF, Hendrey GR, Lewin KF, Nagy J, Nowak RS, Smith SD, Coleman JS, and Seemann JR. 1999. Biotic, abiotic and performance aspects of the Nevada Desert Free-air CO₂ Enrichment (FACE) Facility. *Global Change Biology* **5:** 659-668.

Jordan PW and Nobel PS. 1979. Infrequent establishment of seedlings of *Agave deserti* in the northwestern Sonoran Desert. *American Journal of Botany* **66:** 1079-1084.

Karl TR and Trenberth KE. 2003. Modern global climate change. *Science* **302:** 1719-1723.

Körner Ch. 2003. Carbon limitation in trees. *Journal of Ecology* **91:** 4-17.

Körner Ch, Pelaez-Riedl S, and Bell AJE. 1995. CO₂ responsiveness of plants: a possible link to phloem loading. *Plant, Cell and Environment* **18:** 595-600.

Kozlowski TT. 2000. Responses of woody plants to human-induced environmental stresses: Issues, problems, and strategies for alleviating stress. *Critical Reviews in Plant Sciences* **19:**91-170.

Lewis JD, Olszyk D, and Tingey DT. 1999. Seasonal patterns of photosynthetic light response in Douglas-fir seedlings subjected to elevated CO₂ and temperature. *Tree Physiology* **19:** 243-252.

Loik ME and Nobel PS. 1993. Freezing tolerance and water relations of *Opuntia fragilis* from Canada and the United States. *Ecology* **74:** 1722-1732.

Loik ME, Huxman TE, Hamerlynck EP, and Smith SD. 2000. Low temperature tolerance and cold acclimation for seedlings of three Mojave Desert *Yucca* species exposed to elevated CO_2. *Journal of Arid Environments* **46:** 43-56.

Loik ME, Breshears DD, Lauenroth WK, and Belnap J. 2004a. A multi-scale perspective of water pulses in dryland ecosystems: climatology and ecohydrology of the western USA. *Oecologia* **140:** 11-25.

Loik ME, Still CJ, Huxman TE, and Harte J. 2004b. *In situ* photosynthetic freezing tolerance of plants exposed to a global warming manipulation in the Rocky Mountains, Colorado, U.S.A. *New Phytologist* **162:** 331-341.

Long SP. 1991. Modification of the response of photosynthetic productivity to rising temperature by atmospheric CO_2 concentrations: has its importance been underestimated? *Plant, Cell and Environment* **14:** 729-739.

Luo YQ, Hui DF, and Zhang DQ. 2006. Elevated CO_2 stimulates net accumulations of carbon and nitrogen in land ecosystems: A meta-analysis. *Ecology* **87**:53-63.

Lüscher A, Hendrey GR, and Nösberger J. 1998. Long-term responsiveness to free-air CO2 enrichment of functional types, species and genotypes of plants from fertile permanent grassland. *Oecologia* **113:** 37-45.

Lynch JP, and St. Clair SB. 2004. Mineral stress: the missing link in understanding how global climate change will affect plants in real world soils. *Field Crops Research* **90**:101-115.

Margolis HA and Vezina LP. 1990. Atmospheric CO_2 enrichment and the development of frost hardiness in containerized black spruce seedlings. *Canadian Journal of Forest Research* **20:** 1392-1398.

Medlyn BE, Badeck FW, de Pury DG, Barton CV, Broadmeadow M, Ceulemans R, de Angelis P, Forstreuter M, Jach ME, Kellomaki S, Laitat E, Marek M, Philippot S, Rey A, Strassemeyer J, Laitinen K, Liozon R, Portier B, Roberntz P, Wang K, and Jarvis PG. 1999. Effects of elevated [CO_2] on photosynthesis in European forest species: a meta-analysis of model parameters. *Plant, Cell and Environment* **22:** 1475-1495.

Melillo JM, McGuire AD, Kicklighter DW, Moore B III, Vorosmarty CJ, and Schloss AL. 1993. Global climate change and terrestrial net primary production. *Nature* **363:** 234-240.

Mooney HA, Field C, Gulmon SL, and Bazzaz FA. 1981. Photosynthetic capacity in relation to leaf position in desert versus old-field annuals. *Oecologia* **50:** 109-112.

Moore BD, Cheng SH, Sims D, and Seemann JR. 1999. The biochemical and molecular basis for photosynthetic acclimation to elevated atmospheric CO_2. *Plant, Cell and Environment* **22:** 567-582.

Moore LA and Field CB. 2006. The effects of elevated atmospheric CO2 on the amount and depth distribution of plant water uptake in a California annual grassland. *Global Change Biology* **12**:578-587.

Morgan JA, Mosier AK, Milchunas DG, LeCain DR, Nelson JA, and Parton WJ. 2004a. CO_2 enhances productivity, alters species composition, and reduces digestibility of shortgrass steppe vegetation. *Ecological Applications* **14:** 208-219.

Morgan JA, Pataki DE, Körner C, Clark H, Del Grosso SJ, Grunzweig JM, Knapp AK, Mosier AR, Newton PCD, Niklaus PA, Nippert JB, Nowak RS, Parton WJ, Polley HW, and Shaw MR. 2004b. Water relations in grassland and desert ecosystems exposed to elevated atmospheric CO_2. *Oecologia* **140:** 11-25.

Nagel JM, Huxman TE, Griffin KL, and Smith SD. 2004. CO_2 enrichment reduces the energetic cost of biomass construction in an invasive desert grass. *Ecology* **85:** 100-106.

Naumburg E, Housman DC, Huxman TE, Charlet TN, Loik ME, and Smith SD. 2003. Photosynthetic responses of Mojave Desert shrubs to free-air CO_2 enrichment are greatest during wet years. *Global Change Biology* **9:** 276-285.

Naumburg E, Loik ME, and Smith SD. 2004. Photosynthetic responses of *Larrea tridentata* to seasonal temperature extremes under elevated CO_2. *New Phytologist* **162:** 323-330.

Niklaus PA, Wohlfender M, Siegwolf R, and Körner C. 2001. Effects of six years atmospheric CO_2 enrichment on plant, soil, and soil microbial C of a calcareous grassland. *Plant and Soil* **233:** 189-202.

Nobel PS. 1988. *Environmental Biology of Agaves and Cacti.* Cambridge University Press, Cambridge.

Nobel PS, and Hartsock TL. 1986. Short-term and long-term responses of Crassulacean acid metabolism plants to elevated CO_2. *Plant Physiology* **82:** 604-606.

Nobel PS and Israel AA. 1994. Cladode development, environmental responses of CO_2 uptake, and productivity for *Opuntia ficus-indica* under elevated CO_2. *Journal of Experimental Botany* **45:** 295-303.

Nobel PS, Cui M, Miller PM, and Luo Y. 1994. Light, chlorophyll, carboxylase activity and CO_2 fixation at various depths in the chlorenchyma of *Opuntia ficus-indica* (L.) Miller under current and elevated CO_2. *New Phytologist* **128:** 315-322.

Nobel PS, Israel AA, and Wang N. 1996. Growth, CO_2 uptake, and responses of the carboxylating enzymes to inorganic carbon in two highly productive CAM species at current and doubled CO_2 concentrations. *Plant, Cell and Environment* **19:** 585-592.

Norby RJ and Iverson CM. 2006. Nitrogen uptake, distribution, turnover, and efficiency of use in a CO_2-enriched sweetgum forest. *Ecology* **87:**5-14.

Nowak RS, DeFalco LA, Wilcox CS, Jordan DN, Coleman JS, Seemann JR, and Smith SD. 2001. Leaf conductance decreased under free-air CO_2 enrichment (FACE) for three perennials in the Nevada desert. *New Phytologist* **150:** 449-458.

Nowak RS, Ellsworth DS, and Smith SD. 2004a. *Tansley Review*: Plant functional responses to elevated atmospheric CO_2 – Do data from FACE experiments support early predictions? *New Phytologist* **162:** 253-280.

Nowak RS, Zitzer SF, Babcock D, Smith-Longozo V, Charlet TN, Coleman JS, Seemann JR, and Smith SD. 2004b. Elevated atmospheric CO2 does not conserve soil water in the Mojave Desert. *Ecology* **85:** 889-897.

Osmond CB. 1978. Crassulacean acid metabolism, a curiosity in context. *Annual Review of Plant Physiology* **29:** 379-414.

Osmond CB, Austin MP, Berry JA, Billings WD, Boyer JS, Dacey JWH, Nobel PS, Smith SD, and Winner WE. 1987. Stress physiology and the distribution of plants. *BioScience* **37:** 38-47.

Owensby CE, Ham JM, Knapp AK, and Auen LM. 1999. Biomass production and species composition change in a tallgrass prairie ecosystem after long-term exposure to elevated atmospheric CO_2. *Global Change Biology* **5:** 497-506.

Pake CE and Venable L. 1995. Is coexistence of Sonoran Desert annuals mediated by temporal variability in reproductive success? *Ecology* **76:** 246-261.

Pataki DE, Huxman TE, Jordan DN, Zitzer SF, Coleman JS, Smith SD, Nowak RS, and Seemann JR. 2000. Water use of two Mojave Desert shrubs under elevated CO_2. *Global Change Biology* **6:** 889-897.

Pockman WT and Sperry JS. 1997. Freezing induced xylem cavitation and the northern limit of *Larrea tridentata*. *Oecologia* **109:** 19-27.

Polley HW. 1997. Implications of rising atmospheric carbon dioxide concentration for rangelands. *Journal of Range Management* **50:** 561-577.

Polley HW, Johnson HB, and Tischler CR. 2002. Woody invasion of grasslands: evidence that CO_2 enrichment indirectly promotes establishment of *Prosopis glandulosa*. *Plant Ecology* **164:** 85-94.

Poorter H. 1998. Do slow-growing species and nutrient-stressed plants respond relatively strongly to elevated CO_2? *Global Change Biology* **4:** 693-697.

Poorter H and Navas ML. 2003. Plant growth and competition at elevated CO_2: on winners, losers, and functional groups. *New Phytologist* **157:** 175-198.

Poorter H, Gifford RM, Kriedemann PE, and Wong SC. 1992. A quantitative analysis of dark respiration and carbon content as factors in the growth response of plants to elevated CO_2. *Australian Journal of Botany* **40:** 501-513.

Reich PB, Ellsworth DS, and Walters MB. 2003. Leaf structure (specific leaf area) modulates photosynthesis:nitrogen relations: evidence from within and across species and functional groups. *Functional Ecology* **12:** 948-958.

Reich PB, Tilman D, Craine J, Ellsworth D, Tjoelker MG, Knops J, Wedin D, Naeem S, Bahauddin D, Goth J, Bengtson W, and Lee TD. 2001. Do species and functional groups differ in acquisition and use of C, N and water under varying atmospheric CO_2 and N availability regimes? A field test with 16 grassland species. *New Phytologist* **150:** 435-448.

Reid CD, Maherali H, Johnson HB, Smith SD, Wullschleger SD, and Jackson RB. 2003. On the relationship between stomatal characters and atmospheric CO_2. *Geophysical Research Letters* **30:** 1983-1987.

Robertson EJ, Williams M, Harwood JL, Lindsay JG, Leaver CJ, and Leach RM. 1995. Mitochondria increase three-fold and mitochondrial proteins and lipid change dramatically in post-meristematic cells in young wheat leaves grown in elevated CO_2. *Plant Physiology* **108:** 469-474.

Sage RF. 1994. Acclimation of photosynthesis to increasing atmospheric CO_2: the gas exchange perspective. *Photosynthesis Research* **39:** 351-368.

Sage RF, Sharkey TD, and Seemann JR. 1988. The *in vivo* response of the ribulose-1:5-bisphosphate carboxylase activation state and pool sizes of photosynthetic metabolites to elevated CO_2 in *Phaseolus vulgaris* L. *Planta* **174:** 407-416.

Schneider SH. 1992. The climatic response to greenhouse gases. *Advances in Ecological Research* **22:** 1-32.

Shreve F and Wiggins IL. 1964. *Vegetation and Flora of the Sonoran Desert, Vol. 1.* Stanford University Press, Stanford, CA.

Smith SD and Nowak RS. 1990. Ecophysiology of plants in the Intermountain lowlands. In: Osmond CB, Pitelka LF, and Hidy G (eds.) *Plant Biology of the Basin and Range*. Springer-Verlag, Berlin. Pp. 179-242.

Smith SD, Hartsock TL, and Nobel PS. 1983. Ecophysiology of *Yucca brevifolia*, an arborescent monocot of the Mojave Desert. *Oecologia* **60:** 10-17.

Smith SD, Herr C, Leary K, and Piorkowski J. 1985. Soil-plant water relations in a Mojave Desert mixed shrub community: a comparison of three geomorphic surfaces. *Journal of Arid Environments* **29:** 339-351.

Smith SD, Huxman TE, Zitzer SF, Charlet TN, Housman DC, Coleman JS, Fenstermaker LK, Seemann JR, and Nowak RS. 2000. Elevated CO_2 increases productivity and invasive species success in an arid ecosystem. *Nature* **408:** 79-82.

Smith SD, Jordan DN, and Hamerlynck EP. 1999. Effects of elevated CO_2 and temperature stress on ecosystem processes. In: Luo Y and Mooney HA (eds.) *Carbon Dioxide and Environmental Stress*. Academic Press, San Diego. Pp. 107-137.

Smith SD, Monson RK, and Anderson JE. 1997. *Physiological Ecology of North American Desert Plants*. Springer, Berlin.

Smith SD, Strain BR, and Sharkey TD. 1987. Effects of CO_2 enrichment on four Great Basin grasses. *Functional Ecology* **1:** 139-143.

Steenbergh WF and Lowe CH. 1976. Ecology of the Saguaro: I. The role of freezing weather in a warm-desert population. National Park Service Scientific Monograph Series, Number One. U.S. Government Printing Office, Washington, DC, USA.

Strain BR and Bazzaz FA. 1983. Terrestrial plant communities. In: Lemon E (ed.) *CO$_2$ and Plants: The Response of Plants to Rising Levels of Carbon Dioxide.* AAAS Symposium 84. American Association for the Advancement of Science, Washington, DC. Pp. 177-222.

Taub DR, Seemann JR, and Coleman JS. 2000. Growth in elevated CO2 protects photosynthesis against high-temperature damage. *Plant, Cell and Environment* **23:** 649-656.

Thomas RB, and Griffin KL. 1994. Direct and indirect effects of atmospheric carbon dioxide enrichment on leaf respiration of *Glycine max* (L.) Merr. *Plant Physiology* **104:** 351-361.

Tissue DT and Oechel WC. 1987. Response of *Eriophorum vaginatum* to elevated CO$_2$ and temperature in the Alaskan tussock tundra. *Ecology* **68:** 401-410.

Tissue DT, Griffin KL, and Ball JT. 1999. Photosynthetic adjustment in field-grown ponderosa pine trees after six years of exposure to elevated CO$_2$. *Tree Physiology* **19:** 221-228.

Tissue DT, Griffin KL, Turnbull MH, and Whitehead D. 2001. Canopy position and needle age affect photosynthetic response in field-grown *Pinus radiata* after five years of exposure to elevated CO$_2$ partial pressure. *Tree Physiology* **21:** 915-923.

Tissue DT, Lewis JD, Wullschleger SD, Amthor JS, Griffin KL, and Anderson OR. 2002. Leaf respiration at different canopy positions in sweetgum (*Liquidambar styraciflua*) grown in ambient and elevated concentrations of carbon dioxide in the field. *Tree Physiology* **22:** 1157-1166.

Tissue DT, Thomas RB, and Strain BR. 1993. Long-term effects of elevated CO$_2$ and nutrients on photosynthesis and Rubisco in loblolly pine seedlings. *Plant, Cell and Environment* **16:** 859-865.

Tjoelker MG, Oleksyn J, and Reich PB. 1999. Acclimation of respiration to temperature and CO$_2$ in seedlings of boreal tree species in relation to plant size and relative growth rate. *Global Change Biology* **5:** 679-691.

Turner FB and Randall DC. 1989. Net production by shrubs and winter annuals in southern Nevada. *Journal of Arid Environments* **17:** 23-36.

Venable DL. 1992. Size-number trade-offs and the variation of seed size with plant resource status. *American Naturalist* **140:** 287-304.

Venable DL and Lawlor L. 1980. Delayed germination in desert annuals: Escape in space and time. *Oecologia* **46:** 272-282.

Venable DL, Pake CE, and Caprio AC. 1993. Diversity and coexistence of Sonoran Desert winter annuals. *Plant Species Biology* **8:** 207-216.

Vitousek PM, Mooney HA, Lubchenko J, and Melillo JM. 1997. Human domination of earth's ecosystems. *Science* **277:** 494-499.

Wand SJE, Midgley GF, Jones MH, and Curtis PS. 1999. Responses of wild C$_4$ and C$_3$ grass (Poaceae) species to elevated atmospheric CO$_2$ concentrations: a meta-analytic test of current theories and perceptions. *Global Change Biology* **5:** 723-741.

Wang N and Nobel PS. 1996. Doubling the CO$_2$ concentration enhanced the activity of carbohydrate-metabolism enzymes, source carbohydrate production, photoassimilate transport, and sink strength for *Opuntia ficus-indica*. *Plant Physiology* **110:** 893-902.

Wang XZ and Curtis PS. 2002. A meta-analytical test of elevated CO$_2$ effects on plant respiration. *Plant Ecology* **161:** 251-261.

Wang XZ, Lewis JD, Tissue DT, Seemann JR, and Griffin KL. 2001. Effects of elevated atmospheric CO_2 concentration on leaf dark respiration of *Xanthium strumarium* in light and darkness. *Proceedings of the National Academy of Sciences (USA)* **98:** 2479-2484.

Ward JK, Tissue DT, Thomas RB, and Strain BR. 1999. Comparative responses of model C_3 and C_4 plants to drought in low and elevated CO_2. *Global Change Biology* **5:** 857-867.

Wullschleger SD. 1993. Biochemical limitations to carbon assimilation in C_3 plants – a retrospective analysis of the A-C_i curves from 109 species. *Journal of Experimental Botany* **44:** 907-920.

Wullschleger SD, Norby RJ, and Gunderson CA. 1992. Growth and maintenance respiration in leaves of *Liriodendron tulipifera* L. exposed to long-term carbon dioxide enrichment in the field. *New Phytologist* **121:** 515-523.

Zhang H and Nobel PS. 1996. Photosynthesis and carbohydrate partitioning for the C3 desert shrub *Encelia farinosa* under current and doubled CO_2 concentrations. *Plant Physiology* **110:** 1361-1366.

IV. Perspectives

Chapter 15

EPILOGUE

Erick De la Barrera and William K. Smith

> *Chance favors only the prepared mind*
> Louis Pasteur

The natural world that current environmental scientists study is substantially different from that faced by our predecesors. Unlike the vast unexplored territories from a few centuries ago that favored the expansion of European empires and the largely unknown biodiversity that stimulated the expeditions of explorers of the like of Malaspina, Humboldt, Darwin, and Lewis and Clarke, there is essentially no place on this planet that has not been directly or indirectly altered by human activity (Bazzaz *et al.* 1998). Moreover, our increasing human population, expected to approach 11 billion by 2050 (Pimentel and Pimentel 2006), will continue to exert pressure on our limited natural resources. It is only a question of time before our non-renewable resources are exhausted. Thus environmental sciences are facing a series of ever-emerging challenges, including the uncertain fate of numerous biological species and their associated ecosystem processes, and an alteration of precipitation patterns that often result in desertification of natural and cultivated lands, despite the growing need to secure a greater food supply worldwide. Biophysical plant ecophysiology continues to identify adaptive responses to the environment, a first step in targeting existing genes that might generate greater food production across a broader spectrum of habitat types. The political nature of environmental issues is at an all-time high within the global community. The various scenarios for increased atmospheric CO_2 and global warming have been published by the Intergovernmental Panel for Climate Change (2007), a recent recipient of the Nobel Peace Price. Their outcome is largely dependent on the environmental and energy policies implemented by individual countries. These decisions must be based on the combination of a technological trade-off, considering either the rapid adoption of alternative energy

Perspectives in Biophysical Plant Ecophysiology: A Tribute to Park S. Nobel, pp. 393-397
Edited by: E. De la Barrera and W.K. Smith

sources or that countries will continue to rely heavily on fossil fuels, plus a political dilemma of whether countries will act independently or, instead, by installing common development policies among countries.

Perhaps more severe than the direct effects of higher CO_2 levels with increasing green house gases is the impact of global warming. Associated influences on global precipitation could add substantially to this warming impact. As illustrated in various chapters of this volume, temperature has effects on most physiological processes owing to the temperature-dependence of enzymes. The research approaches of biophysical plant eco-physiology allow the simulation of plant performance under future climate scenarios, as well as provide information for making strategic land management decisions (De la Barrera and Andrade 2005; Wikelski and Cooke 2006). For example, the Environmental Productivity Indices, developed in Park's laboratory, can predict plant productivity under specific field conditions for some twenty species of agaves and cacti (Nobel 1988). However, because ecosystem-level responses are largely unknown, implementation of experimental manipulations of temperature in the field is of great pertinence (Wan *et al.* 2005).

While the effects of changes in mean air temperature are likely to result in departures from productivity maxima for plants, changes in the frequency and severity of episodes of extremely high or low temperatures will most likely dominate these changes. Corresponding changes in the species distribution patterns are also likely. From ecological and evolutionary perspectives, physiological measurements typically reflect rather conservative traits of a species. In this respect, biophysical ecophysiology can improve the accuracy of species distribution models based solely on probabilistic computer simulations by aiding in the characterization of the fundamental niche, which can be described as a core set of attributes for a given taxon that determines the environmental conditions beyond which it is unable to adapt, acclimate, or even survive (Peterson *et al.* 1999). In the midst of a scientific era dominated by molecular biology and genomic explorations, an emphasis on ecophysiological research at the organism level is much needed to provide an evolutionary perspective that acts to couple the genomic and ecosystem levels of the biological hierarchy of complexity (Smith *et al.* 2004).

In addition to higher terrestrial air temperatures, global warming will lead to an increase in ocean surface temperature that will alter precipitation regimes on a global scale (Garduño 1997; Magaña *et al.* 1997; Adem *et al.* 2000; Weltzin *et al.* 2003). In many regions, decreases in total rainfall are expected leading to land aridization, especially for tropical regions (Adeel *et al.* 2005). The declaration of the years 2005 to 2015 as the International Decade of Water by the United Nations is recognition that a better understanding of the global hydrological cycle is needed. For example, multidisciplinary studies of water are being conducted in the arid southern United States, where certain species of mesquite (genus *Prosopis*) are agressively competing for water reserves with nearby municipalities (Grover and Musick 1990; Huxman *et al.* 2005). As illustrated by various chapters in this festschrift for Park, plant adaptations for dealing with water stress have been a traditional and expansive focus of plant biophysical eco-physiology. Our present knowledge of water use by plants, especially for species from arid and semi-arid environments, is now extensive. This will enable the rapid establishment of research programs that explicitly consider the capability to acclimate and addapt to drier environments for plants from mesic environments. At present, integrative studies incorporating multiple environmental factors (*e.g.*, CO_2 concentration, temperature, water, and mineral nutrients) are still lacking (Norby and Luo 2004).

The agricultural sector is a critical user of water, ranging from about 10% of the available potable water in France to 41% in the U.S.A., 77% in Mexico, and over 90% in Uruguay, Afganistan, and Somalia (Comisión Nacional del Agua 2008). A major reason for such large water use in agriculture is an accumulation of inefficient practices at all levels that contribute to substantial water losses via the distribution network, *e.g.*, open canals for water distribution, the irrigation methods such as flooding or aspersion instead of drip irrigation, and even by highly productive crops with low water use efficiency (Hong-Bo *et al.* 2006). Here, ecophysiology can help identify genomes that generate higher water use efficiencies within traditional crops (Condon *et al.* 2002). Moreover, new crops can be developed from plants that combine a high nutritional quality with reasonably low water requirements. This kind of agronomic research has been conducted over the years in Israel, where an important effort for developing new crops is of great importance given their water-limited environment. In addition, there is a rich history in the use of endemic species in many high biodiversity countries. For instance, in the Tehuacán Valley in central-southern Mexico, one of nearly a dozen sites in the world where agriculture originated, most of the twenty something species of columnar cacti are used for food (Casas *et al.* 1999; Smith 2005; Zeder 2005). An example of a crop that recently acquired global importance is that of the prickly pear cactus, *Opuntia ficus-indica*, whose popularity can be credited to combined efforts of scientists and producers. This species is cultivated primarily for forage in nearly fourty countries, although some 10% of the one million hectares covered by this CAM plant are dedicated to fruit production as well (Nobel 2000).

The current environmental crisis gripping the world poses exciting research challenges for natural scientists. Beyond the mere satisfaction of scientific curiosity of researchers, pressing issues such as the effects of global climate change and accelerated human population growth on biodiversity demand timely and comprehensive information. Moreover, the complex question of such research necessitates the participation of various disciplines acting at different spatial and temporal scales. Plant biophysical ecophysiology, based on plant responses driven by physical principles, can provide an underlying framework for such multidisciplinary challenges, acting to bridge between organism responses and ecosystem processes. Paraphrasing one of Park Nobel's favorite quotes (by Louis Pasteur), important discoveries leading to a healthy and sustainable environment will not occur by chance alone, but only if prepared minds are there to recognize opportunities.

Literature cited

Adeel Z, Safriel U, Niemeijer D, and White R. 2005. *Ecosystems and Human Well-being: Desertification Synthesis*. World Resources Institute, Washington, D.C.

Adem J, Mendoza VM, Ruiz A, Villanueva EE, and Garduño R. 2000. Recent numerical experiments on three-months extended and seasonal weather prediction with a thermodynamic model. *Atmósfera* 13: 53-83.

Bazzaz F, Ceballos G, Davis M, Dirzo R, Ehrlich PR, Eisner T, Levin S, Lawton JH, Lubchenco J, Matson PA, Mooney HA, Raven PA, Roughgarden JE, Sarukhán J, Tilman GD, Vitousek P, Walker B, Wall DH, Wilson EO, and Woodwell GM. 1998. Ecological science and the human predicament. *Science* 282: 879.

Casas A, Caballero J, and Valiente-Banuet A. 1999. Use, management, and domestication of columnar cacti in south-central Mexico: a historical perspective. *Journal of Ethnobiology* 19: 97-95.

Comisión Nacional del Agua. 2008. *Programa Nacional Hídrico 2007-2012*. Secretaría del Medio Ambiente y Recursos Naturales, Mexico.

Condon AG, Richards RA, Rebetzke GJ, and Farquar GD. 2002. Improving intrinsic water-use efficiency and crop yield. *Crop Science* 42: 122-131.

De la Barrera E and Andrade JL. 2005. Challenges to plant megadiversity: how environmental physiology can help. *New Phytologist* 167: 5-8.

Garduño R. 1997. Past and future climates simulated with the Adem thermodynamic model. *Quaternary International* 43/44: 19-24.

Grover HD and Musick HB. 1990. Shrubland encroachment in southern New Mexico, U.S.A.: An analysis of desertification processes in the American southwest. *Climatic Change* 17: 305-330.

Magaña V, Conde C, Sánchez O, and Gay C. 1997. Assessment of current and future regional climate scenarios for Mexico. *Climate Research* 9: 107-114.

Hong-Bo S, Li-Ye C, Gang W, Jin-Heng Z, and Zhao-Hua, L. 2006. Where is the road to bio-water saving for the globe? *Colloids and Surfaces B: Biointerfaces* 55: 251-255.

Huxman TE, Wilcox BP, Breshears DD, Scott RL, Snyder KA, Small EE, Hultine K, Pockman WT, and Jackson RB. 2005. Ecohydrological implications of woody plant encroachment. *Ecology* 86: 308-319.

Intergovernmental Panel for Climate Change. 2007. *Climate Change 2007. Impacts, adaptation, and vulnerability. Contribution of Working Group II. 4th. Assessment Report of the International Panel on Climate Change (IPCC)*. Available on-line at: http://www.ipcc.ch/ipccreports/ar4-wg2.htm.

Nobel PS. 1988. *Environmental Biology of Agaves and Cacti*. Cambridge University Press.

Nobel PS. 2000. Crop ecosystem responses to climate change: Crassulacean acid metabolism crops. In: Reddy KR and Hodges HF (eds.) *Climate Change and Global Crop Productivity*. CAB International, Oxford, U.K.

Norby RJ and Luo Y. 2004. Evaluating ecosystem responses to rising atmospheric CO_2 and global warming in a multi-factor world. *New Phytologist* 162: 281-293.

Peterson AT, Soberón J, and Sánchez-Cordero V. 1999. Conservatism of ecological niches in evolutionary time. *Science* 285: 1265-1267.

Pimentel D and Pimentel M. 2006. Global environmental resourses versus world population growth. *Ecological Economics* 59: 195-198.

Smith BD. 2005. Reassessing Coscatlán Cave and the early history of domesticated plants in Mesoamerica. *Proceedings of the National Academy of Sciences* **102:** 9438-9445.

Smith WK, Vogelmann TC, and Chistley C (eds.). 2004. *Photosynthetic Adaptation: Chloroplats to Landscape.* Springer, New York.

Wan S, Hui D, Wallace L, Luo Y. 2005. Direct and indirect effects of experimental warming on ecosystem carbon processes in a tall grass prairie. *Global Biogeochemical Cycles* **19:** GB2014. DOI: 10.1029/2004GB002315.

Weltzin JF, Loik ME, Schwinning S, Williams DG, Fay PA, Haddad BM, Harte J, Huxman TE, Knapp AK, Lin G, Pockman WT, Shaw R, Small E, Smith M, Smith SD, Tissue DT, and Zak JC. 2003. Assessing the response of terrestrial ecosystems to potential changes in precipitation. *Bioscience* **53:** 941-952.

Wikelski M and Cooke SJ. 2006. Conservation physiology. *Trends in Ecology and Evolution* **21:** 38-46.

Zeder MA. 2005. Central questions in the domestication of plants and animals. *Evolutionary Anthropology* **15:** 105-117.

INDEX

www.ingramcontent.com/pod-product-compliance
Lightning Source LLC
Chambersburg PA
CBHW020720180526
45163CB00001B/50